SI Prefixes

Multiple	Exponential Form	Prefix	SI Symbol
1 000 000 000	10^9	giga	G
1 000 000	10^6	mega	M
1 000	10^3	kilo	k
Submultiple			
0.001	10^{-3}	milli	m
0.000 001	10^{-6}	micro	μ
0.000 000 001	10^{-9}	nano	n

6.5

Conversion Factors (FPS) to (SI)

Quantity	Unit of Measurement (FPS)	Equals	Unit of Measurement (SI)
Force	lb		4.4482 N
Mass	slug		14.5938 kg
Length	ft		0.3048 m

Conversion Factors (FPS)

1 ft = 12 in. (inches)

1 mi. (mile) = 5,280 ft

1 kip (kilopound) = 1,000 lb

1 ton = 2,000 lb

Engineering Mechanics
Dynamics

Engineering Mechanics

Dynamics

TH EDITION

R. C. Hibbeler

Macmillan Publishing Company
New York

Collier Macmillan Publishers
London

Macmillan Publishing Company
866 Third Avenue, New York, New York 10022

Collier Macmillan Canada, Inc.

Library of Congress Cataloging in Publication Data

Hibbeler, R. C.
  Engineering mechanics—dynamics.

  Includes index.
  1. Dynamics   I. Title.
TA352.H5   1986   620.1′04   85-21367
ISBN 0-02-354660-3 (Hardcover edition)
ISBN 0-02-946250-9 (International edition)

Printing: 1 2 3 4 5 6 7 8   Year: 6 7 8 9 0 1 2

ISBN 0-02-354660-3

Preface

The purpose of this book is to provide the student with a clear and thorough presentation of the theory and application of the principles of engineering mechanics. Emphasis is placed on developing the student's ability to analyze problems—a most important skill of any engineer. In preparing this revision, presentation and arrangement of some of the topics have been made more effective, and the examples and problem assignments have been improved. Throughout the book, attention has been given to numerical accuracy and clarity in both defining and developing the concepts.

Organization and Approach. The contents of each chapter are organized into well-defined sections. Selected groups of sections contain an explanation of specific topics, illustrative example problems, and a set of homework problems. The topics within each section are placed into subgroups defined by boldface titles. The purpose of this is to present a structured method for introducing each new definition or concept, and to make the book convenient for later reference and review.

A "procedure for analysis" is used at the end of many sections of the book in order to provide the student with a logical and orderly method to follow when applying the theory. As in the previous editions, the example problems are solved using this outlined method in order to clarify its numerical application. It is to be understood, however, that once the relevant principles have been mastered and enough confidence and judgment have been obtained, the student can then develop his own procedures for solving problems. In most cases it is felt that the first step in any procedure should require drawing a diagram. In doing so, the student forms the habit of tabulating the necessary data while focusing on the physical aspects of the problem and its associated geometry. If this step is correctly performed, applying the relevant equations of mechanics becomes somewhat methodical, since the data can be taken directly from the diagram.

Since mathematics provides a systematic means of applying the principles of mechanics, the student is expected to have prior knowledge of algebra, geometry, trigonometry, and, for complete coverage, some calculus. Vector analysis is introduced at points where it is most applicable. Its use often provides a convenient means for presenting concise derivations of the theory, and it makes possible a simple and systematic solution of many complicated three-dimensional problems. Occasionally, the example problems are solved using more than one method of analysis so that the student develops the ability to use mathematics as a tool whereby the solution of any problem may be carried out in the most direct and effective manner.

Problems. Numerous problems in the book depict realistic situations encountered in engineering practice. It is hoped that this realism will both stimulate the student's interest in engineering mechanics and provide a means for developing the skill to reduce any such problem from its physical description to a model or symbolic representation to which the principles of mechanics may be applied. Most of the problems in this edition are new and many problems retained from the previous edition have been changed numerically. Also, in this edition an effort has been made to include some problems which may be solved using a numerical procedure executed on either a microcomputer or programmable pocket calculator. Suitable numerical techniques along with associated computer programs are given in Appendix B. The intent here is to broaden the student's capacity for using other forms of mathematical analysis *without* sacrificing the necessary time needed to focus on the application of the principles of mechanics. Problems of this type which either can or must be solved using numerical procedures are identified by a "square" symbol (■) preceding the problem number.

Throughout the text there is an approximate balance of problems using either SI or FPS units. Furthermore, in any set, an attempt has been made to arrange the problems in order of increasing difficulty. The answers to all but every fourth problem are listed in the back of the book. To alert the user to a problem without a reported answer, an asterisk (*) is placed before the problem number.

Contents. The book is divided into 11 chapters.* In particular, the kinematics of a particle is discussed in Chapter 12, followed by a discussion of particle kinetics in Chapter 13 (equation of motion), Chapter 14 (work and energy), and Chapter 15 (impulse and momentum). The concepts of particle dynamics contained in these four chapters are then summarized in a "review" section and the student is then given the chance to identify and solve a variety

---

*The first eleven chapters of this sequence form the contents of *Engineering Mechanics: Statics*.

of different types of problems. A similar sequence of presentation is given for the planar motion of a rigid body: Chapter 16 (planar kinematics), Chapter 17 (equation of motion), Chapter 18 (work and energy), and Chapter 19 (impulse and momentum), followed by a summary and set of review problems for these chapters. If desired, it is possible to cover Chapters 12 through 19 in the following order with no loss in continuity: Chapters 12 and 16 (kinematics), Chapters 13 and 17 (equation of motion), Chapters 14 and 18 (work and energy), and Chapters 15 and 19 (impulse and momentum).

Time permitting, some of the material involving three-dimensional rigid-body motion may be included in the course. The kinematics and kinetics of this motion are discussed in Chapters 20 and 21, respectively. Chapter 22 (vibrations) may be included if the student has the necessary mathematical background. Sections of the book which are considered to be beyond the scope of the basic dynamics course are indicated by a star (★) and may be omitted. Note, however, that this more advanced material provides a suitable reference for basic principles when it is covered in other courses.

Acknowledgments.   I have endeavored to write this book so that it will appeal to both the student and instructor. Through the years many people have helped in its development and I should like to acknowledge their valued suggestions and comments. Specifically, I wish to personally thank the reviewers who have contributed to this revision, namely, Professors Nicholas J. Altiero, Michigan State University; James G. Andrews, University of Iowa; A. Bedford, University of Texas at Austin; William J. Bottega, Rutgers University; Robert L. Collins, University of Louisville; K. L. DeVries, University of Utah; Ronald C. Dykhuizen, Arizona State University; C. Stuart Ferrell, Louisiana Tech University; Dale Foremann, Colorado School of Mines; John H. Forrester, University of Tennessee; Ray Hagglund, Worcester Polytechnic Institute; Allen Hoffman, Worcester Polytechnic Institute; Edward E. Hornsey, University of Missouri, Rolla; C. C. Hsiao, University of Minnesota; Terry Ishihara, Saginaw Valley State College; Will Liddel, Auburn University; George E. Mase, Michigan State University; John C. McWhorter, Mississippi State University; Benjamin W. Mooring, Texas A & M University; William W. Predebon, Michigan Technological University; Walter F. Rowland, California State University, Fresno; Frank J. Tarantine, Youngstown State University; Youngil Youm, Catholic University; and Warren C. Young, University of Wisconsin, Madison.

Many thanks are also extended to all my students and to members of the teaching profession who have freely taken the time to send me their suggestions and comments. Although the list is too long to mention, it is hoped that those who have given help in this manner will accept this anonymous recognition. Furthermore, I greatly appreciate the personal support given to me by my editors and the staff at Macmillan. Lastly, I should like to acknowledge the assistance of my wife, Conny, who has always been quite helpful in attending to many of the details of preparing the manuscript for publication.

Russell Charles Hibbeler

Contents

Kinematics of a Particle 12
1

Kinetics of a Particle: Force and Acceleration 13
81

Contents

Planar Kinetics of a Rigid Body: Force and Acceleration

309

17

Planar Kinetics of a Rigid Body: Work and Energy

361

18

Planar Kinetics of a Rigid Body: Impulse and Momentum

389

19

Review 2: Planar Kinematics and Kinetics of a Rigid Body

419

Three-Dimensional Kinematics of a Rigid Body

435

20

Three-Dimensional Kinetics of a Rigid Body

461

Vibrations
505

APPENDIXES

Mathematical Expressions
541

Numerical and Computer Analysis
544

Vector Analysis
553

Answers
557

Index
569

Engineering Mechanics
Dynamics

Kinematics of a Particle

Engineering mechanics involves the study of both statics and dynamics. *Statics* is concerned with the equilibrium of bodies at rest or moving with constant velocity, whereas *dynamics* is concerned with bodies having accelerated motion. In this text, the subject of dynamics will be presented in two parts: *kinematics*, which treats only the geometrical aspects of motion, and *kinetics*, which is the analysis of the forces causing the motion. For simplicity particle dynamics will be discussed first, followed by topics in rigid-body dynamics presented in two and then three dimensions.

We will begin our study of dynamics by discussing particle kinematics. Recall that a *particle* has a mass but negligible size and shape. Therefore we must limit application to those objects that have dimensions that are of no consequence in the analysis of the problem. In most problems encountered, one is interested in bodies of finite size, such as rockets, projectiles, or vehicles. Such objects may be considered as particles, provided motion of the body is characterized by motion of its mass center and any rotation of the body is neglected.

This chapter begins with a study of the absolute motion of a particle, which is motion measured with respect to a fixed coordinate system. In this regard, motion along a straight line will be studied before the more general motion along a curved path. Afterward, the relative motion between two particles will be considered, using a translating coordinate system.

Rectilinear Kinematics 12.1

Rectilinear motion occurs when a particle moves along a straight-line path. The kinematics of the motion is characterized by specifying the particle's position, velocity, and acceleration.

1

Fig. 12–1

Position. The straight-line path of the particle can be described using a single coordinate axis s, Fig. 12–1a. The origin O on the path is a fixed point, and from this point the coordinate s is used to define the *position* of the particle P at any given instant. Note that position has the properties of a vector. We can regard its *magnitude* as the *distance* from O to P, usually measured in meters (m) or feet (ft), and its *sense* by the algebraic sign of s. Although the choice is arbitrary, in this case s is positive when the particle is located to the right of the origin and negative when it is located to the left.

Displacement. The *displacement* of the particle is defined as the *change* in its *position*. For example, if the particle moves from P to P', Fig. 12–1b, the displacement is $\Delta s = s' - s$. Here Δs is *positive* since the particle's final position is to the *right* of its initial position, i.e., $s' > s$. Likewise, if the final position is to the *left*, Δs is *negative*.

The displacement of a particle, which is a vector, should be distinguished from the distance the particle travels. Specifically, the *distance traveled* is a positive scalar-quantity which represents the total length of path traversed by the particle. This length is *always positive*.

Velocity. If the particle moves through a displacement Δs from P to P' during the time interval Δt, Fig. 12–1c, the *average velocity* of the particle during this time interval is

$$v_{\text{avg}} = \frac{\Delta s}{\Delta t}$$

If we take smaller and smaller values of Δt, the magnitude of Δs becomes smaller and smaller. Consequently, the *instantaneous velocity* is defined as $v = \lim_{\Delta t \to 0} (\Delta s / \Delta t)$ or

$$v = \frac{ds}{dt} \qquad (12\text{--}1)$$

Since Δt or dt is always positive, the sign used to define the *sense* of the velocity is the same as that of Δs (or ds). For example, if the particle is moving to the *right*, Fig. 12–1c, the velocity is *positive;* whereas if it is moving to the *left,* the velocity is *negative.* The *magnitude* of the velocity is known as the *speed* and is generally expressed in units of m/s or ft/s.

Occasionally, the term "average speed" is used. The *average speed* is a positive scalar and is defined as the total distance traveled by a particle, s_T, divided by the elapsed time Δt, i.e.,

$$(v_{\text{sp}})_{\text{avg}} = \frac{s_T}{\Delta t}$$

Acceleration. Provided that the velocities for the particle are known at the two points P and P', the *average acceleration* for the particle during the time interval Δt is defined as

$$a_{\text{avg}} = \frac{\Delta v}{\Delta t}$$

Here Δv represents the difference in the velocities during the time interval Δt, i.e., $\Delta v = v' - v$, Fig. 12–1d.

The *instantaneous acceleration* at time t is found by taking smaller and smaller values of Δt and corresponding smaller and smaller values of Δv, so that $a = \lim_{\Delta t \to 0} (\Delta v / \Delta t)$ or

$$a = \frac{dv}{dt} \qquad (12\text{--}2)$$

Taking the second time derivative of Eq. 12–1, we can also write

$$a = \frac{d^2 s}{dt^2}$$

Both the average and instantaneous acceleration can be either positive or negative. In particular, when the particle is *slowing down* it is said to be *decelerating.* In this case, v' in Fig. 12–1e is *less* than v and so $\Delta v = v' - v$ will be negative. Consequently, a will also be negative and therefore it will act to the *left,* in the opposite *sense* to v. Also, note that when the *velocity* is *constant,* the *acceleration is zero* since $\Delta v = v - v = 0$. Units commonly used to express the magnitude of acceleration are m/s^2 or ft/s^2.

A differential relation involving the displacement, velocity, and acceleration along the path may be obtained by eliminating the time differential dt between Eqs. 12–1 and 12–2. Show that this yields

$$a\,ds = v\,dv \qquad (12\text{--}3)$$

Constant Acceleration, $a = a_c$. When the acceleration is constant, each of the three kinematic equations $a_c = dv/dt$, $v = ds/dt$, and $a_c\,ds = v\,dv$ may be integrated to obtain formulas that relate a_c, v, s, and t.

Velocity as a Function of Time. Integrate $a_c = dv/dt$, assuming that initially $v = v_0$ when $t = 0$.

$$\int_{v_0}^{v} dv = \int_{0}^{t} a_c\,dt$$

$$v - v_0 = a_c(t - 0)$$

$$\boxed{v = v_0 + a_c t} \tag{12-4}$$

Position as a Function of Time. Integrate $v = ds/dt = v_0 + a_c t$, assuming that initially $s = s_0$ when $t = 0$.

$$\int_{s_0}^{s} ds = \int_{0}^{t} (v_0 + a_c t)\,dt$$

$$s - s_0 = v_0(t - 0) + a_c(\tfrac{1}{2}t^2 - 0)$$

$$\boxed{s = s_0 + v_0 t + \tfrac{1}{2}a_c t^2} \tag{12-5}$$

Velocity as a Function of Position. Either solve for t in Eq. 12–4 and substitute into Eq. 12–5, or integrate $v\,dv = a_c\,ds$, assuming that initially $v = v_0$ at $s = s_0$.

$$\int_{v_0}^{v} v\,dv = \int_{s_0}^{s} a_c\,ds$$

$$\tfrac{1}{2}v^2 - \tfrac{1}{2}v_0^2 = a_c(s - s_0)$$

$$\boxed{v^2 = v_0^2 + 2a_c(s - s_0)} \tag{12-6}$$

The magnitudes and signs of s_0, v_0, and a_c, used in these equations, are determined from the chosen origin and positive direction of the s axis.

It is important to remember that the above equations are useful *only when the acceleration is constant and when $t = 0$, $s = s_0$, $v = v_0$.* A common example of this motion occurs when a body falls freely toward the earth. If air resistance is neglected and the distance of fall is short, then the constant *downward* acceleration of the body when it is close to the earth is approximately 9.81 m/s^2 or 32.2 ft/s^2.*

*The proof is given in Example 13–2.

PROCEDURE FOR ANALYSIS

Often a mathematical relationship between *any two* of the four variables a, v, s, and t can be established by observation or experiment. When this is the case, the relationships between the remaining variables can be obtained by either differentiation or integration, using the kinematic equations $a = dv/dt$, $v = ds/dt$, or $a\,ds = v\,dv$.* Since each of these equations relates *three* variables, then, if a variable is *known* as a function of another variable, a third variable can be determined by *choosing the kinematic equation which relates all three*. For example, suppose that the *acceleration* is known as a function of *position*, $a = f(s)$. The *velocity* can be determined from $a\,ds = v\,dv$ since $f(s)$ can be substituted for a to yield $f(s)\,ds = v\,dv$. Solution for v requires integration.† Note that the velocity *cannot* be obtained by using $a = dv/dt$, since $f(s)\,dt = dv$ *cannot* be integrated.

Whenever the kinematic equations are applied, it is important to specify the positive sense of the position coordinate. By doing so, the *sense* of s, v, and a can then be determined from the *algebraic signs* of their numerical quantities. In the following examples we will indicate the positive sense alongside each kinematic equation as it is applied.

*Some standard differentiation and integration formulas are given in Appendix A.

†The position and velocity must be known at a given instant in order to evaluate either the constant of integration if an indefinite integral is used, or the limits of integration if a definite integral is used.

Example 12–1

The bicyclist in Fig. 12–2 has a constant acceleration of 2 ft/s². If he starts from rest, determine his velocity and position when $t = 5\ s$.

Fig. 12–2

Solution

Velocity. This problem can be solved using the formulas of *constant acceleration* with $a_c = 2$ ft/s². Since the time limit is specified, and initially $v_0 = 0$, the velocity is determined from

$(\overset{+}{\rightarrow})$
$$v = v_0 + a_c t$$
$$v = 0 + 2(5) = 10 \text{ ft/s} \rightarrow \qquad Ans.$$

Position. In a similar manner, the bicyclist's position is determined from

$(\overset{+}{\rightarrow})$
$$s = s_0 + v_0 t + \tfrac{1}{2} a_c t^2$$
$$s = 0 + 0(5) + \tfrac{1}{2}(2)(5)^2 = 25 \text{ ft} \qquad Ans.$$

Since the final velocity has been computed, it is also possible to obtain the position as follows:

$(\overset{+}{\rightarrow})$
$$v^2 = v_0^2 + 2a_c(s - s_0)$$
$$(10)^2 = (0)^2 + 2(2)(s - 0)$$
$$s = 25 \text{ ft} \qquad Ans.$$

Example 12–2

The car in Fig. 12–3 moves in a straight line such that for a short time its velocity is defined by $v = (9t^2 + 2t)$ ft/s, where t is in seconds. Determine its position and acceleration when $t = 3$ s.

Fig. 12–3

Solution

Position. The car's velocity is given as a function of time so that its position can be determined from $v = ds/dt$, since this equation relates v, s, and t. Noting that $s = 0$ when $t = 0$, we have*

$$(\overset{+}{\rightarrow}) \qquad v = \frac{ds}{dt} = (9t^2 + 2t)$$

$$\int_0^s ds = \int_0^t (9t^2 + 2t)\, dt$$

$$s\Big|_0^s = 3t^3 + t^2\Big|_0^t$$

$$s = 3t^3 + t^2$$

When $t = 3$ s,

$$s = 3(3)^3 + (3)^2 = 90 \text{ ft} \qquad\qquad \textit{Ans.}$$

Acceleration. The acceleration is determined from $a = dv/dt$, since this equation relates a, v, and t.

$$(\overset{+}{\rightarrow}) \qquad a = \frac{dv}{dt} = \frac{d}{dt}(9t^2 + 2t)$$

$$= 18t + 2$$

When $t = 3$ s,

$$a = 18(3) + 2 = 56 \text{ ft/s}^2 \rightarrow \qquad\qquad \textit{Ans.}$$

Note that the formulas for constant acceleration cannot be used to solve this problem.

---

*The *same result* can be obtained by evaluating a constant of integration C rather than using definite limits on the integral. For example, integrating $ds = (9t^2 + 2t)\, dt$ yields $s = 3t^3 + t^2 + C$. Using the condition that at $t = 0$, $s = 0$, then $C = 0$.

Example 12–3 ▬▬▬▬

A small projectile is fired vertically *downward* into a fluid medium with an initial velocity of 60 m/s. If fluid resistance causes a deceleration of the projectile which is equal to $a = (-0.4v^3)$ m/s^2, where v is measured in m/s, determine both the velocity v and position s four seconds after the projectile is fired.

Solution

Velocity. The acceleration is given as a function of velocity so that the velocity as a function of time can be obtained from $a = dv/dt$, since this equation relates v, a, and t. (Why not use $v = v_0 + a_c t$?) Integrating, with the initial condition that $v_0 = 60$ m/s when $t = 0$, yields

$$(+\downarrow) \qquad\qquad a = \frac{dv}{dt} = -0.4v^3$$

$$\int_{60}^{v} \frac{dv}{-0.4v^3} = \int_{0}^{t} dt$$

$$\frac{1}{-0.4}\left(\frac{1}{-2}\right)\frac{1}{v^2}\bigg|_{60}^{v} = t - 0$$

$$\frac{1}{0.8}\left[\frac{1}{v^2} - \frac{1}{(60)^2}\right] = t$$

$$v = \left\{\left[\frac{1}{(60)^2} + 0.8t\right]^{-1/2}\right\} \text{ m/s}$$

Here the positive root is taken, since the projectile is moving downward. When $t = 4$ s,

$$v = 0.559 \text{ m/s} \downarrow \qquad\qquad\qquad \textit{Ans.}$$

Position. Knowing the velocity as a function of time, we obtain the position as a function of time from $v = ds/dt$, since this equation relates s, v, and t. Using the initial condition $s = 0$, when $t = 0$, we have

$$(+\downarrow) \qquad v = \frac{ds}{dt} = \left[\frac{1}{(60)^2} + 0.8t\right]^{-1/2}$$

$$\int_{0}^{s} ds = \int_{0}^{t} \left[\frac{1}{(60)^2} + 0.8t\right]^{-1/2} dt$$

$$s = \frac{2}{0.8}\left[\frac{1}{(60)^2} + 0.8t\right]^{1/2}\bigg|_{0}^{t}$$

$$s = \frac{1}{0.4}\left\{\left[\frac{1}{(60)^2} + 0.8t\right]^{1/2} - \frac{1}{60}\right\} \text{ m}$$

When $t = 4$ s,

$$s = 4.43 \text{ m} \qquad\qquad\qquad \textit{Ans.}$$

Example 12–4

A boy tosses a ball in the vertical direction off the side of a cliff, as shown in Fig. 12–4. If the initial velocity of the ball is 15 m/s upward, and the ball is released 40 m from the bottom of the cliff, determine the maximum height s_B reached by the ball and the speed of the ball just before it hits the ground. During the entire time the ball is in motion, it is subjected to a constant downward acceleration of 9.81 m/s^2 due to gravity. Neglect the effect of air resistance.

Fig. 12–4

Solution

Maximum Height. The origin for the position coordinate is taken at ground level. At the maximum height s_B the velocity $v_B = 0$. Since the ball is thrown *upward* when $t = 0$, it is subjected to a velocity $v_A = +15$ m/s (positive since it is in the same sense as positive displacement). For the entire motion, the acceleration is *constant* such that $a_c = -9.81$ m/s^2 (negative since it acts in the *opposite* sense to positive velocity or positive displacement). Since a_c is *constant* throughout the entire motion, the position may be related to the velocity at points A and B as follows:

$$(+\uparrow) \qquad v_B^2 = v_A^2 + 2a_c(s_B - s_A)$$
$$0 = (15)^2 + 2(-9.81)(s_B - 40)$$

so that

$$s_B = 51.5 \text{ m} \qquad\qquad Ans.$$

Velocity. To obtain the velocity of the ball just before it hits the ground, Eq. 12–6 can be applied between points B and C, Fig. 12–4,

$$(+\uparrow) \qquad v_C^2 = v_B^2 + 2a_c(s_C - s_B)$$
$$= 0 + 2(-9.81)(0 - 51.5)$$
$$v_C = -31.8 \text{ m/s} = 31.8 \text{ m/s} \downarrow \qquad\qquad Ans.$$

The negative root was chosen since the ball is moving *downward*.

Similarly, Eq. 12–6 may also be applied between points A and C, i.e.,

$$(+\uparrow) \qquad v_C^2 = v_A^2 + 2a_c(s_C - s_A)$$
$$= 15^2 + 2(-9.81)(0 - 40)$$
$$v_C = -31.8 \text{ m/s} = 31.8 \text{ m/s} \downarrow \qquad\qquad Ans.$$

Note: It should be realized that the ball is subjected to a *deceleration* from A to B of 9.81 m/s^2, and then from B to C it is *accelerated* at this rate. Furthermore, even though the ball momentarily comes to *rest* at B ($v_B = 0$) the acceleration at B is 9.81 m/s^2 downward!

Example 12–5

A metallic particle is subjected to the influence of a magnetic field such that it travels vertically through a fluid that extends from plate A to plate B, Fig. 12–5. If the particle is released from rest at C, $s = 100$mm, and the acceleration is measured as $a = (4s)$ m/s^2, where s is in meters, determine the velocity of the particle when it reaches plate B, $s = 200$ mm, and the time it needs to travel from C to plate B.

Fig. 12–5

Solution

Velocity. Knowing the acceleration as a function of position, the velocity as a function of position can be obtained by using $v\,dv = a\,ds$. Why not use the formulas for constant acceleration? Realizing that $v = 0$ at $s = 100$ mm $= 0.1$ m, we have

$(+\downarrow)$
$$v\,dv = a\,ds$$

$$\int_0^v v\,dv = \int_{0.1}^s 4s\,ds$$

$$\tfrac{1}{2}v^2\Big|_0^v = \frac{4}{2}s^2\Big|_{0.1}^s$$

$$v = 2(s^2 - 0.01)^{1/2} \qquad\qquad (1)$$

At $s = 200$ mm $= 0.2$ m,

$$v_B = 0.346 \text{ m/s} = 346 \text{ mm/s}\downarrow \qquad\qquad \textit{Ans.}$$

The positive root is chosen since the particle is traveling downward, i.e., in the $+s$ direction.

Time. Once the velocity is known as a function of position, Eq. (1), the time for the particle to travel from C to plate B can be obtained using $v = ds/dt$, where $s = 0.1$ m when $t = 0$.

$(+\downarrow)$
$$ds = v\,dt$$
$$= 2(s^2 - 0.01)^{1/2}\,dt$$

$$\int_{0.1}^s \frac{ds}{(s^2 - 0.01)^{1/2}} = \int_0^t 2\,dt$$

$$\ln(s + \sqrt{s^2 - 0.01})\Big|_{0.1}^s = 2t\Big|_0^t$$

$$\ln(s + \sqrt{s^2 - 0.01}) + 2.30 = 2t$$

At $s = 200$ mm $= 0.2$ m,

$$t = \frac{\ln(0.2 + \sqrt{(0.2)^2 - 0.01}) + 2.30}{2} = 0.657 \text{ s} \qquad \textit{Ans.}$$

9

Example 12–6

A particle moves along a horizontal straight line such that its velocity is given by $v = (3t^2 - 6t)$ m/s, where t is the time in seconds. If it is initially located at the origin O, determine the distance traveled during the time interval $t = 0$ to $t = 3.5$ s, the average velocity, and the average speed of the particle during this time interval.

(a)

Solution

Distance Traveled. Since the velocity is related to time, the function that relates position to time may be found by integrating $v = ds/dt$ with the condition that when $t = 0$, $s = 0$.

(b)

Fig. 12–6

$$(\overset{+}{\rightarrow}) \qquad ds = v\, dt$$

$$= (3t^2 - 6t)\, dt$$

$$\int_0^s ds = 3 \int_0^t t^2\, dt - 6 \int_0^t t\, dt$$

$$s = (t^3 - 3t^2)\ \text{m} \qquad (1)$$

In order to determine the distance traveled in 3.5 s, it is necessary to investigate the path of motion during this time. The graph of the velocity function, Fig. 12–6a, reveals that for $0 \le t < 2$ s the velocity is *negative*, which means the particle is traveling to the *left*, and for $t > 2$ s the velocity is *positive*, and hence the particle is traveling to the *right*. Also, since the velocity changes sign when $t = 2$ s, the particle reverses its sense at this instant. The particle's position when $t = 0$, $t = 2$ s, and $t = 3.5$ s can be computed from Eq. (1). This yields

$$s|_{t=0} = 0, \qquad s|_{t=2s} = -4\ \text{m}, \qquad s|_{t=3.5s} = 6.12\ \text{m}$$

The path is shown in Fig. 12–6b. Hence, the distance traveled in 3.5 s is

$$s_T = 4 + 4 + 6.12 = 14.1\ \text{m} \qquad \textit{Ans.}$$

Velocity. The *displacement* from $t = 0$ to $t = 3.5$ s is

$$\Delta s = s|_{t=3.5s} - s|_{t=0} = 6.12 - 0 = 6.12\ \text{m}$$

so that the average velocity is

$$v_{avg} = \frac{\Delta s}{\Delta t} = \frac{6.12}{3.5 - 0} = 1.75\ \text{m/s} \qquad \textit{Ans.}$$

The average speed is defined in terms of the *distance traveled* s_T. Hence,

$$(v_{sp})_{avg} = \frac{s_T}{\Delta t} = \frac{14.12}{3.5 - 0} = 4.03\ \text{m/s} \qquad \textit{Ans.}$$

Problems

12–1. A car starts from rest and reaches a speed of 80 ft/s after traveling 500 ft along a straight road. Determine its constant acceleration and the time of travel.

12–2. A bicyclist starts from rest and after traveling along a straight path a distance of 20 m reaches a speed of 30 km/h. Determine the acceleration if it is *constant*. Also, how long does it take to reach the speed of 30 km/h?

12–3. Traveling with an initial speed of 70 km/h, a car accelerates at 6000 km/h² along a straight road. How long will it take to reach a speed of 120 km/h? Also, through what distance does the car travel during this time?

*12–4. A baseball is thrown downward from a 50-ft tower with an initial speed of 18 ft/s. Determine the speed at which it hits the ground and the time of travel.

12–5. Tests reveal that a normal driver can *react* to a situation in 0.75 s before beginning to avoid a collision. It takes about 3 s for a driver having 0.1% alcohol in his system to do the same. If two such drivers are traveling on a straight road at 30 mph (44 ft/s) and their cars can decelerate at 2 ft/s², determine the shortest stopping distance d for each from the moment they see the pedestrians at A. *Moral:* If you must drink, please don't drive!

$v_1 = 44$ ft/s

A

d

Prob. 12–5

12–6. A particle travels along a straight-line path such that in 2 s it moves from an initial position $s_A = +0.5$ m to a position $s_B = -1.5$ m. Then in another 4 s it moves from s_B to $s_C = +2.5$ m. Determine the particle's average velocity and average speed during the 6-s time interval.

12–7. The position of a particle along a straight line is given by $s = (t^3 - 9t^2 + 15t)$ ft, where t is in seconds. Determine the position of the particle when $t = 6$ s and the total distance it travels during the 6-s time interval. *Hint:* Plot the path to determine the total distance traveled.

*12–8. A go-cart starts from rest and travels on a straight road with a constant acceleration of 6 ft/s² for 4 s, after which it maintains a constant speed for 2 s. Finally it decelerates at 8 ft/s² until it stops. Determine the total distance traveled and the average speed.

12–9. The acceleration of a particle as it moves along a straight-line path is given by $a = (2t - 1)$ m/s², where t is in seconds. If $s = 1$ m and $v = 2$ m/s when $t = 0$, determine the particle's velocity and positions when $t = 6$ s. Also, determine the total distance the particle travels during this time period.

12–10. A particle travels in a straight line such that for a short time $2 \text{ s} \leqslant t \leqslant 6 \text{ s}$ its motion is described by $v = (4/a)$ ft/s, where a is in ft/s². If $v = 6$ ft/s when $t = 2$ s, determine the particle's acceleration when $t = 3$ s.

12–11. A particle travels to the right along a straight path with a velocity $v = [5/(4 + s)]$ m/s, where s is in meters. Determine its position when $t = 6$ s if $s = 5$ m when $t = 0$.

*12–12. When a train is traveling along a straight track at 2 m/s, it begins to accelerate at $a = (60 v^{-4})$ m/s², where v is in m/s. Determine the velocity v and the position of the train 3 s after the acceleration.

12–13. Car B is traveling a distance d ahead of car A. Both cars are traveling at 60 ft/s when the driver of B suddenly applies the brakes, causing his car to decelerate at 12 ft/s². It takes the driver of car A 0.75 s to react (this is the normal reaction time for drivers). When he applies his brakes, he decelerates at 15 ft/s². Determine the minimum distance d between the cars so as to avoid a collision.

A　B

d

Prob. 12–13

12–14. A particle travels in a straight line with accelerated motion such that $a = -ks$, where s is the distance from the starting point and k is a proportionality constant which is to be determined. For $s = 2$ ft the velocity is 4 ft/s, and for $s = 3.5$ ft the velocity is 10 ft/s. What is s for $v = 0$?

12–15. A particle moving along a straight line is subjected to a deceleration $a = (-2v^3)$ m/s^2, where v is in m/s. If it has a velocity $v = 8$ m/s and a position of $s = 10$ m when $t = 0$, determine its velocity and position when $t = 4$ s.

*12–16. The velocity of a particle traveling in a straight line is given by $v = (6t - 3t^2)$ m/s, where t is in seconds. If $s = 0$ when $t = 0$, determine the particle's deceleration and position when $t = 3$ s. How far has the particle traveled during the 3-s time interval, and what is its average speed?

12–17. The acceleration of a rocket traveling upwards is given by $a = (6 + 0.02s)$ m/s^2, where s is in meters. Determine the rocket's velocity when $s = 2$ km and the time needed to reach this elevation. Initially, $v = 0$ and $s = 0$ when $t = 0$.

12–18. A motorcycle starts from rest at $t = 0$ and travels along a straight road with a constant acceleration of 6 ft/s^2 until it reaches a speed of 50 ft/s. Afterwards it maintains this speed. Also, when $t = 0$, a car located 6000 ft down the road is traveling toward the motorcycle at a constant speed of 30 ft/s. Determine the time and distance traveled by the motorcycle when they pass each other.

12–19. The speed of a particle traveling along a straight-line path within a liquid is measured as a function of its position as $v = (100 - s)$ mm/s, where s is in millimeters. Determine (a) the particle's deceleration when it is located at point A, where $s_A = 50$ mm; (b) the distance the particle travels before it stops; and (c) the time needed to stop the particle. When $t = 0$, $s = 0$.

■*12–20. A particle moves along a straight-line path with an acceleration of $a = 5/(3s^{1/3} + s^{5/2})$, where s is in meters. Determine the particle's velocity when $s = 2$ m, if it starts from rest when $s = 1$ m. Use Simpson's rule to evaluate the integral.

12–21. Determine the time required for a car to travel 1 km along a road if it starts from rest, reaches a maximum speed at some intermediate point, and then stops at the end of the road. The car can accelerate at 1.5 m/s^2 and decelerate at 2 m/s^2.

■12–22. A projectile, initially at the origin, moves along a straight-line path through a fluid medium such that its velocity is defined as $v = 3(8e^{-t} + t)^{1/2}$ m/s, where t is in seconds. Plot the position s of the projectile during the first 2 s. Use the Runge-Kutta method to evaluate s with incremental values of $h = 0.25$ s.

12–23. As a body is projected to a high altitude above the earth's *surface*, the variation of the acceleration of gravity with respect to altitude y must be taken into account. Neglecting air resistance, this acceleration is determined from the formula $a = -g_o[R^2/(R + y)^2]$, where g_o is the constant gravitational acceleration at sea level, R is the radius of the earth, and the positive direction is measured upward. If $g_o = 9.81$ m/s^2 and $R = 6356$ km, determine the minimum initial velocity (escape velocity) at which a projectile should be shot vertically from the earth's surface so that it does not fall back to the earth. *Hint:* This requires that $v = 0$ as $y \to \infty$.

*12–24. Accounting for the variation of gravitational acceleration a with respect to altitude y (see Prob. 12–23), derive an equation that relates the velocity of a freely falling particle to its altitude. Assume that the particle is released from rest at an altitude y_o from the earth's surface. With what velocity does the particle strike the earth if it is released from rest at an altitude $y_o = 500$ km? Use the numerical data in Prob. 12–23.

Graphical Solutions **12.2**

When a particle's motion during a time period is erratic, it may be difficult to obtain a continuous mathematical function to describe its position, velocity, or acceleration. Instead, the motion may best be described graphically using a series of curves. If this graph describes the behavior of any two of the variables, a, v, s, t, a graph describing the behavior of the other variables can be established by using the kinematic equations $a = dv/dt$, $v = ds/dt$, $a\,ds = v\,dv$. The following situations frequently occur.

Given the *s-t* Graph, Construct the *v-t* Graph. If the position of a particle can be *experimentally determined* during a time period t, the *s-t* graph for the particle can be plotted, Fig. 12–7a. Since $v = ds/dt$, *the v-t graph can be established by measuring the slope (ds/dt) of the s-t graph at various times and plotting the results*. (In a graphical sense, the slope is measured with a ruler and protractor.) For example, measurement of the slopes v_0, v_1, v_2, and v_3 at the intermediate points (t_0, s_0), (t_1, s_1), (t_2, s_2), and (t_3, s_3) on the *s-t* graph, Fig. 12–7a, gives corresponding points on the *v-t* graph shown in Fig. 12–7b.

It may also be possible to establish the *v-t* graph *mathematically,* provided the curves of the *s-t* graph can be expressed in the form of an equation $s = f(t)$. Corresponding equations describing curves of the *v-t* graph are then determined by *differentiation* since $v = ds/dt = d(f(t))/dt$.

(a)

Fig. 12–7

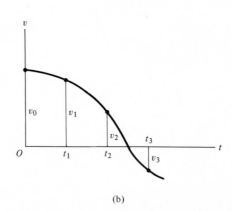

(b)

Given the *v-t* Graph, Construct the *a-t* Graph. When the particle's *v-t* graph is known, as in Fig. 12–8*a*, the acceleration can be determined at any instant using $a = dv/dt$. Hence, *the a-t graph can be established by measuring the slope (dv/dt) of the v-t graph at various times and plotting the results*. For example, measurement of the slopes a_0, a_1, a_2, and a_3 at the intermediate points (t_0, v_0), (t_1, v_1), (t_2, v_2), and (t_3, v_3) on the *v-t* graph, Fig. 12–8*a*, yields corresponding points on the *a-t* graph shown in Fig. 12–8*b*.

The segmented curves of the *a-t* graph can also be determined *mathematically* provided the equations of the corresponding curves of the *v-t* graph are known, $v = g(t)$. This is done by simply taking the *derivative* since $a = dv/dt = d\,(g(t))/dt$.

Since differentiation reduces a polynomial of degree *n* to that of degree *n*-1, then from the above explanation, if the *s-t* graph is parabolic (second-degree curve), the *v-t* graph will be a sloping line (first-degree curve), and the *a-t* graph will be a constant or a horizontal line (zero-degree curve).

(a)

Fig. 12–8

(b)

Example 12–7

A car moves along a straight road such that its position is described by the graph shown in Fig. 12–9a. Construct the *v-t* and *a-t* graphs for the time period $0 \leqslant t \leqslant 30$ s.

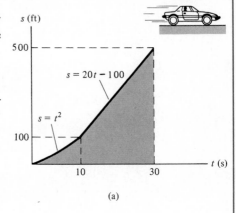

(a)

Solution

v-t Graph. Since $v = ds/dt$, the *v-t* graph can be determined by differentiating the equations defining the *s-t* graph, Fig. 12–9a. We have

$$0 \leqslant t < 10 \text{ s}; \quad s = t^2 \qquad v = \frac{ds}{dt} = 2t$$

$$10 \text{ s} < t \leqslant 30 \text{ s}; \quad s = 20t - 100 \qquad v = \frac{ds}{dt} = 20$$

The results are plotted in Fig. 12–9b. We can also obtain specific values of v by measuring the *slope* of the *s-t* graph at a given instant. For example, $t = 20$ s, the slope of the *s-t* graph is determined from the straight line from 10 s to 30 s, i.e.,

$$t = 20 \text{ s}; \qquad v = \frac{\Delta s}{\Delta t} = \frac{500 - 100}{30 - 10} = 20 \text{ ft/s}$$

(b)

a-t Graph. Since $a = dv/dt$, the *a-t* graph can be determined by differentiating the equations defining the lines of the *v-t* graph. This yields

$$0 \leqslant t < 10 \text{ s}; \qquad v = 2t \qquad a = \frac{dv}{dt} = 2$$

$$10 < t \leqslant 30 \text{ s}; \qquad v = 20 \qquad a = \frac{dv}{dt} = 0$$

The results are plotted in Fig. 12–9c. Show how to determine the specific value of the car's acceleration when $t = 5$ s by measuring the slope of the *v-t* graph.

(c)

Fig. 12–9

15

(a)

(b)

Fig. 12–10

(a)

(b)

Fig. 12–11

Given the *a-t* Graph, Construct the *v-t* Graph. If the *a-t* graph is given, Fig. 12–10*a*, the *v-t* graph may be constructed using the equation $a = dv/dt$, written in integrated form as $\Delta v = \int a \, dt$. In this case, *the change in the particle's speed during a period of time is equal to the area under the a-t graph during the same time period,* Fig. 12–10*b*. (In a graphical sense, any *small* area may be approximated as a trapezoid or rectangle.) Using this method, one begins with knowing the particle's initial velocity v_0 and adding (algebraically) to this small area increments Δv determined from the *a-t* graph. In this manner, one determines successive points, $v_1 = v_0 + \Delta v$, etc., for the *v-t* graph. Notice that an algebraic addition of area is necessary, since areas lying above the *t* axis correspond to an increase in v ("positive" area), whereas those lying below the *t* axis indicate a decrease in v ("negative" area).

If the curves of the *a-t* graph can be described by a series of equations, then each of these equations may be *integrated* to yield equations describing the corresponding curves of the *v-t* graph. Hence, if the *a-t* graph is linear (first-degree curve), the integration will yield a *v-t* graph that is parabolic (second-degree curve), etc.

Given the *v-t* Graph, Construct the *s-t* Graph. When the *v-t* graph is given, Fig. 12–11*a*, it is possible to determine the *s-t* graph using $v = ds/dt$, written in integrated form as $\Delta s = \int v \, dt$. In this case *the particle's displacement during a period of time is equal to the area under the v-t graph during the same time period,* Fig. 12–11*b*. In the same manner as stated above, one begins by knowing the particle's initial position s_0 and adding (algebraically) to this small area increments Δs determined from the *v-t* graph.

If it is possible to describe the segments of the *v-t* graph by a series of equations, then each of these equations may be *integrated* to yield equations describing the *s-t* graph.

Example 12–8

The rocket sled in Fig. 12–12a starts from rest and travels along a straight track such that it accelerates at a constant rate for 10 s and then decelerates at a constant rate. Draw the v-t and s-t graphs and determine the time t' needed to stop the sled. How far has the sled traveled?

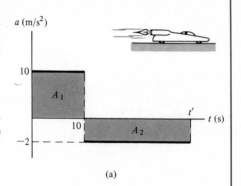

(a)

Solution

v-t Graph. Since $dv = a\,dt$, the v-t graph is determined by integrating the straight-line segments of the a-t graph. Using the initial condition $v = 0$ when $t = 0$, we have

$$0 \leq t < 10 \text{ s}; \quad a = 10; \quad \int_0^v dv = \int_0^t 10\,dt$$

$$v = 10t$$

When $t = 10$ s, $v = 10(10) = 100$ m/s. Using this as the initial condition for the next time period, we have

$$10 \text{ s} < t \leq t'; \quad a = -2; \quad \int_{100}^v dv = \int_{10}^t -2\,dt$$

$$v = -2t + 120$$

When $t = t'$ we require $v = 0$. This yields

$$t' = 60 \text{ s} \qquad \textit{Ans.}$$

The results are shown in Fig. 12–12b.

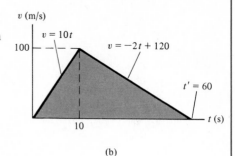

(b)

s-t Graph. Since $ds = v\,dt$, integrating the equations of the v-t graph yields the corresponding equations of the s-t graph. Using the initial conditions $s = 0$ when $t = 0$, we have

$$0 \leq t < 10 \text{ s}; \quad v = 10t; \quad \int_0^s ds = \int_0^t 10t\,dt$$

$$s = 5t^2$$

When $t = 10$ s, $s = 5(10)^2 = 500$ m. Using this initial condition,

$$10 < t \leq 60 \text{ s}; \quad v = -2t + 120;$$

$$\int_{500}^s ds = \int_{10}^t (-2t + 120)\,dt$$

$$s - 500 = -t^2 + 120t - (-(10)^2 + 120(10))$$

$$s = -t^2 + 120t - 600$$

When $t' = 60$ s, the position is

$$s = -(60)^2 + 120(60) - 600 = 3000 \text{ m} \qquad \textit{Ans.}$$

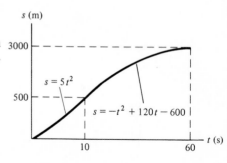

(c)

Fig. 12–12

The s-t graph is shown in Fig. 12–12c. Note that a direct solution for s when $t' = 60$ s is possible, since the *triangular area* under the v-t graph would yield the displacement $\Delta s = s - 0$ from $t = 0$ to $t' = 60$ s. Hence,

$$\Delta s = \tfrac{1}{2}(60)(100) = 3000 \text{ m} \qquad \textit{Ans.}$$

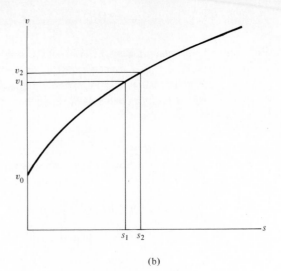

$$\int_{s_1}^{s_2} a \, ds = \tfrac{1}{2} \, (v_2{}^2 - v_1{}^2)$$

(a)

(b)

Fig. 12–13

(a)

(b)

Fig. 12–14

Given the a-s Graph, Construct the v-s Graph. In some cases an a-s graph is known, so that points on the v-s graph can be determined by using $v \, dv = a \, ds$. Integrating this equation between the limits $v = v_1$ at $s = s_1$ and $v = v_2$ at $s = s_2$, we have $\tfrac{1}{2}(v_2^2 - v_1^2) = \int_{s_1}^{s_2} a \, ds$. Thus, small segments of area under the a-s graph, $\int_{s_1}^{s_2} a \, ds$, shown colored in Fig. 12–13a, equal one-half the difference in the squares of the speed, $\tfrac{1}{2}(v_2^2 - v_1^2)$. By approximation of the area, $\int_{s_1}^{s_2} a \, ds$, it is possible to compute the value of v_2 at s_2 if an initial value of v_1 at s_1 is known, i.e., $v_2 = (2 \int_{s_1}^{s_2} a \, ds + v_1^2)^{1/2}$, Fig. 12–13b. The v-s graph can be constructed in this manner starting from the initial velocity v_0.

Another way to construct the v-s graph is to first determine the equations which define the various curves of the a-s graph. Then the corresponding equations defining the curves of the v-s graph can be obtained directly from integration, using $v \, dv = a \, ds$.

Given the v-s Graph, Construct the a-s Graph. If the v-s graph is known, the acceleration a at any position s can be determined using the following graphical procedure. At any point $P(s, v)$, Fig. 12–14a, the slope dv/ds of the v-s graph is determined. The perpendicular line drawn from this slope intersects the s axis at point B. The distance AB represents the magnitude of the particle's acceleration a, Fig. 12–14b, since the two colored triangles shown in Fig. 12–14a are similar. (Recall that $a \, ds = v \, dv$, so that $\tan \theta = dv/ds = a/v$.) Of course, one must work with a consistent set of units when using this procedure; for example, if v is measured in m/s and s in meters, then a will be measured in m/s^2.

We can also determine the curves describing the a-s graph analytically, provided the equations of the corresponding curves of the v-s graph are known. As above, this requires application of $a \, ds = v \, dv$.

Example 12–9

The v-s graph describing the motion of a motorcycle is shown in Fig. 12–15a. Construct the a-s graph of the motion and determine the time needed for the motorcycle to reach the position $s = 400$ ft.

Solution

a-s Graph. Since the equations of the v-s graph are given, the a-s graph can be determined using the equation $a\,ds = v\,dv$, which yields

$0 \le s < 200$ ft; $v = 0.2s + 10$

$$a = v\frac{dv}{ds} = (0.2s + 10)\frac{d}{ds}(0.2s + 10)$$

$$= 0.04s + 2$$

200 ft $< s \le 400$ ft; $v = 50$; $a = v\dfrac{dv}{ds} = (50)\dfrac{d}{ds}(50)$

$$= 0$$

The results are plotted in Fig. 12–15b.

Time. The time can be obtained using the v-s graph and $v = ds/dt$, because this equation relates v, s, and t. For the two segments of motion, we have

$0 \le s < 200$ ft; $v = 0.2s + 10$; $dt = \dfrac{ds}{v} = \dfrac{ds}{0.2s + 10}$

$$\int_0^t dt = \int_0^s \frac{ds}{0.2s + 10}$$

$$t = 5 \ln (0.2s + 10) - 5 \ln 10$$

At $s = 200$ ft, $t = 5 \ln (0.2(200) + 10) - 5 \ln 10 = 8.05$ s.

200 ft $< s \le 400$ ft; $v = 50$; $dt = \dfrac{ds}{v} = \dfrac{ds}{50}$

$$\int_{8.05}^t dt = \int_{200}^s \frac{ds}{50}$$

$$t - 8.05 = \frac{s}{50} - 4$$

$$t = \frac{s}{50} + 4.05$$

Therefore, at $s = 400$ ft,

$$t = \frac{400}{50} + 4.05 = 12.0 \text{ s} \qquad\qquad Ans.$$

v (ft/s)

$v = 0.2s + 10$

$v = 50$

50

10

200 400

s (ft)

(a)

a (ft/s^2)

$a = 0.04s + 2$

10

2

$a = 0$

200 400

s (ft)

(b)

Fig. 12–15

Problems

12–25. The v-t graph for the motion of a train as it moves from station A to station B is shown. Draw the a-t graph and determine the average speed and the distance between the stations.

Prob. 12–25

12–26. The bobsled moves down along the straight course such that its v-t graph is as shown. Construct the s-t and a-t graphs for the same 50-s time interval. When $t = 0$, $s = 0$.

Prob. 12–26

12–27. Two cars start from rest side by side and race along a straight road. Car A accelerates at 4 m/s² for 10 s and then maintains a constant speed. Car B accelerates at 5 m/s² until reaching a constant speed of 25 m/s and then maintains this speed. Construct the a-t, v-t, and s-t graphs for each car until $t = 15$ s. What is the distance between the two cars when $t = 15$ s?

*12–28. The s-t graph for a train has been experimentally determined. From the data, construct the v-t and a-t graphs for the motion; $0 \leqslant t \leqslant 40$ s. For $0 \leqslant t \leqslant 30$ s, the curve is $s = (0.4t^2)$ m, where t is in seconds.

Prob. 12–28

12–29. A car travels along a straight road with the speed shown by the v-t graph. Determine the total distance the car travels until it stops when $t = 48$ s. Also plot the s-t and a-t graphs.

Prob. 12–29

12–30. The v-t graph for particles moving through an electric field from one plate to another has the shape shown in the figure. The acceleration and deceleration that occur are constant and both have a magnitude of 4 m/s². If the plates are spaced 200 mm apart, determine the maximum velocity v_{max} and the time t' to travel from one plate to the other. Also draw the s-t graph.

Prob. 12–30

12–31. A dragster starting from rest travels along a straight road and for 10 s has an acceleration as shown. Construct the v-t graph that describes the motion and find the distance traveled in 10 s.

Prob. 12–31

*12–32. The sports car travels from rest along a straight road and reaches a speed of 26 m/s in 8 s as shown on the v-t graph. The flat part of the graph is caused by shifting gears. Draw the a-t graph and determine the total distance traveled.

Prob. 12–32

12–33. The v-s graph for the car is given for the first 500 ft of its motion. Plot the a-s graph for $0 \leq s \leq 500$ ft. How long does it take to travel the 500-ft distance? The car starts from $s = 0$ when $t = 0$.

Prob. 12–33

12–34. The *a-s* graph for a train traveling along a straight track is given for the first 400 m of its motion. Plot the *v-s* graph. $v = 0$ at $s = 0$.

Prob. 12–34

12–35. A two-stage rocket is fired vertically from rest with an acceleration as shown. After 15 s the first stage *A* burns out and the second stage *B* ignites. Plot the *v-t* and *s-t* graphs which describe the motion of the second stage for $0 \leq t \leq 40$ s.

Prob. 12–35

***12–36.** The *v-s* graph for a go-cart traveling on a straight road is shown. Determine the acceleration of the go-cart at $s = 100$ m and $s = 150$ m. Draw the *a-s* graph.

Prob. 12–36

12–37. The rocket car is originally traveling at a speed of 20 m/s when it is subjected to the acceleration shown in the graph. Determine the car's maximum speed and the time *t* when it stops.

Prob. 12–37

■**12–38.** The *a-s* graph for a rocket moving along a straight horizontal track has been experimentally determined. If the rocket starts at $s = 0$ when $v = 0$, determine its speed when it is at $s = 75$ ft, and 125 ft, respectively. Use Simpson's rule with $n = 100$ to evaluate v at $s = 125$ ft.

$$a = s + 6(\sqrt{s} - 10)^{5/3}$$

Prob. 12–38

12–39. A freight train starts from rest when $t = 0$ and travels on a straight track with a constant acceleration of 0.75 ft/s². When $t = t'$ it maintains a constant speed so that when $t = 160$ s the train has traveled 1500 ft. Determine the time t' and draw the *v-t* graph for the motion.

*12–40. Starting from rest at $s = 0$, a boat travels in a straight line with an acceleration as shown by the a-s graph. Determine the boat's speed when $s = 40$, 90, and 200 ft.

Prob. 12–40

12–41. The rocket sled starts from rest at $s = 0$ and is subjected to an acceleration as shown by the a-s graph. Draw the v-s graph and determine the time needed to travel 500 ft.

Prob. 12–41

12–42. A motorcyclist at A is traveling at 60 ft/s when he wishes to pass the truck T which is traveling at a constant speed of 60 ft/s. To do so the motorcyclist accelerates at 6 ft/s^2 until reaching a maximum speed of 85 ft/s. If he then maintains this speed, determine the time needed for him to reach a point located 100 ft in front of the truck. Draw the v-t graph for both the motorcycle and the truck during this time.

Prob. 12–42

General Curvilinear Motion 12.3

When a particle moves along a curved path, the motion is called *curvilinear motion*. Because the path is often represented in three dimensions, *vector analysis* will be used to formulate the particle's position, velocity, and acceleration.* In this section the general aspects of curvilinear motion are discussed and in subsequent sections three types of coordinate systems often used to describe this motion will be introduced.

---

*A summary of some of the important concepts of vector analysis is given in Appendix B.

Path

Position
(a)

Displacement
(b)

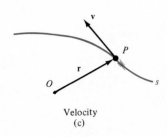

Velocity
(c)

Fig. 12–16a–c

Position. Consider a particle located at point P on a space curve defined by the path function s, Fig. 12–16a. The position of the particle, measured from a fixed point O, will be designated by the *position vector* $\mathbf{r} = \mathbf{r}(t)$. This vector is a function of time since, in general, both its magnitude and direction change as the particle moves along the curve.

Displacement. Suppose that during a small time interval Δt the particle moves a distance Δs along the curve to a new position P', defined by $\mathbf{r}' = \mathbf{r} + \Delta \mathbf{r}$, Fig. 12–16b. The *displacement* $\Delta \mathbf{r}$ represents the change in the particle's position and is determined by vector subtraction; i.e., $\Delta \mathbf{r} = \mathbf{r}' - \mathbf{r}$.

Velocity. During the time Δt, the *average velocity* of the particle is defined as

$$\mathbf{v}_{\text{avg}} = \frac{\Delta \mathbf{r}}{\Delta t}$$

The *instantaneous velocity* is determined from this equation by letting $\Delta t \to 0$ and consequently the direction of $\Delta \mathbf{r}$ approaches the *tangent* to the curve at point P. Hence, $\mathbf{v} = \lim_{\Delta t \to 0} (\Delta \mathbf{r}/\Delta t)$ or

$$\mathbf{v} = \frac{d\mathbf{r}}{dt} \tag{12–7}$$

Since $d\mathbf{r}$ will be tangent to the path at P, the *direction* of $\mathbf{v}$ is *also tangent to the path*, Fig. 12–16c. The *magnitude* of $\mathbf{v}$, which is called the *speed*, may be obtained by noting that the magnitude of the displacement $\Delta \mathbf{r}$ is the length of the straight line segment from P to P', Fig. 12–16b. Realizing that this length, Δr, approaches the arc length Δs as $\Delta t \to 0$, we have $v = \lim_{\Delta t \to 0} (\Delta r/\Delta t) = \lim_{\Delta t \to 0} (\Delta s/\Delta t)$, or

$$v = \frac{ds}{dt} \tag{12–8}$$

Thus, the *speed* may be obtained by differentiating the path function s with respect to time.

Fig. 12–16d–g

Acceleration. If the particle has a velocity **v** at time t and a velocity $\mathbf{v'} = \mathbf{v} + \Delta\mathbf{v}$ at $t + \Delta t$, Fig. 12–16d, then the *average acceleration* of the particle during the time interval Δt is

$$\mathbf{a}_{\text{avg}} = \frac{\Delta\mathbf{v}}{\Delta t}$$

where $\Delta\mathbf{v} = \mathbf{v'} - \mathbf{v}$. To study this time rate of change, the two velocity vectors in Fig. 12–16d are plotted in Fig. 12–16e such that their tails are located at the fixed point O' and their tips reach points on the dashed curve. This curve is called a *hodograph,* and when constructed, it describes the locus of points for the tip of the velocity vector in the same manner as the *path s* describes the locus of points for the tip of the position vector, Fig. 12–16a.

To obtain the *instantaneous acceleration,* let $\Delta t \rightarrow 0$ so that in the limit $\Delta\mathbf{v}$ becomes *tangent to the hodograph* and we have $\mathbf{a} = \lim\limits_{\Delta t \rightarrow 0} (\Delta\mathbf{v}/\Delta t)$, or

$$\mathbf{a} = \frac{d\mathbf{v}}{dt} \qquad\qquad (12\text{–}9)$$

Using Eq. 12–7, we can also write

$$\mathbf{a} = \frac{d^2\mathbf{r}}{dt^2}$$

By definition of the derivative, **a** acts *tangent to the hodograph,* Fig. 12–16f, and therefore, *in general,* **a** *is not tangent to the path of motion,* Fig. 12–16g. To clarify this point, realize that $\Delta\mathbf{v}$ and consequently **a** account for the change made in both the magnitude *and* direction of the velocity **v** as the particle moves from P to P', Fig. 12–16d. The magnitude change increases (or decreases) the "length" of **v**, and in order for the particle to follow the curve, the directional change always swings the velocity vector toward the "inside" or "concave side" of the curve.

12.4 Curvilinear Motion: Rectangular Components

In some cases the motion of a particle can be conveniently expressed in terms of its three rectangular components, which are measured from the origin of a fixed x, y, z frame of reference.

Position. If at a given instant the particle is at point $P(x, y, z)$ on the curved path s, Fig. 12–17a, its location is then defined by the *position vector*

$$\mathbf{r} = x\mathbf{i} + y\mathbf{j} + z\mathbf{k} \qquad (12\text{–}10)$$

Because of the particle motion and the shape of the path, the x, y, z components of $\mathbf{r}$ are generally all functions of time; i.e., $x = x(t)$, $y = y(t)$, and $z = z(t)$, so that $\mathbf{r} = \mathbf{r}(t)$.

In accordance with the discussion in Appendix C, the *magnitude* of $\mathbf{r}$ is defined from Eq. C–3 as

$$r = \sqrt{x^2 + y^2 + z^2}$$

and the *direction* is specified by the components of the unit vector $\mathbf{u}_r = \mathbf{r}/r$.

Velocity. The first time derivative of $\mathbf{r}$ yields the velocity $\mathbf{v}$ of the particle. Hence,

$$\mathbf{v} = \frac{d\mathbf{r}}{dt} = \frac{d}{dt}(x\mathbf{i}) + \frac{d}{dt}(y\mathbf{j}) + \frac{d}{dt}(z\mathbf{k})$$

When taking the derivative, it is necessary to account for changes in *both* the magnitude and direction of the vector's components. From Eq. C–23, the derivative of the $\mathbf{i}$ component of $\mathbf{v}$ is

$$\frac{d}{dt}(x\mathbf{i}) = \frac{dx}{dt}\mathbf{i} + x\frac{d\mathbf{i}}{dt}$$

Position

(a)

Velocity

(b)

Acceleration

(c)

Fig. 12–17

The second term on the right side is zero, since the x, y, z reference frame is *fixed*, and therefore the *direction* (and the *magnitude*) of $\mathbf{i}$ does not change with time. Differentiation of the $\mathbf{j}$ and $\mathbf{k}$ components may be carried out in a similar manner, which yields the final result,

$$\mathbf{v} = \frac{d\mathbf{r}}{dt} = v_x\mathbf{i} + v_y\mathbf{j} + v_z\mathbf{k} \qquad (12\text{--}11)$$

where

$$\begin{aligned} v_x &= \dot{x} \\ v_y &= \dot{y} \\ v_z &= \dot{z} \end{aligned} \qquad (12\text{--}12)$$

The "dot" notation $\dot{x}$, $\dot{y}$, $\dot{z}$ represents the first time derivatives of the parametric equations $x = x(t)$, $y = y(t)$, and $z = z(t)$, respectively.

The velocity has a *magnitude* defined by

$$v = \sqrt{v_x^2 + v_y^2 + v_z^2}$$

and a *direction* that is specified by the components of the unit vector $\mathbf{u}_v = \mathbf{v}/v$. This direction is *always tangent* to the path, as shown in Fig. 12–17b.

Acceleration. The acceleration of the particle is obtained by taking the first time derivative of Eq. 12–11 (or the second time derivative of Eq. 12–10). Using dots to represent the derivatives of the components, we have

$$\mathbf{a} = \frac{d\mathbf{v}}{dt} = a_x\mathbf{i} + a_y\mathbf{j} + a_z\mathbf{k} \qquad (12\text{--}13)$$

where

$$\begin{aligned} a_x &= \dot{v}_x = \ddot{x} \\ a_y &= \dot{v}_y = \ddot{y} \\ a_z &= \dot{v}_z = \ddot{z} \end{aligned} \qquad (12\text{--}14)$$

Hence, a_x, a_y, and a_z represent, respectively, the first time derivatives of the time functions v_x, v_y, and v_z, or the second time derivatives of the time functions x, y, and z.

The acceleration has a *magnitude* defined by

$$a = \sqrt{a_x^2 + a_y^2 + a_z^2}$$

and a *direction* specified by the components of the unit vector $\mathbf{u}_a = \mathbf{a}/a$. Since $\mathbf{a}$ represents the time rate of *change* in velocity, in general $\mathbf{a}$ will *not* be tangent to the path traveled by the particle, Fig. 12–17c.

Example 12–10

At any instant the position of the kite in Fig. 12–18a is defined by the coordinates $x = (30t)$ ft and $y = (9t^2)$ ft, where t is given in seconds. Determine (a) the equation which describes the path and the distance of the kite from the boy when $t = 2$ s, (b) the magnitude and direction of the velocity when $t = 2$ s, and (c) the magnitude and direction of the acceleration when $t = 2$ s.

Solution

Position. The equation of the path is determined by eliminating t from the expressions for x and y, i.e., $t = x/30$ so that $y = 9(x/30)^2$ or

$$y = \frac{1}{100}x^2 \qquad \qquad \text{Ans.}$$

This is the equation of a parabola, Fig. 12–18a. When $t = 2$ s, the kite is located at

$$x = 30(2) = 60 \text{ ft}, \qquad y = 9(2)^2 = 36 \text{ ft}$$

The distance AB is therefore

$$r = \sqrt{(60)^2 + (36)^2} = 70.0 \text{ ft} \qquad \qquad \text{Ans.}$$

Velocity. Using Eqs. 12–12, the components of velocity when $t = 2$ s are

$$v_x = \frac{d}{dt}(30t) = 30 \text{ ft/s} \rightarrow$$

$$v_y = \frac{d}{dt}(9t^2) = 18t \bigg|_{t=2\,s} = 36 \text{ ft/s} \uparrow$$

When $t = 2$ s, the magnitude of velocity is

$$v = \sqrt{(30)^2 + (36)^2} = 46.9 \text{ ft/s} \qquad \qquad \text{Ans.}$$

The direction is tangent to the path, Fig. 12–18b, where

$$\theta_v = \tan^{-1}\frac{36}{30} = 50.2° \qquad \qquad \text{Ans.}$$

Acceleration. The components of acceleration are determined from Eqs. 12–14,

$$a_x = \frac{d}{dt}(30) = 0$$

$$a_y = \frac{d}{dt}(18t) = 18 \text{ ft/s}^2 \uparrow$$

Here the acceleration is constant for all time and thus

$$a = \sqrt{(0)^2 + (18)^2} = 18 \text{ ft/s}^2 \qquad \qquad \text{Ans.}$$

The direction, as shown in Fig. 12–18c, is

$$\theta_a = \tan^{-1}\frac{18}{0} = 90° \qquad \qquad \text{Ans.}$$

(a)

(b)

(c)

Fig. 12–18

Example 12–11

The motion of a bead B sliding down along the spiral path shown in Fig. 12–19 is defined by the position vector $\mathbf{r} = \{0.5 \sin (2t)\mathbf{i} + 0.5 \cos (2t)\mathbf{j} - 0.2t\mathbf{k}\}$ m, where t is given in seconds and the arguments for sine and cosine are given in radians (π rad $= 180°$). Determine the location of the bead when $t = 0.75$ s and the magnitudes of the bead's velocity and acceleration at this instant.

Solution

Position. Evaluating $\mathbf{r}$ when $t = 0.75$ s yields

$$\mathbf{r}|_{t=0.75\,\text{s}} = \{0.5 \sin (1.5 \text{ rad})\mathbf{i} + 0.5 \cos (1.5 \text{ rad})\mathbf{j} - 0.2(0.75)\mathbf{k}\} \text{ m}$$
$$= \{0.499\mathbf{i} + 0.035\mathbf{j} - 0.150\mathbf{k}\} \text{ m} \qquad Ans.$$

The magnitude of this vector represents the distance of the bead from the origin O, which is

$$r = \sqrt{(0.499)^2 + (0.035)^2 + (-0.150)^2} = 0.522 \text{ m} \qquad Ans.$$

The direction of $\mathbf{r}$ is obtained from the components of the unit vector

$$\mathbf{u}_r = \frac{\mathbf{r}}{r} = \frac{0.499}{0.522}\mathbf{i} + \frac{0.035}{0.522}\mathbf{j} - \frac{0.150}{0.522}\mathbf{k}$$
$$= 0.956\mathbf{i} + 0.067\mathbf{j} - 0.287\mathbf{k}$$

Hence, the coordinate direction angles α, β, and γ, shown in Fig. 12–19, are

$$\alpha = \cos^{-1} (0.956) = 17.1° \qquad Ans.$$
$$\beta = \cos^{-1} (0.067) = 86.2° \qquad Ans.$$
$$\gamma = \cos^{-1} (-0.287) = 106° \qquad Ans.$$

Velocity. The velocity is defined by

$$\mathbf{v} = \frac{d\mathbf{r}}{dt} = \{1 \cos (2t)\mathbf{i} - 1 \sin (2t)\mathbf{j} - 0.2\mathbf{k}\} \text{ m/s}$$

Hence, when $t = 0.75$ s the magnitude of velocity, or speed, is

$$v = \sqrt{v_x^2 + v_y^2 + v_z^2}$$
$$= \sqrt{(1 \cos (1.5 \text{ rad}))^2 + (-1 \sin (1.5 \text{ rad}))^2 + (-0.2)^2}$$
$$= 1.02 \text{ m/s} \qquad Ans.$$

The velocity is tangent to the path as shown in Fig. 12–19.

Acceleration. The bead's acceleration $\mathbf{a}$, which is shown in Fig. 12–19, is *not* tangent to the path. Show that

$$\mathbf{a} = \frac{d\mathbf{v}}{dt} = \{-2 \sin (2t)\mathbf{i} - 2 \cos (2t)\mathbf{j}\} \text{ m/s}^2$$

and when $t = 0.75$ s the magnitude is $a = 2$ m/s^2.

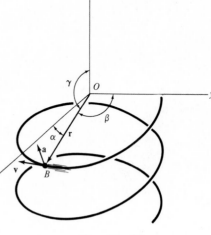

Fig. 12–19

12.5 Motion of a Projectile

Free-flight motion of a projectile is often studied in terms of its rectangular components. To illustrate the concepts involved in the kinematic analysis, consider a projectile launched at point $((s_x)_0, (s_y)_0)$, as shown in Fig. 12–20. The path is defined in the x-y plane such that the initial velocity is $\mathbf{v}_0$, having components $(\mathbf{v}_x)_0$ and $(\mathbf{v}_y)_0$. When air resistance is neglected, the only force acting on the projectile is its weight, which creates a *constant downward acceleration* of approximately $a_c = g = 9.81$ m/s^2 or $g = 32.2$ ft/s^2.*

Fig. 12–20

Horizontal Motion. Since the component of acceleration in the x direction is $a_x = 0$, application of the constant acceleration equations, 12–4 to 12–6, yields

$$(\xrightarrow{+})v = v_0 + a_c t; \qquad v_x = (v_x)_0$$
$$(\xrightarrow{+})s = s_0 + v_0 t + \tfrac{1}{2}a_c t^2; \qquad s_x = (s_x)_0 + (v_x)_0 t$$
$$(\xrightarrow{+})v^2 = v_0^2 + 2a_c(s - s_0); \qquad v_x = (v_x)_0$$

The first and last equations indicate that *the horizontal component of velocity remains constant throughout the motion.*

---

*This assumes that the earth's gravitational field does not vary with altitude (see Example 13–2).

Vertical Motion. Since the positive y axis is directed upward, the component of acceleration in the y direction is $a_y = -g$. Applying Eqs. 12–4 to 12–6, we get

$(+\uparrow)v = v_0 + a_c t;$ $v_y = (v_y)_0 - gt$

$(+\uparrow)s = s_0 + v_0 t + \frac{1}{2}a_c t^2;$ $s_y = (s_y)_0 + (v_y)_0 t - \frac{1}{2}gt^2$

$(+\uparrow)v^2 = v_0^2 + 2a_c(s - s_0);$ $v_y^2 = (v_y)_0^2 - 2g(s_y - (s_y)_0)$

Recall that the last equation can be formulated on the basis of eliminating the time t between the first two equations, and therefore *only two of the above three equations are independent of one another*.

At any instant, the resultant position **r** and velocity **v** are defined as the *vector sum* of their horizontal and vertical components, which are determined from the above equations and shown in Fig. 12–20.

PROCEDURE FOR ANALYSIS

Using the above results, the following procedure provides a method for solving problems concerning free-flight projectile motion.

Kinematic Diagram. Establish the x, y coordinate axes and sketch the trajectory of the particle. Between any *two points* on the path specify the given problem data and the three unknowns. In all cases the acceleration of gravity acts downward. The particle's initial and final velocities should be represented in terms of their x and y components. Remember that positive and negative position, velocity, and acceleration components always act in accordance with the associated coordinate directions.

Kinematic Equations. Depending upon the known data and what is to be determined, a choice should be made as to which three of the following four equations should be applied between the two points on the path to obtain the most direct solution to the problem.

Horizontal Motion. The *velocity* in the horizontal or x direction is *constant*, i.e., $(v_x) = (v_x)_0$, and

$$s_x = (s_x)_0 + (v_x)_0 t$$

Vertical Motion. In the vertical or y direction *only two* of the following three equations can be used for solution.

$$v_y = (v_y)_0 + a_c t$$
$$v_y^2 = (v_y)_0^2 + 2a_c(s_y - (s_y)_0)$$
$$s_y = (s_y)_0 + (v_y)_0 t + \frac{1}{2}a_c t^2$$

The following examples numerically illustrate application of this procedure.

31

Example 12–12

A cannon ball is fired from point A with a horizontal muzzle velocity of 120 m/s, as shown in Fig. 12–21. If the cannon is located at an elevation of 60 m above the ground, determine the time for the cannon ball to strike the ground and the range R.

Fig. 12–21

Solution

Kinematic Diagram. The origin of coordinates is established at the initial point A, Fig. 12–21. The initial velocity of the cannon ball has components $(v_A)_x = 120$ m/s and $(v_A)_y = 0$. Also, between points A and B the acceleration is $a_y = -9.81$ m/s^2. Since $(v_B)_x = (v_A)_x = 120$ m/s, the three unknowns are $(v_B)_y$, R, and the time of flight t.

Vertical Motion. The vertical distance from A to B is known, and therefore we can obtain a direct solution for t by using the equation

$$(+\uparrow) \qquad s_y = (s_y)_0 + (v_y)_0 t + \tfrac{1}{2} a_c t^2$$
$$-60 = 0 + 0 + \tfrac{1}{2}(-9.81) t^2$$
$$t = 3.50 \text{ s} \qquad \qquad \textit{Ans.}$$

This calculation also indicates that if an object is released from rest at A at the instant the cannon is fired (horizontally), it will strike the ground at C the same instant the cannon ball strikes the ground at B.

Horizontal Motion. Since t has been calculated, R is determined as follows:

$$(\xrightarrow{+}) \qquad s_x = (s_x)_0 + (v_x)_0 t$$
$$R = 0 + 120(3.50)$$
$$R = 420 \text{ m} \qquad \qquad \textit{Ans.}$$

Example 12–13

A ball is thrown from a position 5 ft above the ground to the roof of a 40-ft-high building, as shown in Fig. 12–22. If the initial velocity of the ball is 70 ft/s, inclined at an angle of 60° from the horizontal, determine the range or horizontal distance R from the point where the ball is thrown to where it strikes the roof.

Fig. 12–22

Solution

Kinematic Diagram. When the motion is analyzed between points O and A, the three unknowns are represented as the horizontal distance R, time of flight t_{OA}, and vertical component of velocity $(v_y)_A$. With the origin of coordinates at O, Fig. 12–22, the ball's initial velocity has components of

$$(v_x)_O = 70 \cos 60° = 35 \text{ ft/s} \rightarrow$$
$$(v_y)_O = 70 \sin 60° = 60.62 \text{ ft/s} \uparrow$$

Also, $(v_x)_A = (v_x)_O = 35$ ft/s and $a_y = -32.2$ ft/s².

Horizontal Motion

$(\xrightarrow{+})$

$$(s_x)_A = (s_x)_O + (v_x)_O t_{OA}$$
$$R = 0 + 35 t_{OA} \qquad (1)$$

Vertical Motion. Relating t_{OA} to the initial and final elevations of the ball,

$(+\uparrow)$

$$(s_y)_A = (s_y)_O + (v_y)_O t_{OA} + \tfrac{1}{2} a_c t_{OA}^2$$
$$(40 - 5) = 0 + 60.62 t_{OA} + \tfrac{1}{2}(-32.3)t_{OA}^2$$

Solving for the two roots, using the quadratic formula, we have $t_{OB} = 0.712$ s and $t_{OA} = 3.05$ s. The first root (shortest time) is designated as t_{OB}, since it represents the time needed for the ball to reach point B, which has the same elevation as point A, Fig. 12–22. Substituting t_{OA} into Eq. (1) and solving for R yields

$$R = 35(3.05) = 107 \text{ ft} \qquad \textit{Ans.}$$

Using only one equation, can you show that the maximum height reached by the ball, Fig. 12–22, is $h = 62.1$ ft?

Example 12–14

When a ball is kicked from A as shown in Fig. 12–23, it just clears the top of a wall at B as it reaches its maximum height. Knowing that the distance from A to the wall is 20 m and the wall is 4 m high, determine the initial speed at which the ball was kicked. Neglect the size of the ball.

Fig. 12–23

Solution

Kinematic Diagram. The three unknowns are represented by the initial speed v_A, angle of inclination θ, and the time t_{AB} to travel from A to B, Fig. 12–23. At the highest point, B, the velocity $(v_y)_B = 0$, and $(v_x)_B = (v_x)_A = v_A \cos \theta$.

Horizontal Motion

$$(\xrightarrow{+})$$

$$(s_x)_B = (s_x)_A + (v_x)_A t_{AB}$$
$$20 = 0 + (v_a \cos \theta) t_{AB} \tag{1}$$

Vertical Motion

$$(+\uparrow)$$

$$(v_y)_B = (v_y)_A + a_c t_{AB}$$
$$0 = v_A \sin \theta - 9.81 t_{AB} \tag{2}$$

$$(+\uparrow)$$

$$(v_y)_B^2 = (v_y)_A^2 + 2a_c[(s_y)_B - (s_y)_A]$$
$$0 = v_A^2 \sin^2 \theta + 2(-9.81)(4 - 0) \tag{3}$$

To obtain v_A, eliminate t_{AB} from Eqs. (1) and (2), which yields

$$v_A^2 \sin \theta \cos \theta = 196.2 \tag{4}$$

Solve for v_A^2 in Eq. (4) and substitute into Eq. (3), so that

$$\frac{\sin \theta}{\cos \theta} = \tan \theta = \frac{2(9.81)(4)}{196.2} = 0.4$$

$$\theta = \tan^{-1}(0.4) = 21.8°$$

Then using Eq. (4), the required initial speed is

$$v_A = \sqrt{\frac{196.2}{(\sin 21.8°)(\cos 21.8°)}} = 23.9 \text{ m/s} \qquad Ans.$$

Problems

12–43. A particle moves along the path $\mathbf{r} = \{8t^2\mathbf{i} + (t^3 + 5)\mathbf{j}\}$ m, where t is in seconds. Determine the magnitudes of the particle's velocity and acceleration when $t = 3$ s. Also determine the equation $y = f(x)$ of the path.

*12–44. A particle, originally at rest and located at point (3 ft, 2 ft, 5 ft), is subjected to an acceleration $\mathbf{a} = \{6t\mathbf{i} + 12t^2\mathbf{k}\}$ ft/s^2. Determine the particle's position (x, y, z) when $t = 2$ s.

12–45. The curvilinear motion of a particle is defined by $x = 3t^2$, $y = 4t + 2$, and $z = 6t^3 - 8$, where the x, y, z position is given in meters and the time in seconds. Determine the magnitudes and coordinate direction angles α, β, γ of the particle's velocity and acceleration when $t = 2$ s.

12–46. If the velocity of a particle is defined as $\mathbf{v}(t) = \{0.8t^2\mathbf{i} + 12t^{1/2}\mathbf{j} + 5\mathbf{k}\}$ m/s, determine the magnitude and coordinate direction angles α, β, γ of the particle's acceleration when $t = 2$ s.

12–47. A particle moves in the x-y plane such that its position is defined by $\mathbf{r} = \sin^2\theta\mathbf{i} + (\theta^2 + \cos 2\theta)\mathbf{j}$, where θ is in radians and r is in feet. If $\theta = (4t^2)$ rad, where t is in seconds, determine the particle's velocity and acceleration when $t = 2$ s.

■*12–48. A particle is traveling with a velocity of $\mathbf{v} = \{3\sqrt{t}\,e^{-0.2t}\mathbf{i} + 4e^{-0.8t^2}\mathbf{j}\}$ m/s, where t is in seconds. Determine the magnitude of the particle's displacement from $t = 0$ to $t = 3$ s. Use Simpson's rule with $n = 100$ to evaluate the integrals. What is the magnitude of the particle's acceleration when $t = 2$ s?

12–49. A particle travels along a path such that its position is $\mathbf{r} = \{(2 \sin t)\mathbf{i} + 2(1 - \cos t)\mathbf{j}\}$ ft, where t is in seconds and the arguments for the sine and cosine are given in radians. Find the equation $y = f(x)$ which describes the path, and show that the magnitudes of the particle's velocity and acceleration are constant. What is the magnitude of the particle's displacement from $t = 0$ to $t = 2$ s?

12–50. Motion of particles A and B is described by the position vectors $\mathbf{r}_A = \{2t\mathbf{i} + (t^2 - 1)\mathbf{j}\}$ ft and $\mathbf{r}_B = \{(t + 2)\mathbf{i} + (2t^2 - 5)\mathbf{j}\}$ ft, respectively, where t is in seconds. Determine the point where the particles collide and their speeds just before the collision.

12–51. Motion of particles A and B is described by the position vectors $\mathbf{r}_A = \{3t\mathbf{i} + 9t(2 - t)\mathbf{j}\}$ m and $\mathbf{r}_B = \{3(t^2 - 2t + 2)\mathbf{i} + 3(t - 2)\mathbf{j}\}$ m, respectively, where t is in seconds. Determine the point where the particles collide and their speeds just before the collision. How long does it take before the collision occurs?

*12–52. The car travels from A to B, then from B to C as shown in the figure. Compute the magnitude of the displacement of the car and the distance traveled.

Prob. 12–52

12–53. A car traveling along the straight portions of the road has the velocities indicated in the figure when it arrives at points A, B, and C. If it takes 3 s to go from A to B, and then 5 s to go from B to C, determine the average acceleration between points A and B and between points A and C.

Prob. 12–53

12–54. A particle travels along the curve from A to B in 2 s. It takes 4 s for it to go from B to C and then 3 s to go from C to D. Determine its average velocity when it goes from A to D.

Prob. 12–54 / 12–55

12–55. Determine the average speed of the particle in Prob. 12–54 when it travels from A to D.

*12–56. A particle travels along the path $y = x^2$, where x and y are in meters. If the particle's component of velocity in the y direction is always $v_y = 3$ m/s, determine the magnitude of the particle's velocity when $t = 2$ s. When $t = 0$ the particle is at point $(1, 1)$. How far from the origin is the particle when $t = 2$ s?

12–57. When a rocket reaches an altitude of 40 m it begins to travel along the parabolic path $(y - 40)^2 = 160x$, where the coordinates are measured in meters. If the component of velocity in the vertical direction is constant at $v_y = 180$ m/s, determine the magnitudes of the rocket's velocity and acceleration when it reaches an altitude of 80 m.

Prob. 12–57

12–58. The flight path of the helicopter as it takes off from A is defined by the parametric equations $x = (2t^2)$ m and $(y = 0.04t^3)$ m, where t is the time in seconds after takeoff. Determine the distance the helicopter is from point A and the magnitudes of its velocity and acceleration when $t = 10$ s.

Prob. 12–58

12–59. The pitcher throws the baseball horizontally with a speed of 110 ft/s from a height of 5 ft. If the batter is 60 ft away, determine the time needed for the ball to arrive at the batter and the height h at which it passes the batter.

Prob. 12–59

*12–60. The balloon A is ascending at the rate $v_A = 12$ km/h and is being carried horizontally by the wind at $v_w = 20$ km/h. If a ballast bag is dropped from the balloon such that it takes 8 s to reach the ground, determine the balloon's altitude h at the instant the bag was released. Assume that the bag was released from the balloon with the same velocity as the balloon.

Prob. 12–60

12–61. A particle moves with curvilinear motion in the x-y plane such that the y component of motion is described by the equation $y = (7t^3)$ m, where t is in seconds. If the particle starts from rest at the origin when $t = 0$, and maintains a *constant* acceleration in the x direction of 12 m/s^2, determine the particle's speed when $t = 2$ s.

12–62. The snowmobile is traveling at 10 m/s when it leaves the embankment at A. Determine the time of flight from A to B and the range R of the trajectory.

Prob. 12–62

12–63. The girl throws the toy at an angle of 30° from point A as shown. Determine the maximum and minimum speed v_A it can have so that it lands in the pool.

Prob. 12–63

*12–64. It is observed that the skier leaves the ramp A at an angle $\theta = 25°$ with the horizontal. If he strikes the ground at point B, determine his initial speed v_A and the time of flight t_{AB}.

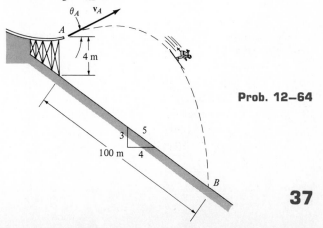

Prob. 12–64

37

12–65. The drinking fountain is designed such that the nozzle is located from the edge of the basin as shown. Determine the maximum and minimum speed at which water can be ejected from the nozzle so that it does not splash over the sides of the basin at B and C.

Prob. 12–65

12–66. The plane is flying horizontally with a constant speed of 250 ft/s at an altitude of 3000 ft. If the pilot drops a package with the same horizontal speed of 250 ft/s, determine the angle θ at which he must sight the target B so that when the package is released it falls and strikes the target. Air resistance neglected, explain why the package appears to remain directly beneath the plane as it falls.

Prob. 12–66

12–67. The water sprinkler, positioned at the base of a hill, releases a stream of water with a velocity of 15 ft/s as shown. Determine the point $B(x, y)$ where the water particles strike the ground on the hill. Assume that the hill is defined by the equation $y = (0.05x^2)$ ft, and neglect the size of the sprinkler.

Prob. 12–67

*12–68. Measurements of a shot which was recorded on a videotape during a basketball game are shown. The ball passed through the hoop even though it barely cleared the hands of player B who attempted to block it. Neglecting the size of the ball, determine the magnitude v_A of its initial velocity and the height h of the ball when it passes over player B.

Prob. 12–68

12–69. A boy at A wishes to throw two balls to the boy at B so that they arrive at the *same time*. If each ball is thrown with a speed of 10 m/s, determine the angles θ_C and θ_D at which each ball should be thrown and the time between each throw. Note that the first ball must be thrown at $\theta_C(>\theta_D)$, then the second ball is thrown at θ_D.

Prob. 12–69

12–70. A girl at A throws an acorn directly at a squirrel perched on a branch at B. If the initial speed of the acorn is 14 m/s and the squirrel, out of fright; happens to fall from rest off the branch at the instant the acorn is released at A, show that the squirrel can still catch the acorn, and determine how far h the squirrel falls before the catch is made.

Prob. 12–70

12–71. A football player runs along the 30-yard line (y axis) with a speed of 6 ft/s. If he passes the ball perpendicular to this direction as shown, and gives it an initial speed of 20 ft/s, inclined at an angle of 30° with the x axis, determine the point x, y where the ball can be caught at the same elevation from which it was thrown.

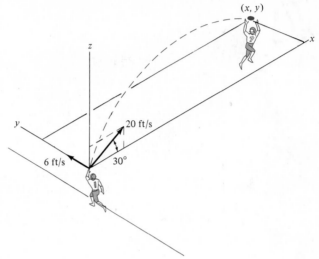

Prob. 12–71

*12–72. The fireman at A uses the hose which ejects a steady stream of water at 10 m/s. Calculate the maximum range R_{max} to which the water particles can be "thrown" and the angle θ at which he must hold the nozzle.

Probs. 12–72 / 12–73

12–73. Show that the fireman at A can "throw" water particles on point B by aiming the nozzle at equal angles measured up or down from a 45° inclination. Determine the range R if this value is 15°, i.e., $\theta_1 = 45° - 15° = 30°$ and $\theta_2 = 45° + 15° = 60°$. The water has a speed of 10 m/s at the nozzle A.

12.6 Curvilinear Motion: Normal and Tangential Components

When the path along which a particle is moving is *known*, it is often convenient to describe the motion using n and t coordinates which act normal and tangent to the path, respectively, and at the instant considered have their *origin located at the particle*.

(a)

Planar Motion. Consider the particle P shown in Fig. 12–24a, which is moving in a plane such that at a given instant it is located at position s on the curve $s = s(t)$. The t axis is *tangent* to the curve at O and is positive in the direction of *increasing* s. We will designate this positive direction with the unit vector $\mathbf{u}_t$. A unique choice for the *normal axis* can be made by considering the fact that geometrically the curve consists of a series of differential arc segments ds. As shown in Fig. 12–24b, each segment is constructed from the arc of an associated circle having a *radius of curvature ρ* (rho) and *center of curvature O'*. The normal axis n which will be chosen is perpendicular to the t axis and is directed from P *toward* the center of curvature O', Fig. 12–24a. This positive direction, which is *always* on the concave side of the curve, will be designated by the unit vector $\mathbf{u}_n$. The plane which contains the n and t axes is referred to as the *osculating plane*, and in this case it is fixed in the plane of motion.*

Velocity
(b)

Fig. 12–24a–b

Velocity. As indicated in Sec. 12.3, $\mathbf{v}$ has a *direction* that is always tangent to the curve, Fig. 12–24b, and a *magnitude* that is determined by taking the time derivative of the path function $s = s(t)$, i.e., $v = ds/dt$ (Eq. 12–8). Hence

$$\mathbf{v} = v\mathbf{u}_t \qquad (12\text{–}15)$$

where

$$v = \dot{s} \qquad (12\text{–}16)$$

Acceleration. The acceleration of the particle is the time rate of change of the velocity. Thus,

$$\mathbf{a} = \dot{\mathbf{v}} = \dot{v}\mathbf{u}_t + v\dot{\mathbf{u}}_t \qquad (12\text{–}17)$$

In order to compute the time derivative $\dot{\mathbf{u}}_t$, note that as the particle moves along the arc ds in time dt, $\mathbf{u}_t$ preserves its magnitude of unity; however, it

---

*The osculating plane may also be defined as that plane which has the greatest contact with the curve at a point. It is the limiting position of a plane contacting both the point and the arc segment ds. As noted above, the osculating plane is always coincident with a plane curve; however, each point on a three-dimensional curve has a unique osculating plane.

(c)

(d)

Fig. 12–24c–e

Acceleration
(e)

changes its *direction*, so that it becomes $\mathbf{u}_t'$, Fig. 12–24c. The change $d\mathbf{u}_t$ in $\mathbf{u}_t$ is shown in Fig. 12–24d. Here $d\mathbf{u}_t$ stretches between the tips of $\mathbf{u}_t$ and $\mathbf{u}_t'$, which lie on an infinitesimal arc of radius $u_t = 1$. Hence, $d\mathbf{u}_t$ has a *magnitude* of $du_t = (1)\, d\theta$ and its *direction* is defined by $\mathbf{u}_n$. Consequently, $d\mathbf{u}_t = d\theta \mathbf{u}_n$ or $\dot{\mathbf{u}}_t = \dot{\theta}\mathbf{u}_n$. Since $ds = \rho\, d\theta$, Fig. 12–24c, $\dot{\theta} = \dot{s}/\rho$, then

$$\dot{\mathbf{u}}_t = \frac{\dot{s}}{\rho}\mathbf{u}_n = \frac{v}{\rho}\mathbf{u}_n$$

Substituting into Eq. 12–17, $\mathbf{a}$ can be written as the sum of its two components,

$$\mathbf{a} = a_t\mathbf{u}_t + a_n\mathbf{u}_n \qquad (12\text{–}18)$$

where

$$a_t = \dot{v} \qquad \text{or} \qquad a_t = v\frac{dv}{ds} \qquad (12\text{–}19)$$

and

$$a_n = \frac{v^2}{\rho} \qquad (12\text{–}20)$$

These two perpendicular components are shown in Fig. 12–24e, in which case the *magnitude* of acceleration is

$$a = \sqrt{a_t^2 + a_n^2} \qquad (12\text{–}21)$$

Each of the acceleration components has a physical meaning which can be explained using two special cases of motion.

1. If the particle moves along a straight line, then $\rho \to \infty$ and from Eq. 12–20, $a_n = 0$. Thus $a = a_t = \dot{v}$, and we can conclude that the *tangential component of acceleration represents the time rate of change in the magnitude of the velocity.*

2. If the particle moves along a curve with a constant speed, then $a_t = \dot{v} = 0$ and $a = a_n = v^2/\rho$. Therefore, the *normal component of acceleration represents the time rate of change in the direction of the velocity*. Since it *always* acts toward the center of curvature, this component is sometimes referred to as the *centripetal acceleration*.

As a result of these interpretations, a particle moving along the curved path shown in Fig. 12–25 will have accelerations directed as shown in the figure.

Fig. 12–25

Three-Dimensional Motion. If the particle is moving along a space curve, Fig. 12–26, then at a given instant the *t* axis is uniquely specified; however, an infinite number of straight lines can be constructed normal to the tangent axis at *P*. As in the case of planar motion, we will choose the positive *n* axis directed from *P* toward the path's center of curvature O'. This axis is referred to as the *principal normal* to the curve at *P*. With the *n* and *t* axes so defined, Eqs. 12–15 to 12–21 can be used to determine **v** and **a.** Since $\mathbf{u}_t$ and $\mathbf{u}_n$ are always perpendicular to one another, and lie in the osculating plane, for spatial motion a third unit vector, $\mathbf{u}_b$, defines a *binormal axis b* which is perpendicular to $\mathbf{u}_t$ and $\mathbf{u}_n$, Fig. 12–26. Using the vector cross product, these three vectors are related by the equation $\mathbf{u}_b = \mathbf{u}_t \times \mathbf{u}_n$.

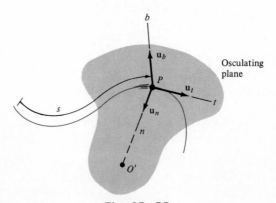

Fig. 12–26

PROCEDURE FOR ANALYSIS

Motion of a particle along a *known* path *s* can be expressed in terms of its normal and tangential components. The following is a summary of the equations necessary for doing this.

Velocity. The particle's *velocity* is tangent to the path and acts in the direction of motion. Its magnitude is found from the time derivative of the path function.

$$v = \dot{s}$$

Tangential Acceleration. The *tangential component of acceleration is the result of the time rate of change in the magnitude of velocity*. This component acts in the direction of motion if the particle's speed is increasing or in the opposite direction if the speed is decreasing. The magnitudes of **v** and **a**$_t$ are related to the path function *s* and time *t* by the equations of rectilinear motion; namely,

$$a_t = \dot{v} \qquad a_t \, ds = v \, dv$$

If a_t is *constant*, $a_t = (a_t)_c$, the above equations, when integrated, yield

$$s = s_0 + v_0 t + \tfrac{1}{2}(a_t)_c t^2$$
$$v = v_0 + (a_t)_c t$$
$$v^2 = v_0^2 + 2(a_t)_c(s - s_0)$$

Normal Acceleration. The *normal component of acceleration is the result of the time rate of change in the direction of the particle's velocity*. This component is *always* directed toward the center of curvature of the path, i.e., along the positive *n* axis. Its magnitude is determined from

$$a_n = \frac{v^2}{\rho}$$

If the path is expressed as $y = f(x)$, the radius of curvature ρ is computed from the equation*

$$\rho = \left| \frac{[1 + (dy/dx)^2]^{3/2}}{d^2y/dx^2} \right|$$

The following examples numerically illustrate application of these equations.

---

*The derivation of this result is given in any standard calculus text.

Example 12–15

A skier travels with a constant speed of 6 m/s along the parabolic path $y = \frac{1}{20}x^2$ shown in Fig. 12–27. Determine his velocity and acceleration at the instant he arrives at A. Neglect the size of the skier in the calculation.

Fig. 12–27

Solution

Although the path has been expressed in terms of its x and y coordinates, we can still establish the n, t axes at A and determine the components of $\mathbf{v}$ and $\mathbf{a}$ along these axes.

Velocity. By definition, the velocity is always directed tangent to the path. Since $y = \frac{1}{20}x^2$, $dy/dx = \frac{1}{10}x$, then $dy/dx|_{x=10} = 1$. Hence, at A, $\mathbf{v}$ makes an angle of $\theta = \tan^{-1} 1 = 45°$ with the x axis, Fig. 12–27. Therefore,

$$v = 6 \text{ m/s} \qquad \text{\small 45°} \qquad \textit{Ans.}$$

Acceleration. The acceleration is computed by using $\mathbf{a} = \dot{v}\mathbf{u}_t + (v^2/\rho)\mathbf{u}_n$. However, it is first necessary to determine the radius of curvature of the path at A (10 m, 5 m). Since $d^2y/dx^2 = \frac{1}{10}$, then

$$\rho = \left| \frac{[1 + (dy/dx)^2]^{3/2}}{d^2y/dx^2} \right| = \left| \frac{[1 + (\frac{1}{10}x)^2]^{3/2}}{\frac{1}{10}} \right|_{x=10\text{m}} = 28.3 \text{ m}$$

The acceleration becomes

$$a_A = \dot{v}\mathbf{u}_t + \frac{v^2}{\rho}\mathbf{u}_n$$

$$= 0\mathbf{u}_t + \frac{(6)^2}{28.3}\mathbf{u}_n$$

$$= \{1.27\mathbf{u}_n\} \text{ m/s}^2$$

Since $\mathbf{a}_A$ acts in the direction of the positive n axis, it makes an angle of $\theta + 90° = 135°$ with the positive x axis. Hence,

$$a_A = 1.27 \text{ m/s}^2 \qquad \text{\small \(\mathbf{a}_A\) 135°} \qquad \textit{Ans.}$$

Example 12–16

A race car C travels around the horizontal circular track that has a radius of 300 ft, Fig. 12–28. If the car increases its speed at a constant rate of 7 ft/s^2 starting from rest, determine the time needed for it to reach an acceleration of 8 ft/s^2. What is its speed at this instant?

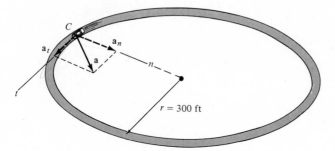

Fig. 12–28

Solution

Acceleration. The magnitude of acceleration can be related to its components using $a = \sqrt{a_t^2 + a_n^2}$. Here $a_t = 7$ ft/s^2. Also $a_n = v^2/\rho$, where $\rho = 300$ ft and

$$v = v_0 + (a_t)_c t$$
$$v = 0 + 7t$$

Thus

$$a_n = \frac{v^2}{\rho} = \frac{7t^2}{300} = 0.163t^2 \text{ ft/s}^2$$

The time needed for the acceleration to reach 8 ft/s^2 is therefore

$$a = \sqrt{a_t^2 + a_n^2}$$
$$8 = \sqrt{(7)^2 + (0.163t^2)^2}$$

Solving for the positive value of t yields

$$0.163t^2 = \sqrt{(8)^2 - (7)^2}$$
$$t = 4.87 \text{ s} \qquad\qquad \textit{Ans.}$$

Velocity. The speed at time $t = 4.87$ s is

$$v = 7t = 7(4.87) = 34.1 \text{ ft/s} \qquad\qquad \textit{Ans.}$$

Example 12–17

A car starts from rest at point A and travels along the horizontal track shown in Fig. 12–29a. During the motion, the increase in speed is $a_t = (0.2t)$ m/s^2, where t is in seconds. Determine the magnitude of the car's acceleration when it arrives at point B.

3 m

Solution

The acceleration is computed from $\mathbf{a} = \dot{v}\mathbf{u}_t + (v^2/\rho)\mathbf{u}_n$. To use this equation, however, it is first necessary to formulate v and $\dot{v}$ so that they may be evaluated at B. Since $v_A = 0$ when $t = 0$, then

$$a_t = \dot{v} = 0.2t \tag{1}$$

$$\int_0^v dv = \int_0^t 0.2t \, dt$$

$$v = 0.1t^2 \tag{2}$$

(a)

The time needed for the car to reach point B can be determined by realizing that the position of B is $s_B = 3 + 2\pi(2)/4 = 6.14$ m, Fig. 12–29a, and since $s_A = 0$ when $t = 0$ we have

$$v = \frac{ds}{dt} = 0.1t^2$$

$$\int_0^{6.14} ds = \int_0^{t_B} 0.1t^2 \, dt$$

$$6.14 = 0.0333t_B^3$$

$$t_B = 5.69 \text{ s}$$

(b)

Fig. 12–29

Substituting into Eqs. (1) and (2) yields

$$(a_t)_B = 0.2(5.69) = 1.14 \text{ m/s}^2$$

$$v_B = 0.1(5.69)^2 = 3.24 \text{ m/s}$$

At B, $\rho_B = 2$ m, so that

$$(a_n)_B = \frac{v_B^2}{\rho_B} = \frac{(3.24)^2}{2} = 5.25 \text{ m/s}^2$$

The magnitude of $\mathbf{a}_B$, Fig. 12–29b, is therefore

$$a_B = \sqrt{(1.14)^2 + (5.25)^2} = 5.37 \text{ m/s}^2 \qquad\qquad Ans.$$

Problems

12–74. A car is traveling along a circular curve having a radius of 50 m. If its speed is 16 m/s and is increasing uniformly at 8 m/s², determine the magnitude of its acceleration at this instant.

12–75. A jet plane flies along the vertical curve having a radius of 800 m. If its speed is uniformly increased from 180 m/s to 230 m/s in 4 s, determine the magnitude of its acceleration at the instant the plane's speed is 200 m/s.

800 m

200 m/s

Prob. 12–75

*12–76. At a given instant the train engine at E has a speed of 20 m/s and an acceleration of 14 m/s² acting in the direction shown. Determine the rate of increase in the train's speed and the radius of curvature ρ of the path.

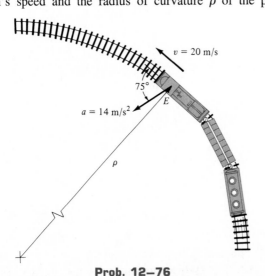

$v = 20$ m/s

75°

E

$a = 14$ m/s²

ρ

Prob. 12–76

12–77. A spiral transition curve is used on railroads to connect a straight portion of the track with a curved portion. If the spiral is defined by the equation $y = (10^{-6})x^3$, where x and y are in feet, determine the magnitude of the acceleration of a train engine moving with a constant speed of 50 ft/s when it is at point $x = 600$ ft.

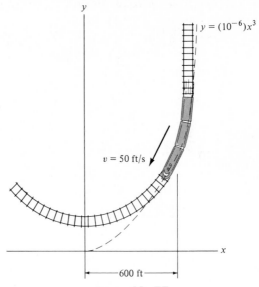

y

$y = (10^{-6})x^3$

$v = 50$ ft/s

x

600 ft

Prob. 12–77

12–78. A boy sits on a merry-go-round so that he is always located at $r = 8$ ft from the center of rotation. The merry-go-round is originally at rest, and then due to rotation the boy's speed is increased at 2 ft/s². Determine the time needed for his acceleration to become 4 ft/s².

12–79. A particle moves along the hyperbolic curve $y = 1/x$ with a constant speed of 6 ft/s. Determine the magnitude and direction of the particle's acceleration at the instant it reaches the point (1 ft, 1 ft).

■*12–80. A go-cart moves along a circular track of radius 100 ft such that its speed for a short period of time $0 \leq t \leq 4$ s is $v = 60(1 - e^{-t})$ ft/s. Determine the magnitude of its acceleration when $t = 2$ s. How far has it traveled in $t = 2$ s? Use Simpson's rule with $n = 50$ to evaluate the integral.

12–81. A particle travels in a circular path of radius $\rho = 3$ m at a speed of 4 m/s. If its speed is increased at $\dot{v} = (0.2s)$ m/s^2, where s is in meters, determine its speed and the magnitude of its acceleration when it has moved through an arc of 90° along its path.

12–82. Starting from rest, the motorboat travels around the circular path, $\rho = 50$ m, at a speed $v = (0.8t)$ m/s, where t is in seconds. Determine the magnitudes of the boat's velocity and acceleration just after it has traveled 20 m.

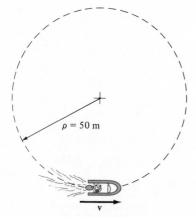

$\rho = 50$ m

Prob. 12–82

12–83. A particle P moves along the curve $y = (x^2 - 4)$ m with a constant speed of 5 m/s. Determine the point on the curve where the maximum magnitude of acceleration occurs, and compute its value.

*12–84. The ball is thrown horizontally with a speed of 8 m/s. Find the equation of the path, $y = f(x)$, and then find the ball's velocity and the normal and tangential components of acceleration when $t = 0.25$ s.

$v_A = 8$ m/s

A

Prob. 12–84

12–85. The Ferris wheel turns such that the speed of the passengers is increased by $\dot{v} = (4t)$ ft/s^2, where t is in seconds. If the wheel starts from rest when $\theta = 0°$, determine the magnitudes of the velocity and acceleration of one of the passengers when the wheel turns $\theta = 30°$.

40 ft

θ

Prob. 12–85

12–86. A race car has an initial speed $v_A = 15$ m/s when it is at A. If it increases its speed along the circular track, $\rho = 150$ m, at the rate $a_t = (0.4s)$ m/s^2, where s is in meters, determine the normal and tangential components of the car's acceleration at $s = 100$ m.

ρ

s

A

Probs. 12–86 / 12–87

12–87. If the race car in Prob. 12–86 has an initial speed $v_A = 15$ m/s at A, compute the time needed for the car to travel 100 m. Assume that $a_t = (0.4s)$ m/s^2 and $\rho = 150$ m.

*12–88. The automobile is originally at rest at $s = 0$. If it then starts to increase its speed according to the equation $\dot{v} = (0.05t^2)$ ft/s^2, where t is in seconds, determine the magnitudes of its velocity and acceleration when $t = 18$ s.

Probs. 12–88 / 12–89

12–89. The automobile is originally at rest at $s = 0$. If it then starts to increase its speed at $\dot{v} = (0.05t^2)$ ft/s^2, where t is in seconds, determine the magnitudes of its velocity and acceleration at $s = 550$ ft.

12–90. Particles A and B are traveling counterclockwise around a circular track at a constant speed of 8 m/s at the instant shown. If the speed of A is increased at $\dot{v}_A = (4s_A)$ m/s^2, where s_A is in meters, determine the distance measured counterclockwise along the track from A to B between the particles when $t = 1$ s. What is the magnitude of the acceleration of each particle at this instant?

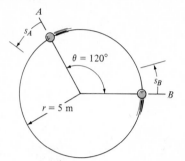

Probs. 12–90 / 12–91

■12–91. Particles A and B are traveling around a circular track at a constant speed of 8 m/s at the instant shown. If the speed of B is increased at $\dot{v}_B = 4$ m/s^2, and at the same instant A has an increase of speed $\dot{v}_A = 0.8t$ m/s^2, determine how long it takes for a collision to occur. What is the magnitude of the acceleration of each particle just before the collision occurs?

12.7 Curvilinear Motion: Cylindrical Components

In some engineering problems it is often convenient to express the path of motion of a particle in terms of cylindrical coordinates, r, θ, and z. If motion is restricted to the plane, the polar coordinates r and θ are used.

Polar Coordinates. We can specify the location of particle P shown in Fig. 12–30a using both the *radial coordinate r*, which extends outward from the fixed origin O to the particle, and a *transverse coordinate θ*, which is the counterclockwise angle between a fixed reference line and the r axis. The angle is generally measured in degrees or radians, where 1 rad = 180°/π. The positive directions of the r and θ coordinates are defined by the unit vectors $\mathbf{u}_r$ and $\mathbf{u}_\theta$, respectively.

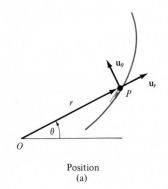

r

Position
(a)

Position. At any instant the position of the particle, Fig. 12–30a, is defined by the position vector

$$\vec{\mathbf{r}} = r\mathbf{u}_r \qquad (12\text{–}22)$$

(b)

Velocity. The instantaneous velocity $\mathbf{v}$ is obtained by taking the time derivative of $\mathbf{r}$. Using a dot to represent time differentiation, we have

$$\vec{\mathbf{v}} = \dot{\vec{\mathbf{r}}} = \dot{r}\mathbf{u}_r + r\dot{\mathbf{u}}_r$$

In order to evaluate $\dot{\mathbf{u}}_r$, notice that $\mathbf{u}_r$ changes only its direction with respect to time, since by definition the magnitude of this vector is always unity. Hence, when θ changes by an amount $\Delta\theta$ during the time Δt, $\mathbf{u}_r$ becomes unit vector $\mathbf{u}_r'$, Fig. 12–30b. The time change in $\mathbf{u}_r$ is then $\Delta\mathbf{u}_r$. For small angles $\Delta\theta$ this vector has a magnitude $\Delta u_r \approx 1(\Delta\theta)$ and acts in the $\mathbf{u}_\theta$ direction, i.e., $\Delta\mathbf{u}_r = \Delta\theta\mathbf{u}_\theta$. Thus,

$$\dot{\mathbf{u}}_r = \lim_{\Delta t \to 0} \frac{\Delta\mathbf{u}_r}{\Delta t} = \left(\lim_{\Delta t \to 0} \frac{\Delta\theta}{\Delta t} \right)\mathbf{u}_\theta$$

$$\dot{\mathbf{u}}_r = \dot{\theta}\mathbf{u}_\theta \qquad (12\text{–}23)$$

Substituting into the above equation, the velocity can be written in component form as

$$\mathbf{v} = v_r\mathbf{u}_r + v_\theta\mathbf{u}_\theta \qquad (12\text{–}24)$$

Velocity
(c)

Fig. 12–30a–c

where

$$v_r = \dot{r}$$
$$v_\theta = r\dot{\theta} \qquad (12\text{–}25)$$

These components are shown graphically in Fig. 12–30c. Note that the radial component of velocity $\mathbf{v}_r$, is a measure of the rate of increase or decrease in the length of the radial coordinate, i.e., $\dot{r}$, whereas the transverse component of velocity, $\mathbf{v}_\theta$, can be interpreted as the rate of motion in the direction of the circumference of a circle having a radius r. In particular, the term $\dot{\theta} = d\theta/dt$ is called the *angular velocity* since it provides a measure of the time rate of change of the angle θ. Common units used for this measurement are rad/s.

Since the components are mutually perpendicular, the *magnitude* of velocity is simply

$$v = \sqrt{(\dot{r})^2 + (r\dot{\theta})^2} \qquad (12\text{--}26)$$

and the *direction* of $\mathbf{v}$ is, of course, tangent to the path at P, Fig. 12–30c.

Acceleration. Taking the time derivatives of Eq. 12–24, using Eqs. 12–25, we obtain the particle's instantaneous acceleration,

$$\mathbf{a} = \dot{\mathbf{v}} = \ddot{r}\mathbf{u}_r + \dot{r}\dot{\mathbf{u}}_r + \dot{r}\dot{\theta}\mathbf{u}_\theta + r\ddot{\theta}\mathbf{u}_\theta + r\dot{\theta}\dot{\mathbf{u}}_\theta$$

To evaluate the term involving $\dot{\mathbf{u}}_\theta$, it is necessary to consider the change made in the direction of $\mathbf{u}_\theta$. As shown in Fig. 12–30d, the new or changed position of $\mathbf{u}_\theta$ is designated as $\mathbf{u}'_\theta$. The time change in $\mathbf{u}_\theta$ is thus $\Delta\mathbf{u}_\theta$. For small angles this vector has a magnitude $\Delta u_\theta \approx 1(\Delta\theta)$ and acts in the $-\mathbf{u}_r$ direction, i.e., $\Delta\mathbf{u}_\theta = -\Delta\theta\mathbf{u}_r$. Thus,

(d)

$$\dot{\mathbf{u}}_\theta = \lim_{\Delta t \to 0} \frac{\Delta\mathbf{u}_\theta}{\Delta t} = -\left(\lim_{\Delta t \to 0} \frac{\Delta\theta}{\Delta t}\right)\mathbf{u}_r$$

$$\dot{\mathbf{u}}_\theta = -\dot{\theta}\mathbf{u}_r \qquad (12\text{--}27)$$

Substituting this result and Eq. 12–23 into the above equation for $\mathbf{a}$, we can write the acceleration in component form as

$$\boxed{\mathbf{a} = a_r\mathbf{u}_r + a_\theta\mathbf{u}_\theta} \qquad (12\text{--}28)$$

where

$$\boxed{\begin{aligned} a_r &= \ddot{r} - r\dot{\theta}^2 \\ a_\theta &= r\ddot{\theta} + 2\dot{r}\dot{\theta} \end{aligned}} \qquad (12\text{--}29)$$

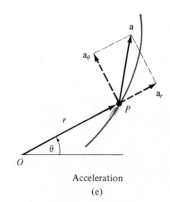

Acceleration
(e)

Fig. 12–30d–e

The term $\ddot{\theta} = d^2\theta/dt^2 = d/dt(d\theta/dt)$ is called the *angular acceleration* since it measures the change made in the rate of change of θ during an instant of time. Common units for this measurement are rad/s^2.

Since $\mathbf{a}_r$ and $\mathbf{a}_\theta$ are always perpendicular, the *magnitude* of acceleration is simply

$$a = \sqrt{(\ddot{r} - r\dot{\theta}^2)^2 + (r\ddot{\theta} + 2\dot{r}\dot{\theta})^2} \qquad (12\text{--}30)$$

The *direction* is determined from the vector addition of its two components. In general, $\mathbf{a}$ will *not* be tangent to the path, Fig. 12–30.

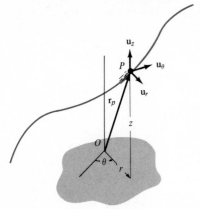

Fig. 12–31

Cylindrical Coordinates. If the particle moves along a space curve as shown in Fig. 12–31, then its location P is specified by the three *cylindrical coordinates, r, θ, z*. The z coordinate is identical to that used for rectangular coordinates. Since the unit vector defining its direction, $\mathbf{u}_z$, is constant, the time derivatives of this vector are zero and therefore the position, velocity, and acceleration of the particle can be written in terms of its cylindrical coordinates as follows:

$$\mathbf{r}_P = r\mathbf{u}_r + z\mathbf{u}_z$$

$$\mathbf{v} = \dot{r}\mathbf{u}_r + r\dot{\theta}\mathbf{u}_\theta + \dot{z}\mathbf{u}_z \tag{12-31}$$

$$\mathbf{a} = (\ddot{r} - r\dot{\theta}^2)\mathbf{u}_r + (r\ddot{\theta} + 2\dot{r}\dot{\theta})\mathbf{u}_\theta + \ddot{z}\mathbf{u}_z \tag{12-32}$$

PROCEDURE FOR ANALYSIS

Application of the preceding polar equations for computing the particle's velocity and acceleration requires a *straightforward substitution* once r and the time derivatives $\dot{r}$, $\ddot{r}$, $\dot{\theta}$, and $\ddot{\theta}$ have been computed and evaluated at the instant considered. Two types of problems generally occur:

1. If the coordinates are specified as time-parametric equations, $r = r(t)$ and $\theta = \theta(t)$, the derivatives are simply $\dot{r} = dr/dt$, $\ddot{r} = d^2r/dt^2$, $\dot{\theta} = d\theta/dt$, and $\ddot{\theta} = d^2\theta/dt^2$ (see Examples 12–18 and 12–19).

2. If the time-parametric equations of motion are not given, it will be necessary to specify the path $r = f(\theta)$ and compute the relationship between the time derivatives using the chain rule of calculus. In this case,

$$\dot{r} = \frac{df(\theta)}{d\theta}\dot{\theta} \qquad \text{and} \qquad \ddot{r} = \left(\frac{d^2f(\theta)}{d\theta^2}\dot{\theta}\right)\dot{\theta} + \frac{df(\theta)}{d\theta}\ddot{\theta}$$

Thus, if two of the *four* time derivatives $\dot{r}$, $\ddot{r}$, $\dot{\theta}$, and $\ddot{\theta}$ are *known*, the other two can be obtained from these equations (see Example 12–20). In some problems, however, two of these time derivatives may *not* be known; instead the magnitude of the particle's velocity or acceleration may be specified. If this is the case, $v^2 = \dot{r}^2 + (r\dot{\theta})^2$ and $a^2 = (\ddot{r} - r\dot{\theta}^2)^2 + (r\ddot{\theta} + 2\dot{r}\dot{\theta})^2$ may be used to obtain the necessary relationships involving $\dot{r}$, $\ddot{r}$, $\dot{\theta}$, and $\ddot{\theta}$ (see Example 12–21).

Motion in three dimensions requires a simple extension of the above procedure to include $\dot{z}$ and $\ddot{z}$.

Besides the examples which follow, further examples involving the calculation of a_r and a_θ can be found in the "kinematics" of Examples 13–11 to 13–13.

Example 12–18

The ball B shown in Fig. 12–32a is rotating in a horizontal circular path of radius r such that the attached cord has an angular velocity $\dot{\theta}$ and angular acceleration $\ddot{\theta}$. Express the velocity and acceleration of the ball in terms of its polar coordinates.

Fig. **12–32**

Solution

Time Derivatives. Since Eqs. 12–31 and 12–32 will be used for the solution, it is first necessary to specify the first and second time derivatives of r and θ. Since r is *constant,* we have

$$r = r, \qquad \dot{r} = 0, \qquad \ddot{r} = 0$$

Velocity

$$v_r = \dot{r} = 0 \qquad\qquad Ans.$$
$$v_\theta = \dot{\theta} \qquad\qquad Ans.$$

Acceleration

$$a_r = \ddot{r} - r\dot{\theta}^2 = -r\dot{\theta}^2 \qquad\qquad Ans.$$
$$a_\theta = r\ddot{\theta} + 2\dot{r}\dot{\theta} = r\ddot{\theta} \qquad\qquad Ans.$$

These results are shown in Fig. 12–32b. Also shown are the n, t axes, which in this special case of circular motion happen to be *collinear* with the r and θ axes, respectively. In particular note that $v = v_r = v_t = r\dot{\theta}$. Also,

$$-a_r = a_n = \frac{v^2}{\rho} = \frac{(r\dot{\theta})^2}{r} = r\dot{\theta}^2$$

and

$$a_\theta = a_t = \frac{dv}{dt} = \frac{d}{dt}(r\dot{\theta}) = \frac{dr}{dt}\dot{\theta} + r\frac{d\dot{\theta}}{dt} = 0 + r\ddot{\theta}$$

53

Example 12–19

The rod OA, shown in Fig. 12–33a, is rotating in the horizontal x-y plane such that at any instant $\theta = (t^3)$ rad. At the same time, the collar B is sliding outward along OA so that $r = (100t^2)$ mm. If in both cases t is measured in seconds, determine the velocity and acceleration of the collar when $t = 1$ s.

(a)

Solution

Time Derivatives. Computing the time derivatives, and evaluating when $t = 1$ s, we have

$$r = 100t^2\Big|_{t=1\,s} = 100 \text{ mm} \qquad \theta = t^3\Big|_{t=1\,s} = 1 \text{ rad} = 57.3°$$

$$\dot{r} = 200t\Big|_{t=1\,s} = 200 \text{ mm/s} \qquad \dot{\theta} = 3t^2\Big|_{t=1\,s} = 3 \text{ rad/s}$$

$$\ddot{r} = 200\Big|_{t=1\,s} = 200 \text{ mm/s}^2 \qquad \ddot{\theta} = 6t\Big|_{t=1\,s} = 6 \text{ rad/s}^2$$

(b)

Velocity. Fig. 12–33b

$$\mathbf{v} = \dot{r}\mathbf{u}_r + r\dot{\theta}\mathbf{u}_\theta$$

$$= 200\mathbf{u}_r + 100(3)\mathbf{u}_\theta$$

$$= \{200\mathbf{u}_r + 300\mathbf{u}_\theta\} \text{ mm/s}$$

The magnitude of $\mathbf{v}$ is

$$v = \sqrt{(200)^2 + (300)^2} = 361 \text{ mm/s} \qquad \textit{Ans.}$$

From Fig. 12–33b,

$$\delta = \tan^{-1}\left(\frac{300}{200}\right) = 56.3° \qquad \delta + 57.3° = 114° \qquad \textit{Ans.}$$

Acceleration. Fig. 12–33c

$$\mathbf{a} = (\ddot{r} - r\dot{\theta}^2)\mathbf{u}_r + (r\ddot{\theta} + 2\dot{r}\dot{\theta})\mathbf{u}_\theta$$

$$= [200 - 100(3)^2]\mathbf{u}_r + [100(6) + 2(200)3]\mathbf{u}_\theta$$

$$= \{-700\mathbf{u}_r + 1800\mathbf{u}_\theta\} \text{ mm/s}^2$$

The magnitude of $\mathbf{a}$ is

$$a = \sqrt{(700)^2 + (1800)^2} = 193 \text{ mm/s}^2 \qquad \textit{Ans.}$$

From Fig. 12–33c,

$$\phi = \tan^{-1}\left(\frac{1800}{700}\right) = 68.7° \qquad \phi - 57.3° = 11.4° \qquad \textit{Ans.}$$

(c)

Fig. 12–33

Example 12–20

The searchlight shown in Fig. 12–34a casts a spot of light along the face of a wall that is located 100 m from the searchlight. Determine the magnitudes of the velocity and acceleration at which the spot appears to travel across the wall at the instant $\theta = 45°$. The searchlight is rotating about the z axis at a constant rate $\dot{\theta} = 4$ rad/s.

Wall

100 m

$\dot{\theta} = 4$ rad/s

(a)

Solution

Time Derivatives. To compute the necessary time derivatives it is first necessary to relate r to θ. From Fig. 12–34a, this relation is

$$r = \frac{100}{\cos \theta} = 100 \sec \theta$$

100 m

(b)

Using the chain rule of calculus, noting that $d(\sec \theta) = \sec \theta \tan \theta \, d\theta$, and $d(\tan \theta) = \sec^2 \theta \, d\theta$, we have

$$\dot{r} = 100(\sec \theta \tan \theta)\dot{\theta}$$
$$\ddot{r} = 100(\sec \theta \tan \theta)\dot{\theta}(\tan \theta)\dot{\theta} + 100 \sec \theta(\sec^2 \theta)\dot{\theta}(\dot{\theta}) + 100 \sec \theta \tan \theta(\ddot{\theta})$$
$$= 100 \sec \theta \tan^2 \theta(\dot{\theta})^2 + 100 \sec^3 \theta(\dot{\theta})^2 + 100(\sec \theta \tan \theta)\ddot{\theta}$$

Since $\dot{\theta} = 4$ rad/s = constant, then $\ddot{\theta} = 0$, and the above equations, when $\theta = 45°$, become

$$r = 100 \sec 45° = 141.4$$
$$\dot{r} = 400 \sec 45° \tan 45° = 565.7$$
$$\ddot{r} = 1600(\sec 45° \tan^2 45° + \sec^3 45°) = 6788.2$$

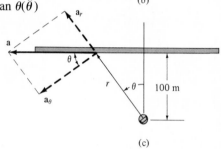

100 m

(c)

Fig. 12–34

Velocity. Fig. 12–34b

$$\mathbf{v} = \dot{r}\mathbf{u}_r + r\dot{\theta}\mathbf{u}_\theta$$
$$= 565.7\mathbf{u}_r + 141.4(4)\mathbf{u}_\theta$$
$$= 565.7\mathbf{u}_r + 565.7\mathbf{u}_\theta$$
$$v = \sqrt{v_r^2 + v_\theta^2} = \sqrt{(565.7)^2 + (565.7)^2}$$
$$= 800 \text{ m/s} \qquad\qquad Ans.$$

Acceleration. Fig. 12–34c

$$\mathbf{a} = (\ddot{r} - r\dot{\theta}^2)\mathbf{u}_r + (r\ddot{\theta} + 2\dot{r}\dot{\theta})\mathbf{u}_\theta$$
$$= [6788.2 - 141.4(4)^2]\mathbf{u}_r + [141.4(0) + 2(565.7)4]\mathbf{u}_\theta$$
$$= 4525.8\mathbf{u}_r + 4525.8\mathbf{u}_\theta$$
$$a = \sqrt{a_r^2 + a_\theta^2} = \sqrt{(4525.8)^2 + (4525.8)^2}$$
$$= 6400 \text{ m/s}^2 \qquad\qquad Ans.$$

Example 12–21

Due to the rotation of the forked rod, the cylindrical peg A in Fig. 12–35a travels around the slotted path, a portion of which is in the shape of a cardioid, $r = 0.5(1 - \cos \theta)$ ft, where θ is in radians. If the peg's velocity is $v = 4$ ft/s and its acceleration is $a = 30$ ft/s^2 at the instant $\theta = 180°$, determine the angular velocity $\dot{\theta}$ and angular acceleration $\ddot{\theta}$ of the fork.

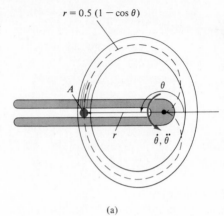

$r = 0.5 (1 - \cos \theta)$

(a)

Solution

Time Derivatives. Computing the time derivatives of r using the chain rule of calculus yields

$$r = 0.5(1 - \cos \theta)$$
$$\dot{r} = 0.5(\sin \theta)\dot{\theta}$$
$$\ddot{r} = 0.5(\cos \theta)\dot{\theta}(\dot{\theta}) + 0.5(\sin \theta)\ddot{\theta}$$

Evaluating these results at $\theta = 180°$, we have

$$r = 1 \text{ ft} \qquad \dot{r} = 0 \qquad \ddot{r} = -0.5\dot{\theta}^2$$

Velocity. Using Eq. 12–26 to determine $\dot{\theta}$ yields

$$v = \sqrt{(\dot{r})^2 + (r\dot{\theta})^2}$$
$$4 = \sqrt{(0)^2 + (1\dot{\theta})^2}$$
$$\dot{\theta} = 4 \text{ rad/s} \qquad\qquad\qquad \textbf{\textit{Ans.}}$$

$v = 4$ ft/s $\qquad a = 30$ ft/s^2

(b)

Fig. 12–35

Acceleration. In a similar manner, $\ddot{\theta}$ can be found using Eq. 12–30.

$$a = \sqrt{(\ddot{r} - r\dot{\theta}^2)^2 + (r\ddot{\theta} + 2\dot{r}\dot{\theta})^2}$$
$$30 = \sqrt{[-0.5(4)^2 - 1(4)^2]^2 + [1\ddot{\theta} + 2(0)(4)]^2}$$
$$(30)^2 = (-24)^2 + \ddot{\theta}^2$$
$$\ddot{\theta} = 18 \text{ rad/s}^2 \qquad\qquad\qquad \textbf{\textit{Ans.}}$$

Vectors **a** and **v** are shown in Fig. 12–35b.

Problems

*12–92. A particle is moving along a circular path having a radius of 0.4 m. Its angular position as a function of time is given by $\theta = (2t^2)$ rad, where t is in seconds. Determine the magnitude of the particle's acceleration when $\theta = 30°$. The particle starts from rest when $\theta = 0°$.

12–93. The time rate of change of acceleration is referred to as the *jerk*, which is often used as a means of measuring passenger discomfort. Calculate this vector, **ȧ**, in terms of its cylindrical components, using Eq. 12–32.

12–94. A train is traveling along the circular curve of radius $r = 600$ ft. At the instant shown, its angular rate of rotation is $\dot{\theta} = 0.02$ rad/s, which is decreasing at the rate $\ddot{\theta} = -0.001$ rad/s². Determine the magnitudes of the train's velocity and acceleration at this instant.

Prob. 12–94

12–95. A truck is traveling along the horizontal circular curve of radius $r = 60$ m with a constant speed $v = 20$ m/s. Determine the angular rate of rotation $\dot{\theta}$ of the radial line r and the magnitude of the truck's acceleration.

Prob. 12–95

*12–96. Starting from rest, the boy runs outward in the *radial direction* from the center of the platform with a constant acceleration of 0.5 m/s². If the platform is rotating at a constant rate $\dot{\theta} = 0.2$ rad/s, determine the radial and transverse components of the velocity and acceleration of the boy when $t = 3$ s.

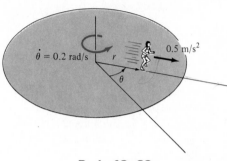

Prob. 12–96

12–97. Because of telescopic action, the end of the industrial robotic arm extends along the path of the limaçon $r = (1 + 0.5 \cos \theta)$ m. At the instant $\theta = \pi/4$, the arm has an angular rotation $\dot{\theta} = 0.6$ rad/s, which is increasing at $\ddot{\theta} = 0.25$ rad/s². Determine the radial and transverse components of the velocity and acceleration of the object A held in its grip at this instant.

Prob. 12–97

12–98. The airplane on the amusement park ride moves along a path defined by the equations $r = 4$ m, $\theta = (0.2t)$ rad, and $z = (0.5 \cos \theta)$ m, where t is in seconds. Determine the cylindrical components of the velocity and acceleration of the airplane when $t = 6$ s.

Prob. 12–98

12–99. The slotted link is pinned at O, and as a result of the constant angular velocity $\dot{\theta} = 3$ rad/s it drives the peg P for a short distance along the spiral guide $r = (4\theta)$ m, where θ is measured in radians. Compute the radial and transverse components of the velocity and acceleration of P at the instant $\theta = \pi/3$ rad.

Probs. 12–99 / 12–100

*12–100. Solve Prob. 12–99 if the slotted link has an angular acceleration $\ddot{\theta} = 8$ rad/s² when $\dot{\theta} = 3$ rad/s at $\theta = \pi/3$ rad.

12–101. The ceiling fan is turning at a constant rate of 20 rad/s. If vibrations of the paddle cause the tip A to describe a path $z = (15 \sin \theta)$ mm, determine the magnitudes of the velocity and acceleration of point A at the instant $\theta = 90°$.

Prob. 12–101

12–102. At the instant shown, the man is twirling a hose over his head with an angular speed $\dot{\theta} = 2$ rad/s and an angular acceleration $\ddot{\theta} = 3$ rad/s². If it is assumed that the hose lies in a horizontal plane, and water is flowing through it at a constant rate of 3 m/s, determine the magnitudes of the velocity and acceleration of a water particle as it exits the open end, $r = 1.5$ m.

Prob. 12–102

12–103. A particle travels along the "four-leaf rose" defined by the equation $r = (5 \cos 2\theta)$ m. If the angular velocity of the radial coordinate line is $\dot{\theta} = (3t^2)$ rad/s, where t is in seconds, determine the radial and transverse components of the particle's velocity and acceleration at the instant $\theta = 30°$. When $t = 0$, $\theta = 0$.

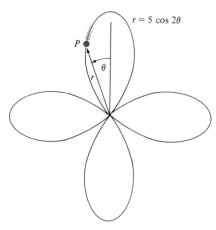

$r = 5 \cos 2\theta$

P

θ

r

Prob. 12–103

12–105. For a short time the jet plane moves along a path in the shape of a lemniscate, $r^2 = (2500 \cos 2\theta)$ km². At the instant $\theta = 30°$, the radar tracking device is rotating at $\dot{\theta} = 5(10^{-3})$ rad/s with $\ddot{\theta} = 2(10^{-3})$ rad/s². Determine the radial and transverse components of velocity and acceleration of the plane at this instant.

$r^2 = 2500 \cos 2\theta$

r

θ

Prob. 12–105

*12–104. The car travels around the circular track with a constant speed of 20 m/s. Determine the car's radial and transverse components of velocity and acceleration at the instant $\theta = \pi/4$ rad.

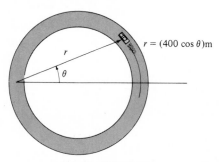

$r = (400 \cos \theta)$m

r

θ

Prob. 12–104

12–106. For a short distance the train travels along a track having the shape of a spiral, $r = (1000/\theta)$ m, where θ is in radians. If it maintains a constant speed $v = 20$ m/s, determine the radial and transverse components of its velocity when $\theta = 9\pi/4$ rad.

r

θ

$r = \dfrac{1000}{\theta}$

Prob. 12–106

12–107. The rocket is traveling upwards such that when $\theta = 30°$ the radar tracking device is turning at $\dot{\theta} = 0.6$ rad/s and $\ddot{\theta} = 0.2$ rad/s². Determine the velocity and acceleration of the rocket at this instant.

Probs. 12–107 / 12–108

*12–108.** As the rocket travels upwards, the radar tracking device has an angular position $\theta = (0.03t^2)$ rad, where t is in seconds. Determine the velocity and acceleration of the rocket at the instant $t = 3$ s.

■**12–109.** The double collar C is pin-connected together such that one collar slides over a fixed rod and the other slides over a rotating rod AB. If the angular velocity of AB is given as $\dot{\theta} = (e^{0.5t^2})$ rad/s, where t is in seconds, and the path defined by the fixed rod is $r = |(0.4 \sin \theta + 0.2)|$ m, determine the radial and transverse components of the collar's velocity and acceleration when $t = 1$ s. When $t = 0$, $\theta = 0$. Use Simpson's rule with $n = 50$ to determine θ at $t = 1$ s.

Prob. 12–109

12–110. A cameraman standing at A is following the movement of a race car, B, which is traveling around a curved track at a constant speed of 30 m/s. Determine the angular rate $\dot{\theta}$ at which the man must turn in order to keep the camera directed on the car at the instant $\theta = 30°$.

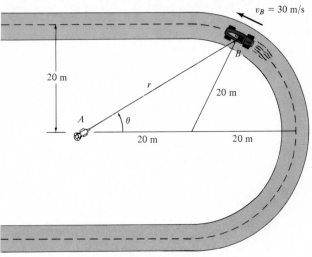

Prob. 12–110

12–111. The mechanism of a machine is constructed so that the roller at A follows the surface of the cam described by the equation $r = (300 + 200 \cos \theta)$ mm. If $\dot{\theta} = 0.5$ rad/s and $\ddot{\theta} = 0$, determine the magnitudes of the roller's velocity and acceleration when $\theta = 30°$. Neglect the size of the roller. Also compute the velocity components $(v_A)_x$ and $(v_A)_y$ of the roller at this instant. The rod to which the roller is attached remains vertical and can slide up or down along the guides while the guides translate horizontally to the left.

Prob. 12–111

*12–112. The roller coaster is traveling down along the spiral ramp with a constant speed $v = 6$ m/s. If the track descends a distance of 10 m for every full revolution, $\theta = 2\pi$ rad, determine the magnitude of the roller coaster's acceleration as it moves along the track, $r = 5$ m. *Hint:* For part of the solution, note that the tangent to the ramp at any point is at an angle $\phi = \tan^{-1}[10/2\pi(5)] = 17.66°$ from the horizontal. Use this to determine the velocity components v_θ and v_z, which in turn are used to determine $\dot{\theta}$ and $\dot{z}$.

$r = 5$ m

10 m

Prob. 12–112

12–113. The motion of a particle along a fixed path is defined by the parametric equations $r = 1.5$ m, $\theta = (2t)$ rad, and $z = (t^2)$ m, where t is in seconds. Determine the unit vector that specifies the direction of the binormal axis to the osculating plane with respect to a set of fixed x, y, z coordinate axes when $t = 0.25$ s. *Hint:* Formulate the particle's velocity $\mathbf{v}_P$ and acceleration $\mathbf{a}_P$ in terms of their $\mathbf{i}$, $\mathbf{j}$, $\mathbf{k}$ components. Note that $x = r \cos \theta$ and $y = r \sin \theta$. The binormal is parallel to $\mathbf{v}_P \times \mathbf{a}_P$. Why?

12–114. A particle P travels along an elliptical spiral path such that its position vector $\mathbf{r}$ is defined by $\mathbf{r} = \{2 \cos (0.1t)\mathbf{i} + 1.5 \sin (0.1t)\mathbf{j} + (2t)\mathbf{k}\}$ m, where t is in seconds and the arguments for the sine and cosine are given in radians. When $t = 10$ s, determine the coordinate direction angles α, β, and γ, which the binormal axis to the osculating plane makes with the x, y, and z axes. *Hint:* Solve for the velocity $\mathbf{v}_P$ and acceleration $\mathbf{a}_P$ of the particle in terms of their $\mathbf{i}$, $\mathbf{j}$, $\mathbf{k}$ components. The binormal is parallel to $\mathbf{v}_P \times \mathbf{a}_P$. Why?

z

P

$\mathbf{r}$

y

x

Prob. 12–114

12.8 Absolute Dependent Motion Analysis of Two Particles

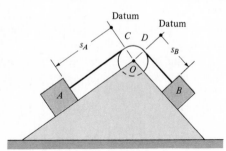

Fig. 12–36

In some types of problems the motion of one particle will *depend* upon the corresponding motion of another particle. This dependency commonly occurs if the particles are interconnected by inextensible cords which are wrapped around pulleys. For example, the movement of block A downward along the inclined plane in Fig. 12–36 will cause a corresponding movement of block B up the other incline. We can show this mathematically by first specifying the positions of the blocks using the path coordinates s_A and s_B. Note that these coordinates are (1) referenced from a fixed point or datum, (2) measured along the plane in the direction of motion, and (3) have a positive sense from C to A and D to B. Since the total cord length l_T is constant, the coordinates are related by the equation

$$s_A + s_B = l_T - l_{CD} = l'$$

where l_{CD} is the length of cord passing over arc CD.* Taking the time derivative of this expression yields a relation between the velocities of the blocks, i.e.,

$$\frac{ds_B}{dt} + \frac{ds_A}{dt} = 0$$

or

$$v_B = -v_A$$

The negative sign indicates that when the block A is moving downward, i.e., in the direction of positive s_A, it causes a corresponding upward motion of block B; i.e., B moves in the negative s_B direction.

In a similar manner, time differentiation of the velocities yields the relation between the accelerations, i.e.,

$$a_B = -a_A$$

A more complicated example involving dependent motion of two blocks is shown in Fig. 12–37a. In this case, the position of block A is specified by s_A, and the position of the *end* of the cord from which block B is suspended is defined by s_B. Here we have chosen coordinates which are (1) referenced from fixed points or datums, (2) measured in the direction of motion of each block, and (3) positive to the right (s_A) and downward (s_B). The *constant* cord length l, which represents the total length of cord *minus* the constant cord

---

*Note that when the blocks move, the position coordinates s_A and s_B measure the lengths of the *changing segments* of the cord. The cord length l_{CD} always remains *constant*.

segments which are colored, can be related to the position coordinates by the equation

$$2s_B + h + s_A = l$$

Since l and h are constants, the two time derivatives yield

$$2v_B = -v_A \qquad 2a_B = -a_A$$

Hence, when B moves downward $(+s_B)$, A moves to the left $(-s_A)$ at two times the motion.

This example can also be worked by defining the position of block B from the center of the bottom pulley (a fixed point), Fig. 12–37b. In this case

$$2(h - s_B) + h + s_A = l$$

Time differentiation yields

$$2v_B = v_A \qquad 2a_B = a_A$$

Here the signs are the same. Why?

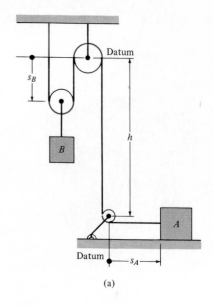

(a)

PROCEDURE FOR ANALYSIS

The dependent motion of particle A located on a cord may be related to the motion of another particle B located on the *same cord* by using the following procedure.

Position-Coordinate Equation. Establish position coordinates which have their origin referenced to a *fixed* point or datum and extend from this point to each of the particles. It is *not necessary* that the *origin* be the *same* for both of these coordinates; however, it is *important* that each coordinate axis selected be directed along the *path of motion* of the particle.

Using geometry or trigonometry, relate the coordinates to the total length of the cord, l_T, or to that portion of cord, l, which *excludes* the segments that do not change length as the particles move—such as the segments wrapped over pulleys.

Time Derivatives. The two successive time derivatives of the position-coordinate equation yield the required velocity and acceleration equations which relate the motions of the two particles. The signs of the terms in these equations will be consistent with those that specify the positive and negative sense of the position coordinates.

If a problem involves a *system* of two or more cords wrapped around pulleys, then the motion of a point on one cord must be related to the motion of a point on another cord using the above procedure. Separate equations are written for corresponding cords of the system and the positions of the two particles are then related by these equations (see Example 12–24).

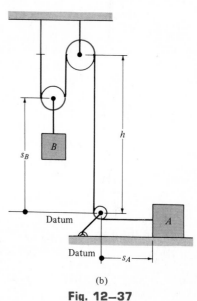

(b)

Fig. 12–37

Example 12–22

Determine the speed of block A in Fig. 12–38. Block B has an upward speed of 6 ft/s.

Fig. 12–38

Solution

Position-Coordinate Equation. There is *one* cord in this system having segments which are changing length. Position coordinates s_A and s_B will be used since each is measured from a fixed point (C or D) and extends along the block's *path of motion*. In particular, s_B is directed to point E since motion of B and E is the *same*.

The colored segments of the cord in Fig. 12–38 remain at a constant length and do not have to be considered as the blocks move. The remaining length of cord, l, is also constant and is related to the changing position coordinates s_A and s_B by the equation

$$s_A + 3s_B = l$$

Time Derivative. Taking the time derivative yields

$$v_A + 3v_B = 0$$

so that when $v_B = -6$ ft/s (upward), then

$$v_A = 18 \text{ ft/s} \downarrow \qquad\qquad \textit{Ans.}$$

Example 12–23

Determine the speed of block A in Fig. 12–39. Block B has an upward speed of 6 ft/s.

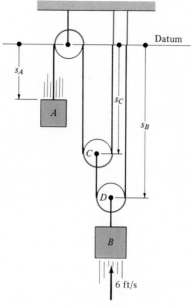

Fig. 12–39

Solution

Position-Coordinate Equation. This system has two cords which have segments that change length. Motion of the end of the cord attached to A can be related to the motion of the end of the cord attached to the pulley at C using coordinates s_A and s_C. Likewise, the motion of the end of the cord at C can be related to the motion of the end of the cord at D using position coordinates s_C and s_B.

The colored segments of the cords in Fig. 12–39 do not have to be considered in the analysis. Why? For the remaining cord lengths, say l_1 and l_2, we have

$$s_A + 2s_C = l_1 \qquad s_B + (s_B - s_C) = l_2$$

Eliminating s_C yields an equation defining the positions of both blocks, i.e.,

$$s_A + 4s_B = 2l_2 + l_1$$

Time Derivative. The time derivative gives

$$v_A + 4v_B = 0$$

so that when $v_B = -6$ ft/s (upward), then

$$v_A = 24 \text{ ft/s} \downarrow \qquad\qquad \textit{Ans.}$$

Example 12–24

Determine the speed with which block B rises in Fig. 12–40 if the end of the cord at A is pulled down with a speed of 2 m/s.

Fig. 12–40

Solution

Position-Coordinate Equation. By inspection this system has two cords. Three position coordinates have been selected. These are all measured from the horizontal datum passing through the fixed point D. The position of the end of the cord at A is defined by s_A, the position of the end of the cord at C is defined by s_C, and the position of the end of the cord at E (block B) is defined by s_E.

Excluding the colored segments of the cords in Fig. 12–40, the remaining constant cord lengths l_1 and l_2 (with the hook dimensions) can be expressed in terms of the three position coordinates. We have

$$(s_A - s_C) + (s_E - s_C) + s_E = l_1$$

$$s_C + s_E = l_2$$

Eliminating s_C yields

$$s_A + 4s_E = l_1 + 2l_2$$

This equation relates the position s_E of block B to the position s_A of point A.

Time Derivative. The time derivative gives

$$v_A + 4v_E = 0$$

so that when $v_A = 2$ m/s (downward), then

$$v_E = -0.5 \text{ m/s} \quad \text{or} \quad v_E = 0.5 \text{ m/s} \uparrow \qquad \textit{Ans.}$$

Example 12-25

A man at A is hoisting a safe S as shown in Fig. 12–41 by walking to the right with a constant velocity $v_A = 0.5$ m/s. Determine the velocity and acceleration of the safe when it reaches the window elevation at E. The rope is 30 m long and passes over a small pulley D.

Fig. 12–41

Solution

Position-Coordinate Equation. This problem is unlike the previous examples, since rope segment DA changes both direction and magnitude. However, the ends of the rope, which define the positions of S and A, are specified by means of the x and y coordinates measured from fixed points and *directed along the paths of motion* of the ends of the rope.

The x and y coordinates may be related since the rope has a fixed length $l = 30$ m, which at all times is equal to the length of segment DA plus CD. Using the Pythagorean theorem to determine l_{DA}, we have $l_{DA} = \sqrt{(15)^2 + x^2}$; also, $l_{CD} = 15 - y$. Hence,

$$l = l_{DA} + l_{CD}$$
$$30 = \sqrt{(15)^2 + x^2} + (15 - y)$$
$$y = \sqrt{225 + x^2} - 15 \qquad (1)$$

Time Derivatives. Taking the time derivative, where $v_S = dy/dt$ and $v_A = dx/dt$, yields

$$v_S = \frac{dy}{dt} = \frac{1}{2} \frac{2x}{\sqrt{225 + x^2}} \frac{dx}{dt}$$

$$= \frac{x}{\sqrt{225 + x^2}} v_A \qquad (2)$$

At $y = 10$ m, x is determined from Eq. (1), i.e., $x = 20$ m. Hence, from Eq. (2) with $v_A = 0.5$ m/s,

$$v_S = \frac{20}{\sqrt{225 + (20)^2}}(0.5) = 0.4 \text{ m/s} = 400 \text{ mm/s} \qquad \textit{Ans.}$$

The acceleration is determined by taking the time derivative of Eq. (2). Since v_A is constant, then $a_A = dv_A/dt = 0$, and we have

$$a_S = \frac{d^2y}{dt^2} = \left[\frac{-x^2}{(225 + x^2)^{3/2}} \frac{dx}{dt} + \frac{1}{(225 + x^2)^{1/2}} \frac{dx}{dt} \right] v_A$$

$$= \frac{225 v_A^2}{(225 + x^2)^{3/2}}$$

At $x = 20$ m, with $v_A = 0.5$ m/s, the acceleration becomes

$$a_S = \frac{225(0.5)^2}{(225 + (20)^2)^{3/2}} = 0.00360 \text{ m/s}^2 = 3.60 \text{ mm/s}^2 \qquad \textit{Ans.}$$

12.9 Relative-Motion Analysis of Two Particles Using Translating Axes

Throughout this chapter the absolute motion of a particle has been determined using a single fixed reference frame for measurement. There are many cases, however, where the path of motion for a particle is complicated, so that it may be feasible to analyze the motion in parts by using two or more frames of reference. For example, the motion of a particle located at the tip of an airplane propeller, while the plane is in flight, is more easily described if one observes first the motion of the airplane from a fixed reference and then superimposes (vectorially) the circular motion of the particle measured from a reference attached to the airplane. Any type of coordinates—rectangular, cylindrical, etc.—may be chosen to describe these two different motions.

In this section only *translating frames of reference* will be considered for the analysis. Relative-motion analysis of particles using rotating frames of reference will be treated in Secs. 16.10 and 20.4, since such an analysis depends upon prior knowledge of the kinematics of line segments.

Position. Consider particles A and B, which move along the arbitrary paths aa and bb, respectively, as shown in Fig. 12–42a. The *absolute position* of each particle, $\mathbf{r}_A$ and $\mathbf{r}_B$, is measured from the common origin O of the *fixed x, y, z* reference frame. The origin of a second frame of reference x', y', z' is attached to and moves with particle A. The axes of this frame do not rotate; rather they are *only permitted to translate* relative to the fixed frame. The *relative position* of "B with respect to A" is observed from this moving frame and is designated by $\mathbf{r}_{B/A}$, called a *relative-position vector*. Using vector addition, the three vectors shown in Fig. 12–42a can be related by the equation*

$$\mathbf{r}_B = \mathbf{r}_A + \mathbf{r}_{B/A} \qquad (12\text{–}33)$$

Fig. 12–42a

(a)

---

*An easy way to remember the setup of this equation, and others like it, is to note the "cancellation" of the subscript A between the two terms, i.e., $\mathbf{r}_B = \mathbf{r}_{\cancel{A}} + \mathbf{r}_{B/\cancel{A}}$.

Velocity. An equation that relates the velocities of the particles can be determined by taking the time derivative of Eq. 12–33, i.e.,

$$\mathbf{v}_B = \mathbf{v}_A + \mathbf{v}_{B/A} \qquad (12\text{–}34)$$

Here $\mathbf{v}_B = d\mathbf{r}_B/dt$ and $\mathbf{v}_A = d\mathbf{r}_A/dt$ refer to *absolute velocities,* since they are measured from the fixed frame of reference; whereas the *relative velocity* $\mathbf{v}_{B/A} = d\mathbf{r}_{B/A}/dt$ is observed from the translating reference frame. It is important to note that since the x', y', z' axes translate, the *components* of $\mathbf{r}_{B/A}$ will *not* change direction and therefore the time derivative of this vector will only have to account for changes in the vector's magnitude. The above equation therefore states that the velocity of B is equal to the velocity of A plus (vectorially) the relative velocity of "B with respect to A," as measured by a *translating observer* fixed in the x', y', z' reference, Fig. 12–42b.

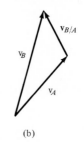

(b)

Acceleration. The time derivative of Eq. 12–34 yields a similar vector relationship between the *absolute* and *relative accelerations* of particles A and B,

$$\mathbf{a}_B = \mathbf{a}_A + \mathbf{a}_{B/A} \qquad (12\text{–}35)$$

Here $\mathbf{a}_{B/A}$ is the acceleration of B as seen by an observer located at A and translating with the x', y', z' reference frame. The vector addition is shown in Fig. 12–42c.

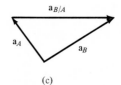

(c)

Fig. 12–42b, c

PROCEDURE FOR ANALYSIS

When applying the relative-position equation, $\mathbf{r}_B = \mathbf{r}_A + \mathbf{r}_{B/A}$, it is first necessary to specify the locations of the fixed x, y, z and translating x', y', z' axes. Usually, the origin A of the translating axes is located at a point having a *known position*, $\mathbf{r}_A$, Fig. 12–42a. A graphical representation of the vector addition $\mathbf{r}_B = \mathbf{r}_A + \mathbf{r}_{B/A}$ can be shown, and both the known and unknown quantities labeled on this sketch. Since vector addition forms a triangle there can be at most *two unknowns*, represented by the magnitudes and/or directions of the vector quantities. These unknowns can be solved for either graphically, using trigonometry (law of sines, law of cosines), or by resolving each of the three vectors $\mathbf{r}_B$, $\mathbf{r}_A$, and $\mathbf{r}_{B/A}$ into Cartesian components, thereby generating two scalar equations. The latter method is illustrated in the example problems which follow.

The relative-motion equations $\mathbf{v}_B = \mathbf{v}_A + \mathbf{v}_{B/A}$ and $\mathbf{a}_B = \mathbf{a}_A + \mathbf{a}_{B/A}$ are applied in the same manner as explained above, except in this case the origin O of the fixed x, y, z axes does not have to be specified, Fig. 12–42b and c.

Example 12–26

Water drips from a faucet at the rate of five drops per second as shown in Fig. 12–43. Determine the vertical separation between two consecutive drops after the lower drop has attained a velocity of 3 m/s.

Fig. 12–43

Solution

If the first and second drops of water are denoted as B and A, respectively, Fig. 12–43, the separation $s_{B/A}$ (position of B with respect to A) can be determined by the equation

$$s_{B/A} = s_B - s_A \qquad (1)$$

Here the origin of the fixed frame of reference is located at the head of the faucet (datum) with positive s downward. The origin of the moving frame s' is at drop A. Since each drop is subjected to rectilinear motion, having a *constant* downward acceleration $a_c = g = 9.81$ m/s^2, Eq. 12–4 may be applied with the initial condition $v_0 = 0$ to determine the time needed for drop B to attain a velocity of 3 m/s.

$(+\downarrow)$
$$v = v_0 + a_c t$$
$$3 = 0 + 9.81\, t_B$$
$$t_B = 0.306\ \text{s}$$

Since one drop falls every fifth of a second (0.2 s), drop A falls for a time

$$t_A = 0.306\ \text{s} - 0.2\ \text{s} = 0.106\ \text{s}$$

before drop B attains a velocity of 3 m/s.

The position of each drop can be found by means of Eq. 12–5, with the condition $s_0 = 0$, $v_0 = 0$.

$(+\downarrow)$
$$s = s_0 + v_0 t + \tfrac{1}{2} a_c t^2$$
$$s_B = 0 + 0 + \tfrac{1}{2}(9.81)(0.306)^2$$
$$= 0.459\ \text{m}$$
$$s_A = 0 + 0 + \tfrac{1}{2}(9.81)(0.106)^2$$
$$= 0.055\ \text{m}$$

Applying Eq. (1) yields

$$s_{B/A} = s_B - s_A = 0.459 - 0.055 = 0.404\ \text{m}$$
$$= 404\ \text{mm} \qquad\qquad \textit{Ans.}$$

Example 12–27

A train, traveling at a constant speed of 60 mi/h, crosses over a road as shown in Fig. 12–44a. If an automobile A is traveling at 45 mi/h along the road, determine the relative velocity of the train with respect to the automobile.

$v_T = 60$ mi/h

$v_A = 45$ mi/h

(a)

Solution

The relative velocity $\mathbf{v}_{T/A}$ is measured from the translating x', y' axes attached to the automobile, Fig. 12–44a. It is determined from $\mathbf{v}_T = \mathbf{v}_A + \mathbf{v}_{T/A}$. Since $\mathbf{v}_T$ and $\mathbf{v}_A$ are known in *both* magnitude and direction, the unknowns become the components of $\mathbf{v}_{T/A}$. Using the x, y axes in Fig. 12–44b and a Cartesian vector analysis, we have

$$\mathbf{v}_T = \mathbf{v}_A + \mathbf{v}_{T/A}$$
$$60\mathbf{i} = (45 \cos 45° \, \mathbf{i} + 45 \sin 45° \, \mathbf{j}) + \mathbf{v}_{T/A}$$
$$\mathbf{v}_{T/A} = \{28.2\mathbf{i} - 31.8\mathbf{j}\} \text{ mi/h}$$

The components of $\mathbf{v}_{T/A}$ can also be determined by applying a scalar analysis as follows:

$$\mathbf{v}_T = \mathbf{v}_A + \mathbf{v}_{T/A}$$
$$60 \text{ mi/h} = 45 \text{ mi/h} + (v_{T/A})_x + (v_{T/A})_y$$
$$\rightarrow \qquad \measuredangle 45° \qquad \rightarrow \qquad \downarrow$$

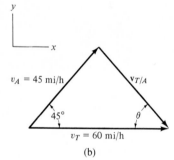

(b)

Fig. 12–44

Resolving each vector into its horizontal and vertical components yields

$(\xrightarrow{+})$ $\qquad\qquad 60 = 45 \cos 45° + (v_{T/A})_x + 0$
$(+\uparrow)$ $\qquad\qquad 0 = 45 \sin 45° + 0 - (v_{T/A})_y$

Solving, we obtain the previous results,

$$(v_{T/A})_x = 28.2 \text{ mi/h} \rightarrow$$
$$(v_{T/A})_y = 31.8 \text{ mi/h} \downarrow$$

The magnitude of $\mathbf{v}_{T/A}$ is thus

$$v_{T/A} = \sqrt{(28.2)^2 + (31.8)^2} = 42.5 \text{ mi/h} \qquad \textit{Ans.}$$

The direction, defined from the x axis, is

$$\tan \theta = \frac{(v_{T/A})_y}{(v_{T/A})_x} = \frac{31.8}{28.2}$$
$$\theta = 48.4° \quad \measuredangle\theta \qquad \textit{Ans.}$$

Note that the vector addition shown in Fig. 12–44b indicates the correct sense for $\mathbf{v}_{T/A}$. This figure anticipates the answer and can be used to check it.

Example 12–28

Two jet planes are flying horizontally at the same elevation, as shown in Fig. 12–45a. Plane A is flying along a straight-line path, and at the instant shown it has a speed of 700 km/h and an acceleration of 50 km/h². Plane B is flying along a circular path at 600 km/h and decreasing its speed at the rate of 100 km/h². Determine the velocity and acceleration of B as measured by the pilot in A.

(a)

Solution

Velocity. Plane A is traveling with rectilinear motion and a *translating frame of reference x′, y′* will be attached to it. Applying the relative-velocity equation in scalar form since the velocity vectors of both planes are parallel at the instant shown, Fig. 12–45b, we have

$(+\uparrow)$
$$v_B = v_A + v_{B/A}$$
$$600 = 700 - v_{B/A}$$
$$v_{B/A} = 100 \text{ km/h} \downarrow \qquad \textit{Ans.}$$

Acceleration. Plane B has both tangential and normal components of acceleration, since it is flying along a *curved path*. From Eq. 12–20, the magnitude of the normal component is

$$(a_B)_n = \frac{v_B^2}{\rho} = \frac{(600)^2}{400} = 900 \text{ km/h}^2$$

(b)

Applying the relative-acceleration equation, Fig. 12–45c, we have

$$\mathbf{a}_B = \mathbf{a}_A + \mathbf{a}_{B/A}$$
$$100 + 900 = 50 + (a_{B/A})_x + (a_{B/A})_y$$
$$\downarrow \qquad \rightarrow \qquad \uparrow \qquad \rightarrow \qquad \downarrow$$

(c)

Fig. 12–45

Thus,

$(\xrightarrow{+})$
$$0 + 900 = 0 + (a_{B/A})_x$$
$(+\uparrow)$
$$-100 + 0 = 50 - (a_{B/A})_y$$

Solving, we obtain

$$(a_{B/A})_x = 900 \text{ km/h}^2 \rightarrow \qquad (a_{B/A})_y = 150 \text{ km/h}^2 \downarrow$$

The magnitude and direction of $\mathbf{a}_{B/A}$ are therefore

$$a_{B/A} = 912 \text{ km/h}^2 \qquad \theta = 9.46° \quad \searrow_\theta \qquad \textit{Ans.}$$

Notice that the solution to this problem is possible using a translating frame of reference, since the pilot in plane A is "translating." Observation of plane A with respect to plane B, however, must be obtained using a *rotating* set of axes attached to plane B. (This assumes, of course, that the pilot of B does not turn his eyes to follow the motion of A.) The analysis for this case is given in Example 16–20.

Problems

12–115. The pulley system supports three blocks as shown. If *A* is moving downward with a speed of 4 ft/s while *C* is moving up at 2 ft/s, determine the speed of block *B*.

Probs. 12–115 / 12–116

*12–116. If block *A* of the pulley system is moving downward at 6 ft/s while block *C* is moving down at 18 ft/s, determine the relative velocity of block *B* with respect to *C*.

12–117. Block *B* is moving downward with a constant acceleration $a_B = 0.2$ m/s^2. If both blocks start from rest and are at the same elevation $h = 0$ when $t = 0$, determine (a) the velocities of blocks *A* and *B* when $t = 3$ s, and (b) the elevation h between the blocks when $t = 3$ s.

Prob. 12–117

12–118. Determine the constant speed at which the cable at *A* must be drawn in by the motor in order to hoist the load at *B* 6 m in 1.5 s.

Prob. 12–118

12–119. The cable system is used to tow the log *B*. Determine the displacement of the log if the man at *C* pulls the cable 4 ft to the right.

Prob. 12–119

73

*12–120. The hook B on the oil rig is supported by a cable which is connected to the drum at A, passes over a pulley at E, down around the hook's pulley at B, up and around another pulley at E, and is then attached to the drum at C. Determine the time needed to hoist the hook from $h = 0$ to $h = 100$ ft if (a) drum A draws the cable in at a constant speed of 4 ft/s and drum C is stationary, and (b) drum A draws in the cable at a constant speed of 3 ft/s and drum C draws in the cable at a constant speed of 2 ft/s.

Prob. 12–120

12–121. The pipe P is being lifted using the cable and pulley system shown. If point A on the cable is being drawn toward the drum with a speed of 2 m/s, determine the speed of the pipe.

Probs. 12–121 / 12–122

12–122. The pipe P is being lifted using the cable and pulley system shown. Starting from rest, point A on the cable is being drawn in with an acceleration $a_A = (5t^2)$ m/s^2, where t is in seconds. If $s = 0$ when $t = 0$, determine the height s of the pipe when $t = 2$ s.

12–123. Determine the velocity of the block at B if the cord at A is pulled down with a speed of 8 ft/s. What is the relative velocity of B with respect to A?

Prob. 12–123

*12–124. The pulley arrangement shown is used to obtain a high mechanical advantage. If supports *B* and *C remain fixed* while the plunger *P* is pushed downward with a speed of 4 ft/s, determine the speed of the hook *A* located at the end of the cable.

Prob. 12–124

12–125. The motor at *C* pulls in the cable with an acceleration $a_C = (3t^2)$ m/s^2, where *t* is in seconds. The motor at *D* draws in its cable at $a_D = 5$ m/s^2. If both motors start at the same instant from rest when *d* = 3 m, determine (a) the time needed for *d* = 0, and (b) the relative velocity of block *A* with respect to block *B* when this occurs.

Prob. 12–125

12–126. If block *B* is moving down with a velocity v_B and has an acceleration a_B, determine the velocity and acceleration of block *A* in terms of the parameters shown.

Prob. 12–126

12–127. The cable at *A* is drawn in at a constant rate $v_A = 6$ m/s. Determine the speed at which the load at *B* rises as a function of its vertical position *y*.

Probs. 12–127 / 12–128

*12–128. If the cable at *A* is drawn in with an acceleration $a_A = 2$ m/s^2 when it has a speed $v_A = 6$ m/s, determine the velocity and acceleration of the load at *B* at this instant, which occurs when *y* = 3 m.

75

12–129. The man pulls the boy up to the tree limb C by walking backward at a constant speed of 1.5 m/s. Determine the speed at which the boy is being lifted at the instant $x_A = 4$ m. Neglect the size of the limb. When $x_A = 0$, $y_B = 8$ m, so that points A and B are coincident, i.e., the rope is 16 m long.

Probs. 12–129 / 12–130

12–130. The man pulls the boy up to the tree limb C by walking backward. If he starts from rest when $x_A = 0$ and moves backward with a constant acceleration $a_A = 0.2$ m/s^2, determine the speed of the boy at the instant $y_B = 4$ m. Neglect the size of the limb. When $x_A = 0$, $y_B = 8$ m, so that points A and B are coincident, i.e., the rope is 16 m long.

12–131. The girl at C stands near the edge of the pier and pulls in the rope *horizontally* at a constant speed of 6 ft/s. Determine how fast the boat approaches the pier when the rope length AB is 50 ft.

Prob. 12–131

***12–132.** The crate C is being lifted by moving the roller at A downward with a constant speed of $v_A = 2$ m/s along the guide. Determine the velocity and acceleration of the crate at the instant $s = 1$ m. When the roller is at B, the crate rests on the ground. Neglect the size of the pulley in the calculation. *Hint:* Relate the coordinates x_C and x_A using the problem geometry, then take the first and second time derivatives.

Prob. 12–132

12–133. The motorboat can travel with a speed $v_B = 15$ ft/s in still water. If it heads for C at the opposite side of the river, and the river flows with a speed $v_R = 5$ ft/s, determine the resultant velocity of the boat. How far downstream, d, is the boat carried when it reaches the other side at D?

Prob. 12–133

12–134. A man walks at 5 km/h in the direction of a horizontal 15-km/h wind. If raindrops fall vertically at 7 km/h in *still air,* determine the direction in which the drops appear to fall with respect to the man.

$v_w = 15$ km/h

$v_m = 5$ km/h

Prob. 12–134

12–135. A man is riding on a mail car traveling forward at 10 mi/h. Determine the direction θ, measured in the horizontal plane, in which he must throw a package P horizontally at 20 mi/h, measured relative to himself, so that the package has a velocity $\mathbf{v}_P$ which is perpendicular to his direction of travel.

$\mathbf{v}_P$ P

θ 20 mi/h

$v_c = 10$ mi/h

Prob. 12–135

*12–136. A passenger in an automobile observes that raindrops make an angle of 30° with the horizontal as the auto travels forward with a speed of 60 km/h. Compute the terminal (constant) velocity $\mathbf{v}_r$ of the rain if it is assumed to fall vertically.

$\mathbf{v}_r$ $v_a = 60$ km/h

Prob. 12–136

12–137. The airplane can fly at a speed of 650 km/h in still air. The point of destination is located at a bearing of north 60° east. If a 50-km/h wind is blowing from north to south, determine the bearing angle θ at which the pilot must direct the plane to stay on course.

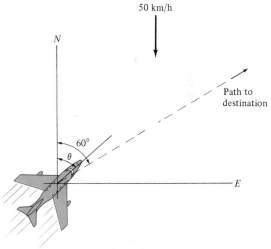

50 km/h

N

Path to destination

60°

θ

E

Prob. 12–137

12–138. At the instant shown, cars A and B are traveling at speeds of 30 mi/h and 20 mi/h, respectively. If B is increasing its speed by 1200 mi/h², while A maintains a constant speed, determine the velocity and acceleration of B with respect to A.

Prob. 12–138

12–139. At the instant shown, the bicyclist at A is traveling at 7 m/s around the curve on the race track while increasing his speed at 0.5 m/s². The bicyclist at B is traveling at 8.5 m/s along the straight-a-way and increasing his speed at 0.7 m/s². Determine the relative velocity and relative acceleration of A with respect to B at this instant.

Prob. 12–139

*12–140. An aircraft carrier is traveling forward with a velocity of 50 km/h. At the instant shown, the plane at A has just taken off and has attained a forward horizontal air speed of 200 km/h, measured from still water. If the plane at B is traveling along the runway of the carrier at 175 km/h in the direction shown, determine the velocity of A with respect to B.

Prob. 12–140

12–141. Two planes, A and B, are flying at the same altitude. If their velocities are $v_A = 600$ km/h and $v_B = 500$ km/h such that the angle between their straight-line courses is $\theta = 75°$, determine the velocity of plane B with respect to plane A.

Prob. 12–141

12–142. Two boats leave the pier P at the same time and travel in the directions shown. If $v_A = 40$ ft/s and $v_B = 30$ ft/s, determine the speed of boat A relative to boat B. How long after leaving the pier will the boats be 1500 ft apart?

Prob. 12–142

12–143. The boy at A is running in a straight line away from the building at a constant speed of 4 ft/s. The boy at C throws the ball B horizontally toward him when he is 10 ft away. With what speed must the boy at C throw the ball so that the boy at A can catch it? Also determine the relative speed of the ball with respect to the boy at A at the instant the catch is made.

Prob. 12–143

*12–144. At a given instant the football player at A throws a football C with a velocity of 20 m/s in the direction shown. Determine the constant speed at which the player at B must run so that he can catch the football at the same elevation at which it was thrown. Also calculate the relative velocity and relative acceleration of the football with respect to B at the instant the catch is made. Player B is 15 m away from A when A starts to throw the football.

Prob. 12–144

12–145. The boy B is running down the slope at a constant speed of 12 ft/s. At the instant he reaches A, his friend (?) throws an egg horizontally off the side of the slope, which later strikes him on the head. Determine the time of flight before the egg gets scrambled, the launch velocity $\mathbf{v}_0$, and the speed with which the egg strikes B. Experimental verification is not recommended.

12–146. At a given instant, each of the two particles A and B are moving with a speed of 8 m/s along the paths shown. If B is decelerating at 6 m/s^2 and the speed of A is increasing at 5 m/s^2, determine the relative acceleration of A with respect to B at this instant.

Prob. 12–145

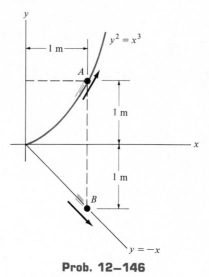

Prob. 12–146

Kinetics of a Particle: Force and Acceleration

In Chapter 12 we developed the methods needed to formulate the acceleration of a particle in terms of its velocity and position. In this chapter we will use these concepts when applying Newton's second law of motion, $\mathbf{F} = m\mathbf{a}$, to study the effects caused by an unbalanced force acting on a particle. Depending upon the geometry of the path, the analysis of problems will be performed using rectangular, normal and tangential, or cylindrical coordinates. In the last part of the chapter, Newton's second law of motion will be used to study problems in space mechanics.

Newton's Laws of Motion 13.1

Many of the earlier notions about dynamics were dispelled after 1590 when Galileo performed experiments to study the motions of pendulums and falling bodies. The conclusions drawn from these experiments gave some insight as to the effects of forces acting on bodies in motion. The general motion of a body subjected to forces was not known, however, until 1687, when Isaac Newton first presented three basic laws governing the motion of a particle. In a slightly reworded form, Newton's three laws of motion can be stated as follows:

First Law. A particle originally at rest, or moving in a straight line with a constant velocity, will remain in this state provided the particle is not subjected to an unbalanced force.

Second Law. A particle acted upon by an unbalanced force **F** experiences an acceleration **a** that has the same direction as the force and a magnitude that is directly proportional to the force.*

Third Law. For every force acting on a particle, the particle exerts an equal, opposite, and collinear reactive force.

The first and third laws were used extensively in developing the concepts of statics. Although these laws are also considered in dynamics, Newton's second law of motion forms the basis for most of this study, since this law relates the accelerated motion of a particle to the forces that act on it. Measurements of force and acceleration can be recorded in a laboratory so that in accordance with the second law, if a known unbalanced force F_1 is applied to a particle, the acceleration a_1 of the particle may be measured. Since the force and acceleration are directly proportional, the constant of proportionality, m, may be determined from the ratio $m = F_1/a_1$. Provided the units of measurement are consistent, a different unbalanced force F_2 applied to the particle will create an acceleration a_2, such that $F_2/a_2 = m$.† In both cases the ratio will be the same and the acceleration and the force, both being vector quantities, will have the same direction. The scalar m is called the *mass* of the particle. Being constant during any acceleration, m provides a quantitative measure of the resistance of the particle to a change in its velocity.

If the mass of the particle is m, Newton's second law of motion may be written in mathematical form as

$$\mathbf{F} = m\mathbf{a}$$

This equation, which is referred to as the *equation of motion,* is one of the most important formulations in mechanics.‡ As stated above, its validity has been based solely upon *experimental evidence.* In 1905, however, Albert Einstein developed the theory of relativity and placed limitations on the use of Newton's second law for describing general particle motion. Through experiments it was proven that *time* is not an absolute quantity as assumed by Newton; and as a result, the equation of motion fails to predict the exact behavior of a particle, especially when the particle's speed approaches the

---

*Stated another way, the unbalanced force acting on the particle is proportional to the time rate of change of the particle's linear momentum. See footnote ‡ below.

†Throughout this discussion, the units of force, mass, length, and time used to measure F, m, and a must be chosen such that the units for one of these quantities are defined in terms of all the others. Such is the case for the SI and FPS units, where it may be recalled that N = kg · m/s² and slug = lb · s²/ft (see Sec. 1.4 of *Statics*). Note, however, that if units of force, mass, length, and time are *all* selected arbitrarily, then $F = kma$ and k (a dimensionless constant) would have to be experimentally determined in order to preserve the equality.

‡Since m is constant, we can also write $\mathbf{F} = \dfrac{d}{dt}(m\mathbf{v})$, where $m\mathbf{v}$ is the particle's linear momentum.

speed of light [0.3 Gm/s]. Developments of the theory of quantum mechanics by Schrödinger and others indicate further that conclusions drawn from using this equation are also invalid when particles move within an atomic distance of one another. For the most part, however, these requirements regarding particle speed and size are not encountered in engineering problems, so that their effects will not be considered in this book.

Newton's Law of Gravitational Attraction. Shortly after formulating his three laws of motion, Newton postulated a law governing the mutual attraction between any two particles. In mathematical form this law can be expressed as

$$F = G \frac{m_1 m_2}{r^2} \qquad (13-1)$$

where
F = force of attraction between the two particles

G = universal constant of gravitation; according to experimental evidence $G = 6.673(10^{-11})$ m^3/(kg · s^2)

m_1, m_2 = mass of each of the two particles

r = distance between the centers of the two particles

Any two particles or bodies have a mutually attractive (gravitational) force acting between them. In the case of a particle located at or near the surface of the earth, however, the only gravitational force having any sizable magnitude is that between the earth and the particle. This force is termed the "weight" and, for our purpose, it will be the only gravitational force considered.

Mass and Weight. *Mass* is a property of matter by which we can compare the action of one body with that of another. As shown above, this property manifests itself as a gravitational attraction between two bodies and provides a quantitative measure of the resistance of matter to a change in velocity. It is an *absolute* quantity since the measurement of mass can be made at any location. The weight of a body, however, is *not absolute* since it is measured in a gravitational field, and hence its magnitude depends upon where the measurement is made. From Eq. 13–1, we can develop a general expression for finding the weight W of a particle having a mass $m = m_1$. Let m_2 be the mass of the earth and r the distance between the earth's center and the particle. Then if $g = Gm_2/r^2$, we have

$$W = mg$$

By comparison with $F = ma$, we term g the acceleration due to gravity. For most engineering calculations we measure g at a point on the surface of the earth at sea level, and at a latitude of 45°, considered the "standard location."

The mass and weight of a body are measured differently in the SI and FPS systems of units, and the method of defining the units should be thoroughly understood.

Fig. 13–1

SI System of Units. In the SI system the mass of the body is specified in kilograms and the weight must be calculated using the equation of motion, $F = ma$. Hence, if a body has a mass m (kg) and is located at a point where the acceleration due to gravity is g (m/s^2), then the weight is expressed in *newtons* as $W = mg$ (N), Fig. 13–1a. In particular, if the body is located at the "standard location," the acceleration due to gravity is $g = 9.806\,65$ m/s^2. For calculations, the value $g = 9.81$ m/s^2 will be used, so that

$$W = mg \text{ (N)} \quad (g = 9.81 \text{ m/s}^2) \tag{13–2}$$

Therefore, a body of mass 1 kg has a weight of 9.81 N; a 2-kg body weighs 19.62 N; and so on.

FPS System of Units. In the FPS system the weight of the body is specified in pounds and the mass must be calculated from $F = ma$. Hence, if a body has a weight W (lb) and is located at a point where the acceleration due to gravity is g (ft/s^2), then the mass is expressed in *slugs* as $m = W/g$ (slug), Fig. 13–1b. Since the acceleration of gravity at the standard location is approximately 32.2 ft/s^2 ($= 9.81$ m/s^2), the mass of the body measured in slugs is

$$m = \frac{W}{g} \text{(slug)} \quad (g = 32.2 \text{ ft/s}^2) \tag{13–3}$$

Therefore, a body weighing 32.2 lb has a mass of 1 slug; a 64.4-lb body has a mass of 2 slugs; and so on.

13.2 The Equation of Motion

When more than one force acts on a particle, the resultant force is determined by a vector summation of all the forces; i.e., $\mathbf{F}_R = \Sigma\mathbf{F}$. For this more general case, the equation of motion may be written as

$$\Sigma\mathbf{F} = m\mathbf{a} \tag{13–4}$$

To illustrate application of this equation, consider the particle P shown in Fig. 13–2a, which has a mass m and is subjected to the action of two forces, $\mathbf{F}_1$ and $\mathbf{F}_2$. We can graphically account for the magnitude and direction of each force acting on the particle by drawing the particle's *free-body diagram*, Fig. 13–2b. Since the *resultant* of these forces *produces* the vector $m\mathbf{a}$, its magnitude and direction can be represented graphically on the *kinetic diagram*, shown in Fig. 13–2c. The equal sign written between the diagrams symbolizes the *graphical* equivalency between the free-body diagram and the kinetic diagram; i.e., $\Sigma\mathbf{F} = m\mathbf{a}$. In particular, note that if $\mathbf{F}_R = \Sigma\mathbf{F} = \mathbf{0}$, then the acceleration is also zero, so that the particle will either remain at *rest* or move along a straight-line path with *constant velocity*. Such are the conditions of *static equilibrium*, Newton's first law of motion.

Dynamic Equilibrium. The equation of motion can be rewritten in the form

$$\Sigma\mathbf{F} - m\mathbf{a} = \mathbf{0}$$

and the result illustrated on a diagram, which includes the "$m\mathbf{a}$ vector" acting in its reverse sense ($-m\mathbf{a}$), Fig. 13–2d. If $-m\mathbf{a}$ is treated in the same way as a "force vector," then the state of "equilibrium" created in Fig. 13–2d is referred to as *dynamic equilibrium*. This method for application of the equation of motion is often referred to as the *D'Alembert principle*, named after the French mathematician Jean le Rond d'Alembert (1717–1783).

The vector $-m\mathbf{a}$ is referred to as the *inertia force vector*. The word "inertia" is incorporated in this terminology, because by definition *mass* is a quantitative measure of the inertia or resistance of a body to a *change* in its velocity (acceleration). It is important to realize, however, that the inertia force vector is actually not the same as a force. The inertia of a body manifests itself as a force whenever an unbalanced force acts on the body and thereby causes an acceleration. For example, consider the passenger riding in a train that is accelerating, Fig. 13–3a. The forward motion of the train creates a horizontal force $\mathbf{F}$ which the seat exerts on her back, Fig. 13–3b. By the equation of motion, it is this unbalanced *force* which gives her a forward acceleration ($\mathbf{F} = m\mathbf{a}$). No force exists which pushes her back toward the seat ($-m\mathbf{a}$), although this is the sensation she receives, Fig. 13–3c.

(a)

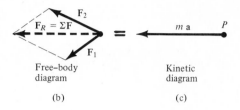

Free–body diagram   Kinetic diagram

(b)   (c)

Dynamic Equilibrium

(d)

Fig. 13–2

(a)   (b)   (c)

Fig. 13–3

Inertial Frame of Reference. Whenever the equation of motion is applied, it is required that measurements of the acceleration be made from a *Newtonian* or *inertial frame of reference*. Such a *coordinate system does not rotate and is either fixed or translates in a given direction with a constant velocity (zero acceleration)*. This definition ensures that the particle's *acceleration* measured by observers in two different inertial frames of reference will always be the *same*. When studying the motions of rockets and satellites it is justifiable to consider the inertial reference frame as fixed to the stars, whereas dynamics problems concerned with motions on or near the surface of the earth may be solved by using an inertial frame which is assumed fixed to the earth. Even though the earth *rotates* both about its own axis and about the sun, the acceleration created by these rotations can be neglected in most computations.

13.3 Equation of Motion for a System of Particles

The equation of motion will now be extended to include a system of n particles isolated within an enclosed region in space, as shown in Fig. 13–4a. In particular, there is no restriction in the way the particles are connected, and as a result the following analysis will apply equally well to the motion of a solid, liquid, or gas system. At the instant considered, the arbitrary ith particle, having a mass m_i, is subjected to a set of internal forces and a resultant external force. The *internal forces,* represented symbolically as $\sum\limits_{j=1(j\neq i)}^{n} \mathbf{f}_{ij} = \mathbf{f}_i$, are forces which the other particles exert on the ith particle. The summation extends over all n particles; however, it is meaningless for $j = i$ since the ith particle cannot exert a force on itself. The *resultant external force* $\mathbf{F}_i$ represents, for example, the effect of gravitational, electrical, magnetic, or contact forces between adjacent bodies or particles *not* included within the system.

The free-body and kinetic diagrams for the ith particle are shown in Fig. 13–4b. Applying the equation of motion yields

$$\Sigma\mathbf{F} = m\mathbf{a}; \qquad\qquad \mathbf{F}_i + \mathbf{f}_i = m_i\mathbf{a}_i$$

When the equation of motion is applied to each of the other particles of the system, similar equations will result. If all these equations are added together *vectorially,* we obtain

$$\Sigma\mathbf{F}_i + \Sigma\mathbf{f}_i = \Sigma m_i\mathbf{a}_i$$

The summation of the internal forces, if carried out, will equal zero since the internal forces between particles all occur in equal, but opposite collinear pairs. Consequently, only the sum of the external forces will remain and therefore the equation of motion, written for the system of particles, becomes

$$\Sigma\mathbf{F}_i = \Sigma m_i\mathbf{a}_i \qquad\qquad (13\text{–}5)$$

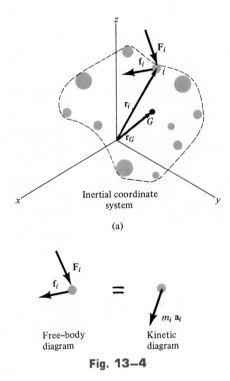

(a)

Free-body diagram $\quad=\quad$ Kinetic diagram

Fig. 13–4

If $\mathbf{r}_G$ is a position vector which locates the *center of mass G* of the particles, Fig. 13–4a, then by definition of the center of mass $m\mathbf{r}_G = \Sigma m_i\mathbf{r}_i$, where $m = \Sigma m_i$ is the total mass of all the particles. Differentiating this equation twice with respect to time, assuming no mass is entering or leaving the system,* yields

$$m\mathbf{a}_G = \Sigma m_i\mathbf{a}_i$$

Substituting this result into the above equation, we obtain

$$\Sigma\mathbf{F} = m\mathbf{a}_G \qquad\qquad (13–6)$$

This equation states that the sum of the external forces acting on the system of particles is equal to the mass $m = \Sigma m_i$ of a single "fictitious" particle times its acceleration. This fictitious particle is located at the center of mass G of all the particles.

Since, in reality, all particles must have finite size to possess mass, Eq. 13–6 justifies application of the equation of motion to a *body* that is represented as a single particle.

*A case in which m is a function of time (variable mass) is discussed in Sec. 15.9.

13.4 Equations of Motion: Rectangular Coordinates

When a particle is moving relative to an inertial x, y, z frame of reference, the forces acting on the particle, as well as its acceleration, may be expressed in terms of their **i, j, k** components, Fig. 13–5. Applying the equation of motion, we have

$$\Sigma \mathbf{F} = m\mathbf{a}$$

$$\Sigma F_x \mathbf{i} + \Sigma F_y \mathbf{j} + \Sigma F_z \mathbf{k} = m(a_x \mathbf{i} + a_y \mathbf{j} + a_z \mathbf{k})$$

For this equation to be satisfied, the respective **i, j,** and **k** components on the left side must equal the corresponding components on the right side. Consequently, we may write the following three scalar equations:

$$\begin{aligned} \Sigma F_x &= ma_x \\ \Sigma F_y &= ma_y \\ \Sigma F_z &= ma_z \end{aligned} \qquad (13\text{–}7)$$

In particular, if the particle is constrained to move in the x-y plane, then only the first two of these equations are used to specify the motion.

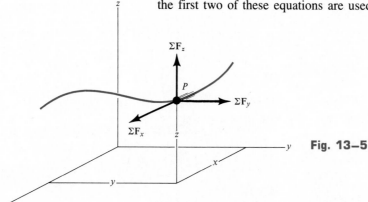

Fig. 13–5

PROCEDURE FOR ANALYSIS

The following procedure provides a method for solving problems using the equations of motion:

Free-Body Diagram. Establish the x, y, z inertial coordinate system and draw the particle's free-body diagram. Drawing the free-body diagram is *very important* since it provides a graphical representation of *all the forces* ($\Sigma \mathbf{F}$) which act on the particle and thereby makes it easy to resolve these forces into their components. The direction and sense of the particle's acceleration **a** should also be established. If the sense of its components is unknown assume that they are in the *direction* of the *positive* inertial coordinate(s)*. The acceleration may be sketched near, *but not on,* the free-body diagram, or it may be represented as the *ma* vector on the kinetic diagram.†

Equations of Motion.   If the forces can be resolved directly from the free-body diagram, apply the equations of motion in their scalar component form. If the geometry of the problem appears complicated, which often occurs in three dimensions, Cartesian vector analysis can be used for the solution.

Friction: If the particle contacts a rough surface, it may be necessary to use the *frictional equation,* which relates the coefficient of kinetic friction μ_k to the magnitudes of the frictional and normal forces $\mathbf{F}_f$ and $\mathbf{N}$ acting at the surfaces of contact, i.e., $F_f = \mu_k N$.

Spring: If the particle is connected to an *elastic spring* having negligible mass, the magnitude of spring force $\mathbf{F}_s$ can be related to the stretch or compression s of the spring by the equation $F_s = ks$, where k is the spring's stiffness measured as a force per unit length.

Kinematics.   Use the equations of kinematics if a complete solution cannot be obtained strictly from the equation of motion. In this regard, if the velocity or position of the particle is to be found, it will be necessary to apply the proper kinematic equations once the particle's acceleration is determined from $\Sigma\mathbf{F} = m\mathbf{a}.$

If the *acceleration is a function of time,* use $a = dv/dt$ and $v = ds/dt$ which, when integrated, yield the particle's velocity and position.

If the *acceleration is a function of displacement,* integrate $a\, ds = v\, dv$ to obtain the velocity as a function of position.

If the *acceleration is constant,* use

$$v = v_0 + a_c t \qquad s = s_0 + v_0 t + \tfrac{1}{2} a_c t^2 \qquad v^2 = v_0^2 + 2a_c(s - s_0)$$

to determine the position or velocity of the particle.

If the problem involves the dependent motion of several particles, use the method outlined in Sec. 12.8 to relate their accelerations. In all cases, make sure the positive coordinate directions used for writing the kinematic equations are the *same* as those used for writing the equations of motion.

The following examples numerically illustrate application of this procedure.

---

*This provides a mathematical convenience to account for the proper sign of **a** if the equations of kinematics are used in the solution.

†It is a convention in this text to always use the kinetic diagram as a graphical aid when developing the proofs and theory. The particle's acceleration or its components will be shown as colored vectors near the free-body diagram in the examples.

Example 13–1

The 50-kg crate shown in Fig. 13–6a rests on a horizontal plane for which the coefficient of kinetic friction is $\mu_k = 0.3$. If the crate is subjected to a 400-N towing force as shown, determine the velocity of the crate in 5 s starting from rest.

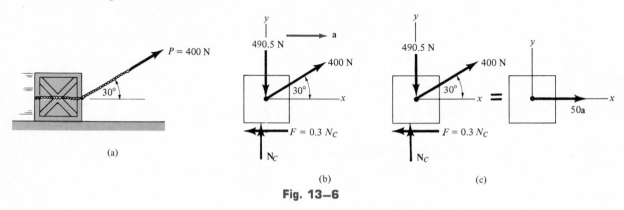

Fig. 13–6

Solution

Free-Body Diagram. The weight of the crate is $W = mg = 50\ \text{kg}(9.81\ \text{m/s}^2) = 490.5\ \text{N}$. As shown in Fig. 13–6b, the frictional force has a magnitude $F = \mu_k N_C$ and acts to the left, since it opposes the motion of the crate. Note that **a** acts horizontal, in the positive x direction.

Equations of Motion. Using the data shown on the free-body diagram, we have

$$\xrightarrow{+} \Sigma F_x = ma_x; \qquad 400 \cos 30° - 0.3\,N_C = 50a \qquad (1)$$
$$+\uparrow \Sigma F_y = ma_y; \qquad N_C - 490.5 + 400 \sin 30° = 0 \qquad (2)$$

Solving Eq. (2) for N_C, substituting the result into Eq. (1), and solving for a yields

$$N_C = 290.5\ \text{N}$$
$$a = 5.19\ \text{m/s}^2 \qquad \Sigma F_x = ma \atop a = 5.19\ \text{m/s}^2$$

Kinematics. Since the acceleration is *constant*, and the initial velocity is zero, the velocity of the crate in 5 s is

$$(\xrightarrow{+}) \qquad v = v_0 + a_c t$$
$$= 0 + 5.19(5)$$
$$= 26.0\ \text{m/s} \rightarrow \qquad \textit{Ans.}$$

Note the alternative procedure of drawing the crate's free-body *and* kinetic diagrams, Fig. 13–6c, prior to applying the equations of motion.

Example 13–2

A 10-kg projectile is fired vertically upward from the ground, with an initial velocity of 50 m/s, Fig. 13–7a. Determine the maximum height to which it will travel if (a) atmospheric resistance is neglected; and (b) the atmospheric resistance is measured as $F_D = (0.01v^2)$ N, where v is the speed at any instant, measured in m/s.

(a)

Solution
Part (a)
Free-Body Diagram. As shown in Fig. 13–7b, the projectile's weight is $W = mg = 10(9.81) = 98.1$ N. We assume **a** acts upward in the *positive z* direction.

(b)

Equation of Motion
$$+\uparrow \Sigma F_z = ma_z; \qquad -98.1 = 10a$$
$$a = -9.81 \text{ m/s}^2$$

The result indicates that the projectile, like every object having free-flight motion near the earth's surface, is subjected to a constant downward acceleration of 9.81 m/s².

Kinematics. Initially, $z_0 = 0$ and $v_0 = 50$ m/s, and at the maximum height $z = h$, $v = 0$. Since the acceleration is *constant,* then

$$(+\uparrow) \qquad v^2 = v_0^2 + 2a_c(z - z_0)$$
$$0 = (50)^2 + 2(-9.81)(h - 0)$$
$$h = 127 \text{ m} \qquad\qquad Ans.$$

Part (b)
Free-Body Diagram. Since the force $F_D = (0.01v^2)$ N tends to retard the upward motion of the projectile, it acts downward as shown on the free-body diagram, Fig. 13–7c.

(c)

Fig. 13–7

Equation of Motion*
$$+\uparrow \Sigma F_z = ma_z; \qquad -0.01v^2 - 98.1 = 10a$$
$$a = -0.001v^2 - 9.81$$

Kinematics. Here the acceleration is *not constant;* however, it can be related to the velocity and displacement by using $a\,dz = v\,dv$.

$$(+\uparrow)\; a\,dz = v\,dv; \quad (-0.001v^2 - 9.81)dz = v\,dv$$

Separating the variables and integrating, realizing that initially $z_0 = 0$, $v_0 = 50$ m/s (positive upward), and at $z = h$, $v = 0$, we have

$$\int_0^h dz = -\int_{50}^0 \frac{v\,dv}{0.001v^2 + 9.81}$$

$$h = -500 \ln (v^2 + 9810)\Big|_{50}^0 = 113 \text{ m} \qquad Ans.$$

The answer indicates a lower elevation than that obtained in Part (a). Why?

*Note that if the projectile were fired downward, the equation of motion would then be $-0.01v^2 + 98.1 = 10a$.

Example 13–3

The crate shown in Fig. 13–8a has a weight of 50 lb and is acted upon by a force having a variable magnitude $P = 20t$, where P is in pounds and t is in seconds. Compute the crate's velocity 2 s after **P** has been applied. The crate's initial velocity is $v_0 = 3$ ft/s down the plane, and the coefficient of kinetic friction between the crate and the plane is $\mu_k = 0.3$.

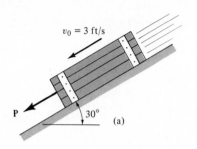

$v_0 = 3$ ft/s

P

30°

(a)

Fig. 13–8

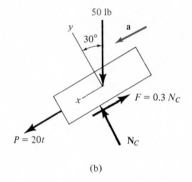

50 lb

y

30°

a

x

$F = 0.3\,N_C$

$P = 20t$

N_C

(b)

Solution

Free-Body Diagram.   As shown in Fig. 13–8b, the frictional force is directed opposite to the crate's sliding motion and has a magnitude $F = \mu_k N_C = 0.3 N_C$. The mass is $m = W/g = 50/32.2 = 1.55$ slug.

Equations of Motion.   Using the data shown on the free-body diagram, we have

$$+\swarrow \Sigma F_x = ma_x; \quad 20t - 0.3 N_C + 50 \sin 30° = 1.55a \qquad (1)$$
$$+\nwarrow \Sigma F_y = ma_y; \quad\quad N_C - 50 \cos 30° = 0 \qquad (2)$$

Solving for N_C in Eq. (2) ($N_C = 43.3$ lb), substituting into Eq. (1), and simplifying yields

$$a = 12.88t + 7.73 \qquad (3)$$

Kinematics.   Since the acceleration is a function of time, the velocity of the crate is obtained by using $a = dv/dt$ with the initial condition that $v_0 = 3$ ft/s at $t = 0$. We have

$$(+\swarrow) \qquad\qquad dv = a\, dt$$
$$\int_3^v dv = \int_0^t (12.88t + 7.73)dt$$
$$v = 6.44t^2 + 7.73t + 3$$

When $t = 2$ s,

$$v = 44.2 \text{ ft/s} \swarrow \qquad\qquad Ans.$$

Example 13–4

A smooth 2-kg collar, shown in Fig. 13–9a, is attached to a spring having a stiffness $k = 3$ N/m and an unstretched length of 0.75 m. If the collar is released from rest at A, determine its acceleration and the normal force of the rod on the collar at the instant $y = 1$ m.

Fig. 13–9

(a)

(b)

Solution

Free-Body Diagram. The free-body diagram of the collar when it is located at the arbitrary position y is shown in Fig. 13–9b. Note that the weight is $W = 2(9.81) = 19.62$ N. Furthermore, the collar is *assumed* to be accelerating so that '**a**' acts downward in the *positive* y direction.

Equations of Motion. Using the data in Fig. 13–9b,

$$\xrightarrow{+} \Sigma F_x = ma_x; \qquad\qquad -N_C + F_s \cos \theta = 0 \qquad\qquad (1)$$

$$+ \downarrow \Sigma F_y = ma_y; \qquad\qquad 19.62 - F_s \sin \theta = 2a \qquad\qquad (2)$$

From Eq. (2) it is seen that the acceleration is not constant; rather, it depends upon the magnitude and direction of the spring force. Solution is possible once F_s and θ are known.

The magnitude of the spring force is a function of the stretch s of the spring; i.e., $F_s = ks$. Here the unstretched length is $AB = 0.75$ m, Fig. 13–9a; therefore, $s = CB - AB = \sqrt{y^2 + (0.75)^2} - 0.75$. Since $k = 3$ N/m, then

$$F_s = ks = 3(\sqrt{y^2 + (0.75)^2} - 0.75) \qquad\qquad (3)$$

From Fig. 13–9a, the angle θ is related to y by trigonometry.

$$\sin \theta = \frac{y}{\sqrt{y^2 + (0.75)^2}} \qquad\qquad (4)$$

Substituting $y = 1$ m into Eqs. (3) and (4) yields $F_s = 1.50$ N and $\theta = 53.1°$. Substituting these results into Eqs. (1) and (2), we obtain

$$N_C = 0.90 \text{ N} \qquad\qquad\qquad\qquad Ans.$$

$$a = 9.21 \text{ m/s}^2 \qquad\qquad\qquad\qquad Ans.$$

Example 13–5

The 100-kg block A shown in Fig. 13–10a is released from rest. If the mass of the pulleys and the cord is neglected, determine the speed of the 20-kg block B in 2 s.

(a)

Solution

Motion of blocks A and B will be analyzed separately.

Free-Body Diagram. Since the mass of the pulleys is neglected, the equilibrium condition for pulley C is shown in Fig. 13–10b. The free-body diagrams for blocks A and B are shown in Fig. 13–10c and d, respectively. Here we *assume* both blocks accelerate downward, in the direction of $+s_A$ and $+s_B$.

Equations of Motion

Block A (Fig. 13–10c):

$$+\downarrow \Sigma F_y = ma_y; \qquad 981 - 2T = 100a_A \qquad (1)$$

Block B (Fig. 13–10d):

$$+\downarrow \Sigma F_y = ma_y; \qquad 196.2 - T = 20a_B \qquad (2)$$

(b)

Kinematics. The necessary third equation is obtained by studying the kinematics of the pulley arrangement in order to relate a_A to a_B. Using the technique developed in Sec. 12.8, the coordinates s_A and s_B measure the positions of A and B from the fixed datum, Fig. 13–10a. It is seen that

$$2s_A + s_B = l$$

where l is constant and represents the total vertical length of cable. Differentiating this expression twice with respect to time yields

$$2a_A = -a_B \qquad (3)$$

Hence when block A accelerates *downward*, block B accelerates *upward* at twice the amount. Notice, however, that in writing Eqs. (1) to (3), the *positive direction was always assumed downward*. It is important to be *consistent* in this assumption since we are seeking a simultaneous solution of equations. The solution yields

$$T = 327.0 \text{ N}$$
$$a_A = 3.27 \text{ m/s}^2$$
$$a_B = -6.54 \text{ m/s}^2$$

Since a_B is constant, the velocity of block B in 2 s is thus

$$(+\downarrow) \qquad v = v_0 + a_B t$$
$$= 0 + (-6.54)(2)$$
$$= -13.1 \text{ m/s} \qquad \qquad Ans.$$

The negative sign indicates that block B is moving upward.

(c)

(d)

Fig. 13–10

Problems

Except when stated otherwise, throughout this chapter assume that the coefficients of static and kinetic friction are equal, i.e., $\mu = \mu_s = \mu_k$.

13–1. Determine (only) the "gravitational attraction" between an 80-kg man and a 50-kg woman. The distance between their centers of mass is 0.5 m.

13–2. The moon has a mass of $73.5(10^{21})$ kg and the earth's mass is $5.98(10^{24})$ kg. If their centers are $384(10^6)$ m apart, determine the gravitational attractive force between the two bodies.

13–3. The planets Mars and Earth have diameters of 6775 and 12 755 km, respectively. The mass of Mars is 0.107 times that of the earth. If a body weighs 200 N on the surface of the earth, what would its weight be on Mars? Also, what is the mass of the body and the acceleration of gravity on Mars?

*13–4. The 500-kg fuel assembly for a nuclear reactor is being lifted out from the core of the nuclear reactor using the pulley system shown. It is hoisted upward with a constant acceleration such that $s = 0$ when $t = 0$ and $s = 2.5$ m when $t = 1.5$ s. Determine the tension in the cable at A during the motion.

Prob. 13–4

13–5. The rocket belt and man have a total weight of 210 lb. If the engines provide a constant thrust having x and y components of $T_x = 35$ lb and $T_y = 300$ lb, determine the total distance the man travels when $t = 5$ s after takeoff. What is his velocity at this instant? Neglect air resistance and the loss of fuel mass.

$T_x = 35$ lb

$T_y = 300$ lb

Prob. 13–5

13–6. By using an inclined plane to retard the motion of a falling object, and thus make the observations more accurate, Galileo was able to determine experimentally that the distance through which an object moves in free fall is proportional to the square of the time for travel. Prove that this is the case, i.e., $s \propto t^2$, by determining the time t_B, t_C, and t_D needed for a block of mass m to slide from rest at A to points B, C, and D, respectively. Neglect the effects of friction.

$s_D = 9$ m

$s_C = 4$ m

$s_B = 2$ m

$20°$

Prob. 13–6

13–7. The conveyor belt C is moving at 4 m/s. If the static coefficient of friction between the conveyor and the 10-kg box B is $\mu = 0.2$, determine the shortest time the conveyor can stop so that the box does not shift or move on the belt.

B

$v_C = 4$ m/s

C

Prob. 13–7

95

*13–8. The man weighs 180 lb and supports the barbells which have a weight of 100 lb. If he lifts them 2 ft in the air in 1.5 s starting from rest, determine the reaction of *both* of his feet on the ground during the lift. Assume the motion is with uniform acceleration.

2 ft

Prob. 13–8

13–9. The boy having a weight of 80 lb hangs uniformly from the bar. Determine the force in each of his arms in $t = 2$ s if the bar is moving upward with (a) a constant velocity of 3 ft/s, and (b) a speed of $v = (4t^2)$ ft/s, where t is in seconds.

Prob. 13–9

13–10. The man pushes on the 60-lb crate with a force **F** until it begins to slide. The force is always directed down at 30° from the horizontal as shown, and its magnitude is increased until the crate begins to slide. Determine the crate's initial acceleration if the static coefficient of friction is $\mu_s = 0.6$ and the kinetic coefficient of friction is $\mu_k = 0.3$.

F

30°

Prob. 13–10

13–11. A freight elevator, including its load, has a mass of 250 kg. It is prevented from rotating by using the track and wheels mounted along its sides. If the motor M develops a constant tension $T = 1.20$ kN in its attached cable, determine the velocity of the elevator when it has moved upward 10 m starting from rest. Neglect the mass of the pulleys and cables. The counterweight C has a mass of 150 kg.

C

M

Prob. 13–11

*13–12. Determine the time needed to pull the cord at B down 4 ft starting from rest when a force of 10 lb is applied to the cord. Block A weighs 20 lb. Neglect the mass of the pulleys and cords.

B

10 lb

A

Prob. 13–12

13–13. The motor M exerts a force on the cord of $F = (8t + 60)$ N, where t is in seconds. If the block has a mass of 20 kg and is traveling downward at 20 m/s at the instant the force is applied, determine the time needed to stop the block. Neglect the mass of the pulley and cord.

Prob. 13–13

13–14. At a given instant the 10-lb block A is moving downward with a speed of 6 ft/s. Determine its speed 2 s later. Block B has a weight of 4 lb, and the coefficient of kinetic friction between it and the horizontal plane is $\mu_k = 0.2$. Neglect the mass of the pulleys and cord.

Prob. 13–14

13–15. As a point of historical interest, the acceleration of gravity g used to be determined using an "Atwood Machine," which is simply the arrangement of the blocks and pulley shown. In an experiment of this type, consider block A to have a mass of 2 kg and block B a mass of 2.6 kg. If it is observed that B descends 3 m in 2.2 s starting from rest, determine the value of g. Discuss what effects must be considered in order to increase the accuracy of this experiment.

Prob. 13–15

*13–16. The two boxcars A and B have a weight of 20,000 lb and 30,000 lb, respectively. If they are freely coasting down the incline when the brakes are applied to all the wheels of car A, determine the force in the coupling C between the two cars. The coefficient of friction between the wheels of A and the tracks is $\mu = 0.5$. The wheels of car B are free to roll. Neglect their mass in the calculation. *Suggestion:* Solve the problem by representing single resultant normal forces acting on A and B, respectively.

Prob. 13–16

13–17. The conveyor belt delivers each 12-kg crate to the ramp at A such that the crate's velocity is $v_A = 2.5$ m/s, directed down *along* the ramp. If the kinetic coefficient of friction between each crate and the ramp is $\mu_k = 0.3$, determine the speed at which each crate slides off the ramp at B. Assume that no tipping occurs.

Prob. 13–17

13–18. The 800-kg car at B is connected to the 350-kg car at A by a spring coupling. Determine the stretch in the spring if (a) the wheels of both cars are free to roll and (b) car B applies brakes to all four wheels. Take $\mu_B = 0.4$. Neglect the mass of the wheels.

Prob. 13–18

13–19. The 100-kg crate is hoisted up the incline using the cable and motor M. For a short time, the force in the cable is $F = 800t^2$ N, where t is in seconds. If the crate has an initial velocity $v_1 = 2$ m/s when $t = 0$, determine its velocity when $t = 2$ s. The coefficient of kinetic friction between the crate and the incline is $\mu_k = 0.3$.

Probs. 13–19 / 13–20

*13–20. The 100-kg crate is hoisted up the incline using the cable and motor M. For a short time, the force in the cable is $F = 800t^2$ N, where t is in seconds. If the crate has an initial velocity $v_1 = 2$ m/s at $s = 0$ and $t = 0$, determine the distance the crate moves up the plane when $t = 2$ s. The coefficient of kinetic friction between the crate and the incline is $\mu_k = 0.3$.

13–21. The safe S has a weight of 200 lb and is supported by the rope and pulley arrangement shown. If the end of the rope is given to a boy B of weight 90 lb, determine his acceleration if in the confusion he doesn't let go of the rope. Neglect the mass of the pulleys and rope.

Prob. 13–21

13–22. Cylinder B has a mass m and is hoisted using the cord and pulley system shown. Determine the magnitude of force **F** as a function of the block's vertical position y so that when **F** is applied to the end of the cord the block rises with a constant acceleration $\mathbf{a}_B$. Neglect the mass of the cord and pulleys.

Prob. 13–22

13–23. In the event of an emergency, a nuclear reactor is shut down by dropping the control rod R, having a weight of 20 lb, into the reactor core C. If the rod is immersed in ''heavy'' water, which offers a drag resistance to downward motion $F_D = (0.25v)$ lb, where v is the instantaneous velocity of the rod in ft/s, determine the distance s it must descend into the reactor core before it attains a speed of 12 ft/s. The rod is released from rest when $s = 0$. Neglect the effects of buoyancy.

Prob. 13–23

*13–24. A parachutist having a mass of 70 kg is falling at 10 m/s when he opens his parachute at a very high altitude. If the atmospheric drag resistance is $F_D = (35v^2)$ N, where v is in m/s, determine his velocity when he has fallen 4 m. What is his velocity when he lands on the ground? This velocity is referred to as the *terminal velocity,* which is found by letting the distance of fall $y \rightarrow \infty$.

Prob. 13–24

13–25. A fast-moving automobile is subjected to air resistance which is proportional to the square of the car's speed; i.e., $R = kv^2$. In an experiment to determine k, it is found that when a 4000-lb car is coasting at 88 ft/s it decreases its speed to 76 ft/s in $t = 4$ s. Determine the coefficient of drag, k, for the car. Neglect the effects of mechanical friction, rolling resistance, and the mass of the wheels. The engine is turned off and the wheels are free to roll.

Prob. 13–25

13–26. The tanker has a weight of $800(10^6)$ lb and is traveling forward at $v_0 = 3$ ft/s in still water when the engines are shut off. If the drag resistance of the water is proportional to the speed of the tanker at any instant and can be approximated by $F_D = (400(10^3)v)$ lb, where v is in ft/s, determine the time needed for the tanker's speed to become 1 ft/s.

Probs. 13–26 / 13–27

13–27. How far must the tanker in Prob. 13–26 travel in order to reduce its speed from 3 ft/s to 1.5 ft/s?

*13–28. The spring mechanism is used as a shock absorber for railroad cars. Determine the maximum compression of spring HI if the fixed bumper R of a 5-Mg railroad car, rolling freely at 1.5 m/s, strikes the plate P. Bar AB slides along the guide paths CE and DF. The ends of all springs are attached to their respective members and the springs are originally unstretched. Assume that P and R remain in contact during collision.

Prob. 13–28

13–29. Determine the tension developed in the cords attached to each block and the accelerations of the blocks. Neglect the mass of the pulleys and cords.

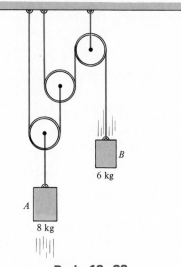

Prob. 13–29

■13–30. The 10-kg crate rests on the cart for which the coefficient of static friction is $\mu_s = 0.3$ between the crate and cart. Determine the largest angle θ of the plane so that the crate does not slide on the cart when the cart is given an acceleration of 6 m/s².

Prob. 13–30

13–31. Determine the speed of the collar in Example 13–4 at the instant $y = 1$ m. *Hint:* Combine Eqs. (2), (3), and (4) to determine $a = f(y)$. Then substitute into the equation $v\,dv = a\,dy$ and integrate.

*13–32. If a horizontal force $P = 12$ lb is applied to block A, determine the acceleration of block B. Neglect friction. *Hint:* Show that $a_B = a_A \tan 15°$.

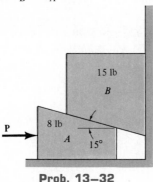

Prob. 13–32

13–33. Determine the acceleration of cylinder B when the system is released from rest. Neglect the mass of the pulleys and cords.

Prob. 13–33

13–34. Block A has a mass of 60 kg and rests on block B, which has a mass of 30 kg. If the coefficients of friction are indicated in the figure, determine the largest horizontal force P which can be applied to block B so that block A does not slip on block B.

Probs. 13–34 / 13–35

13–35. If a horizontal force $P = 600$ N is applied to the 30-kg block B, determine the acceleration of the 60-kg block A. The coefficients of friction are indicated in the figure.

*13–36. In the cathode-ray tube, electrons having a mass m are emitted from a source point S and begin to travel horizontally with an initial velocity $\mathbf{v}$. While passing between the grid plates a distance l, they are subjected to a vertical force having a magnitude eV/w, where e is the charge of an electron, V the applied voltage acting across the plates, and w the distance between the plates. After passing clear of the plates, the electrons then travel in straight lines and strike the screen at A. Determine the deflection s of the electrons in terms of the dimensions of the voltage plate and tube.

Prob. 13–36

101

13.5 Equations of Motion: Normal and Tangential Coordinates

When a particle moves over a curved path which is known, the equation of motion for the particle may be written in the normal and tangential directions. We have

$$\Sigma \mathbf{F} = m\mathbf{a}$$
$$\Sigma F_t \mathbf{u}_t + \Sigma F_n \mathbf{u}_n + \Sigma F_b \mathbf{u}_b = m\mathbf{a}_t + m\mathbf{a}_n$$

Here ΣF_n, ΣF_t, and ΣF_b represent the sums of all the force components acting on the particle in the normal, tangential, and binormal directions, respectively, Fig. 13–11. Note that there is no motion of the particle in the binormal direction, since the particle is constrained to move along the path. The above equation is satisfied provided

$$\Sigma F_t = ma_t$$
$$\Sigma F_n = ma_n \qquad (13\text{–}8)$$
$$\Sigma F_b = 0$$

Fig. 13–11

Recall that $a_t (= dv/dt)$ represents the time rate of change in the magnitude of velocity. Consequently, if $\Sigma \mathbf{F}_t$ acts in the direction of motion, the particle's speed will increase, whereas if it acts in the opposite direction, the particle will slow down. Likewise, $a_n (= v^2/\rho)$ represents the time rate of change in the direction of velocity. Since this vector *always* acts in the positive n direction, then $\Sigma \mathbf{F}_n$, which causes $\mathbf{a}_n$, also acts in this direction. In particular, when the particle is traveling on a constrained circular path with a constant speed, there is a normal force exerted on the particle by the constraint. Since this force is always directed towards the center of the path it is often referred to as the *centripetal force*.

PROCEDURE FOR ANALYSIS

The procedure given in Sec. 13.4 may be stated as follows when applied to problems involving n, t, b coordinates:

Free-Body Diagram. Establish the inertial n, t, b coordinate system at the particle and draw the particle's free-body diagram. Assume the particle's acceleration components $\mathbf{a}_n$ and $\mathbf{a}_t$ act in the positive direction of n and t if they are unknown.

Equations of Motion. Apply the equations of motion, Eqs. 13–8.

Kinematics. Formulate the tangential and normal components of acceleration; i.e., $a_t = dv/dt$ or $a_t = v \, dv/ds$ and $a_n = v^2/\rho$. If the path is defined as $y = f(x)$, the radius of curvature can be obtained from $\rho = |[1 + (dy/dx)^2]^{3/2}/(d^2y/dx^2)|$.

The following examples numerically illustrate application of this procedure.

Example 13–6

The ball B is fastened to the end of a 1-m-long string, Fig. 13–12a. If air resistance is neglected, then the ball, as it moves with constant speed, describes a horizontal circular path in which the string OB generates a locus of points defining the surface of a cone. Determine the speed of the ball along its circular path if $\theta = 45°$.

Fig. 13–12

(a)   (b)

Solution

Free-Body Diagram. As shown in Fig. 13–12b, the weight of the ball is represented as $W = mg$, where $g = 9.81$ m/s². Since the ball moves around the circle with constant speed, only a normal component of acceleration $\mathbf{a}_n$ (directed toward the center of curvature) is present. Why? This component has a magnitude $a_n = v^2/\rho = v^2/(1 \cos 45°)$.

Equations of Motion. Using the n, b axes shown, we have

$$+\uparrow \Sigma F_b = 0; \qquad T \sin 45° - m(9.81) = 0 \qquad (1)$$

$$\xrightarrow{+}\Sigma F_n = ma_n; \qquad T \cos 45° = m\frac{v^2}{1 \cos 45°} \qquad (2)$$

Application of the equation of motion in the tangential direction is of no consequence to the solution, since there are no forces in this direction. Solving Eq. (1) for T and substituting into Eq. (2), it is found that m cancels and the speed v of the ball is then

$$v = 2.63 \text{ m/s} \qquad\qquad \textit{Ans.}$$

Example 13–7

Determine the banking angle θ of the circular track so that the wheels of the sports car shown in Fig. 13–13a will not have to depend upon friction to prevent the car from sliding either up or down the curve. The car travels at a constant speed of 100 ft/s. The radius of the track is 600 ft.

(a)

Fig. 13–13

(b)

Solution

Free-Body Diagram. As shown in Fig. 13–13b, the car is assumed to have a mass m. As stated in the problem, no frictional force acts on the car.

Equations of Motion. Using the n, b axes shown,

$$+\uparrow \Sigma F_b = 0; \qquad N_C \cos\theta - mg = 0 \qquad\qquad (1)$$

$$\xrightarrow{+}\Sigma F_n = ma_n; \qquad N_C \sin\theta = m\frac{v^2}{\rho} \qquad\qquad (2)$$

Eliminating N_C and m from these equations by dividing Eq. (2) by Eq. (1), we obtain

$$\tan\theta = \frac{v^2}{g\rho} = \frac{(100)^2}{32.2(600)}$$

$$\theta = \tan^{-1}(0.518)$$

$$= 27.4° \qquad\qquad Ans.$$

A force summation in the tangential direction of motion is of no consequence to the solution. If it were considered, note that $a_t = dv/dt = 0$, since the car moves with *constant speed*.

Example 13–8

The 3-kg disk D is attached to the end of a cord as shown in Fig. 13–14a. The other end of the cord is attached to a ball-and-socket joint located at the center of a platform. If the platform is rotating rapidly, and the disk is placed on it and released from rest as shown, determine the time it takes for the disk to reach a speed great enough to break the cord. The maximum tension the cord can sustain is 100 N, and the coefficient of kinetic friction between the disk and the platform is $\mu_k = 0.1$.

Fig. 13–14

(a)

(b)

Solution

Free-Body Diagram. As shown in Fig. 13–14b, the disk has *both* normal and tangential components of acceleration as a result of the unbalanced forces **T** and **F**. Since sliding occurs, the frictional force has a magnitude $F = \mu_k N_D = 0.1 N_D$ and a direction that opposes the *relative motion* of the disk with respect to the platform. The weight of the disk is $W = 3(9.81) = 29.43$ N.

Equations of Motion

$$\Sigma F_b = 0; \qquad\qquad N_D - 29.43 = 0 \qquad\qquad (1)$$

$$\Sigma F_t = ma_t; \qquad\qquad 0.1N_D = 3a_t \qquad\qquad (2)$$

$$\Sigma F_n = ma_n; \qquad\qquad T = 3\left(\frac{v^2}{1}\right) \qquad\qquad (3)$$

Since the maximum tension sustained by the cord is $T = 100$ N, Eq. (3) can be solved for the critical speed v_{cr} of the disk needed to break the cord. Solving all the equations, we obtain

$$N_D = 29.43 \text{ N}$$
$$a_t = 0.981 \text{ m/s}^2$$
$$v_{cr} = 5.77 \text{ m/s}$$

Kinematics. Since a_t is *constant*, the time needed to break the cord is

$$v_{cr} = v_0 + a_t t$$
$$5.77 = 0 + (0.981)t$$
$$t = 5.89 \text{ s} \qquad\qquad Ans.$$

Example 13-9

The skier in Fig. 13–15a descends the smooth slope, which may be approximated by a parabola. If she has a weight of 120 lb, determine the normal force she exerts on the ground at the instant she arrives at point A, where her velocity is 30 ft/s. Also compute her acceleration at A.

(a)

(b)

Fig. 13–15

Solution

Free-Body Diagram. The free-body diagram for the skier when she is at A is shown in Fig. 13–15b. Since the path is *curved*, there are two components of acceleration, $\mathbf{a}_n$ and $\mathbf{a}_t$.

Equations of Motion

$$+\uparrow \Sigma F_n = ma_n; \qquad N_A - 120 = \frac{120}{32.2}\left(\frac{(30)^2}{\rho}\right) \qquad (1)$$

$$\xleftarrow{+} \Sigma F_t = ma_t; \qquad 0 = \frac{120}{32.2}a_t \qquad (2)$$

The radius of curvature ρ for the path must be computed at point $A(0, -15 \text{ ft})$. Since $y = \frac{1}{60}x^2 - 15$, $dy/dx = \frac{1}{30}x$, $d^2y/dx^2 = \frac{1}{30}$, then at $x = 0$,

$$\rho = \left| \frac{[1 + (dy/dx)^2]^{3/2}}{d^2y/dx^2} \right|_{x=0} = \left| \frac{[1 + (0)^2]^{3/2}}{\frac{1}{30}} \right| = 30 \text{ ft}$$

Substituting into Eq. (1) and solving for N_A, we have

$$N_A = 231.8 \text{ lb} \qquad\qquad \textit{Ans.}$$

Kinematics. From Eq. (2),

$$a_t = 0$$

Thus,

$$a_n = \frac{v^2}{\rho} = \frac{(30)^2}{30} = 30 \text{ ft/s}^2$$

$$a = a_n = 30 \text{ ft/s}^2 \uparrow \qquad\qquad \textit{Ans.}$$

Example 13–10

A block having a mass of 2 kg is given an initial velocity of 1 m/s when it is at the top surface of the smooth cylinder shown in Fig. 13–16a. If the block slides along a path of radius 0.5 m, determine the angle $\theta = \theta_{max}$ at which it begins to leave the cylinder's surface.

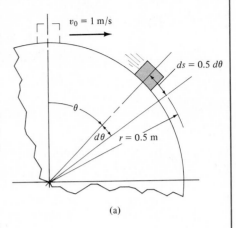

(a)

Solution
Free-Body Diagram. The free-body diagram for the block, when the block is located at the *general position* θ, is shown in Fig. 13–16b. The block must have a tangential acceleration $\mathbf{a}_t$, since its *speed* is always *increasing* as it slides downward. The weight is $W = 2(9.81) = 19.62$ N.

Equations of Motion

$$+\swarrow\Sigma F_n = ma_n; \qquad -N_B + 19.62 \cos\theta = 2\frac{v^2}{0.5} \qquad (1)$$

$$+\searrow\Sigma F_t = ma_t; \qquad 19.62 \sin\theta = 2a_t \qquad (2)$$

These two equations contain four unknowns, N_B, v, a_t, and θ. At the instant $\theta = \theta_{max}$, however, the block leaves the surface of the cylinder so that $N_B = 0$.

(b)

Fig. 13–16

Kinematics. A third equation for the solution may be obtained by noting that the magnitude of tangential acceleration a_t may be related to the speed of the block v and the angle θ. Since $a_t\, ds = v\, dv$ and $ds = r\, d\theta = 0.5\, d\theta$, Fig. 13–16a, we have

$$a_t = \frac{v\, dv}{0.5\, d\theta} \qquad (3)$$

Substituting Eq. (3) into Eq. (2) and separating the variables, we have

$$v\, dv = 4.905 \sin\theta\, d\theta$$

Integrating both sides, realizing that when $\theta = 0°$, $v_0 = 1$ m/s, yields

$$\int_1^v v\, dv = 4.905 \int_{0°}^{\theta} \sin\theta\, d\theta$$

$$\left.\frac{v^2}{2}\right|_1^v = -4.905 \cos\theta \left.\vphantom{\frac{v^2}{2}}\right|_{0°}^{\theta}$$

$$v^2 = 9.81(1 - \cos\theta) + 1$$

Substituting into Eq. (1) with $N_B = 0$ and solving for $\cos\theta_{max}$ gives

$$19.62 \cos\theta_{max} = \frac{2}{0.5}[9.81(1 - \cos\theta_{max}) + 1]$$

$$\cos\theta_{max} = \frac{43.24}{58.86}; \qquad \theta_{max} = 42.7° \qquad \textit{Ans.}$$

Problems

13–37. Compute the mass of the sun, knowing that the distance from the earth to the sun is $149.6(10^6)$ km. *Hint:* Use Eq. 13–1 to represent the force of gravity acting on the earth.

13–38. Determine the maximum constant speed at which the pilot can travel around the vertical curve having a radius of curvature $\rho = 800$ m, so that he experiences a maximum acceleration $a_n = 8\,g = 78.5$ m/s². If he has a mass of 70 kg, determine the normal force he exerts on the seat of the airplane when the plane is traveling at this speed and is at its lowest point.

$\rho = 800$ m

Prob. 13–38

13–39. The 4-Mg helicopter maneuvers a horizontal turn having a radius of curvature $r = 400$ m. Determine the lift force $\mathbf{F}_L$ required and the angle of bank θ when it is flying with a constant speed of $v = 40$ m/s. Show that θ increases if v increases by also calculating θ when $v = 60$ m/s.

F_L

θ

$r = 400$ m

G

Prob. 13–39

*13–40. The 150-lb man lies against the cushion for which the coefficient of friction is $\mu = 0.5$. Determine the normal and frictional force he exerts on the cushion if, due to rotation about the z axis, he has a constant speed $v = 20$ ft/s.

z

8 ft

G

60°

Prob. 13–40

13–41. The device shown is used to produce the experience of weightlessness in a passenger when he reaches point A, $\theta = 90°$, along the path. If the passenger has a mass of 75 kg, determine the minimum speed he should have when he reaches A so that he does not exert a normal reaction on the seat. The chair is pin-connected to the frame BC so that he is always seated in an upright position. During the motion his speed remains constant.

A

B

10 m

C

θ

Probs. 13–41 / 13–42

13–42. A man having a mass of 75 kg sits in the chair which is pin-connected to the frame BC. If the man is always seated in an upright position, determine the horizontal and vertical reactions he exerts on the chair at the instant $\theta = 45°$. At this instant he has a speed of 6 m/s, which is increasing at 0.5 m/s².

13–43. When crossing an intersection, a motorcyclist encounters the slight bump or crown caused by the intersecting road. If the crest of the bump has a radius of curvature $\rho = 50$ ft, determine the maximum constant speed at which he can travel without leaving the surface of the road. Neglect the size of the motorcycle and rider in the calculation. The rider and his motorcycle have a total weight of 450 lb.

$\rho = 50$ ft

Prob. 13–43

*13–44. A girl having a mass of 25 kg sits at the edge of the merry-go-round so her center of mass G is at a distance of 1.5 m from the axis of rotation. If the angular motion of the platform is *slowly* increased so that the girl's tangential component of acceleration can be neglected, determine the maximum speed which she can have before she begins to slip off the merry-go-round. The coefficient of friction between the girl and the merry-go-round is $\mu = 0.3$.

Probs. 13–44 / 13–45

13–45. Solve Prob. 13–44 assuming the platform starts rotating from rest so that the girl's speed is increased uniformly at $\dot{v} = 0.5$ m/s^2.

13–46. Determine the constant speed of the chairs on the amusement park ride if it is observed that the supporting cables are directed at $\theta = 30°$ from the vertical. Each chair including its passenger has a mass of 80 kg. Also, what are the components of force in the n, t, and b directions which a 50-kg passenger exerts on the chair during the motion?

Prob. 13–46

13–47. Prove that if the block is released from rest at the top point B of a smooth path of *arbitrary shape*, the speed it attains when it reaches point A is equal to the speed it attains when it falls freely through a distance h; i.e., $v = \sqrt{2gh}$.

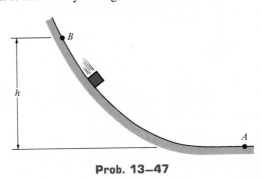

Prob. 13–47

*13–48. A skier starts from rest at $A(10 \text{ m}, 0)$ and descends the smooth slope, which may be approximated by a parabola. If she has a mass of 52 kg, determine the normal force she exerts on the ground at the instant she arrives at point B. Neglect the size of the skier. *Hint:* Use the result of Prob. 13–47.

$$y = \frac{1}{20}x^2 - 5$$

Prob. 13–48

13–49. The 600-kg wrecking ball is suspended from the crane by a cable having a negligible mass. If the ball has a speed $v = 8$ m/s at the instant it is at its lowest point, $\theta = 0°$, determine the tension in the cable at this instant. Also, determine the angle θ to which the ball swings before it stops.

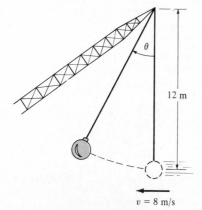

$v = 8$ m/s

Prob. 13–49

13–50. A toboggan and rider have a total mass of 90 kg and travel down along the (smooth) slope defined by the equation $y = 0.08x^2$. At the instant $x = 10$ m, the toboggan's speed is 5 m/s. At this point, determine the rate of increase in speed and the normal force which the toboggan exerts on the slope. Neglect the size of the toboggan and rider for the calculation.

$v = 5$ m/s

$y = 0.08\,x^2$

10 m

Prob. 13–50

13–51. The roller coaster is traveling at 40 ft/s when it reaches its lowest point A. Determine the normal force which it exerts on the tracks at this instant and the rate of increase in its speed. Assume that the car is free to roll and its total weight, including the passenger, is 800 lb. Neglect the mass of the wheels and the size of the car.

$$y = 50 \cos \frac{\pi}{50} x$$

50 ft

50 ft

25 ft   25 ft

Prob. 13–51

*13–52. The 0.8-Mg car is traveling over the hill having the shape of a parabola. If the driver maintains a constant speed of 9 m/s, determine both the resultant normal force and the resultant frictional force that all the wheels of the car exert on the road at the instant it reaches point A. Neglect the size of the car.

$$y = 20\left(1 - \frac{x^2}{6400}\right)$$

80 m

Probs. 13–52 / 13–53

13–53. The 0.8-Mg car is traveling over the hill having the shape of a parabola. When the car is at point A, it is traveling at 9 m/s and increasing its speed at 3 m/s^2. Determine both the resultant normal force and the resultant frictional force that all the wheels of the car exert on the road at this instant. Neglect the size of the car.

13–54. The 2-lb block is released from rest at A and slides down along the smooth cylindrical surface. If the attached spring has a stiffness $k = 2$ lb/ft, determine its unstretched length so that it does not allow the block to leave the surface until $\theta = 60°$.

A

2 ft

θ

$k = 2$ lb/ft

Prob. 13–54

13–55. The cyclist is freely coasting down the hill defined by the shape $y = 0.2e^x$. If he has a speed of 15 m/s when $y = 0.2$ m, determine the resultant normal reaction he exerts on the ground and the rate of increase in his speed at this instant. He and his bicycle have a total mass of 80 kg. Neglect friction, the mass of the wheels, and the size of the bicycle.

$y = 0.2e^x$

Prob. 13–55

*13–56. The block has a weight of 2 lb and it is free to move along the smooth slot of the rotating disk. The spring has a stiffness of 2.5 lb/ft and an unstretched length of 1.25 ft. Determine the force of the spring on the block and the tangential component of force which the slot exerts on the side of the block, when the block is at rest with respect to the disk and is traveling with a constant speed of 12 ft/s.

$k = 2.5$ lb/ft

Prob. 13–56

13–57. The collar has a mass of 2 kg and is free to slide along the smooth rod which rotates in the *horizontal plane*. The attached spring has a stiffness $k = 1$ kN/m, and when no force is applied to the spring the length $d = 0.5$ m. Determine the force which the spring exerts on the collar, if the collar is at rest with respect to the rod yet due to the rod's rotation is moving with a constant speed $v = 6$ m/s.

Probs. 13–57 / 13–58

13–58. The collar has a mass of 2 kg and is free to slide along the smooth rod which rotates in the *horizontal plane*. The attached spring has a stiffness $k = 1$ kN/m, and when no force is applied to the spring the length $d = 0.5$ m. Determine the length d, if the collar is at rest with respect to the rod yet due to the rod's rotation is moving with a constant speed $v = 6$ m/s.

13–59. The collar has a mass of 5 kg and is confined to move along the smooth rod which lies in the vertical plane. The attached spring has an unstretched length of 200 mm. If at the instant $\theta = 30°$ the collar has a speed $v = 2$ m/s, determine the normal force of the rod on the collar and the collar's acceleration.

Prob. 13–59

*13–60. The 2-kg spool S fits loosely on the inclined rod for which $\mu = 0.2$. If the spool is located 0.25 m from A, determine the minimum constant speed the spool can have so that it does not slip down the rod.

Probs. 13–60 / 13–61

13–61. The 2-kg spool S fits loosely on the inclined rod for which $\mu = 0.2$. If the spool is located 0.25 m from A, determine the maximum constant speed the spool can have so that it does not slip up the rod.

13–62. Determine the constant speed of the satellite S so that it circles the earth with an orbit of radius $r = 20$ Mm. The mass of the earth is $5.976(10^{24})$ kg. *Hint:* Use Eq. 13–1.

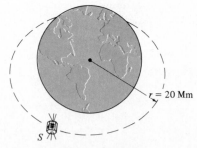

Prob. 13–62

■**13–63.** Packages having a weight of 6 lb ride on the surface of the conveyor belt. If the belt travels at a constant speed of 2 ft/s and the coefficient of friction between the belt and each package is $\mu = 0.4$, determine the angle θ at which the packages first begin to slip off the surface of the belt. Neglect the size of the packages.

Prob. 13–63

*13–64.** A sled and rider, weighing 30 lb and 150 lb respectively, travel down a smooth slope that has the shape shown. If the rider does not hold on to the sides of the sled, determine the normal force which he exerts on the sled when the sled is (a) at point B, where it has a velocity of 15 ft/s, and (b) at some point C in mid-air.

Prob. 13–64

Equations of Motion: Cylindrical Coordinates **13.6**

When all the forces acting on a particle are resolved into components along the unit-vector directions $\mathbf{u}_r$, $\mathbf{u}_\theta$, and $\mathbf{u}_z$, Fig. 13–17, the equation of motion may be expressed as

$$\Sigma \mathbf{F} = m\mathbf{a}$$

$$\Sigma F_r \mathbf{u}_r + \Sigma F_\theta \mathbf{u}_\theta + \Sigma F_z \mathbf{u}_z = m a_r \mathbf{u}_r + m a_\theta \mathbf{u}_\theta + m a_z \mathbf{u}_z$$

To satisfy this equation, the respective $\mathbf{u}_r$, $\mathbf{u}_\theta$, and $\mathbf{u}_z$ components on the left side must equal the corresponding components on the right side. Consequently, we may write the following three scalar equations:

$$\begin{aligned} \Sigma F_r &= m a_r \\ \Sigma F_\theta &= m a_\theta \\ \Sigma F_z &= m a_z \end{aligned} \qquad (13\text{–}9)$$

If the particle is constrained to move only in the r-θ plane, then only the first two of Eqs. 13–9 are used to specify the motion.

Fig. 13–17

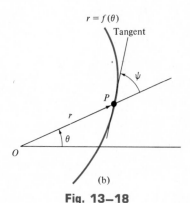

Fig. 13—18

Tangent and Normal Forces. The most straightforward type of problem involving cylindrical coordinates requires the determination of the resultant force components ΣF_r, ΣF_θ, and ΣF_z causing a particle to move with a *known* acceleration. If, however, the particle's accelerated motion is not completely specified at the given instant, then perhaps some information regarding the directions or magnitudes of the forces acting on the particle must be known or computed in order to solve Eqs. 13–9. For example, the force **T** causes the particle in Fig. 13–18a to move along a path defined in polar coordinates; $r = f(\theta)$. The *normal force* **N** which the path exerts on the particle is always *perpendicular to the tangent of the path*, whereas the frictional force **F** always acts along the tangent in the opposite direction of motion. The *directions* of **N** and **F** can be specified by first computing the angle ψ (psi), Fig. 13–18b, which is defined between the *extended* radial line $r = OP$ and the tangent to the curve. This angle, *always measured counterclockwise*, or in the *positive direction of* θ, is determined from the equation*

$$\tan \psi = r \left/ \frac{dr}{d\theta} \right. \qquad (13\text{--}10)$$

Application is illustrated numerically in Example 13–13.

PROCEDURE FOR ANALYSIS

The procedure given in Sec. 13.4 may be stated as follows when applied to problems involving cylindrical coordinates.

Free-Body Diagram. Establish the r, θ, z inertial coordinate system and draw the particle's free-body diagram. Assume that $\mathbf{a}_r$, $\mathbf{a}_\theta$, and $\mathbf{a}_z$ act in the *positive directions* of r, θ, and z if they are unknown.

Equations of Motion. Apply the equations of motion, Eqs. 13–9.

Kinematics. Use the methods of Sec. 12.7 to determine r and the time derivatives $\dot{r}$, $\ddot{r}$, $\dot{\theta}$, $\ddot{\theta}$, and $\ddot{z}$, and evaluate the acceleration components $a_r = \ddot{r} - r\dot{\theta}^2$, $a_\theta = r\ddot{\theta} + 2\dot{r}\dot{\theta}$, and $a_z = \ddot{z}$. If any of these three components are computed as negative quantities, it indicates that they act in their negative coordinate directions.

The following examples numerically illustrate application of this procedure. Further application of Eqs. 13–9 is given in Sec. 13.7, where problems involving rockets, satellites, and planetary bodies subjected to gravitational forces are considered.

---

*The derivation is given in any standard calculus text.

Example 13–11

The 2-lb block in Fig. 13–19a moves on the smooth horizontal plane, such that its path is specified in polar coordinates by the parametric equations $r = (10t^2)$ ft and $\theta = (0.5t)$ rad, where t is in seconds. Determine the magnitude of the *resultant* force **F** causing the motion at the instant $t = 1$ s.

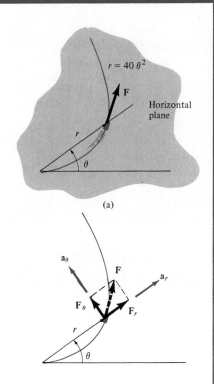

(a)

Solution

Free-Body Diagram. Note that if the parameter t is eliminated between r and θ, the equation of the path becomes $r = 40\theta^2$. The block's free-body diagram when it is located at an arbitrary point along the path is shown in Fig. 13–19b. Here the r and θ components of the resultant force are represented. They, and $\mathbf{a}_r$ and $\mathbf{a}_\theta$, are assumed to act in the *positive* coordinate directions.

Equations of Motion

$$\Sigma F_r = ma_r; \qquad F_r = \frac{2}{32.2}(\ddot{r} - r\dot{\theta}^2) \qquad (1)$$

$$\Sigma F_\theta = ma_\theta; \qquad F_\theta = \frac{2}{32.2}(r\ddot{\theta} + 2\dot{r}\dot{\theta}) \qquad (2)$$

(b)

Fig. 13–19

Kinematics. Since the motion is specified, the coordinates and the required time derivatives can be computed and evaluated at $t = 1$ s.

$$r = 10t^2 \Big|_{t=1\text{s}} = 10 \text{ ft} \qquad \theta = 0.5t \Big|_{t=1\text{s}} = 0.5 \text{ rad}$$

$$\dot{r} = 20t \Big|_{t=1\text{s}} = 20 \text{ ft/s} \qquad \dot{\theta} = 0.5 \text{ rad/s}$$

$$\ddot{r} = 20 \text{ ft/s}^2 \qquad \ddot{\theta} = 0$$

Substituting these results into Eqs. (1) and (2) yields

$$F_r = \frac{2}{32.2}(20 - 10(0.5)^2) = 1.087 \text{ lb}$$

$$F_\theta = \frac{2}{32.2}(0 + 2(20)(0.5)) = 1.242 \text{ lb}$$

Hence, the resultant force has a magnitude of

$$F = \sqrt{(1.087)^2 + (1.242)^2}$$
$$= 1.65 \text{ lb} \qquad\qquad Ans.$$

Example 13–12

The smooth 2-kg cylinder C in Fig. 13–20a has a peg P through its center which passes through the slot in arm OA. If the arm rotates in the *vertical plane* at a constant rate $\dot{\theta}=0.5$ rad/s, determine the force that the arm exerts on the peg at the instant $\theta = 60°$.

(a)

Solution

Free-Body Diagram. The free-body diagram for the cylinder is shown in Fig. 13–20b. The force of the peg, $\mathbf{F}_P$, acts perpendicular to the slot in the arm. As usual, $\mathbf{a}_r$ and $\mathbf{a}_\theta$ are assumed to act in the directions of *positive r* and θ, respectively.

Equations of Motion. Using the data in Fig. 13–20b, we have

$$+\swarrow \Sigma F_r = ma_r; \qquad 19.62 \sin\theta - N_C \sin\theta = 2a_r \tag{1}$$

$$+\searrow \Sigma F_\theta = ma_\theta; \quad 19.62 \cos\theta + F_P - N_C \cos\theta = 2a_\theta \tag{2}$$

Kinematics. From Fig. 13–20a, r can be related to θ by the equation

$$r = \frac{0.4}{\sin\theta} = 0.4 \csc\theta$$

Since $d(\csc\theta) = -(\csc\theta \cot\theta)d\theta$ and $d(\cot\theta) = -(\csc^2\theta)\,d\theta$, then r and the necessary time derivatives become

$$\dot{\theta} = 0.5 \qquad r = 0.4 \csc\theta$$
$$\ddot{\theta} = 0 \qquad \dot{r} = -0.4(\csc\theta \cot\theta)\dot{\theta}$$
$$= -0.2 \csc\theta \cot\theta$$
$$\ddot{r} = -0.2(-\csc\theta \cot\theta)(\dot{\theta}) \cot\theta - 0.2 \csc\theta(-\csc^2\theta)\dot{\theta}$$
$$= 0.1 \csc\theta(\cot^2\theta + \csc^2\theta)$$

Evaluating these formulas at $\theta = 60°$, we get

$$\dot{\theta} = 0.5 \qquad r = 0.462$$
$$\ddot{\theta} = 0 \qquad \dot{r} = -0.133$$
$$\ddot{r} = 0.192$$

$$a_r = \ddot{r} - r\dot{\theta}^2 = 0.192 - 0.462(0.5)^2 = 0.0765$$
$$a_\theta = r\ddot{\theta} + 2\dot{r}\dot{\theta} = 0 + 2(-0.133)(0.5) = -0.133$$

Substituting these results into Eqs. (1) and (2) with $\theta = 60°$ and solving yields

$$N_C = 19.4 \text{ N} \qquad F_P = -0.355 \text{ N} \qquad\qquad \textit{Ans.}$$

The negative sign indicates that $\mathbf{F}_P$ acts opposite to that shown in Fig. 13–20b.

19.62 N

(b)

Fig. 13–20

Example 13–13

A can C, having a mass of 0.5 kg, moves along a grooved horizontal slot shown in Fig. 13–21a. The slot is in the form of a spiral, which is defined by the equation $r = (0.1\theta)$ m, where θ is measured in radians. If the arm OA is rotating at a constant rate $\dot{\theta} = 4$ rad/s in the horizontal plane, determine the force it exerts on the can at the instant $\theta = \pi$ rad. Neglect friction and the size of the can.

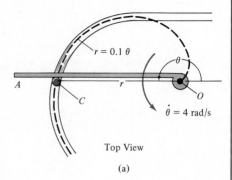

Top View

(a)

Solution

Free-Body Diagram. The driving force $\mathbf{F}_C$ acts perpendicular to the arm OA, whereas the normal force of the wall of the slot on the can, $\mathbf{N}_C$, acts perpendicular to the tangent to the curve at $\theta = \pi$ rad, Fig. 13–21b. As usual, $\mathbf{a}_r$ and $\mathbf{a}_\theta$ act in the *positive directions* of r and θ, respectively. Since the path is specified, the angle ψ which the extended radial line r makes with the tangent, Fig. 13–21c, can be determined from Eq. 13–10. We have $r = 0.1\theta$, so that $dr/d\theta = 0.1$, and therefore

$$\tan \psi = r \left/ \frac{dr}{d\theta} \right. = \frac{0.1\theta}{0.1} = \theta$$

When $\theta = \pi$, $\psi = \tan^{-1} \pi = 72.3°$, so that $\phi = 90° - \psi = 17.7°$, as shown in Fig. 13–21$b$ and c.

(b)

Equations of Motion. Using the data shown in Fig. 13–21b, we have

$$\xleftarrow{+} \Sigma F_r = ma_r; \qquad N_C \cos 17.7° = 0.5a_r \qquad (1)$$
$$+ \downarrow \Sigma F_\theta = ma_\theta; \qquad F_C - N_C \sin 17.7° = 0.5a_\theta \qquad (2)$$

Kinematics. Computing the time derivatives of r and θ gives

$$\dot{\theta} = 4 \text{ rad/s} \qquad r = 0.1\theta$$
$$\ddot{\theta} = 0 \qquad \dot{r} = 0.1\dot{\theta} = 0.1(4) = 0.4 \text{ m/s}$$
$$\ddot{r} = 0.1\ddot{\theta} = 0$$

At the instant $\theta = \pi$ rad,

$$a_r = \ddot{r} - r\dot{\theta}^2 = 0 - 0.1(\pi)(4)^2 = -5.03 \text{ m/s}^2$$
$$a_\theta = r\ddot{\theta} + 2\dot{r}\dot{\theta} = 0 + 2(0.4)(4) = 3.20 \text{ m/s}^2$$

Substituting these results into Eqs. (1) and (2) and solving yields

$$N_C = -2.64 \text{ N}$$
$$F_C = 0.797 \text{ N} \qquad\qquad \textit{Ans.}$$

(c)

Fig. 13–21

Problems

13–65. A particle, having a mass of 3 kg, moves along a three-dimensional path defined by the equations $r = (4 + 3t)$ m, $\theta = (t^2 + 2)$ rad, and $z = (6 - t^2)$ m, where t is in seconds. Determine the r, θ, and z components of force which the path exerts on the particle when $t = 2$ s.

13–66. The boy of mass 40 kg is sliding down the spiral slide at a constant speed such that his position, measured from the top of the chute, has components $r = 1.5$ m, $\theta = (0.7t)$ rad, and $z = (-0.5t)$ m, where t is in seconds. Determine the components of force $\mathbf{F}_r$, $\mathbf{F}_\theta$, and $\mathbf{F}_z$ which he exerts on the chute at the instant $t = 2$ s. Neglect the size of the boy.

Prob. 13–66

13–67. The spring-held follower AB has a weight of 0.75 lb and moves back and forth as it bears on the smooth contoured surface of the cam, where $r = 0.2$ ft and $z = (0.5 \sin \theta)$ ft. If the cam is rotating at a constant rate of 6 rad/s, determine the force at the end A when $\theta = 90°$. In this position the spring is compressed 0.2 ft. Neglect friction at the bearing C.

Prob. 13–67

*13–68. Rod OA rotates counterclockwise with a constant angular rate $\dot{\theta} = 4$ rad/s. The double collar B is pin-connected together such that one collar slides over the rotating rod and the other collar slides over the *horizontal* circular rod described by the equation $r = (1.6 \cos \theta)$ m. If *both* collars have a mass of 0.5 kg, determine the force which the circular rod exerts on one of the collars and the force that OA exerts on the other collar at the instant $\theta = 45°$.

Probs. 13–68 / 13–69

13–69. Solve Prob. 13–68 if the path is *vertical*.

13–70. A boy standing on firm ground spins the girl sitting on a circular "dish" or sled in a circular path of radius $r_0 = 3$ m such that the angular rate of rotation is $\dot{\theta}_0 = 0.1$ rad/s. If the attached cable OC is drawn inward at a constant speed of $\dot{r} = -0.5$ m/s, determine the tension it exerts on the sled at the instant $r = 2$ m. The sled and girl have a total mass of 50 kg. Neglect the size of the boy and sled and the effects of friction between the sled and ice. *Hint:* First show that the equation of motion in the θ direction yields $a_\theta = r\ddot{\theta} + 2\dot{r}\dot{\theta} = d/dt(r^2\dot{\theta}) = 0$. When integrated, $r^2\dot{\theta} = C$, where the constant C is determined from the problem data.

Probs. 13–70 / 13–71

13–71. Compute the angular velocity of the sled in Prob. 13–70 at the instant $r = 2$ m.

*13–72. A car of a roller coaster travels along a track which for a short distance is defined by a conical spiral, $r = \frac{3}{4}z$, $\theta = -1.5z$, where r and z are in meters and θ is in radians. If the angular motion $\dot{\theta} = 1$ rad/s is always maintained, determine the r, θ, z components of reaction exerted on the car by the tracks at the instant $z = 6$ m. The car and passenger have a total mass of 300 kg. Neglect the mass of the wheels and the size of the car.

Prob. 13–72

13–73. The 0.5-lb particle is guided along the circular path using the slotted arm guide. If the arm has an angular velocity $\dot{\theta} = 4$ rad/s and an angular acceleration $\ddot{\theta} = 8$ rad/s² at the instant $\theta = 30°$, determine the force of the guide on the particle. Motion occurs in the *horizontal plane*.

Probs. 13–73 / 13–74

13–74. Solve Prob. 13–73 if motion occurs in the *vertical plane*.

13–75. The pilot of an airplane executes a vertical loop which in part follows the path of a "four-leaved rose," $r = (-600 \cos 2\theta)$ ft, where θ is in radians. If his speed at A is a constant $v_P = 80$ ft/s, determine the vertical reaction he exerts on the seat of the plane when the plane is at A. He weighs 150 lb. *Hint:* To determine the time derivatives necessary to compute the plane's acceleration components a_r and a_θ, take the first and second time derivatives of $r = -600 \cos 2\theta$. Then, for further information, use Eq. 12–26 to determine $\dot{r}$ and $\dot{\theta}$. Also, take the time derivative of Eq. 12–26, noting that $\dot{v}_P = 0$, to determine $\ddot{r}$ and $\ddot{\theta}$.

Prob. 13–75

*13–76. Using air pressure, the 0.5-kg ball is forced to move through the tube lying in the horizontal plane and having the shape of a logarithmic spiral. If the tangential force exerted on the ball due to the air is 6 N, determine the magnitude and direction of the rate of increase in the ball's speed at the instant $\theta = \pi/2$.

Prob. 13–76

119

13–77. The particle has a mass of 0.5 kg and is confined to move along the smooth vertical slot due to the rotation of the arm OA. Determine the force of the rod on the particle and the normal force of the slot on the particle when $\theta = 30°$. The rod is rotating with a constant angular velocity of $\dot{\theta} = 2$ rad/s. Assume the particle contacts only one side of the slot at any instant.

Prob. 13–77

13–78. The smooth particle has a mass of 80 g. It is attached to an elastic cord extending from O to P and due to the slotted arm guide moves along the *horizontal* circular path $r = (0.8 \sin \theta)$ m. If the cord has a stiffness $k = 30$ N/m and an unstretched length of 0.25 m, determine the force of the guide on the particle when $\theta = 60°$. The guide has a constant angular velocity $\dot{\theta} = 5$ rad/s.

Prob. 13–78

13–79. The collar, which has a weight of 3 lb, slides along the smooth rod lying in the *horizontal plane* and having the shape of a parabola $r = 4/(1 - \cos \theta)$, where θ is in radians. If the collar's angular rate is constant and equals $\dot{\theta} = 4$ rad/s, determine the tangential force P needed to cause the motion and the normal force that the collar exerts on the rod at the instant $\theta = 90°$.

Probs. 13–79 / 13–80

*13–80. Solve Prob. 13–79 if the parabolic path (rod) lies in the *vertical plane*.

13–81. The 4.20-Mg car C is traveling along a portion of a horizontal road defined by the cardioid $r = 400(1 + \cos \theta)$ m. If the car maintains a constant speed $v_C = 4$ m/s, determine the radial and transverse components of the frictional force which must be exerted by the road on all the wheels in order to maintain the motion when $\theta = 0$. *Hint:* To determine the time derivatives necessary to compute the car's acceleration components a_r and a_θ, take the first and second time derivatives of $r = 400(1 + \cos \theta)$. Then, for further information, use Eq. 12–26 to determine $\dot{\theta}$. Also, take the time derivative of Eq. 12–26, noting that $\dot{v}_C = 0$, to determine $\ddot{r}$ and $\ddot{\theta}$.

13–82. Using a forked rod, a smooth can C having a mass of 0.5 kg is forced to move along the *vertical slotted* path $r = (0.6\theta)$ m, where θ is in radians. If the can has a constant speed $v_C = 2$ m/s, determine the force of the forked rod on the can and the normal force of the slot on the can at the instant $\theta = \pi$ rad. Assume the can is in contact with only *one edge* of the rod and slot at any instant. *Hint:* To obtain the time derivatives necessary to compute the can's acceleration components a_r and a_θ, take the first and second time derivatives of $r = 0.6\theta$. Then, for further information, use Eq. 12–26 to determine $\dot{r}$ and $\dot{\theta}$. Also, take the time derivative of Eq. 12–26, noting that $\dot{v}_C = 0$, to determine $\ddot{r}$ and $\ddot{\theta}$.

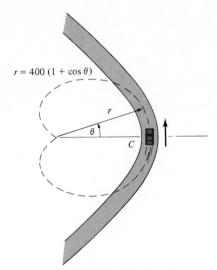

$r = 400 (1 + \cos \theta)$

Prob. 13–81

$\theta = \pi$

$r = 0.6\,\theta$

Prob. 13–82

*13.7 Central-Force Motion and Space Mechanics

$dA = \frac{1}{2}r^2\,d\theta$

(a)

(b)

Fig. 13-22

If a particle is moving only under the influence of a force having a line of action which is always directed toward a fixed point, the motion is called *central-force motion*. This type of motion is commonly caused by electrostatic and gravitational forces.

In Fig. 13–22a, the particle P having a mass m is acted upon only by the central force **F**. The free-body diagram for the particle is shown in Fig. 13–22b. Using polar coordinates (r, θ), the equations of motion, Eqs. 13–9, become

$$-F = m\left[\frac{d^2 r}{dt^2} - r\left(\frac{d\theta}{dt}\right)^2\right]$$

$$0 = m\left(r\frac{d^2\theta}{dt^2} + 2\frac{dr}{dt}\frac{d\theta}{dt}\right)$$

(13–11)

The second of these equations may be written in the form

$$\frac{1}{r}\left[\frac{d}{dt}\left(r^2\frac{d\theta}{dt}\right)\right] = 0$$

so that integrating yields

$$r^2\frac{d\theta}{dt} = h$$

(13–12)

where h is a constant of integration. From Fig. 13–22a it can be seen that the shaded area described by the radius r, as r moves through an angle dθ, is $dA = \frac{1}{2}r^2\,d\theta$. If the *areal velocity* is defined as

$$\frac{dA}{dt} = \frac{1}{2}r^2\frac{d\theta}{dt}$$

(13–13)

then, by comparison with Eq. 13–12, it is seen that the areal velocity for a particle subjected to central-force motion is *constant*. To obtain the *path of motion*, $r = f(\theta)$, the independent variable t must be eliminated from Eqs. 13–11. Using the chain rule of calculus and Eq. 13–12, the time derivatives of Eqs. 13–11 may be replaced by

$$\frac{dr}{dt} = \frac{dr}{d\theta}\frac{d\theta}{dt} = \frac{h}{r^2}\frac{dr}{d\theta}$$

$$\frac{d^2r}{dt^2} = \frac{d}{dt}\left(\frac{h}{r^2}\frac{dr}{d\theta}\right) = \frac{d}{d\theta}\left(\frac{h}{r^2}\frac{dr}{d\theta}\right)\frac{d\theta}{dt} = \left[\frac{d}{d\theta}\left(\frac{h}{r^2}\frac{dr}{d\theta}\right)\right]\frac{h}{r^2}$$

Substituting a new dependent variable (xi) $\xi = 1/r$ into the second equation, it is seen that

$$\frac{d^2r}{dt^2} = -h^2\xi^2\frac{d^2\xi}{d\theta^2}$$

Also, the square of Eq. 13–12 becomes

$$\left(\frac{d\theta}{dt}\right)^2 = h^2\xi^4$$

Substituting these last two equations into the first of Eqs. 13–11, we have

$$-h^2\xi^2\frac{d^2\xi}{d\theta^2} - h^2\xi^3 = -\frac{F}{m}$$

or

$$\frac{d^2\xi}{d\theta^2} + \xi = \frac{F}{mh^2\xi^2} \tag{13–14}$$

This differential equation defines the path over which the particle travels when it is subjected to the central force* **F.**

For application the force of gravitational attraction will be considered. Some common examples of central-force systems which depend upon gravitation include the motion of the moon and artificial satellites about the earth, and the motion of the planets about the sun. As a typical problem in space mechanics, consider the trajectory of a space satellite or space vehicle launched into orbit with an initial velocity $\mathbf{v}_0$, Fig. 13–23. It will be assumed that this velocity is initially *parallel* to the tangent at the surface of the earth, as shown in the figure.† Just after the satellite is released into free flight the only force acting upon it is the gravitational force of the earth. (Gravitational attractions involving other bodies such as the moon or sun will be neglected, since for orbits close to the earth their effect is small in comparison with the gravitational force of the earth.) According to Newton's law of gravitation, force **F** will always act between the mass centers of the earth and the satellite, Fig. 13–23. From Eq. 13–1, this force of attraction has a magnitude of

$$F = G\frac{M_e m}{r^2}$$

where M_e and m represent the mass of the earth and the satellite, respectively, G is the gravitational constant, and r is the distance between the mass centers. Setting $\xi = 1/r$ in the above equation and substituting the result into Eq. 13–14, we obtain

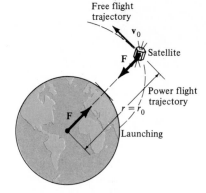

Fig. 13–23

---

*In the derivation, **F** is considered positive when it is directed toward point O. If **F** is oppositely directed, the right side of Eq. 13–14 should be negative.

†The case where $\mathbf{v}_0$ acts at some initial angle θ to the tangent is best described using the conservation of angular momentum (see Prob. 15–70).

$$\frac{d^2\xi}{d\theta^2} + \xi = \frac{GM_e}{h^2} \tag{13–15}$$

This second-order ordinary differential equation has constant coefficients and is nonhomogeneous. The solution is represented as the sum of the complementary and particular solutions. The complementary solution is obtained when the term on the right is equal to zero. It is,

$$\xi_c = C \cos(\theta - \phi)$$

where C and ϕ are constants of integration. The particular solution is

$$\xi_p = \frac{GM_e}{h^2}$$

Thus, the complete solution to Eq. 13–15 is

$$\xi = \xi_c + \xi_p$$

$$= \frac{1}{r} = C \cos(\theta - \phi) + \frac{GM_e}{h^2} \tag{13–16}$$

The validity of this result may be checked by substitution into Eq. 13–15.

Equation 13–16 represents the *free-flight trajectory* of the satellite. It is the equation of a conic section expressed in terms of polar coordinates. As shown in Fig. 13–24, a *conic section* is defined as the locus of point P which moves in a plane in such a way that the ratio of its distance from a fixed point F to its distance from a fixed line is constant. The fixed point is called the *focus*, and the fixed line DD is called the *directrix*. The constant ratio is called the *eccentricity* of the conic and is denoted by e. Thus,

$$e = \frac{FP}{PA}$$

which may be written in the form

$$FP = r = e(PA) = e[p - r\cos(\theta - \phi)]$$

Fig. 13–24

or

$$\frac{1}{r} = \frac{1}{p}\cos(\theta - \phi) + \frac{1}{ep}$$

Comparing this equation with Eq. 13–16, it is seen that the eccentricity of the conic section for the trajectory is

$$e = \frac{Ch^2}{GM_e} \tag{13–17}$$

and the fixed distance from the focus to the directrix is

$$p = \frac{1}{C} \tag{13–18}$$

Provided the polar angle θ is measured from the x axis (an axis of symmetry since it is perpendicular to the directrix DD), the angle ϕ is zero, Fig. 13–24, and therefore Eq. 13–16 reduces to

$$\frac{1}{r} = C \cos \theta + \frac{GM_e}{h^2} \qquad (13\text{–}19)$$

The constants h and C are determined from the data obtained for the position and velocity of the satellite at the end of the *power-flight trajectory*. For example, if the initial height or distance to the space vehicle is r_0 (measured from the center of the earth) and its initial speed is v_0 at the beginning of its free flight, Fig. 13–25, then the constant h may be obtained from Eq. 13–12. When $\theta = \phi = 0°$, the velocity $\mathbf{v}_0$ has no radial component; therefore, from Eq. 12–19, $v_0 = r_0(d\theta/dt)$, so that

$$h = r_0^2 \frac{d\theta}{dt}$$

or

$$h = r_0 v_0 \qquad (13\text{–}20)$$

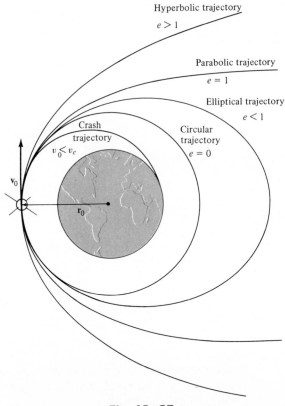

Fig. 13–25

To determine C, use Eq. 13–19 with $\theta = 0°$, $r = r_0$, and substitute Eq. 13–20 for h:

$$C = \frac{1}{r_0}\left(1 - \frac{GM_e}{r_0 v_0^2}\right) \qquad (13\text{–}21)$$

The equation for the free-flight trajectory therefore becomes

$$\frac{1}{r} = \frac{1}{r_0}\left(1 - \frac{GM_e}{r_0 v_0^2}\right)\cos\theta + \frac{GM_e}{r_0^2 v_0^2} \qquad (13\text{–}22)$$

The type of path taken by the satellite is determined from the value of the eccentricity of the conic section as given by Eq. 13–17. If

$$\begin{array}{ll} e = 0, & \text{free-flight trajectory is a circle} \\ e = 1, & \text{free-flight trajectory is a parabola} \\ e < 1, & \text{free-flight trajectory is an ellipse} \\ e > 1, & \text{free-flight trajectory is a hyperbola} \end{array} \qquad (13\text{–}23)$$

A graph of each of these trajectories is shown in Fig. 13–25. From the curves it is seen that when the satellite follows a parabolic path, it is "on the border" of not returning to its initial starting point. The initial velocity at launching, v_0, required for the satellite to follow a parabolic path is called the *escape velocity*. This velocity, v_e, can be determined by using the second of Eqs. 13–23 with Eqs. 13–17, 13–20, and 13–21. It is left as an exercise to show that

$$v_e = \sqrt{\frac{2GM_e}{r_0}} \qquad (13\text{–}24)$$

The velocity v_c required to launch a satellite into a *circular orbit* can be found using the first of Eqs. 13–23. Since e is related to h and C, Eq. 13–17, C must be zero to satisfy this equation (from Eq. 13–20, h cannot be zero); and therefore, using Eq. 13–21, we have

$$v_c = \sqrt{\frac{GM_e}{r_0}} \qquad (13\text{–}25)$$

Provided r_0 represents a minimum height for launching, in which frictional resistance from the atmosphere is neglected, velocities of launching which are less than v_c will cause the satellite to reenter the earth's atmosphere and either burn up or crash, Fig. 13–25.

All the trajectories attained by planets and most satellites are elliptical, Fig. 13–26. For an orbit about the earth, the *minimum distance* from the orbit to the center of the earth O (which is located at one of the foci of the ellipse) is r_p and can be found using Eq. 13–22 with $\theta = 0°$. Therefore,

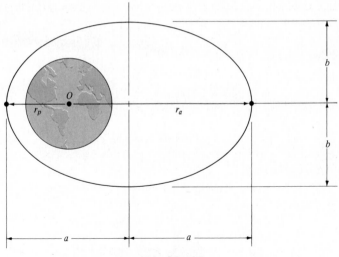

Fig. 13–26

$$r_p = r_0 \tag{13-26}$$

This minimum distance is called the *perigee* of the orbit. The *apogee* or maximum distance r_a can be found using Eq. 13–22 with $\theta = 180°$.*
Thus,

$$r_a = \frac{r_0}{\dfrac{2GM_e}{r_0 v_0^2} - 1} \tag{13-27}$$

With reference to Fig. 13–26, the semimajor axis a of the ellipse is

$$a = \frac{r_p + r_a}{2} \tag{13-28}$$

Using analytical geometry it can be shown that the minor axis b is determined from the equation

$$b = \sqrt{r_p r_a} \tag{13-29}$$

Furthermore, by direct integration, the area of an ellipse is

$$A = \pi ab = \frac{\pi}{2}(r_p + r_a)\sqrt{r_p r_a} \tag{13-30}$$

*Actually, the terminology perigee and apogee pertains only to orbits about the *earth*. If any other body is located at the focus of an elliptical orbit, the minimum and maximum distances are referred to respectively as the *periapsis* and *apoapsis* of the orbit.

The areal velocity has been defined by Eq. 13–13. Substituting this equation into Eq. 13–12, we have $dA/dt = h/2$.

Integrating yields $A = hT/2$, where T is the *period* of time required to make one orbital revolution. From Eq. 13–30, the period is

$$T = \frac{\pi}{h}(r_p + r_a)\sqrt{r_p r_a} \qquad (13\text{–}31)$$

In addition to predicting the orbital trajectory of earth satellites, the theory developed in this section is valid, as a surprisingly close approximation, in predicting the actual motion of the planets traveling around the sun. In this case the mass of the sun, M_s, should be substituted for M_e when the appropriate formulas are used.

The fact that the planets do indeed follow elliptic orbits about the sun was discovered by the German astronomer Johannes Kepler (1571–1630). His discovery was made *before* Newton had developed the laws of motion and the law of gravitation. Kepler's laws, stipulated after 20 years of planetary observation, are summarized by the following three statements:

1. Every planet moves in its orbit such that the line joining it to the sun sweeps over equal areas in equal intervals of time, whatever the line's length.
2. The orbit of every planet is an ellipse with the sun placed at one of its foci.
3. The square of the period of any planet is directly proportional to the cube of the semimajor axis of its orbit.

A mathematical statement of the first and second laws is given by Eqs. 13–13 and 13–22. The third law can be shown from Eq. 13–31 (see Prob. 13–84).

Example 13-14

A satellite is launched 600 km from the surface of the earth, with an initial velocity of 30 Mm/h acting parallel to the tangent at the surface of the earth, Fig. 13–27. Assuming that the radius of the earth is 6378 km and that its mass is $5.976(10^{24})$ kg, determine (a) the eccentricity of the orbital path, and (b) the velocity of the satellite at apogee.

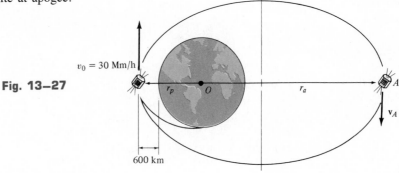

Fig. 13–27

Solution

Part (a). The eccentricity of the orbit is obtained using Eq. 13–17. The constants h and C are first determined from Eqs. 13–20 and 13–21. Since

$$r_p = 6378 \text{ km} + 600 \text{ km} = 6.978(10^6) \text{ m}$$

$$v_0 = 30 \text{ Mm/h} = 8333.3 \text{ m/s}$$

then

$$h = r_p v_0 = 6.978(10^6)(8333.3) = 5.815(10^{10}) \text{ m}^2/\text{s}$$

$$C = \frac{1}{r_p}\left(1 - \frac{GM_e}{r_p v_0^2}\right)$$

$$= \frac{1}{6.978(10^6)}\left(1 - \frac{6.673(10^{-11})[5.976(10^{24})]}{6.978(10^6)(8333.3)^2}\right) = 2.54(10^{-8}) \text{ m}^{-1}$$

Hence,

$$e = \frac{Ch^2}{GM_e} = \frac{2.54(10^{-8})[5.815(10^{10})]^2}{6.673(10^{-11})[5.976(10^{24})]} = 0.215 < 1 \qquad Ans.$$

From Eq. 13–23, observe that the orbit is an *ellipse*.

Part (b). If the satellite were launched at the apogee A shown in Fig. 13–27, with a velocity v_A, the same orbit would be maintained provided

$$h = r_p v_0 = r_a v_A = 5.815(10^{10}) \text{ m}^2/\text{s}$$

Using Eq. 13–27, we have

$$r_a = \frac{r_p}{\dfrac{2GM_e}{r_p v_0^2} - 1} = \frac{6.978(10^6)}{\left(\dfrac{2[6.673(10^{-11})][5.976(10^{24})]}{6.978(10^6)(8333.3)^2} - 1\right)} = 10.804(10^6)$$

Thus,

$$v_A = \frac{5.815(10^{10})}{10.804(10^6)} = 5382.3 \text{ m/s} = 19.4 \text{ Mm/h} \qquad Ans.$$

Problems

In the following problems, assume that the radius of the earth is 6378 km, the earth's mass is $5.976(10^{24})$ kg, the mass of the sun is $1.99(10^{30})$ kg, and the gravitational constant is $G = 6.673(10^{-11})$ m^3/(kg $\cdot$ s^2).

13–83. A communications satellite is to be placed into an equatorial circular orbit around the earth so that it always remains directly over a point on the earth's surface. If this requires the period to be 24 hours (approximately), determine the radius of the orbit and the satellite's velocity.

*13–84. Prove Kepler's third law of motion. *Hint:* Use Eqs. 13–19, 13–28, 13–29, and 13–31.

13–85. A satellite is placed into an elliptical orbit over the earth such that its perigee is 8 Mm and apogee 10 Mm. Determine its speed at both of these points and the eccentricity of its orbit.

13–86. The rocket is in a free-flight elliptical orbit about the earth such that the eccentricity of its orbit is e and its perigee is r_0. Determine the minimum increment of speed it should have in order to escape the earth's gravitational field. Where along the path should this occur?

Prob. 13–86

13–87. A satellite is to be placed into an elliptical orbit about the earth such that at the perigee of its orbit it has an *altitude* of 800 km and at apogee its *altitude* is 2400 km. Determine its required launch velocity tangent to the earth's surface at perigee and the period of its orbit.

*13–88. Show that the velocity of a satellite launched into a circular orbit about the earth is given by Eq. 13–25. Determine the velocity of a satellite launched parallel to the surface of the earth so that it travels in a circular orbit 800 km from the earth's surface.

13–89. The eccentricity for the earth's orbit about the sun is $e = 0.0167$. If its farthest distance from the sun is $151.3(10^6)$ km, determine its speed at this point. Also compute how close the earth approaches the sun and its speed at this point.

13–90. An asteroid is in an elliptical orbit about the sun such that its periapsis is $9.30(10^9)$ km. If the eccentricity of the orbit is $e = 0.073$, determine the apoapsis of the orbit.

13–91. The elliptical path of a satellite has an eccentricity $e = 0.130$. If it has a speed of 15 Mm/h when it is at perigee, P, determine its speed when it arrives at apogee, A. Also, how far is it from the earth's surface when it is at A?

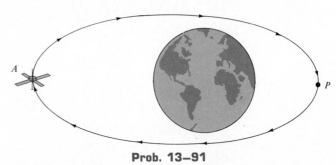

Prob. 13–91

*13–92. A satellite is launched with an initial velocity $v_0 = 4000$ km/h, parallel to the surface of the earth. Determine the required altitude (or range of altitudes) above the earth's surface for launching if the free-flight trajectory is to be (a) circular, (b) parabolic, (c) elliptical, and (d) hyperbolic.

13–93. A satellite is placed into orbit at a velocity of 6 km/s, parallel to the surface of the earth. Determine the proper altitude of the satellite above the earth's surface such that its orbit remains circular. What will happen to the satellite if its initial velocity is 5 km/s when placed tangentially into orbit at the calculated altitude?

13–94. The planet Jupiter travels around the sun in an elliptical orbit such that the eccentricity is $e = 0.048$. If the periapsis between Jupiter and the sun is $r_p = 704(10^6)$ km, determine (a) Jupiter's speed at periapsis, and (b) the apoapsis of the orbit.

13–95. A satellite is in an elliptic orbit around the earth such that $e = 0.156$. If its perigee is 5 Mm, determine its velocity at this point and also the distance OB when it is at point B, located $135°$ away as shown.

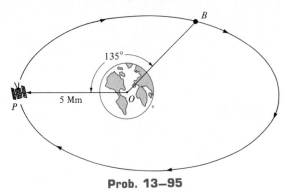

Prob. 13–95

*13–96. A satellite S travels in a circular orbit around the earth. A rocket is located at the apogee of its elliptical orbit for which $e = 0.58$. Determine the sudden change in velocity that must occur at A so that the rocket can enter the satellite's orbit while in free flight along the dashed elliptical trajectory. When it arrives at B, determine the sudden adjustment in speed that must be given to the rocket in order to maintain the circular orbit.

Prob. 13–96

13–97. The rocket is traveling in a free-flight elliptical orbit about the earth such that $e = 0.76$ as shown. Determine its velocity when it is at point A. Also determine the sudden change in speed the rocket must experience at B in order to travel in free flight along the orbit indicated by the dashed path.

Prob. 13–97

13–98. The rocket shown is originally in a free-flight circular orbit 6 Mm above the surface of the earth. It is required that it travel in another circular orbit having an altitude of 12 Mm. To do this, the rocket is given a short pulse of power at A so that it travels in free flight along the dashed elliptical path from the first orbit to the second orbit. Determine the necessary speed it must have at A just after the power pulse, and the time required to get to the outer orbit along the path AA'. What sudden adjustment in speed must be made at A' to maintain the second circular orbit?

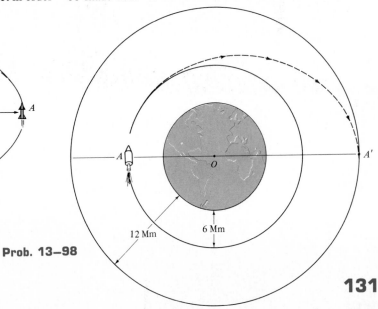

Prob. 13–98

131

Kinetics of a Particle: Work and Energy

14

In this chapter we shall integrate the equation of motion and thereby obtain the principle of work and energy. This principle is useful for solving problems which involve force, velocity, and displacement. Later in the chapter the concept of power is discussed and a method is presented for solving kinetic problems using the theorem of conservation of energy. Before presenting these topics, however, it is first necessary to define the work done by various types of forces.

The Work of a Force 14.1

In mechanics a force $\mathbf{F}$ does work only when it undergoes a displacement in the direction of the force. For example, consider the force $\mathbf{F}$ having a location on the path s which is specified by the position vector $\mathbf{r}$, Fig. 14–1. If the force moves along the path to a new position $\mathbf{r}' = \mathbf{r} + d\mathbf{r}$, the displacement is then $d\mathbf{r}$. The work dU which is done by $\mathbf{F}$ is a *scalar quantity,* defined by the dot product

$$dU = \mathbf{F} \cdot d\mathbf{r}$$

The magnitude of $d\mathbf{r}$ is represented by ds, the differential segment along the path. If the angle between the tails of $d\mathbf{r}$ and $\mathbf{F}$ is θ, Fig. 14–1, then by definition of the dot product (see Eq. C–14) the above equation may also be written as

$$dU = F \, ds \cos \theta$$

Work as expressed by this equation may be interpreted in one of two ways:

Fig. 14–1

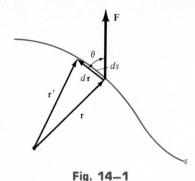

Fig. 14–1

either as the product of F and the component of displacement in the direction of the force, i.e., $ds \cos \theta$, or as the product of ds and the component of force in the direction of displacement, i.e., $F \cos \theta$. Note that if $0° \leq \theta < 90°$, then the vectors having magnitudes $F \cos \theta$ and ds have the *same sense* so that the work is *positive;* whereas if $90° < \theta \leq 180°$, these vectors have an *opposite sense* and therefore the work is *negative*. Also, $dU = 0$ if the force is *perpendicular* to displacement, since $\cos 90° = 0$, or if the force is applied at a *fixed point,* in which case the displacement is zero.

The basic unit for work in the SI system is called a joule (J). This unit combines the units of force and displacement. Specifically, 1 *joule* of work is done when a force of 1 newton moves 1 meter along its line of action (1 J = 1 N · m). The moment of a force has this same combination of units (N · m); however, the concepts of moment and work are in no way related. A moment is a vector quantity, whereas work is a scalar. In the FPS system work is generally defined in ft · lb, which is distinguished from the units for a moment written as lb · ft.

Work of a Variable Force. If a force undergoes a finite displacement along its path from s_1 to s_2, Fig. 14–2a, the work is determined by integration. If $\mathbf{F}$ is expressed as a function of position, $\mathbf{F} = \mathbf{F}(s)$, we have

$$U_{1-2} = \int_{s_1}^{s_2} F \cos \theta \, ds \qquad (14–1)$$

If the working component of the force, $F \cos \theta$, is plotted versus s, Fig. 14–2b, the integral represented in this equation can be interpreted as the *area under the curve* between the points s_1 and s_2.

(a)

(b)

Fig. 14–2

Work of a Constant Force Moving Along a Straight Line. If the force $\mathbf{F}_c$ has a constant magnitude and acts at a constant angle θ from its straight-line path, Fig. 14–3a, then the component of $\mathbf{F}_c$ in the direction of displacement is $F_c \cos \theta$. The work done by $\mathbf{F}_c$ when it is displaced from s_1 to s_2 is determined by Eq. 14–1, in which case

(a)

$$U_{1-2} = F_c \cos \theta \int_{s_1}^{s_2} ds$$

or

$$U_{1-2} = F_c \cos \theta \, (s_2 - s_1) \qquad (14\text{–}2)$$

Here the work of $\mathbf{F}_c$ represents the *area under the rectangle* in Fig. 14–3b.

(b)

Fig. 14–3

Work of a Weight. Consider a particle which moves up along the path s shown in Fig. 14–4 from point 1 to point 2. At an intermediate point, the displacement $d\mathbf{r} = dx\,\mathbf{i} + dy\,\mathbf{j}$. Since $\mathbf{W} = -W\mathbf{j}$, applying Eq. 14–1 yields

$$U_{1-2} = \int \mathbf{F} \cdot d\mathbf{r} = \int_{\mathbf{r}_1}^{\mathbf{r}_2} (-W\mathbf{j}) \cdot (dx\,\mathbf{i} + dy\,\mathbf{j})$$

$$= \int -W\,dy = -W \int dz = -W(y_2 - y_1)$$

or

$$U_{1-2} = -W(\Delta y) \qquad (14\text{–}3)$$

Thus, the work done is equal to the magnitude of the particle's weight times its vertical displacement. In the case shown in Fig. 14–4 the work is *negative,* since W is downward and Δy is upward. Note, however, that if the particle is displaced *downward* $(-\Delta y)$ the work of the weight is *positive.* Why?

Fig. 14–4

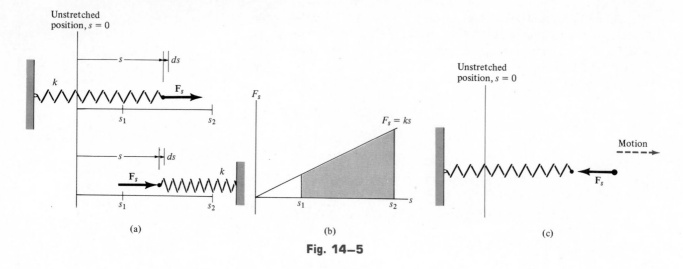

Fig. 14–5

Work of a Spring Force. The magnitude of force developed in a linear elastic spring when the spring is displaced a distance s from its unstretched position is $F_s = ks$, where k is the spring stiffness. If the spring is elongated or compressed from a position s_1 to a further position s_2, Fig. 14–5a, the work done *on the spring* by $\mathbf{F}_s$ is *positive*, since in each case the force and displacement are in the *same direction*. We require

$$U_{1-2} = \int_{s_1}^{s_2} F_s \, ds = \int_{s_1}^{s_2} ks \, ds$$
$$= \tfrac{1}{2}ks_2^2 - \tfrac{1}{2}ks_1^2$$

This equation represents the trapezoidal area under the line $F_s = ks$ versus s, Fig. 14–5b.

If a particle (or body) is attached to a spring, then the force $\mathbf{F}_s$ exerted on the particle is *opposite* to that exerted on the spring, Fig. 14–5c. Consequently, the force will do *negative work* on the particle when the particle is moving so as to further elongate (or compress) the spring. Hence, the above equation becomes

$$U_{1-2} = -(\tfrac{1}{2}ks_2^2 - \tfrac{1}{2}ks_1^2) \tag{14–4}$$

When this equation is used, a mistake in sign can be eliminated if one simply notes the direction of the spring force acting on the particle and compares it with the direction of displacement of the particle—if both are in the *same direction, positive work* results; if they are *opposite* to one another, the *work is negative*.

Example 14–1

The 10-kg block shown in Fig. 14–6*a* rests on the smooth incline. If the spring is originally unstretched, determine the total work done by all the forces acting on the block when a horizontal force $P = 400$ N pushes the block up the plane $s = 2$ m.

Fig. 14–6

Unstretched position of spring

$s = 2$ m

$2 \sin 30°$ m

$P = 400$ N

$k = 30$ N/m

$2 \cos 30°$ m

$30°$

(a)

Solution

The free-body diagram of the block is shown in Fig. 14–6*b*.

Horizontal Force **P.**   Since the force is *constant,* the work is computed using Eq. 14–2. The result can be calculated as the force times the component of displacement in the direction of the force, i.e.,

$$U_P = 400(2 \cos 30°) = 692.8 \text{ J}$$

or, the displacement times the component of force in the direction of displacement, i.e.,

$$U_P = 400 \cos 30°(2) = 692.8 \text{ J}$$

Spring Force **F$_s$.**   Since the spring is originally unstretched and in the final position is stretched 2 m, the work of **F$_s$** is

$$U_s = -\tfrac{1}{2}(30)(2)^2 = -60 \text{ J}$$

Why is the work negative?

Weight **W.**   Since the weight acts in the opposite direction to its vertical displacement the work is negative; i.e.,

$$U_W = -98.1(2 \sin 30°) = -98.1 \text{ J}$$

Note that it is also possible to consider the component of weight in the direction of displacement; i.e.,

$$U_W = -(98.1 \sin 30°)2 = -98.1 \text{ J}$$

Normal Force **N$_B$.**   This force does *no work* since it is *always* perpendicular to the displacement.

The work of all the forces when the block is displaced 2 m is thus

$$U_T = 692.8 - 60 - 98.1 = 535 \text{ J} \qquad \textit{Ans.}$$

14.2 Principle of Work and Energy

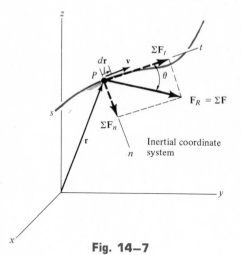

Fig. 14–7

Consider a particle P which is located at some arbitrary point on its path as shown in Fig. 14–7. At the instant considered, P is subjected to a system of external forces, represented by the resultant $\mathbf{F}_R = \Sigma\mathbf{F}$. If the forces are resolved into their normal and tangential components, then during an infinitesimal displacement ds along the path, the normal components, $\Sigma\mathbf{F}_n$, *do no work*, since they do not move in the normal direction; instead, only the tangential components, $\Sigma\mathbf{F}_t$, do work. As a result, consider writing the equation of motion for the particle in the tangential direction.

$$\Sigma F_t = ma_t; \qquad\qquad \Sigma F \cos\theta = ma_t$$

Since $d\mathbf{r}$ has a magnitude which is equal to ds measured along the path, then $a_t = v\,dv/ds$, Eq. 12–3, and therefore

$$\Sigma F \cos\theta\, ds = mv\, dv$$

Integrating both sides, assuming that initially the particle has a position $s = s_1$ and a speed $v = v_1$, and later $s = s_2$, $v = v_2$, we have

$$\Sigma \int_{s_1}^{s_2} F \cos\theta\, ds = \int_{v_1}^{v_2} mv\, dv$$

or

$$\Sigma \int_{s_1}^{s_2} F \cos\theta\, ds = \tfrac{1}{2}mv_2^2 - \tfrac{1}{2}mv_1^2 \qquad (14\text{–}5)$$

Using Eq. 14–1, the final result may be written as

$$\Sigma U_{1-2} = \tfrac{1}{2}mv_2^2 - \tfrac{1}{2}mv_1^2 \qquad (14\text{–}6)$$

This equation represents the *principle of work and energy* for the particle. The term on the left is the sum of the work done by *all* the forces acting on the particle as the particle moves from point 1 to point 2. The two terms on the right side, which are of the form $T = \tfrac{1}{2}mv^2$, define the particle's final and initial *kinetic energy*, respectively. These terms are *positive* scalar quantities since they do not depend on the direction of the particle's velocity. Furthermore, Eq. 14–6 is dimensionally homogeneous so that the kinetic energy has the same units as work, e.g., joules (J) or ft · lb.

When applying Eq. 14–6 it is convenient to rewrite it in the form

$$T_1 + \Sigma U_{1-2} = T_2 \qquad (14\text{–}7)$$

which states that the particle's initial kinetic energy plus the work done by all the forces acting on the particle as it moves from its initial to its final position is equal to the particle's final kinetic energy.

As noted from the derivation, the principle of work and energy represents an integrated form of $\Sigma F_t = ma_t$, acquired by using the kinematic equation $a_t = v \, dv/ds$. As a result, this principle will provide a convenient substitution for $\Sigma F_t = ma_t$ when solving those types of kinetic problems which involve force, velocity, and displacement, since these variables are involved in the terms of Eq. 14–7. For example, if a particle's initial speed is known, and the work of all the forces acting on the particle can be computed, then Eq. 14–7 provides a *direct means* of obtaining the final speed v_2 of a particle after it undergoes a specified displacement. If instead v_2 is determined by means of the equation of motion, a two-step process is necessary; i.e., apply $\Sigma F_t = ma_t$ to obtain a_t, then integrate $a_t = v \, dv/ds$ to obtain v_2. Note that the principle of work and energy cannot be used, for example, to determine forces directed *normal* to the path of motion, since these forces do no work on the particle. Instead $\Sigma F_n = ma_n$ must be applied. For curved paths, however, the magnitude of the normal force is a function of velocity. Hence, it is generally easier to obtain the velocity using the principle of work and energy, and then substitute this quantity into the equation of motion $\Sigma F_n = mv^2/\rho$ to obtain the normal force.

PROCEDURE FOR ANALYSIS

As stated above, the principle of work and energy is used to solve kinetic problems that involve *velocity, force,* and *displacement,* since these variables are involved in the equation. For application it is suggested that the following procedure be used.

Work (Free-Body Diagram).   Draw a free-body diagram of the particle when it is located at an intermediate point along its path in order to account for all the forces that do work on the particle.

Principle of Work and Energy.   Apply the principle of work and energy, $T_1 + \Sigma U_{1-2} = T_2$. The kinetic energy at the initial and final points is always positive, since it involves the speed squared ($T = \frac{1}{2}mv^2$). For the calculation v must be measured from the same inertial reference frame. The work done by each force shown on the free-body diagram is computed by using the appropriate equations developed in Sec. 14.1. Since *algebraic addition* of the work terms is required, it is important that the proper sign of each term be specified. Specifically, work is positive when the force is in the same direction as its displacement, otherwise it is negative.

Numerical application of this procedure is illustrated in the examples following Sec. 14.3.

14.3 Principle of Work and Energy for a System of Particles

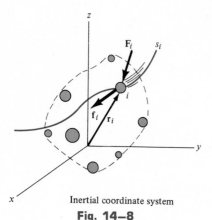

z

$\mathbf{F}_i$ s_i

$\mathbf{f}_i$ $\mathbf{r}_i$

i

y

x

Inertial coordinate system

Fig. 14–8

The principle of work and energy can be extended to include a system of n particles isolated within an enclosed region of space as shown in Fig. 14–8. Here the arbitrary ith particle, having a mass m_i, is subjected to a resultant external force $\mathbf{F}_i$ and a set of internal forces $\displaystyle\sum_{j=1(j\neq i)}^{n} f_{ij} = f_i$ which the other particles exert on the ith particle. The principle of work and energy written for the ith particle is thus

$$\tfrac{1}{2}m_i v_{i1}^2 + \int_{\mathbf{r}_{i1}}^{\mathbf{r}_{i2}} \mathbf{F}_i \cdot d\mathbf{r}_i + \int_{\mathbf{r}_{i1}}^{\mathbf{r}_{i2}} \mathbf{f}_i \cdot d\mathbf{r}_i = \tfrac{1}{2}m_i v_{i2}^2$$

Similar equations result if the principle of work and energy is applied to each of the other particles of the system. Since both work and kinetic energy are scalars, the results may be added together algebraically, so that

$$\Sigma\tfrac{1}{2}m_i v_{i1}^2 + \Sigma\int_{\mathbf{r}_{i1}}^{\mathbf{r}_{i2}} \mathbf{F}_i \cdot d\mathbf{r}_i + \Sigma\int_{\mathbf{r}_{i1}}^{\mathbf{r}_{i2}} \mathbf{f}_i \cdot d\mathbf{r}_i = \Sigma\tfrac{1}{2}m_i v_{i2}^2$$

We can write this equation symbolically as

$$\Sigma T_1 + \Sigma U_{1-2} = \Sigma T_2 \tag{14–8}$$

Hence, the system's initial kinetic energy (ΣT_1) plus the work done by all the external and internal forces acting on the particles of the system (ΣU_{1-2}) is equal to the system's final kinetic energy (ΣT_2). Note that although the internal forces on adjacent particles occur in equal but opposite collinear pairs, the total work done by each of these forces will, in general, *not cancel* out since the paths over which corresponding particles travel will be *different*. There are, however, two important exceptions to this rule which often occur in practice. If the particles are contained within the boundary of a *translating rigid body,* the internal forces all undergo displacement by an equal amount, and therefore the internal work will be zero. Also, particles connected by inextensible cables make up a system which has internal forces which are displaced by an equal amount. In this case, adjacent particles exert equal but opposite internal forces which have components that undergo the same displacement, and therefore the work of these forces cancels.

The procedure for analysis outlined in Sec. 14.2 provides a method for applying Eq. 14–8; however, only one equation applies for the entire system. If the particles are connected by cords, other equations can generally be obtained by using the kinematic principles outlined in Sec. 12.8 in order to relate the particles' speeds. When doing this, assume that the particles are displaced in the *positive coordinate directions* when writing *both* the kinematic equation and the principle of work and energy (see Example 14–6).

Fig. 14-9

Work of Friction Caused by Sliding. A special class of problems will now be investigated which requires a careful application of Eq. 14–8. These problems all involve cases where a body is sliding over the surface of another body in the presence of friction. Consider, for example, a block which is translating a distance d over a rough surface as shown in Fig. 14–9. If the applied force **P** just balances the *resultant* frictional force μN, then due to equilibrium a constant velocity **v** is maintained and one would expect Eq. 14–8 to be applied as follows:

$$\tfrac{1}{2}mv^2 + P(d) - \mu N(d) = \tfrac{1}{2}mv^2$$

Indeed this equation is satisfied if $P = \mu N$; however, as one realizes from experience, the sliding motion will *generate heat*, a form of energy which seems not to be accounted for in the work-energy equation. In order to explain this paradox and thereby more accurately represent the nature of friction, we should actually consider a model of the surfaces of contact which are *deformable* (nonrigid).* Recall that the rough portions at the bottom of the block act as "teeth," and when the block slides these teeth *deform slightly* and break off or vibrate due to interlocking effects and pulling away from "teeth" at the contacting surface. Consequently, the frictional forces acting on the block at the points of contact *are not* displaced the same distance d as the applied force **P**. Instead, the actual displacement d' of these frictional forces will be less ($d' < d$), and therefore the *external work* done by the resultant frictional force will be $\mu N d'$. The remaining amount of work, $\mu N(d - d')$, manifests itself as an increase in internal energy, which in part causes the block's temperature to rise.

In summary then, Eq. 14–8 can be applied to problems involving sliding friction; however, *it should be fully realized that the work of any external frictional force is not represented by $\mu N d$; instead, this term represents both the external work of friction ($\mu N d'$) and internal work [$\mu N(d - d')$] which is converted into various forms of internal energy.*†

*See Chapter 9 of *Engineering Mechanics: Statics*.
†See B. A. Sherwood and W. H. Bernard, "Work and Heat Transfer in the Presence of Sliding Friction," *Am. J. Phys*. 52, 1001 (1984).

Example 14–2

The 3500-lb automobile shown in Fig. 14–10a is traveling down the 10° inclined road at a speed of 20 ft/s. If the driver wishes to stop his car, determine how far s his tires skid on the road if he jams on the brakes causing his wheels to lock. The coefficient of kinetic friction between the wheels and the road is $\mu_A = 0.5$.

20 ft/s

s

A

10°

(a)

s

3500 lb

10°

10°

F_A

N_A

(b)

Fig. 14–10

Solution

Work (Free-Body Diagram). As shown in Fig. 14–10b, the normal force N_A does no work since it never undergoes displacement along its line of action. The weight, 3500 lb, is displaced $s \sin 10°$ and does positive work, whereas the frictional force F_A does both external and internal work when it is *thought* to undergo a displacement s. Applying the equation of equilibrium normal to the road, we have

$$+\nwarrow \Sigma F_y = 0; \quad N_A - 3500 \cos 10° = 0 \quad N_A = 3446.8 \text{ lb}$$

Thus,

$$F_A = 0.5 N_A = 1723.4 \text{ lb}$$

Principle of Work and Energy

$$\tfrac{1}{2}mv^2 \quad \{T_1\} + \{\Sigma U_{1-2}\} = T_2 \quad \tfrac{1}{2}mv^2$$

$$\left\{\frac{1}{2}\left(\frac{3500}{32.2}\right)(20)^2\right\} \overset{F\delta s}{+} \{3500(s \sin 10°) - 1723.4s\} = \{0\}$$

Solving for s yields

$$s = 19.5 \text{ ft} \qquad\qquad \textit{Ans.}$$

Note that if this problem is solved by using the equation of motion, *two steps* are involved. First, from the free-body diagram, Fig. 14–10b, the equation of motion is applied along the incline. This yields

$$+\swarrow \Sigma F_s = ma_s; \quad 3500 \sin 10° - 1723.4 = \frac{3500}{32.2}a$$

$$a = -10.3 \text{ ft/s}^2$$

Then, using the integrated form of $a\,ds = v\,dv$ (kinematics) since a is constant, we have

$$+\swarrow v^2 = v_0^2 + 2a_c(s - s_0) \qquad (0)^2 = (-20)^2 + 2(-10.3)(s - 0)$$

$$s = 19.5 \text{ ft} \qquad\qquad \textit{Ans.}$$

Example 14–3

A 10-kg block rests on the horizontal surface shown in Fig. 14–11a. The spring, which is not attached to the block, has a stiffness $k = 500$ N/m and is initially compressed 0.2 m from C to A. After the block is released from rest at A, determine its velocity when it passes point D. The coefficient of kinetic friction between the block and the plane is $\mu_k = 0.2$.

(a)

(b)

Fig. 14–11

Solution

Work (Free-Body Diagrams). Two free-body diagrams for the block are shown in Fig. 14–11b. The block moves under the influence of the spring force $\mathbf{F}_s$ along the 0.2-m-long path AC, after which it continues to slide along the plane to point D. With reference to either free-body diagram, $\Sigma F_y = 0$; hence, $N_B = 98.1$ N. Only the spring and friction forces do work during the displacement—the spring force does positive work from A to C, whereas the frictional force does negative work. Why?

Principle of Work and Energy

$$\{T_A\} + \{\Sigma U_{A-D}\} = \{T_D\}$$

$$\{\tfrac{1}{2}m(v_A)^2\} + \{\tfrac{1}{2}ks_{AC}^2 - (0.2N_B)s_{AD}\} = \{\tfrac{1}{2}m(v_D)^2\}$$

$$\{0\} + \left\{\frac{1}{2}(500)(0.2)^2 - 0.2(98.1)(0.4)\right\} = \left\{\frac{1}{2}(10)(v_D)^2\right\}$$

Solving for v_D, we get

$$v_D = 0.656 \text{ m/s} \rightarrow \qquad\qquad \textit{Ans.}$$

Example 14–4

The platform P, shown in Fig. 14–12a, has negligible mass and is tied down so that the 0.4-m-long cords keep the spring compressed 0.6 m when *nothing* is on the platform. If a 2-kg block is placed on the platform and released when the platform is pushed down 0.1 m, determine the maximum height h the block rises in the air, measured from the ground.

Fig. 14–12

Solution

Work (Free-Body Diagram). Since the block is released from rest and later reaches its maximum height, the initial and final velocities are zero. The free-body diagram of the block when it is still in contact with the platform is shown in Fig. 14–12b. Note that the weight does negative work and the spring force does positive work. Why? In particular, the *initial compression* in the spring is 0.6 m + 0.1 m = 0.7 m. Due to the cords, the spring's *final compression* is 0.6 m (after the block leaves the platform).

Principle of Work and Energy

$$\{T_1\} + \{\Sigma U_{1-2}\} = \{T_2\}$$

$$\{0\} + \left\{-\left(\frac{1}{2}(200)(0.6)^2 - \frac{1}{2}(200)(0.7)^2\right) - (19.62)[h - (0.4 - 0.1)]\right\} = \{0\}$$

Solving yields

$$h = 0.963 \text{ m} \qquad\qquad \textit{Ans.}$$

Example 14–5

A block having a mass of 2 kg is given an initial velocity of $v_0 = 1$ m/s when it is at the top surface of the smooth cylinder shown in Fig. 14–13a. If the block travels along a path of 0.5-m radius, determine the angle $\theta = \theta_{max}$ at which it begins to leave the cylinder's surface.

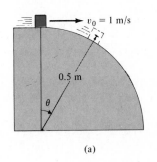

(a)

Solution

Work (Free-Body Diagram). Inspection of Fig. 14–13b reveals that only the weight $W = 2(9.81) = 19.62$ N does work during the displacement. If the block is assumed to leave the surface when $\theta = \theta_{max}$ then the weight moves through a vertical displacement of $0.5(1 - \cos \theta_{max})$ m, as shown in the figure.

Principle of Work and Energy

$$\{T_1\} + \{\Sigma U_{1-2}\} = \{T_2\}$$

$$\left\{ \frac{1}{2}(2)(1)^2 \right\} + \{19.62(0.5)(1 - \cos \theta_{max})\} = \left\{ \frac{1}{2}(2)v_2^2 \right\}$$

or

$$v_2^2 = 9.81(1 - \cos \theta_{max}) + 1 \qquad (1)$$

(b)

Fig. 14–13

Equation of Motion. There are two unknowns in Eq. (1), θ_{max} and v_2. A second equation relating these two variables may be obtained by applying the equation of motion in the *normal direction* to the forces acting on the free-body diagram. Thus,

$$+\swarrow \Sigma F_n = ma_n; \qquad -N_B + 19.62 \cos \theta = 2 \left(\frac{v^2}{0.5} \right)$$

When the block leaves the surface of the cylinder at $\theta = \theta_{max}$, $N_B = 0$ and $v = v_2$; hence,

$$\cos \theta_{max} = \frac{v_2^2}{4.905} \qquad (2)$$

Eliminating the unknown v_2^2 between Eqs. (1) and (2) gives

$$4.905 \cos \theta_{max} = 9.81(1 - \cos \theta_{max}) + 1$$

Solving for $\cos \theta_{max}$, we have

$$\cos \theta_{max} = 0.735$$

Thus,

$$\theta_{max} = 42.7° \qquad\qquad \textit{Ans.}$$

This problem has also been solved in Example 13–10. If the two methods of solution are compared it will be apparent that a work-energy approach yields a more direct solution.

Example 14–6

The blocks *A* and *B* shown in Fig. 14–14*a* have a mass of 10 and 100 kg, respectively. Determine the distance *B* travels from the point where it is released from rest to the point where its speed becomes 2 m/s.

Solution

This problem may be solved by considering the blocks separately and applying the principle of work and energy to each block. However, the work of the (unknown) cable tension can be eliminated from the analysis by considering blocks *A* and *B* together as a *system*. The solution will require simultaneous solution of the equations of work and energy *and* kinematics. To be consistent with our sign convention, we will assume both blocks move in the positive *downward* direction.

Work (Free-Body Diagram). As shown on the free-body diagram of the system, Fig. 14–14*b*, the cable force **T** and reactions **R**₁ and **R**₂ do *no work,* since these forces represent the reactions at the supports and consequently do not move while the blocks are being displaced.

Principle of Work and Energy. Realizing the blocks are released from rest, and assuming *both* move *downward,* we have

$$\{\Sigma T_1\} + \{\Sigma U_{1-2}\} = \{\Sigma T_2\}$$

$$\{\tfrac{1}{2}m_A(v_A)_1^2 + \tfrac{1}{2}m_B(v_B)_1^2\} + \{W_B\Delta s_B + W_A\Delta s_A\} = \{\tfrac{1}{2}m_A(v_A)_2^2 + \tfrac{1}{2}m_B(v_B)_2^2\}$$

$$\{0 + 0\} + \{981(\Delta s_B) + 98.1(\Delta s_A)\} = \left\{\frac{1}{2}(10)(v_A)_2^2 + \frac{1}{2}(100)(2)^2\right\} \quad (1)$$

Kinematics. Using the methods of kinematics discussed in Sec. 12.8, it may be seen from Fig. 14–14*a* that at any given instant the total length *l* of all the vertical segments of cable may be expressed in terms of the position coordinates s_A and s_B as

$$s_A + 4s_B = l$$

Hence, a change in position yields the displacement equation

$$\Delta s_A + 4\,\Delta s_B = 0$$

$$\Delta s_A = -4\,\Delta s_B$$

Taking the time derivative yields

$$v_A = -4v_B = -4(2) = -8 \text{ m/s}$$

Retaining the negative sign and substituting into Eq. (1) yields

$$\Delta s_B = 0.883 \text{ m} \downarrow \qquad\qquad Ans.$$

(a)

(b)

Fig. 14–14

Problems

Except when stated otherwise, throughout this chapter assume that the coefficients of static and kinetic friction are equal, i.e., $\mu = \mu_s = \mu_k$.

14–1. A freight car having a mass of 12 Mg is towed along a horizontal track. If the car starts from rest and attains a speed of 15 m/s after traveling a distance of 100 m, determine the constant horizontal towing force applied to the car. Neglect friction and the mass of the wheels.

14–2. The 20-lb crate has a velocity $v_A = 12$ ft/s when it is at A. Determine its velocity after it slides $s = 6$ ft down the plane. The coefficient of kinetic friction between the crate and the plane is $\mu_k = 0.2$.

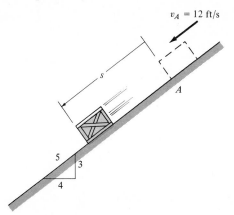

Prob. 14–2

14–3. Water having a specific weight $\gamma = 62.4$ lb/ft^3 is pumped from a lake into the bottom of a tank. If the bottom of the tank is elevated 20 ft from the lake's surface, determine the work required to fill the tank. Neglect any friction losses. The tank is cylindrical with a radius of 5 ft at the base and is 12 ft high.

*14–4. A woman having a mass of 70 kg stands in an elevator which has a downward acceleration of 4 m/s^2 starting from rest. Determine the work done by her weight and the work of the normal force which the floor exerts on her as the elevator descends 6 m. Explain why the work of these forces is different.

14–5. A crate has a weight of 1500 lb. If it is pulled along the ground at a constant speed for a distance of 25 ft, and the towing cable makes an angle of 15° with the horizontal, determine the tension in the cable and the work done by the towing force. The coefficient of friction between the ground and the crate is $\mu = 0.55$.

14–6. The 1500-lb "wrecking" ball, suspended from a 50-ft cable, swings from rest at point A, $\theta = 30°$, toward a wall. After striking the wall it rises to point B, $\phi = 15°$, before stopping. Determine the amount of energy which is lost during the collision, and compute the tension in the cable just before the collision, $\theta = 0°$. Neglect the size of the ball for the calculation. Point C doesn't move, and when $\theta = 0°$ the ball is at its lowest point and about to strike the wall.

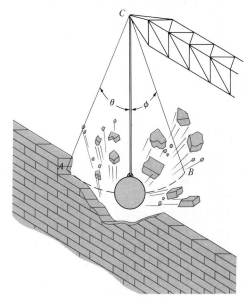

Prob. 14–6

14–7. The collar has a mass of 30 kg and is supported on the rod having a coefficient of kinetic friction $\mu_k = 0.4$. The attached spring has an unstretched length of 0.2 m and a stiffness $k = 50$ N/m. Determine the speed of the collar after the applied force $F = 200$ N causes it to be displaced $x = 1.5$ m from point A. When $x = 0$ the collar is held at rest. Neglect the size of the collar in the calculation.

Prob. 14–7

*14–8. The "air spring" A is used to protect the support structure B and prevent damage to the conveyor belt tensioning weight C in the event of a belt failure D. The force developed by the spring as a function of its deflection is shown by the graph. Determine the suspended height d of the 50-lb counterweight C so that if the conveyor belt fails the counterweight will deform the spring $s_{max} = 0.2$ ft. Neglect the mass of the pulley and belt.

Prob. 14–8

14–9. The ball has a mass of 0.5 kg and is suspended from a rubber band having an unstretched length of 1 m and a stiffness $k = 50$ N/m. If the support at A to which the rubber band is attached is held 2 m from the floor, determine the greatest speed the ball can have at A so that it does not touch the floor when it reaches its lowest point B. Neglect the size of the ball and the mass of the rubber band.

Prob. 14–9

14–10. The 5-lb block is released from rest at A and slides down the smooth circular surface AB. It then continues to slide along the horizontal rough surface until it strikes the spring. Determine how far it compresses the spring before stopping.

Prob. 14–10

14–11. The bumper B on the 5-Mg train car will deflect in accordance with the load-deflection graph shown. Determine the maximum deflection s that will occur in order to stop the car if it is traveling at $v = 5$ m/s when it collides with the rigid stop. Neglect the mass of the car wheels.

Prob. 14–11

■*14–12.** The force $\mathbf{F}$ acting in a constant direction on the 20-kg block has a magnitude which varies with the position s of the block. Determine how far the block slides before its velocity becomes 5 m/s. When $s = 0$ the block is moving to the right at 2 m/s. The coefficient of friction between block and surface is $\mu = 0.3$.

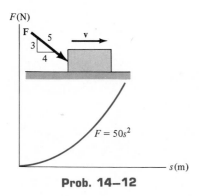

Prob. 14–12

14–13. The 120-lb man acts as a human cannon ball by being "fired" from the spring-loaded cannon shown. If the greatest acceleration he can experience is $a = 10g = 322$ ft/s^2, determine the required stiffness of the spring if it is compressed 2 ft at the moment of firing. With what velocity will he exit the cannon barrel when the cannon is fired? When the spring is compressed $s = 2$ ft then $d = 8$ ft. Neglect friction and assume the man holds himself in a rigid position throughout the motion.

Prob. 14–13

14–14. The two blocks have weights $W_A = 60$ lb and $W_B = 10$ lb. If the kinetic coefficient of friction between the incline and block A is $\mu_k = 0.2$, determine the speed of block A after it moves 3 ft down the plane starting from rest. Neglect the mass of the cord and pulleys.

Prob. 14–14

14–15. The motion of a truck is arrested using a bed of loose stone AB and a set of crash barrels BC. If the stones provide a rolling resistance of 160 lb per wheel and the crash barrels provide a resistance as shown in the graph, determine the distance the 4500-lb truck penetrates the barrels if the truck is moving at 60 ft/s when it approaches A. Neglect the size of the truck.

Prob. 14–15

*14–16.** The 2-kg block is subjected to a force having a constant direction and a magnitude $F = (300/(1 + s))$ N, where s is measured in meters. When $s = 4$ m, the block is moving to the left with a speed of 8 m/s. Determine its speed when $s = 12$ m. The coefficient of kinetic friction between the block and the ground is $\mu_k = 0.25$.

Prob. 14–16

14–17. The 100-kg stone is being dragged across the smooth surface by means of a truck T. If the towing cable passes over a small pulley at A, determine the amount of work that must be done by the truck in order to increase the cable angle θ from $\theta_1 = 30°$ to $\theta_2 = 45°$. The truck exerts a constant force $F = 500$ N on the cable at B. Neglect the mass of the pulley and cable.

Prob. 14–17

14–18. The collar has a mass of 20 kg and rests on the smooth rod. Two springs are attached to it and the ends of the rod as shown. Each spring has an uncompressed length of 1 m. If the collar is displaced $s = 0.5$ m and released from rest, determine its velocity at the instant it returns to the point $s = 0$.

Prob. 14–18 / 14–19

14–19. The collar has a mass of 20 kg and rests on the smooth rod. Two springs are attached to it and the ends of the rod as shown. Each spring has an uncompressed length of 1.5 m. If the collar is displaced $s = 0.5$ m and released from rest, determine its velocity at the instant it returns to the point $s = 0$.

*14–20. Packages having a weight of 15 lb are transferred horizontally from one conveyor to the next using a ramp for which $\mu_k = 0.15$. The top conveyor is moving at 6 ft/s and the packages are spaced 3 ft apart. Determine the required speed of the bottom conveyor so no sliding occurs when the packages come horizontally in contact with it. What is the spacing s between the packages on the bottom conveyor?

Prob. 14–20

14–21. The spring has a stiffness $k = 50$ lb/ft and an *unstretched length* of 2 ft. As shown, it is confined by the plate and wall using cables so that its length is 1.5 ft. A 4-lb block is given a speed v_A when it is at A, and it slides down the incline having a coefficient of friction $\mu_k = 0.2$. If it strikes the plate and pushes it forward 0.25 ft before stopping, determine its speed at A. Neglect the mass of the plate and spring.

Prob. 14–21

14–22. The skier starts from rest at A and travels down the ramp. If friction and air resistance can be neglected, determine his speed v_B when he reaches B. Also, compute the distance s to where he strikes the ground at C, if he makes the jump traveling horizontally at B. Neglect the skier's size. He has a mass of 70 kg.

Prob. 14–22

14–23. The block has a mass of 0.8 kg and moves within the smooth vertical slot. If it starts from rest when the *attached* spring is in the unstretched position at A, determine the *constant* vertical force $\mathbf{F}$ which must be applied to the cord so that the block attains a speed $v_B = 2.5$ m/s when it reaches B; $s_B = 0.15$ m. Neglect the size and mass of the pulley. *Hint:* The work of $\mathbf{F}$ can be determined by finding the difference Δl in cord lengths AC and BC and using $U_F = F\,\Delta l$.

Prob. 14–23

151

*14–24. The catapulting mechanism is used to propel the 10-kg slider A to the right along the smooth track. The propelling action is obtained by drawing the pulley attached to rod BC rapidly to the left by means of a piston P. If the piston applies a constant force $F = 20$ kN to rod BC such that it moves it 0.2 m, determine the speed attained by the slider if it was originally at rest. Neglect the mass of the pulleys, cable, piston, and rod BC.

Prob. 14–24

14–25. A rocket of mass m is fired vertically from the surface of the earth, i.e., at $r = r_1$. Assuming that no mass is lost as it travels upward, determine the work it must do against gravity to reach a distance r_2. The force of gravity is $F = GM_e m/r^2$ (Eq. 13–1), where M_e is the mass of the earth and r the distance between the rocket and the center of the earth.

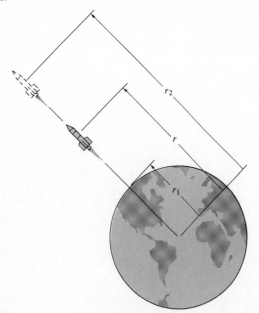

Prob. 14–25

14–26. The 2-kg collar is attached to a spring that has an unstretched length of 3 m. If it is drawn to point B and released from rest, determine its speed when it arrives at point A.

Prob. 14–26

14–27. The cyclist travels to point A pedaling until he reaches a speed $v_A = 8$ m/s. He then coasts freely up the curved surface. Determine the normal force he exerts on the surface when he reaches point B. The total mass of the bike and man is 75 kg. Neglect friction, the mass of the wheels, and the size of the bicycle.

Prob. 14–27 / 14–28

*14–28. Determine the smallest speed at which the man at A must be traveling in order to reach point C if he coasts freely up the curved surface. The total mass of the bike and man is 75 kg. Neglect friction, the mass of the wheels, and the size of the bicycle.

14–29. A 2-lb block rests on the smooth cylindrical surface. An elastic cord having a stiffness $k = 2$ lb/ft is attached to the block at B and to the base of the cylinder at point C. If the block is released from rest at A, determine the longest unstretched length of the cord so the block begins to leave the cylinder at the instant $\theta = 45°$. Neglect the size of the block.

Prob. 14–29

Power and Efficiency 14.4

Power. *Power* is defined as the amount of work performed per unit of time. Hence, the *average power* generated by a machine or engine that performs an amount of work ΔU within the time interval Δt is

$$P_{avg} = \frac{\Delta U}{\Delta t} \qquad (14\text{--}9)$$

If the time $\Delta t \to dt$, and consequently $\Delta U \to dU$, then the *instantaneous power* is defined as

$$P = \frac{dU}{dt} \qquad (14\text{--}10)$$

Provided that the work dU is expressed by $dU = \mathbf{F} \cdot d\mathbf{r}$, then it is also possible to write

$$P = \frac{dU}{dt} = \frac{\mathbf{F} \cdot d\mathbf{r}}{dt}$$

or

$$P = \mathbf{F} \cdot \mathbf{v} \qquad (14\text{--}11)$$

Hence, power is a *scalar,* where in the formulation $\mathbf{v}$ represents the velocity of the point which is acted upon by the force $\mathbf{F}$.

The basic units of power used in the SI and FPS systems are the watt (W) and horsepower (hp), respectively. These units are defined as

$$1 \text{ W} = 1 \text{ J/s} = 1 \text{ N} \cdot \text{m/s}$$
$$1 \text{ hp} = 550 \text{ ft} \cdot \text{lb/s}$$

For conversion between the two systems of units, 1 hp = 746 W.

The term "power" provides a useful basis for determining the type of motor or machine which is required to do a certain amount of work in a given time. For example, two pumps may each be able to empty a reservoir if given enough time; however, the pump having the larger power will complete the job sooner.

Efficiency. The *mechanical efficiency* of a machine is defined as the ratio of the output of useful power produced by the machine to the input of power supplied to the machine. Hence,

$$\varepsilon = \frac{\text{power output}}{\text{power input}} \tag{14-12}$$

If energy applied to the machine occurs during the *same time interval* at which it is removed, then the efficiency may also be expressed in terms of the ratio of output energy to input energy; i.e.

$$\varepsilon = \frac{\text{energy output}}{\text{energy input}} \tag{14-13}$$

Since machines consist of a series of moving parts, frictional forces will always be developed within the machine, and as a result, extra energy or power is needed to overcome these forces. Consequently, *the efficiency of a machine is always less than 1*.

PROCEDURE FOR ANALYSIS

When computing the power supplied to a body, one must first determine the unbalanced external force $\mathbf{F}$ acting on the body which causes the motion. This force is usually developed by a machine or engine placed either within or external to the body. If the body is accelerating, it may be necessary to draw the proper free-body diagram and apply the equation of motion ($\Sigma \mathbf{F} = m\mathbf{a}$) to determine $\mathbf{F}$. Once $\mathbf{F}$ and the velocity $\mathbf{v}$ of the point where $\mathbf{F}$ is applied have been computed, the power is determined by multiplying the force magnitude by the component of velocity acting in the direction of $\mathbf{F}$, i.e., $P = \mathbf{F} \cdot \mathbf{v} = Fv \cos \theta$.

In some problems the power may also be computed by calculating the work done per unit of time ($P_{\text{avg}} = \Delta U / \Delta t$, or $P = dU/dt$). Depending upon the problem, this work can be done either by the external or internal force of a machine or engine, by the weight of the body, or by an elastic spring force acting on the body.

Example 14-7

The motor M of the hoist shown in Fig. 14–15a operates with an efficiency of 0.85. Determine the power that must be supplied to the motor to lift the crate C having a weight of 75 lb. The cable is being drawn in with an acceleration of 4 ft/s^2, and at the instant shown its speed is 2 ft/s. Neglect the mass of the pulley and cable.

Solution

In order to compute the power output of the motor it is first necessary to determine the tension force in the cable. Since the crate is subjected to an acceleration, this requires application of the equation of motion.

From the free-body diagram, Fig. 14–15b, we have

$$+\uparrow \Sigma F_y = ma_y; \qquad 2T - 75 = -\frac{75}{32.2}a_C \qquad (1)$$

The acceleration of the crate can be obtained by using kinematics to relate the motion of the crate to the known motion of a point P located on the cable, Fig. 14–15a. Hence, by the methods of Sec. 12.8, the coordinates s_C and s_P in Fig. 14–15a can be related to a constant portion of cable length l which is changing in the vertical and horizontal directions. We have $2s_C + s_P = l$. Taking the second time derivative of this equation yields

$$2a_C = -a_P \qquad (2)$$

From the problem data, $a_P = +4$ ft/s^2; thus, $a_C = (-4 \text{ ft/s}^2)/2 = -2 \text{ ft/s}^2$. Substituting this result into Eq. (1) and *retaining* the negative sign since the acceleration in *both* Eqs. (1) and (2) is considered positive downward, we have

$$2T - 75 = -\frac{75}{32.2}(-2)$$

$$T = 39.8 \text{ lb}$$

The power output, measured in units of horsepower, required to draw the cable in at a rate of 2 ft/s is therefore

$$P = \mathbf{T} \cdot \mathbf{v} = (39.8 \text{ lb})(2 \text{ ft/s})(1 \text{ hp}/550 \text{ lb} \cdot \text{ft/s})$$

$$= 0.145 \text{ hp}$$

This *power output* requires that the motor provide a *power input* of

$$\text{power input} = \frac{1}{\varepsilon}(\text{power output})$$

$$= \frac{1}{0.85}(0.145) = 0.170 \text{ hp} \qquad \textit{Ans.}$$

Since the velocity of the crate is constantly changing, notice that this power requirement is *instantaneous*.

(a)

(b)

Fig. 14–15

Problems

14–30. The diesel engine of a 400-Mg train increases the train's speed uniformly from rest to 10 m/s in 100 s along a horizontal track. Determine the average power developed.

14–31. To dramatize the loss of energy in an automobile, consider a car having a weight of 5,000 lb that is traveling at 35 mi/h. If the car is brought to a stop, determine how long a 100-W light bulb must burn to expend the same amount of energy. (1 mi = 5280 ft.)

*14–32. A man having a weight of 150 lb is able to run up a 12-ft-high flight of stairs in 4 s. Determine the power generated. How long would a 100-W light bulb have to burn to expend the same amount of energy? *Conclusion:* Turn the lights off when not in use!

Prob. 14–32

14–33. An automobile having a mass of 2 Mg travels up a 7° slope at a constant speed $v = 100$ km/h. If mechanical friction and wind resistance are neglected, determine the power developed by the engine if the automobile has an efficiency $\varepsilon = 0.65$.

14–34. The crate having a weight of 50 lb is hoisted by the pulley system and motor M. If the crate starts from rest and, by constant acceleration, attains a speed of 12 ft/s after rising $s = 10$ ft, determine the power that must be supplied to the motor at the instant $s = 10$ ft. The motor has an efficiency $\varepsilon = 0.76$. Neglect the mass of the pulley and cable.

Prob. 14–34

14–35. A spring having a stiffness of 5 kN/m is compressed 400 mm. The stored energy in the spring is used to drive a machine which requires 90 W of power. Determine how long the spring can supply energy at the required rate.

*14–36. The motor is used to hoist the loaded 500-kg elevator upward with a constant velocity $v_E = 8$ m/s. If the motor draws 60 kW of electrical power, determine the motor's efficiency. Neglect the mass of the pulleys and cable.

Prob. 14–36 / 14–37

14–37. The 500-kg elevator starts from rest and travels upward with a constant acceleration $a_c = 2$ m/s². Determine the power output of the motor M when $t = 3$ s. Neglect the mass of the pulleys and cable.

14–38. An electric streetcar has a weight of 15,000 lb and accelerates along a horizontal straight road from rest such that the power is always 100 hp. Determine how far it must travel to reach a speed of 40 ft/s.

14–39. The rocket sled has a mass of 4 Mg and travels from rest along the horizontal track for which $\mu_k = 0.20$. If the engine provides a constant thrust $T = 150$ kN, determine the power output of the engine as a function of time. Neglect the loss of fuel mass and air resistance.

Prob. 14–39

*14–40. The jeep has a weight of 2,500 lb and an engine which transmits a power of 100 hp to *all* the wheels. Assuming the wheels do not slip on the ground, determine the angle θ of the largest incline the jeep can climb at a constant speed $v = 30$ ft/s.

Prob. 14–40

14–41. The block and tackle arrangement for the oil rig is shown schematically in the figure. Determine the power output of the draw-work motor M necessary to lift the 600-lb drill pipe upward with a constant speed of 4 ft/s.

Prob. 14–41

157

14–42. The plate has a mass of 150 kg and rests on a surface for which $\mu = 0.5$. If a force $F = (60t^2)$ N, where t is in seconds, is applied to the cable, determine the power developed by the force when $t = 5$ s. *Hint:* First determine the time needed for the force to cause motion.

$\mu = 0.5$

Prob. 14–42

14–43. The 50-lb block rests on the rough surface for which $\mu = 0.2$. A force $F = (40 + s^2)$ lb, where s is in ft, acts on the block in the direction shown. If the spring is originally unstretched ($s = 0$) and the block is at rest, determine the power developed by the force when the block has moved $s = 1.5$ ft.

30°

$k = 20$ lb/ft

$\mu = 0.2$

Prob. 14–43

*14–44. The block has a weight of 80 lb and rests on the floor for which $\mu_k = 0.4$. If the motor draws in the cable at a constant rate of 6 ft/s, determine the output of the motor at the instant $\theta = 30°$. Neglect the mass of the cable and pulleys.

3 ft

θ

θ

3 ft

Prob. 14–44

14.5 Conservative Forces and Potential Energy

Conservative Force. A particularly simple type of force acting on a particle is one that depends *only* on the particle's position and is independent of the particle's velocity and acceleration. Furthermore, if the work done by this force in moving the particle from one point to another is *independent of the path* followed by the particle, this force is called a *conservative force*. The weight of a particle and the force of an elastic spring are two examples of conservative forces often encountered in mechanics.

Weight. The work done by the weight of a particle is independent of the path; rather, it depends only on the particle's *vertical displacement*. If this displacement is Δy (positive upward), then from Eq. 14–3,

$$U = -W(\Delta y)$$

Elastic Spring. The work done by a spring force *acting on a particle* is independent of the path of the particle, but depends only on the extension or compression s of the spring. If the spring is elongated or compressed from position s_1 to a further position s_2, then from Eq. 14–4,

$$U = -(\tfrac{1}{2}ks_2^2 - \tfrac{1}{2}ks_1^2)$$

Friction. In contrast to a conservative force, consider the force of friction exerted *on a moving object* by a fixed surface. The work done by the frictional force *depends upon the path*—the longer the path, the greater the work. Consequently, *frictional forces are nonconservative*. The work is dissipated from the body in the form of heat.

Potential Energy. When a conservative force acts on a particle, the force has the capacity to do work. This capacity, measured as *potential energy, V,* depends *only* upon the *location* of the particle when acted upon by the force.

Gravitational Potential Energy. If a particle is located a distance y *above* a datum, as shown in Fig. 14–16, the particle's weight has positive *gravitational potential energy, V_g,* since **W** has the capacity of doing positive work when the particle is moved back down to the datum. Likewise, if the particle is located a distance y *below* the datum, V_g is negative since the weight does negative work when the particle is moved back up to the datum. At the datum $V_g = 0$.

Gravitational Potential Energy

Fig. 14–16

Measuring y as *positive upward,* the gravitational potential energy of the particle of weight W is thus*

$$V_g = Wy \qquad\qquad (14\text{–}14)$$

*Here the weight is assumed to be *constant*. This assumption is suitable for small differences in elevation Δy. If the elevation change is significant, however, a variation of weight with elevation must be taken into account (see Prob. 14–52).

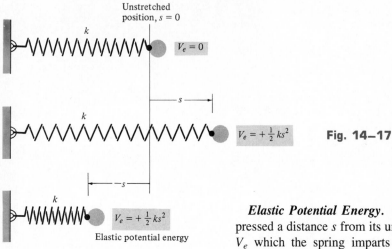

Unstretched
position, $s = 0$

$V_e = 0$

$V_e = +\frac{1}{2} ks^2$

Fig. 14–17

$V_e = +\frac{1}{2} ks^2$

Elastic potential energy

Elastic Potential Energy. When an elastic spring is elongated or compressed a distance s from its unstretched position, the elastic potential energy V_e which the spring imparts to an attached particle can be expressed as

$$V_e = +\tfrac{1}{2}ks^2 \tag{14–15}$$

Here V_e is *always positive* since, in the deformed position, the force of the spring has the *capacity* for doing positive work on the particle when the spring is returned to its unstretched position, Fig. 14–17.

Potential Function. In the general case, if a particle is subjected to both gravitational and elastic forces, the particle's potential energy can be expressed as a *potential function,* which is the algebraic sum

$$V = V_g + V_e \tag{14–16}$$

Measurement of V depends upon the location of the particle with respect to a selected datum in accordance with Eqs. 14–14 and 14–15.

In general, if a particle is located at an arbitrary point (x, y, z) in space, its potential energy or capacity to do work can be defined using a potential function $V = V(x, y, z)$. The work done by a conservative force in moving the particle from point (x_1, y_1, z_1) to point (x_2, y_2, z_2) is measured by the *difference* of this function; i.e.,

$$U_{1-2} = V_1 - V_2 \tag{14–17}$$

For example, the potential function for a block of weight W suspended from a spring can be expressed in terms of its position, s, measured from a datum located at the unstretched length of the spring, Fig. 14–18. We have

$$V = V_g + V_e$$
$$= -Ws + \tfrac{1}{2}ks^2$$

If the block moves from s_1 to a further downward position s_2, then applying Eq. 14–17 it can be seen that the work of $\mathbf{W}$ and $\mathbf{F}_s$ is

Datum

s_1

s

s_2

k F_s

W

Fig. 14–18

$$U_{1-2} = V_1 - V_2 = (-Ws_1 + \tfrac{1}{2}ks_1^2) - (-Ws_2 + \tfrac{1}{2}ks_2^2)$$
$$= W(s_2 - s_1) - (\tfrac{1}{2}ks_2^2 - \tfrac{1}{2}ks_1^2)$$

When the displacement path is infinitesimal, i.e., from point (x, y, z) to $(x + dx, y + dy, z + dz)$, Eq. 14–17 becomes

$$dU = V(x, y, z) - V(x + dx, y + dy, z + dz)$$
$$= -dV(x, y, z) \tag{14–18}$$

Provided both the force and displacement are defined using rectangular coordinates, then the work can also be expressed as

$$dU = \mathbf{F} \cdot d\mathbf{r} = (F_x\mathbf{i} + F_y\mathbf{j} + F_z\mathbf{k}) \cdot (dx\mathbf{i} + dy\mathbf{j} + dz\mathbf{k})$$
$$= F_x dx + F_y dy + F_z dz$$

Substituting this result into Eq. 14–18 and expressing the total derivative of $dV(x, y, z)$ in terms of its partial derivatives yields

$$F_x dx + F_y dy + F_z dz = -\left(\frac{\partial V}{\partial x}dx + \frac{\partial V}{\partial y}dy + \frac{\partial V}{\partial z}dz\right)$$

Since changes in x, y, and z are all independent of one another, this equation is satisfied provided

$$F_x = -\frac{\partial V}{\partial x}, \qquad F_y = -\frac{\partial V}{\partial y}, \qquad F_z = -\frac{\partial V}{\partial z} \tag{14–19}$$

Thus,

$$\mathbf{F} = -\frac{\partial V}{\partial x}\mathbf{i} - \frac{\partial V}{\partial y}\mathbf{j} - \frac{\partial V}{\partial z}\mathbf{k}$$
$$= -\left(\frac{\partial}{\partial x}\mathbf{i} + \frac{\partial}{\partial y}\mathbf{j} + \frac{\partial}{\partial z}\mathbf{k}\right)V$$

or

$$\boxed{\mathbf{F} = -\nabla V} \tag{14–20}$$

where ∇ (del) represents the vector operator $\nabla = (\partial/\partial x)\mathbf{i} + (\partial/\partial y)\mathbf{j} + (\partial/\partial z)\mathbf{k}$.

Equation 14–20 relates a force $\mathbf{F}$ to its potential function V and thereby provides a mathematical criterion for proving that $\mathbf{F}$ is conservative. For example, the gravitational potential function for a weight located a distance y above a datum is $V_g = Wy$, Eq. 14–14. To prove that $\mathbf{W}$ is conservative, it is necessary to show that it satisfies Eq. 14–20 (or Eq. 14–19), in which case

$$F_y = -\frac{\partial V}{\partial y}; \qquad\qquad F = -\frac{\partial}{\partial y}(Wy) = -W$$

The negative sign indicates that $\mathbf{W}$ acts downward, opposite to positive y which is upward.

161

14.6 Conservation of Energy

When a particle is acted upon by a system of *both* conservative and nonconservative forces, the portion of the work done by the *conservative forces* can be written in terms of the difference in their potential energies using Eq. 14–17, i.e., $(\Sigma U_{1-2})_{\text{cons.}} = V_1 - V_2$. As a result, the principle of work and energy can be written as

$$T_1 + V_1 + (\Sigma U_{1-2})_{\text{noncons.}} = T_2 + V_2 \qquad (14\text{–}21)$$

Here $(\Sigma U_{1-2})_{\text{noncons.}}$ represents the work of the nonconservative forces acting on the particle. If *only conservative forces* are applied to the body, this term is zero and then we have

$$T_1 + V_1 = T_2 + V_2 \qquad (14\text{–}22)$$

Potential Energy (max)
Kinetic Energy (zero)

Potential Energy and
Kinetic Energy

Potential Energy (zero)
Kinetic Energy (max)

Fig. 14–19

This equation is referred to as the *conservation of mechanical energy* or simply the *conservation of energy*. It states that during the motion the sum of the particle's kinetic and potential energies remains *constant*. For this to occur, kinetic energy must be transformed into potential energy, and vice versa. For example, if a ball of weight **W** is dropped from a height h above the ground (datum), Fig. 14–19, the potential energy of the ball is maximum before it is dropped, at which time its kinetic energy is zero. The total mechanical energy of the ball in its initial position is thus

$$E = T_1 + V_1 = 0 + Wh = Wh$$

When the ball has fallen a distance $h/2$, its speed can be determined by using $v^2 = v_0^2 + 2a_c(y - y_0)$, which yields $v = \sqrt{2g(h/2)} = \sqrt{gh}$. The energy of the ball at the mid-height position is therefore

$$E = V_2 + T_2 = W\frac{h}{2} + \frac{1}{2}\frac{W}{g}(\sqrt{gh})^2 = Wh$$

Just before the ball strikes the ground, its potential energy is zero and its speed is $v = \sqrt{2gh}$. Here, again, the total energy of the ball is

$$E = V_3 + T_3 = 0 + \frac{1}{2}\frac{W}{g}(\sqrt{2gh})^2 = Wh$$

Note that when the ball comes in contact with the ground, it deforms somewhat, and provided the ground is hard enough, the ball will rebound off the surface, reaching a new height h', which will be less than the height h from which it was first released. The difference in height accounts for an energy loss, $E_l = W(h - h')$, occurring at the moment of collision. Portions of this loss produce noise, deformation of the ball and ground, vibrations, and heat.

System of Particles. If a system of particles is *subjected only to conservative forces*, then an equation similar to Eq. 14–22 can be written for the particles. Applying the ideas of the preceding discussion, Eq. 14–8 ($\Sigma T_1 + \Sigma U_{1-2} = \Sigma T_2$) becomes

$$\Sigma T_1 + \Sigma V_1 = \Sigma T_2 + \Sigma V_2 \qquad (14\text{–}23)$$

Here, the sum of the system's initial kinetic and potential energies is equal to the sum of the system's final kinetic and potential energies. In other words, the system's kinetic energy and potential energy, caused by *both* the internal and external forces, remain constant; i.e., $\Sigma T + \Sigma V = $ const.

PROCEDURE FOR ANALYSIS

The conservation of energy equation is used to solve problems involving *velocity, displacement,* and *conservative force systems*. It is generally *easier to apply* than the principle of work and energy. This is because the energy equation just requires specifying the particle's kinetic and potential energies at only *two points* along the path, rather than determining the work when the particle moves through a *displacement*. For application it is suggested that the following procedure be used.

Potential Energy.   Draw two diagrams showing the particle located at its initial and final points along the path. If the particle is subjected to a vertical displacement, establish the fixed horizontal datum from which to measure the particle's gravitational potential energy V_g. Although this position can be selected arbitrarily, it is best to locate the datum either at the initial or final point of the path, since at the datum $V_g = 0$. Data pertaining to the elevation y of the particle from the datum and the extension or compression s of any connecting springs can be determined from the geometry associated with the two diagrams. Recall that the potential energy $V = V_g + V_e$, where $V_g = \pm Wy$ and $V_e = +\frac{1}{2}ks^2$.

Conservation of Energy.   Apply the equation $T_1 + V_1 = T_2 + V_2$. When computing the kinetic energy, $T = \frac{1}{2}mv^2$, the particle's speed v must be measured from an inertial reference frame.

It is important to remember that only problems involving conservative force systems may be solved by using the conservation of energy theorem. As stated previously, friction or other drag-resistant forces, which depend upon velocity or acceleration, are nonconservative. A portion of the work done by such forces is transformed into thermal energy used to heat up the surfaces of contact, and consequently this energy dissipates into the surroundings and may not be recovered. Therefore, problems involving frictional forces can either be solved by using the principle of work and energy written in the form of Eq. 14–21, if it applies, or the equation of motion. The following example problems numerically illustrate application of the procedure described above.

Example 14–8

The boy and his bicycle shown in Fig. 14–20a have a total weight of 125 lb and a center of mass at G. If he is coasting, i.e., not pedaling, with a speed of 10 ft/s at the top of the hill A, determine the normal force exerted on the wheels of the bicycle when he arrives at B, where the radius of curvature of the road is $\rho = 50$ ft. Neglect friction.

(a)

(b)

Solution

Since the normal force does *no work*, it must be obtained using the equation of motion, $\Sigma F_n = m(v^2/\rho)$. We can, however, determine the bicycle's speed at B using the conservation of energy equation.

Fig. 14–20

Potential Energy. Figure 14–20b shows the bicycle at points A and B. For convenience, the potential-energy datum has been established through the center of mass when the bicycle is located at B.

Conservation of Energy

$$\{T_A\} + \{V_A\} = \{T_B\} + \{V_B\}$$

$$\left\{\frac{1}{2}\left(\frac{125}{32.2}\right)(10)^2\right\} + \{(125)(30)\} = \left\{\frac{1}{2}\left(\frac{125}{32.2}\right)(v_B)^2\right\} + \{0\}$$

$$v_B = 45.1 \text{ ft/s}$$

Equation of Motion. Using the data tabulated on the free-body diagram when the bicycle is at B, Fig. 14–20c, we have

$$+\uparrow \Sigma F_n = ma_n; \qquad N_B - 125 = \frac{125}{32.2}\frac{(45.1)^2}{50}$$

$$N_B = 283 \text{ lb} \qquad\qquad \textit{Ans.}$$

(c)

Example 14–9

The ram R shown in Fig. 14–21a has a mass of 100 kg and is released from rest 0.75 m from the top of a spring A that has a stiffness $k_A =$ 12 kN/m. If a second spring B, having a stiffness $k_B = 15$ kN/m, is "nested" in A, determine the maximum deflection of A needed to stop the downward motion of the ram. The unstretched length of each spring is indicated in the figure. Neglect the mass of the springs.

Solution

Potential Energy. We will *assume* that the ram compresses *both* springs at the instant it comes to rest. The datum is located through the center of gravity of the ram at its initial position, Fig. 14–21b. When the kinetic energy is reduced to zero ($v_2 = 0$), A is compressed a distance s_A so that B compresses $s_B = s_A - 0.1$ m.

Conservation of Energy.

$$T_1 + V_1 = T_3 + V_3$$

$$\{0\} + \{0\} = \{0\} + \{\tfrac{1}{2}k_A s_A^2 + \tfrac{1}{2}k_B(s_A - 0.1)^2 - Wh\}$$

$$\{0\} + \{0\} = \{0\} + \left\{\frac{1}{2}(12000)s_A^2 + \frac{1}{2}(15000)(s_A - 0.1)^2 - 981(0.75 + s_A)\right\}$$

Rearranging the terms,

$$13500 s_A^2 - 2481 s_A - 660.75 = 0$$

Using the quadratic formula, and solving for the positive root,* we have

$$s_A = 0.331 \text{ m} \qquad\qquad \textit{Ans.}$$

Since $s_B = 0.331 - 0.1 = 0.231$ m, which is positive, indeed the assumption that *both* springs are compressed by the ram is correct.

---

*The second root, $s_A = -0.148$ m, does not represent the physical situation. Since positive s is measured downward, the negative sign indicates that spring A would have to be "extended" by an amount of 0.148 m to stop the ram.

(a)

(b)

Fig. 14–21

Example 14–10

A smooth 2-kg collar C, shown in Fig. 14–22a, fits loosely on the vertical shaft. If the spring is unstretched when the collar is in the dashed position A, determine the speed at which the collar is moving when $y = 1$ m, if (a) it is released from rest at A, and (b) it is released at A with an *upward* velocity $v_A = 2$ m/s.

(a)

Solution

Part (a)

Potential Energy. For convenience, the datum is established through AB, Fig. 14–22b. When the collar is at C, the gravitational potential energy is $-(mg)y$, since the collar is *below* the datum, and the elastic potential energy is $\frac{1}{2}ks_{CB}^2$. Here $s_{CB} = 0.5$ m, which represents the *stretch* in the spring as computed in the figure.

Conservation of Energy.

$$\{T_A\} + \{V_A\} = \{T_C\} + \{V_C\}$$

$$\{0\} + \{0\} = \{\tfrac{1}{2}mv_C^2\} + \{\tfrac{1}{2}ks_{CB}^2 - mgy\}$$

$$\{0\} + \{0\} = \left\{\frac{1}{2}(2)v_C^2\right\} + \left\{\frac{1}{2}(3)(0.5)^2 - 2(9.81)(1)\right\}$$

$$v_C = 4.39 \text{ m/s} \qquad \textit{Ans.}$$

This problem can also be solved by using the equation of motion or the principle of work and energy. Note that in *both* of these methods the variation of the magnitude and direction of the spring force must be taken into account (see Example 13–4). Here, however, the above method of solution is clearly advantageous since the calculations depend *only* on data calculated at the initial and final points of the path.

(b)

Fig. 14–22

Part (b)

Conservation of Energy. If $v_A = 2$ m/s, using the data in Fig. 14–22b, we have

$$\{T_A\} + \{V_A\} = \{T_C\} + \{V_C\}$$

$$\{\tfrac{1}{2}mv_A^2\} + \{0\} = \{\tfrac{1}{2}mv_C^2\} + \{\tfrac{1}{2}ks_{CB}^2 - mgy\}$$

$$\left\{\frac{1}{2}(2)(2)^2\right\} + \{0\} = \left\{\frac{1}{2}(2)v_C^2\right\} + \left\{\frac{1}{2}(3)(0.5)^2 - 2(9.81)(1)\right\}$$

$$v_C = 4.82 \text{ m/s} \qquad \textit{Ans.}$$

Note that the kinetic energy of the collar depends only on the *magnitude* of velocity, and therefore it is immaterial if the collar is moving up or down at 2 m/s when released at A.

Problems

14-45. Solve Prob. 14-9 using the conservation of energy equation.

14-46. Solve Prob. 14-19 using the conservation of energy equation.

14-47. Solve Prob. 14-26 using the conservation of energy equation.

*14-48. The collar has a weight of 8 lb. If it is pushed down so as to compress the spring 2 ft and then released from rest, determine its speed after it rises from this point; $h = 4.5$ ft. The spring is not attached to the collar. Neglect friction.

Prob. 14-48

14-49. Tarzan has a mass of 100 kg and from rest swings from the cliff by rigidly holding on to the 10-m-long tree vine, measured from the supporting limb A to his center of mass. Determine his speed just after the vine strikes the lower limb at B. Also, with what force must he hold on to the vine just before and just after the vine contacts the limb at B? *Hint:* Note that the radius of curvature of his path changes from 10 m to 3 m at this point.

Prob. 14-49

14-50. The roller-coaster car has a mass of 800 kg, including its passenger, and moves from the top of the hill A with a speed $v_A = 3$ m/s. Determine the minimum height h of the hill crest so that the car travels around both inside loops without leaving the track. Neglect friction, the mass of the wheels, and the size of the car. What is the normal reaction on the car when the car is at B and when it is at C?

Prob. 14-50

14-51. Four inelastic cables C are attached to a plate P and hold the 20-in. spring 6 in. in compression when *no force* acts on the plate. If a block B, having a weight of 7 lb, is placed on the plate and the plate is pushed down $y = 8$ in. and released from rest, determine how high the block rises from the point where it was released. Neglect the mass of the plate.

Prob. 14-51

*14-52. If the mass of the earth is M_e, show that the gravitational potential energy of a body of mass m located a distance r from the center of the earth is $V_g = -GM_em/r$. Recall that the gravitational force acting between the earth and the body is $F = G(M_em/r^2)$, Eq. 13-1. For the calculation, locate the datum at $r \rightarrow \infty$. Also, prove that **F** is a conservative force.

14–53. A 60-kg satellite is traveling in free flight along an elliptical orbit such that at A, where $r_A = 20$ Mm, it has a speed $v_A = 40$ Mm/h. What is the speed of the satellite when it reaches point B, where $r_B = 80$ Mm? *Hint:* See Prob. 14–52, where $M_e = 5.976(10^{24})$ kg and $G = 6.673(10^{-11})$ m³/(kg · s²).

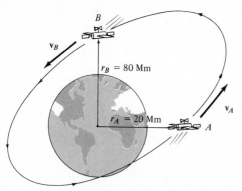

Prob. 14–53

14–54. Two equal-length springs having stiffnesses $k_A = 300$ N/m and $k_B = 200$ N/m are "nested" together in order to form a shock absorber. If a 2-kg block is dropped from an at-rest position 0.6 m above the top of the springs, determine their maximum deformation required to stop the motion of the block.

Prob. 14–54

14–55. The gantry structure is used to test the response of an airplane during a crash. The plane, having a mass of 8 Mg, is hoisted back until $\theta = 60°$, and then the pull-back cable AC is released when the plane is at rest. Determine the speed of the plane just before crashing into the ground, $\theta = 15°$. What is the maximum tension developed in the supporting cable during the motion? Neglect the effects of lift caused by the wings during the motion.

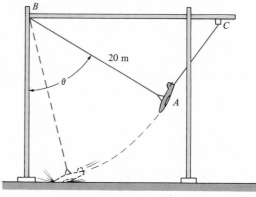

Prob. 14–55

*14–56. The ride at an amusement park consists of a gondola which is lifted to a height of 120 ft at A. If it is released from rest and falls along the parabolic track, determine its speed at the instant $y = 20$ ft. Also determine the normal reaction of the tracks on the gondola at this instant. The gondola and passenger have a total weight of 500 lb. Neglect the effects of friction and the mass of the wheels.

Prob. 14–56

14–57. If the roller-coaster car has a speed $v_A = 5$ ft/s when it is at A and coasts freely down the track, determine the speed v_B it attains when it reaches point B. Also determine the normal force a 150-lb passenger exerts on the car when it is at B. At this point the track follows a path defined by $y = x^2/200$. Neglect the effects of friction, the mass of the wheels, and the size of the car.

Prob. 14–57

14–58. The toboggan and passenger have a total mass of 60 kg. When the toboggan reaches point A it has a speed $v_A = 4$ m/s. Determine the angle θ at which it leaves the smooth circular curve and the distance s where it plunges in the snow. Neglect friction.

Prob. 14–58

14–59. The "slingshot" consists of a 0.2-kg funnel attached to two rubber bands each having a stiffness $k = 150$ N/m. The rubber bands are unstretched when $\theta = 0°$. If a balloon filled with 0.7 kg of water is placed in the funnel and the funnel is pulled back $d = 2$ m, and aimed at 45° from the horizontal, determine the launch velocity v_1 of the balloon when $\theta = 0°$. Also calculate the height h from the point where the balloon was launched to where it strikes the wall of the building. Neglect the change in elevation of the balloon while it is contained in the funnel and the mass of the rubber band.

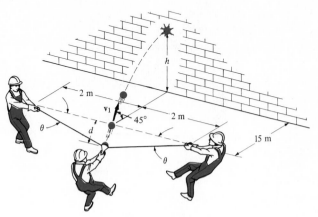

Prob. 14–59

***14–60.** The spring has a stiffness $k = 3$ lb/ft and an unstretched length of 2 ft. If it is attached to the 5-lb smooth collar and the collar is released from rest at A, determine the speed of the collar just before it strikes the end of the rod at B. Neglect the size of the collar.

Prob. 14–60

14–61. The man on the bicycle attempts to coast around the ellipsoidal loop without falling off the track. Determine the speed he must attain at A just before entering the loop in order to perform the stunt. The bicycle and man have a total mass of 85 kg and a center of mass at G. Neglect the mass of the wheels.

Prob. 14–61

14–62. The 7-kg cylinder is attached to a series of two springs as shown. If the cylinder is released from rest when the springs are both unstretched, determine the maximum stretch in each spring.

Prob. 14–62 / 14–63

14–63. The 7-kg cylinder is attached to a series of two springs as shown. If the cylinder is released from rest when the springs are both unstretched, determine the stretch in each spring at the instant the cylinder first attains a speed of 2 ft/s.

*14–64. The double-spring bumper is used to stop the 1500-lb steel billet in the rolling mill. Determine the maximum deflection of the plate A caused by the billet if it strikes the plate with a speed of 8 ft/s. Neglect the mass of the rollers and the plates A and B.

14–65. The block has a mass of 20 kg and is released from rest when $s = 0.5$ m. If the mass of the bumpers A and B is neglected, determine the maximum deflection of each spring due to the collision.

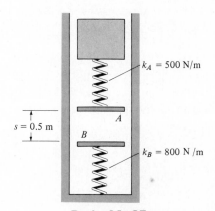

Prob. 14–65

Kinetics of a Particle: Impulse and Momentum

In this chapter we shall integrate the equation of motion and thereby obtain the principle of impulse and momentum. This principle is useful for solving problems involving force, velocity, and time. It will be shown that impulse and momentum principles also provide the necessary means for analyzing problems of impact, steady fluid flow, and systems which gain or lose mass.

Principle of Linear Impulse and Momentum   15.1

The equation of motion for a particle of mass m can be written as

$$\Sigma \mathbf{F} = m\mathbf{a} = m\frac{d\mathbf{v}}{dt} \tag{15–1}$$

where $\mathbf{a}$ and $\mathbf{v}$ are both measured from an inertial frame of reference. Rearranging the terms and integrating between the limits $\mathbf{v} = \mathbf{v}_1$ at $t = t_1$ and $\mathbf{v} = \mathbf{v}_2$ at $t = t_2$, we have

$$\Sigma \int_{t_1}^{t_2} \mathbf{F}\, dt = m \int_{\mathbf{v}_1}^{\mathbf{v}_2} d\mathbf{v}$$

or

$$\Sigma \int_{t_1}^{t_2} \mathbf{F}\, dt = m\mathbf{v}_2 - m\mathbf{v}_1 \tag{15–2}$$

This equation, which is referred to as the *principle of linear impulse and momentum*, provides a *direct means* of obtaining the particle's final velocity $\mathbf{v}_2$ after a specified time period when the particle's initial velocity is known

Variable Impulse

Fig. 15–1

Constant Impulse

Fig. 15–2

and the forces acting on the particle are either constant or can be expressed as functions of time. Notice from the derivation that if $\mathbf{v}_2$ is determined using the equation of motion, a two-step process is necessary; i.e., apply $\Sigma \mathbf{F} = m\mathbf{a}$ to obtain $\mathbf{a}$, then integrate $\mathbf{a} = d\mathbf{v}/dt$ to obtain $\mathbf{v}_2$.

Linear Impulse. The integral $\mathbf{I} = \int \mathbf{F}\, dt$ in Eq. 15–2 is defined as the *linear impulse*. This term is a vector quantity which measures the effect of a force during the time the force acts. The impulse vector acts in the same direction as the force, and its magnitude has units of force–time, e.g., $N \cdot s$ or $lb \cdot s$. If the force is expressed as a function of time, the impulse may be determined by direct evaluation of the integral. In particular, if $\mathbf{F}$ acts in a *constant direction* during the time period t_1 to t_2, the magnitude of the impulse $\mathbf{I} = \int_{t_1}^{t_2} \mathbf{F}\, dt$ can be represented experimentally by the shaded area under the curve of force versus time, Fig. 15–1. However, if the force is constant in magnitude and direction, the resulting impulse becomes $\mathbf{I} = \int_{t_1}^{t_2} \mathbf{F}_c\, dt = \mathbf{F}_c(t_2 - t_1)$, which represents the shaded rectangular area shown in Fig. 15–2.

Linear Momentum. Each of the two vectors of the form $\mathbf{L} = m\mathbf{v}$ in Eq. 15–2 is defined as the *linear momentum* of the particle. Since m is a scalar, the linear-momentum vector has the same direction as $\mathbf{v}$, and its magnitude mv has units of mass–velocity, e.g., $kg \cdot m/s$, $slug \cdot ft/s$.*

Principle of Linear Impulse and Momentum. For problem solving, Eq. 15–2 will be rewritten in the form

$$m\mathbf{v}_1 + \Sigma \int_{t_1}^{t_2} \mathbf{F}\, dt = m\mathbf{v}_2 \qquad (15\text{–}3)$$

which states that the initial momentum of the particle at t_1 plus the vector sum of all the impulses applied to the particle during the time interval t_1 to t_2 is equivalent to the final momentum of the particle at t_2. These three terms are illustrated graphically on the *impulse and momentum diagrams* shown in Fig. 15–3. Each of these diagrams graphically accounts for all the vectors in the equation $m\mathbf{v}_1 + \Sigma \int_{t_1}^{t_2} \mathbf{F}\, dt = m\mathbf{v}_2$. The two *momentum diagrams* are simply outlined shapes of the particle which indicate the direction and magnitude of the particle's initial and final momenta, $m\mathbf{v}_1$ and $m\mathbf{v}_2$, respectively, Fig. 15–3. Similar to the free-body diagram, the *impulse diagram* is an outlined shape of the particle showing all the impulses that act on the particle when it is located at some intermediate point along its path. In general, whenever the magnitude or direction of a force *varies,* the impulse of the force is determined by integration and represented on the impulse diagram as $\mathbf{I} = \int_{t_1}^{t_2} \mathbf{F}\, dt$. If the force is *constant* for the time interval $(t_2 - t_1)$, the impulse applied to the particle is $\mathbf{I} = \mathbf{F}_c(t_2 - t_1)$, acting in the same direction as $\mathbf{F}_c$.

Initial momentum diagram

Impulse diagram

Final momentum diagram

Fig. 15–3

*Although the units for impulse and momentum are defined differently, show that Eq. 15–2 is dimensionally homogeneous.

Scalar Equations. If each of the vectors in Eq. 15–3 or on Fig. 15–3 is resolved into its x, y, and z components, we can write symbolically the following three scalar equations:

$$m(v_x)_1 + \Sigma \int_{t_1}^{t_2} F_x \, dt = m(v_x)_2$$

$$m(v_y)_1 + \Sigma \int_{t_1}^{t_2} F_y \, dt = m(v_y)_2 \qquad (15\text{–}4)$$

$$m(v_z)_1 + \Sigma \int_{t_1}^{t_2} F_z \, dt = m(v_z)_2$$

These equations represent the principle of linear impulse and momentum for the particle in the x, y, and z directions, respectively.

PROCEDURE FOR ANALYSIS

The principle of linear impulse and momentum is used to solve problems involving *force, time,* and *velocity,* since these terms are involved in the formulation. For application it is suggested that the following procedure be used.

Free-Body Diagram. Establish the x, y, z inertial frame of reference and draw the particle's free-body diagram in order to account for all the forces that produce impulses on the particle. The direction and sense of the particle's initial and final velocities should also be established. If any of these are unknown, assume that the sense of its components is in the direction of the positive inertial coordinate(s). The velocities may be sketched near, *but not on,* the free-body diagram.*

As an alternative procedure draw the impulse and momentum diagrams for the particle as discussed above regarding Fig. 15–3.†

Principle of Impulse and Momentum. Apply the principle of linear impulse and momentum, $m\mathbf{v}_1 + \Sigma \int_{t_1}^{t_2} \mathbf{F} \, dt = m\mathbf{v}_2$. If motion occurs in the x-y plane, the two scalar component equations can be formulated by either resolving the vector components of $\mathbf{F}$ from the free-body diagram, or by using the data on the impulse and momentum diagrams.

If the problem involves the dependent motion of several particles, use the method outlined in Sec. 12.8 to relate their velocities. Make sure the positive coordinate directions used for writing these kinematic equations are the *same* as those used for writing the equations of impulse and momentum.

The following examples numerically illustrate application of this procedure.

*We will use this convention in the solution of the example problems.
†This procedure will be followed when developing the proofs and theory in the text.

Example 15–1

The 100-kg crate shown in Fig. 15–4a is originally at rest on the smooth horizontal surface. If a force of 200 N, acting at an angle of 45°, is applied to the crate for 10 s, determine the final velocity of the crate and the normal force which the surface exerts on the crate during the time interval.

200 N

45°

(a)

200 N

45°

v_2

981 N

x

y

N_C

Fig. 15–4a, b

(b)

Solution

Free-Body Diagram. This diagram is shown in Fig. 15–4b. Here it has been assumed that during the motion the crate remains on the surface and after 10 s the crate moves to the left with a velocity v_2. Since all the forces acting on the crate are *constant*, the respective impulses are simply the product of the force magnitude and 10 s ($\mathbf{I} = \mathbf{F}_c(t_2 - t_1)$).

Principle of Impulse and Momentum. Resolving the vectors in Fig. 15–4b along the x,y axes, and applying Eqs. 15–4, yields

$(\xrightarrow{+})$
$$m(v_x)_1 + \Sigma \int_{t_1}^{t_2} F_x \, dt = m(v_x)_2$$

$$0 - 200(10) \cos 45° = -100v_2$$

$$v_2 = 14.1 \quad \text{m/s} \leftarrow \qquad\qquad Ans.$$

$(+\uparrow)$
$$m(v_y)_1 + \Sigma \int_{t_1}^{t_2} F_y \, dt = m(v_y)_2$$

$$0 + N_C(10) - 981(10) + 200(10) \sin 45° = 0$$

$$N_C = 840 \text{ N} \uparrow \qquad\qquad Ans.$$

Since no motion occurs in the y direction, direct application of $\Sigma F_y = 0$ gives the same result for N_C.

Note the alternative procedure of drawing the crate's impulse and momentum diagrams, Fig. 15–4c, prior to applying Eqs. 15–4.

y

x

$+$

200 (10)

45°

y

981 (10)

x

$N_C(10)$

$=$

y

x

$100v_2$

Fig. 15–4c

Example 15–2

The crate shown in Fig. 15–5a has a weight of 50 lb and is acted upon by a force having a variable magnitude $P = (20t)$ lb, where t is in seconds. Compute the crate's velocity 2 s after **P** has been applied. The crate has an initial velocity $v_1 = 3$ ft/s down the plane, and the coefficient of kinetic friction between the crate and the plane is $\mu_k = 0.3$.

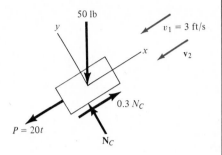

Fig. 15–5

Solution

Free-Body Diagram (Fig. 15–5b). Since the magnitude of force $P = 20t$ *varies* with time, the impulse it creates must be determined by *integrating* over the 2-s time interval. The weight, normal force, and frictional force (which acts opposite to the direction of motion) are all *constant,* so that the impulse created by each of these forces is simply the magnitude of the force times 2 s.

Principle of Impulse and Momentum. Applying Eqs. 15–4 in the x direction, we have

$$(+\nearrow) \qquad m(v_x)_1 + \Sigma \int_{t_1}^{t_2} F_x \, dt = m(v_x)_2$$

$$-\frac{50}{32.2}(3) - \int_0^2 20t \, dt + 0.3 N_C(2) - (50)(2) \sin 30° = -\frac{50}{32.2} v_2$$

$$-4.66 - 40 + 0.6 N_C - 50 = -1.55 v_2$$

The equation of equilibrium can be applied in the y direction. Why?

$$+\nwarrow \Sigma F_y = 0; \qquad N_C - 50 \cos 30° = 0$$

Solving,

$$N_C = 43.3 \text{ lb} \nwarrow \qquad v_2 = 44.2 \text{ ft/s} \swarrow \qquad \textit{Ans.}$$

This problem has also been solved using the equation of motion in Example 13–3. The two methods of solution should be compared. Since *force, velocity,* and *time* are involved in the problem, application of the principle of impulse and momentum eliminates the need for using kinematics ($a = dv/dt$) and thereby yields an easier method for solution.

Example 15–3

Blocks A and B shown in Fig. 15–6a have a mass of 3 kg and 5 kg, respectively. If the system is released from rest, determine the velocity of block B in 6 s. Neglect the mass of the pulleys and cord.

Solution
Free-Body Diagram (Fig. 15–6b). Since the impulses of the blocks' weights are constant, the impulses of the cord tensions will also be constant. Furthermore, since the mass of pulley D is neglected, the cord tension $T_A = 2T_B$, Fig. 15–6c.

Principle of Impulse and Momentum
Block A:

$(+\uparrow)$
$$m(v_A)_1 + \Sigma \int_{t_1}^{t_2} F_y \, dt = m(v_A)_2$$
$$0 + 2T_B(6) - 3(9.81)(6) = -3(v_A)_2 \qquad (1)$$

Block B:

$(+\downarrow)$
$$m(v_B)_1 + \Sigma \int_{t_1}^{t_2} F_y \, dt = m(v_B)_2$$
$$0 + 5(9.81)(6) - T_B(6) = 5(v_B)_2 \qquad (2)$$

Kinematics. Since the blocks are subjected to dependent motion, the velocity of A may be related to that of B by using the kinematic analysis discussed in Sec. 12.8. A horizontal datum is established through the fixed point at C, Fig. 15–6a, and the changing positions of the blocks, s_A and s_B, are related to the constant total length l of the vertical segments of the cord by the equation

$$2s_A + s_B = l$$

Taking the time derivative yields

$$2v_A = -v_B \qquad (3)$$

As indicated by the negative sign, when B moves downward A moves upward.* Substituting this result into Eq. (1) and solving Eqs. (1) and (2) yields

$$(v_B)_2 = 35.8 \text{ m/s} \downarrow \qquad \qquad Ans.$$
$$T_B = 19.2 \text{ N}$$

*Note that the *positive* (downward) directions for $(\mathbf{v}_A)_2$ and $(\mathbf{v}_B)_2$ are *consistent* in Fig. 15–6a and b and in Eqs. (1) to (3). Why is this important?

Datum

(a)

(b)

(c)

Fig. 15–6

Problems

Except when stated otherwise, throughout this chapter assume that the coefficients of static and kinetic friction are equal, i.e., $\mu = \mu_s = \mu_k$.

15–1. A hammer head H having a weight of 0.25 lb is moving vertically downward at 40 ft/s when it strikes the head of a nail of negligible mass and drives it into a block of wood. Find the impulse on the nail if it is assumed that the grip at A is loose, the handle has a negligible mass, and the hammer stays in contact with the nail while it comes to rest. Neglect the impulse caused by the weight of the hammer head during contact with the nail.

Prob. 15–1

15–2. A cannon ball having a weight of 20 lb is fired upward, in the vertical direction, with a muzzle velocity of 1,200 ft/s. Determine how long it takes before its velocity is reduced to zero, which occurs when it reaches its maximum height. Use the principle of impulse and momentum.

15–3. When the 5-kg block is 6 m from the wall, it is sliding at $v_1 = 14$ m/s. If the coefficient of friction between the block and the horizontal plane is $\mu = 0.3$, determine the impulse of the wall on the block necessary to stop the block. Neglect the friction impulse acting on the block during the collision.

Prob. 15–3

*15–4. The uniform beam has a weight of 5,000 lb. Determine the average tension in each of the two cables AB and AC if the beam is given an upward speed of 8 ft/s in 1.5 s. Originally the beam is at rest.

Prob. 15–4

15–5. A man hits the 50-g golf ball such that it leaves the tee at an angle of 40° with the horizontal and strikes the ground at the same elevation a distance of 20 m away. Determine the impulse of the club C on the ball. Neglect the impulse caused by the ball's weight while the club is striking the ball.

Prob. 15–5

15-6. A car is traveling at 14 m/s when it is 5 m away from a wall. The driver applies the brakes to all four wheels at this point and skids toward the wall. If the crash into the wall takes 0.2 s, determine the average net impulsive force of the wall on the car. The car has a mass of 1.5 Mg, and the coefficient of friction between the road and the tires is $\mu = 0.6$. For the solution *include* the friction impulse acting on the car during the collision.

Prob. 15-6

15-7. From experiments, the time variation of the vertical force on a runner's foot as he strikes and pushes off the ground is shown in the graph. These results are reported for a 1-lb *static* load, i.e., in terms of unit weight. If a runner weighs 175 lb, determine the approximate vertical impulse he exerts on the ground if the impulse occurs in 210 ms.

Prob. 15-7

*15-8. The 12-Mg "jump jet" is capable of taking off vertically from the deck of a ship. If its jets exert a constant vertical force of 150 kN on the plane, determine how high it goes in $t = 6$ s, starting from rest. Neglect the loss of fuel during the flight.

Prob. 15-8

15-9. A train consists of a 50-Mg engine and three cars, each having a mass of 30 Mg. If it takes 80 s for the train to increase its speed uniformly to 40 km/h, starting from rest, determine the force **T** developed at the coupling between the engine E and the first car A. The wheels of the engine provide the resultant frictional tractive force **F** which gives the train forward motion, whereas the car wheels roll freely. Determine **F** acting on the engine wheels. Neglect the mass of the wheels.

Prob. 15-9

15–10. A solid-fueled rocket can be made using a fuel grain with either a hole (a), or starred cavity (b), in the cross section. From experiment the engine thrust-time curves (T vs. t) for the same amount of propellent using these geometries are shown. Determine the total impulse in both cases. Realizing that the mass of the fuel is expelled as it is burned, explain which geometry for the propellent is best for obtaining the maximum range when used in a rocket and why.

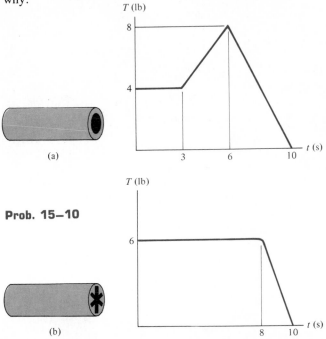

(a)

(b)

Prob. 15–10

15–11. The baseball has a horizontal speed of 35 m/s when it is struck by the bat B. If it then travels away at an angle of 60° from the horizontal and reaches a maximum altitude of 50 m, measured from the height of the bat, determine the magnitude of the net impulse of the bat on the ball. The ball has a mass of 400 g. Neglect the weight of the ball during the time the bat strikes the ball.

Prob. 15–11

***15–12.** Packages having a mass of 6 kg slide down a smooth chute and land horizontally with a speed of 3 m/s on the surface of a conveyor belt. If the coefficient of friction between the belt and a package is $\mu = 0.15$, determine the time needed to bring the package to rest on the belt if the belt is moving in the same direction as the package with a speed $v = 1$ m/s.

Prob. 15–12

15–13. The jet plane has a mass of 250 Mg and a horizontal velocity of 100 m/s when $t = 0$. If *both* engines provide a horizontal thrust which varies as shown in the graph, determine the plane's velocity in $t = 5$ s. Neglect air resistance and the loss of fuel during the motion.

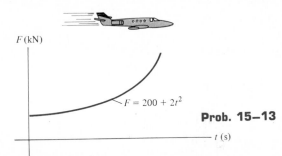

$F = 200 + 2t^2$

Prob. 15–13

15–14. The tanker has a mass of 130 Gg. If it is originally at rest, determine its speed when $t = 10$ s, if the horizontal thrust provided by its propeler varies with time as shown in the graph. Neglect the effect of water resistance.

$F = 3\,(1 - e^{-0.1t})$

Prob. 15–14

15–15. A tankcar has a mass of 20 Mg and is freely rolling to the right with a speed of 0.75 m/s. If it strikes the barrier, determine the horizontal impulse needed to stop the car if the spring in the bumper B has a stiffness (a) $k \to \infty$ (bumper is rigid), and (b) $k = 15$ kN/m.

$v = 0.75$ m/s

k

B

Prob. 15–15

*15–16.** Blocks A and B have masses $m_A = 10$ kg and $m_B = 3$ kg, respectively. If block A is moving down the inclined plane with an initial velocity $(v_A)_1 = 2$ m/s, determine the velocity of B when $t = 2$ s. Assume that the inclined plane is smooth. Neglect the mass of the pulleys and cord.

$(v_A)_1 = 2$ m/s

A

$30°$

B

Prob. 15–16 / 15–17

15–17. Solve Prob. 15–16 if the coefficient of friction between the incline and block A is $\mu_A = 0.15$.

15–18. The force acting on a projectile having a mass m as it passes horizontally through the barrel of the cannon is $F = C \sin(\pi t/t')$. Determine the projectile's velocity when $t = t'$. If the projectile reaches the end of the barrel at this instant, determine the length s.

s

Prob. 15–18

15–19. The log has a mass of 500 kg and rests on the ground for which the coefficients of static and kinetic friction are $\mu_s = 0.5$ and $\mu_k = 0.4$, respectively. The winch delivers a horizontal towing force **T** to its cable at A which varies as shown in the graph. Determine the speed of the log when $t = 5$ s. Originally the cable tension is zero. *Hint:* First determine the force needed to begin moving the log.

T (N)

1800

$T = 200 \ t^2$

3

t (s)

A T

Prob. 15–19 / 15–20

*15–20.** The log has a mass of 500 kg and rests on the ground for which the coefficients of static and kinetic friction are $\mu_s = 0.5$ and $\mu_k = 0.4$, respectively. The winch delivers a towing force **T** to its cable at A which varies as shown in the graph. How much time is required to give the log a speed of 10 m/s? Originally the cable tension is zero. *Hint:* First determine the force needed to begin moving the log.

15–21. The motor pulls on the cable at A with a force $F = (30 + t^2)$ lb, where t is in seconds. If the 17-lb crate is originally at rest at $t = 0$, determine its speed in $t = 4$ s. Neglect the mass of the cable and pulleys. *Hint:* First find the time needed to begin lifting the crate.

Prob. 15–21

15–22. The 50-kg block is hoisted up the incline using the cable and motor arrangement shown. The coefficient of friction between the block and the surface is $\mu = 0.4$. If the block is moving up the plane at $v_A = 2$ m/s when it is at A, and at this instant ($t = 0$) the motor develops a tension in the cord of $T = (300 + 120\sqrt{t})$ N, determine the velocity of the block when $t = 2$ s.

Prob. 15–22

15–23. A cake and plate weighing 2 lb rest at the center of a circular table. Without touching the cake a boy attempts to remove the 2-ft-radius tablecloth by quickly pulling on it horizontally. If slipping between the cake plate and tablecloth occurs at all times, determine the longest time the tablecloth can remain in contact with the plate without having the cake wind up on the floor. The coefficient of friction between the cake plate and the tablecloth is $\mu = 0.3$ and between the cake plate and table surface $\mu' = 0.4$. *Hint:* Apply the principles of impulse and momentum and work and energy to the cake for x ft, the distance the cake plat remains on the tablecloth, and for $(2 - x)$ ft when it remain on the table.

Prob. 15–23

15.2 Principle of Linear Impulse and Momentum for a System of Particles

The principle of linear impulse and momentum for a system of particles moving relative to an inertial reference, Fig. 15–7, may be obtained from the equation of motion $\Sigma \mathbf{F}_i = \Sigma m_i \mathbf{a}_i$, Eq. 13–5, which may be rewritten as

$$\Sigma \mathbf{F}_i = \Sigma m_i \frac{d\mathbf{v}_i}{dt} \tag{15–5}$$

The term on the left side represents only the sum of all the *external forces* acting on the system of particles. Recall that the internal forces between particles do not appear with this summation, since by Newton's third law they occur in equal but opposite collinear pairs and therefore cancel out. Multiplying both sides of Eq. 15–5 by dt and integrating between the limits $t = t_1$, $\mathbf{v}_i = (\mathbf{v}_i)_1$ and $t = t_2$, $\mathbf{v}_i = (\mathbf{v}_i)_2$ yields

$$\Sigma m_i(\mathbf{v}_i)_1 + \Sigma \int_{t_1}^{t_2} \mathbf{F}_i \, dt = \Sigma m_i(\mathbf{v}_i)_2 \tag{15–6}$$

This equation states that the initial linear momenta of the system added vectorially to the impulses of all the *external forces* acting on the system during the time period t_1 to t_2 are equal to the system's final linear momenta.

By definition, the location of the mass center G of the system is determined from $m\mathbf{r}_G = \Sigma m_i \mathbf{r}_i$, where $m = \Sigma m_i$ is the total mass of all the particles, and $\mathbf{r}_G$ and $\mathbf{r}_i$ are defined in Fig. 15–7. Taking the time derivatives, we have

$$m\mathbf{v}_G = \Sigma m_i \mathbf{v}_i$$

which states that the total linear momentum of the system of particles is equivalent to the linear momentum of a "fictitious" aggregate particle of

Inertial coordinate system

Fig. 15–7

mass $m = \Sigma m_i$ moving with the velocity of the mass center G of the system. Substituting into Eq. 15–6 yields

$$m(\mathbf{v}_G)_1 + \Sigma \int_{t_1}^{t_2} \mathbf{F}\, dt = m(\mathbf{v}_G)_2 \qquad (15\text{–}7)$$

This equation states that the initial linear momentum of the aggregate particle plus (vectorially) the external impulses acting on the system of particles during the time interval t_1 to t_2 is equal to the aggregate particle's final linear momentum. Since in reality all particles must have finite size to possess mass, the above equation justifies application of the principle of linear impulse and momentum to a rigid body represented as a single particle.

Conservation of Linear Momentum for a System of Particles 15.3

When the sum of the *external impulses* acting on a system of particles is *zero*, Eq. 15–6 reduces to a simplified form, namely,

$$\Sigma m_i(\mathbf{v}_i)_1 = \Sigma m_i(\mathbf{v}_i)_2 \qquad (15\text{–}8)$$

This equation is referred to as the *conservation of linear momentum*. It states that the linear momenta for a system of particles remain constant throughout the time period t_1 to t_2. Since $m\mathbf{v}_G = \Sigma m_i \mathbf{v}_i$, we can also write

$$(\mathbf{v}_G)_1 = (\mathbf{v}_G)_2 \qquad (15\text{–}9)$$

which indicates that the velocity $\mathbf{v}_G$ of the mass center for the system of particles does not change when no external impulses are applied to the system.

A careful study of the free-body diagram for the entire system of particles should be made before applying the above equations. By doing this, one will be able to identify the forces as creating either external or internal impulses and thereby determine in what direction(s) linear momentum is conserved. As stated earlier, for the system the *internal impulses* will always cancel out since they occur in equal but opposite collinear pairs. If the time period over which the motion is studied is *very short,* some of the external impulses may also be neglected or considered approximately equal to zero. The forces causing these negligible impulses are called *nonimpulsive forces*. By comparison, forces which are very large, act for a very short period of time, and yet produce a significant change in momentum are called *impulsive forces*. They, of course, cannot be neglected in the impulse-momentum analysis.

Impulsive forces normally occur due to an explosion or the striking of one body against another, whereas nonimpulsive forces may include the weight of a body, the force imparted by a slightly deformed spring having a relatively small stiffness, or for that matter, any force that is small compared to other larger (impulsive) forces.

PROCEDURE FOR ANALYSIS

Generally, the principle of linear impulse and momentum or the conservation of linear momentum is applied to a *system of particles* in order to determine the final velocities of the particles *just after* the time period considered. By applying these equations to the entire system, the internal impulses acting within the system, which may be unknown, are *eliminated* from the analysis. For application it is suggested that the following procedure be used.

Free-Body Diagram. Draw the free-body diagram for the system of particles in order to identify the internal and external forces. The conservation of linear momentum applies to the system in a given direction when *no external impulsive forces* act on the system in that direction. Also, establish the direction and sense of the particle's initial and final velocities.

As an alternative procedure, draw the impulse and momentum diagrams for the system of particles just before, during, and just after the impulsive forces are applied. Then investigate the impulse diagram in order to clearly distinguish the external impulsive and nonimpulsive forces from the system's internal impulses.

Momentum Equations. Apply the principle of linear impulse and momentum or the conservation of linear momentum in the appropriate directions. If the particles are subjected to dependent motion using cables and pulleys, kinematics as discussed in Sec. 12.8 can be used to relate the velocities.

The following examples numerically illustrate application of this procedure.

Example 15–4

The 15-Mg boxcar A is coasting freely at 1.5 m/s on the horizontal track when it encounters a tank car B having a mass of 12 Mg and coasting at 0.75 m/s toward it as shown in Fig. 15–8a. If the cars meet and couple together, determine (a) the speed of both cars just after the coupling, and (b) the average force between them if the coupling takes place in 0.8 s.

1.5 m/s   0.75 m/s

A B **Fig. 15–8**

(a)

Solution

Part (a)

Free-Body Diagram.* As shown in Fig. 15–8b, we have considered *both* cars as a single system. Hence the coupling force F is *internal* to the system and therefore will *not* be included in the momentum-impulse analysis. Also, it is assumed both cars, when coupled, move to the right at v_2. By inspection, momentum is conserved in the horizontal direction.

$(v_A)_1 = 1.5$ m/s $(v_B)_1 = 0.75$ m/s

(b)

$(v_A)_1 = 1.5$ m/s

(c)

Conservation of Linear Momentum

$(\xrightarrow{+})$
$$m_A(v_A)_1 + m_B(v_B)_1 = (m_A + m_B)v_2$$
$$15\ 000(1.5) - 12\ 000(0.75) = 27\ 000v_2$$
$$v_2 = 0.50 \text{ m/s} \qquad\qquad Ans.$$

Part (b). The average (impulsive) coupling force, F_{avg}, can be determined by applying the principle of linear momentum to *either one* of the cars.

Free-Body Diagram. As shown in Fig. 15–8c, by isolating the boxcar the coupling force is *external* to the car.

Principle of Impulse and Momentum. Since $\int F\,dt = F_{avg}\,\Delta t = F_{avg}(0.8)$, we have

$(\xrightarrow{+})$
$$m_A(v_A)_1 + \Sigma \int F\,dt = m_A v_2$$
$$15\ 000(1.5) - F_{avg}(0.8) = 15\ 000(0.50)$$
$$F_{avg} = 18.8 \text{ kN} \qquad\qquad Ans.$$

Solution was possible since the boxcar's final velocity was obtained in Part (a). Try solving for F_{avg} by applying the principle of impulse and momentum to the tank car.

*Only horizontal forces are shown on the free-body diagram.

Example 15–5

The 1200-lb cannon shown in Fig. 15–9a fires an 8-lb projectile with a muzzle velocity of 1500 ft/s. If firing takes place in 0.03 s, determine (a) the recoil velocity of the cannon just after firing, and (b) the average impulsive force acting on the projectile. The cannon support is firmly fixed to the ground and the horizontal recoil of the cannon is absorbed by two springs.

Recoil spring

(a)

Solution

Part (a)

Free-Body Diagram.* As shown in Fig. 15–9b, we have considered the projectile and cannon as a single system, since then the impulsive forces, **F**, between the cannon and projectile are *internal* to the system and are therefore not included in the momentum-impulse analysis of the system. Furthermore, during the time $\Delta t = 0.03$ s, the two recoil springs which are attached to the support each exert a *nonimpulsive force* $\mathbf{F}_s$ on the cannon. This is because Δt is very short, so that during this time the cannon only moves through a very small distance† s. Consequently, $F_s = ks \approx 0$. Hence it may be concluded that momentum for the system is conserved in the *horizontal direction*.

(b)

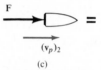

(c)

Fig. 15–9

Conservation of Linear Momentum.

$(\overset{+}{\rightarrow})$ $m_c(v_c)_1 + m_p(v_p)_1 = -m_c(v_c)_2 + m_p(v_p)_2$

$$0 + 0 = -\frac{1200}{32.2}(v_c)_2 + \frac{8}{32.2}(1500)$$

$$(v_c)_2 = 10 \text{ ft/s} \qquad \qquad Ans.$$

Part (b). The average impulsive force exerted by the cannon on the projectile can be determined by applying the principle of linear impulse and momentum to the projectile.

Principle of Impulse and Momentum. Using the data on the free-body diagram, Fig. 15–9c, noting that $\int F \, dt = F_{avg} \Delta t = F_{avg}(0.03)$, we have

$(\overset{+}{\rightarrow})$ $m(v_p)_1 + \Sigma \int F \, dt = m(v_p)_2$

$$0 + F_{avg}(0.03) = \frac{8}{32.2}(1500)$$

$$F_{avg} = 12.4(10^3) \text{ lb} = 12.4 \text{ kip} \qquad Ans.$$

Obtain this same answer by applying the principle of linear impulse and momentum to the cannon.

---

*Only horizontal forces are shown on the free-body diagram.

†If the cannon is firmly fixed to its support (no springs), the reactive force of the support on the cannon must be considered as an external impulse to the system, since the support would allow no movement of the cannon.

Example 15–6

The 350-Mg tugboat T shown in Fig. 15–10a is used to pull the 50-Mg barge B with a rope R. If the initial velocity of the tugboat is $(v_T)_1 = 3$ m/s while the rope is *slack*, determine the velocity of the tugboat *directly after* towing starts to occur. Assume the rope does not stretch. Neglect the frictional effects of the water.

(a)

(b)

Fig. 15–10a, b

Solution

Free-Body Diagram.* As shown in Fig. 15–10b, we have considered the entire system (tugboat and barge). Hence, the impulsive force created by the rope and the barge is *internal* to the system, and therefore momentum of the system is conserved during the instant of towing.

Conservation of Momentum. Noting that $(v_B)_2 = (v_T)_2$, we have

$(\xleftarrow{+})$ $\quad m_T(v_T)_1 + m_B(v_B)_1 = m_T(v_T)_2 + m_B(v_B)_2$

$\qquad 350(10^3)(3) + 0 = 350(10^3)(v_T)_2 + 50(10^3)(v_T)_2$

Solving,

$$(v_T)_2 = 2.63 \text{ m/s} \qquad\qquad \textit{Ans.}$$

This value represents the tugboat's velocity *just after* the towing impulse. Use this result and show that the towing impulse is 131 kN · s.

The alternative procedure of drawing the system's impulse and momentum diagrams is shown in Fig. 15–10c. The conservation of momentum is formulated as above by writing the vector terms directly from the diagrams.

*Only horizontal forces are shown on the free-body diagram.

(c)

Fig. 15–10c

Example 15–7

A rigid pile P shown in Fig. 15–11a has a mass of 800 kg and is driven into the ground using a hammer H that has a mass of 300 kg. The hammer falls from rest from a height $y_0 = 0.5$ m and strikes the top of the pile. Determine the impulse which the hammer imparts on the pile if the pile is surrounded entirely by loose sand so that after striking the hammer does *not* rebound off the pile.

(a)

Solution

Conservation of Energy. The velocity at which the hammer strikes the pile can be determined using the conservation of energy equation applied to the hammer. With the datum at the top of the pile, Fig. 15–11a, we have

$$T_0 + V_0 = T_1 + V_1$$
$$\tfrac{1}{2}m_H(v_H)_0^2 + W_H y_0 = \tfrac{1}{2}m_H(v_H)_1^2 + W_H y_1$$
$$0 + 300(9.81)(0.5) = \frac{1}{2}(300)(v_H)_1^2 + 0$$
$$(v_H)_1 = 3.13 \text{ m/s}$$

Free-Body Diagram. From the physical aspects of the problem, the free-body diagram of the hammer and pile, Fig. 15–11b, indicates that during the *short time* occurring just before to just after the *collision* the weights of the hammer and pile and the resistance force $\mathbf{F}_s$ of the soil are all *nonimpulsive*. Furthermore, the impulsive force $\mathbf{R}$ is internal to the system and therefore cancels. Consequently, momentum is conserved in the vertical direction.

(b)

Conservation of Momentum

$$(+\downarrow) \qquad m_H(v_H)_1 + m_P(v_P)_1 = m_H v_2 + m_P v_2$$
$$300(3.13) + 0 = 300v_2 + 800v_2$$
$$v_2 = 0.854 \text{ m/s}$$

Principle of Impulse and Momentum. The impulse which the hammer imparts to the pile can now be determined since $\mathbf{v}_2$ is known. From the free-body diagram for the hammer, Fig. 15–11c, we have

$$(+\downarrow) \qquad m_H(v_H)_1 + \Sigma \int_{t_1}^{t_2} F_y \, dt = m_H v_2$$
$$300(3.13) - \int R \, dt = 300(0.854)$$
$$\int R \, dt = 683 \text{ N} \cdot \text{s} \qquad\qquad \textit{Ans.}$$

(c)

Fig. 15–11

Try finding the impulse by applying the principle of impulse and momentum to the pile.

Example 15–8

A boy having a mass of 40 kg stands on the back of a 15-kg toboggan which is originally at rest, Fig. 15–12a. If he walks to the front B and stops, determine the distance the toboggan moves. Neglect friction between the bottom of the toboggan and the ground (ice).

Fig. 15–12

Solution I

Free-Body Diagram.* The unknown frictional force of the boy's shoes on the bottom of the toboggan can be *excluded* from the analysis if the toboggan and boy on it are considered as a single system. In this way the frictional force **F** becomes internal and the conservation of momentum applies. Since both the initial and final momenta of the system are zero (because the initial and final velocities are zero), the system's momentum must also be zero when the boy is at some intermediate point between A and B, Fig. 15–12b.

Conservation of Momentum

$$(\xrightarrow{+}) \qquad -m_b v_b + m_t v_t = 0 \qquad (1)$$

Here the two unknowns v_b and v_t represent the velocities of the boy and the toboggan measured from a *fixed inertial reference* on the ground. The *positions* of point A on the toboggan and the boy must be determined by integration. Assuming the initial position of point A to be at the origin, then at the intermediate position we have

$$-m_b s_b + m_t s_t = 0$$

If in the final position point A on the toboggan is located $s_t = +s$ to the right (in the same direction as $\mathbf{v}_t$), Fig. 15–12c, then when the boy is at B this point has the position $s_b = (2 - s)$ m to the left (in the same direction as $\mathbf{v}_b$). Hence,

$$-m_b(2 - s) + m_t s = 0 \qquad (2)$$

$$s = \frac{2m_b}{m_b + m_t} = \frac{2(40)}{40 + 15} = 1.45 \text{ m} \qquad Ans.$$

Solution II

The problem may also be solved by considering the relative motion of the boy with respect to the toboggan, $\mathbf{v}_{b/t}$. This velocity is related to the velocities of the boy and toboggan by the equation $\mathbf{v}_b = \mathbf{v}_t + \mathbf{v}_{b/t}$, Eq. 12–34. Since positive motion is assumed to be to the right in Eq. (1), $\mathbf{v}_b$ and $\mathbf{v}_{b/t}$ are negative, because the boy's motion is to the left. Hence, in scalar form, $-v_b = v_t - v_{b/t}$, and Eq. (1) then becomes $m_b(v_t - v_{b/t}) + m_t v_t = 0$. Integrating gives

$$m_b(s_t - s_{b/t}) + m_t s_t = 0$$

Assuming point A on the toboggan moves a distance s to the right, Fig. 15–12c, realizing $s_{b/t} = 2$ m, we have $m_b(s - 2) + m_t s = 0$, which is the same as Eq. (2).

*Only horizontal forces are shown on the free-body diagram.

Problems

*15–24. A railroad car having a mass of 15 Mg is coasting at 1.5 m/s on a horizontal track. At the same time another car having a mass of 12 Mg is coasting at 0.75 m/s in the opposite direction. If the cars meet and couple together, determine the speed of both cars just after the coupling. Compute the difference between the total kinetic energy before and after coupling has occurred, and explain qualitatively what happened to this energy.

15–25. The bus *B* has a weight of 15,000 lb and is traveling to the right at 5 ft/s. Meanwhile a 3,000-lb car *A* is traveling at 4 ft/s to the left. If the vehicles crash head-on and become entangled, determine their common velocity just after the collision. Assume that the brakes are not applied during collision.

Prob. 15–25

15–26. The 10-kg package is released from rest at *A* and slides down the smooth ramp onto a cart having a mass of 25 kg. If the coefficient of kinetic friction between the cart and package is $\mu = 0.8$, determine the final velocity of the cart after the package stops sliding on the cart. Also, calculate the time the package slides on the cart. Neglect the mass of the wheels.

Prob. 15–26

15–27. A 100-lb boy walks forward over the surface of the 60-lb cart with a constant speed of 3 ft/s relative to the cart. Determine the cart's speed and its displacement at the moment he is about to step off. Neglect the mass of the wheels.

Prob. 15–27

*15–28. The boy and girl are each standing on ice skates and holding on to the ends of the rope. If they begin to pull in the rope, such that their relative speed of approach is $v_{rel} = 2$ m/s, determine their velocities at the instant they meet. If the rope is 5 m long, how long will it take before they embrace? The boy has a mass of 60 kg, and the girl has a mass of 50 kg.

Prob. 15–28

15–29. The boy B jumps off the canoe at A with a velocity of 5 m/s relative to the canoe as shown. If he lands in the second canoe C, determine the final speed of both canoes after the motion. Each canoe has a mass of 40 kg. The boy's mass is 30 kg, and the girl D has a mass of 25 kg. Both canoes are originally at rest.

$v_{B/A} = 5$ m/s

30°

Prob. 15–29

15–30. A 70-kg man on ice skates throws the 8-kg block with an initial horizontal velocity of 3 m/s, measured relative to himself, in the direction shown. If he is originally at rest, determine the distance s he is from the block at the instant the block strikes the ground. Neglect friction, the motion of his arms and the size of the block and man.

$v_{B/m} = 3$ m/s

1.5 m

s

Prob. 15–30

15–31. A tugboat T having a mass of 19 Mg is tied to a barge B having a mass of 75 Mg. If the rope is "elastic" such that it has a stiffness $k = 600$ kN/m, determine the maximum stretch in the rope during the initial towing. Originally both the tugboat and barge are moving in the same direction with speeds $(v_T)_1 = 15$ km/h and $(v_B)_1 = 10$ km/h, respectively. Neglect the resistance of the water.

$\mathbf{v}_T$

$\mathbf{v}_B$

Prob. 15–31

*15–32. A boy having a weight of 75 lb stands at the end of the 50-lb canoe. If the canoe is not tied to the pier P and the boy runs at 4 ft/s relative to the canoe, determine how far d the canoe has moved away from the pier at the instant he leaps off at A. Neglect water resistance. Originally $d = 0$.

8 ft — d

Prob. 15–32

15–33. A boy A having a weight of 80 lb and a man B having a weight of 170 lb stand motionless at the ends of the dolly, which has a weight of 40 lb. If they exchange positions, A going to B and then B going to A's original position, determine the final position of the dolley just after the motion. Assume that the dolly is free to roll. Neglect the mass of the wheels.

5 ft

Prob. 15–33 / 15–34

15–34. Solve Prob. 15–33 if B first moves to A, then A moves to B's original position.

15–35. Two men A and B, each having a weight of 160 lb, stand on the 500-lb raft. Each can run with a speed of 3 ft/s measured relative to the raft. Determine the final speed of the raft if (a) A runs and jumps off, then B runs and jumps off, and (b) both run at the same time and jump off. Neglect turning of the raft and water resistance.

Prob. 15–35

*15–36. Blocks A and B have masses of 40 kg and 60 kg, respectively. They are placed on a smooth surface and the spring connected between them is stretched 2 m. If they are released from rest, determine the speeds of both blocks the instant the spring becomes unstretched.

$k = 180$ N/m

A B

Prob. 15–36

15–37. The free-rolling ramp has a weight of 120 lb. If the boy whose weight is 80 lb slides from rest at the top A, determine the ramp's speed when the boy reaches the bottom at B. Assume that the ramp is smooth, and neglect the mass of the wheels. The boy slides 15 ft down the ramp to go from A to B.

15 ft

A

5

3

4

B

Prob. 15–37

15.4 Impact

Plane of contact

v_A A B v_B Line of impact

Central impact

(a)

Plane of contact

A B Line of impact

θ ϕ

Oblique impact

v_A v_B

(b)

Fig. 15–13

Impact occurs when two bodies collide with each other during a very *short* interval of time, causing relatively large (impulsive) forces to be exerted between the bodies. The striking of a hammer and nail, or a golf club and ball, are common examples of impact loadings.

In general, there are two types of impact. *Central impact* occurs when the direction of motion of the mass centers of the two colliding particles is along a line passing through the mass centers of the particles. This line is called the *line of impact*, Fig. 15–13a. When the motion of one or both of the particles is at an angle with the line of impact, Fig. 15–13b, the impact is said to be *oblique impact*.

Central Impact. To illustrate the method for analyzing the mechanics of impact, consider the case involving the central impact of two *smooth* particles A and B shown in Fig. 15–14.

1. The particles have the initial momenta shown in Fig. 15–14a. Provided that $(v_A)_1 > (v_B)_1$, collision will eventually occur.
2. During the collision the particles will undergo a *period of deformation* such that they exert an equal but opposite deformation impulse $\int \mathbf{P}\, dt$ on each other, Fig. 15–14b.
3. Only at the instant of *maximum deformation* will both particles move with a common velocity $\mathbf{v}$, Fig. 15–14c.
4. Afterward a *period of restitution* occurs, in which case the particles will either return to their original shape or remain permanently deformed. The equal but opposite *restitution impulse* $\int \mathbf{R}\, dt$ pushes the particles apart from one another, Fig. 15–14d. In reality, the physical properties of any two bodies are such that the deformation impulse is *always greater* than that of restitution, i.e., $\int P\, dt > \int R\, dt$.
5. Just after separation the particles will have the final momenta shown in Fig. 15–14e, where $(v_B)_2 > (v_A)_2$.

Before impact

(a)

In most problems the initial velocities of the particles will be *known* and it will be necessary to determine their final velocities $(v_A)_2$ and $(v_B)_2$. In this regard, *momentum* for the *system of particles* is *conserved* since during collision the internal impulses of deformation and restitution *cancel*. Hence, referring to Fig. 15–14a and e,

$(\xrightarrow{+})$ $m_A(v_A)_1 + m_B(v_B)_1 = m_A(v_A)_2 + m_B(v_B)_2$ (15–10)

In order to obtain a second equation, necessary to solve for $(v_A)_2$ and $(v_B)_2$, we must apply the principle of impulse and momentum to *each particle*. For example, during the deformation phase for particle A, Fig. 15–14a, b, and c, we have

$(\xrightarrow{+})$ $m_A(v_A)_1 - \displaystyle\int P\, dt = m_A(v)$

Deformation Impulse

(b)

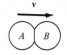

Maximum deformation

(c)

For the restitution phase, Fig. 15–14c, d, and e,

$(\xrightarrow{+})$ $m_A v - \displaystyle\int R\, dt = m_A(v_A)_2$

Restitution Impulse

(d)

The ratio of the restitution impulse to the deformation impulse is called the *coefficient of restitution, e.* From the above equations, this value for particle A is

$$e = \frac{\displaystyle\int R\, dt}{\displaystyle\int P\, dt} = \frac{v - (v_A)_2}{(v_A)_1 - v}$$

After impact

(e)

Fig. 15–14

In a similar manner, we can establish e by considering particle B, Fig. 15–14. This yields

$$e = \frac{\displaystyle\int R\,dt}{\displaystyle\int P\,dt} = \frac{(v_B)_2 - v}{v - (v_B)_1}$$

If the unknown v is eliminated from the above two equations, the coefficient of restitution can be expressed in terms of the particles' initial and final velocities as

$$e = \frac{(v_B)_2 - (v_A)_2}{(v_A)_1 - (v_B)_1} \qquad (15\text{--}11)$$

Provided a value for e is specified, Eqs. 15–10 and 15–11 may be solved simultaneously to obtain $(v_A)_2$ and $(v_B)_2$. In doing so, however, it is important to carefully establish a sign convention for defining the positive directions for both $\mathbf{v}_B$ and $\mathbf{v}_A$, and then use it *consistently* when writing *both* equations. As noted from the application above, we have defined the positive direction to the right when referring to the motions of both A and B. Consequently, if a negative value results from the solution of either $(v_A)_2$ or $(v_B)_2$, it indicates motion to the left.

Coefficient of Restitution. With reference to Fig. 15–14a and e, it is seen that Eq. 15–11 states that the coefficient of restitution is equal to the ratio of the relative velocity of the particles' separation *just after impact,* $(v_B)_2 - (v_A)_2$, to the relative velocity of the particles' approach *just before impact,* $(v_A)_1 - (v_B)_1$. By measuring these relative velocities experimentally, it has been found that e varies appreciably with impact velocity as well as with the size and shape of the colliding bodies. Differences in measurement occur because some of the initial kinetic energy of the bodies is transformed into heat energy as well as creating sound and elastic shock waves when the collision occurs.* For these reasons the coefficient of restitution is reliable only when used under conditions which closely approximate those which were known to exist when measurements were made. In general e has a value between zero and one, although one should be aware of the physical meaning of its two limits.

Elastic Impact (e = 1). If the collision between the two particles is *perfectly elastic,* the deformation impulse ($\int \mathbf{P}\,dt$) is equal and opposite to the restitution impulse ($\int \mathbf{R}\,dt$). Although in reality this can never be achieved, $e = 1$ for an elastic collision. Under these conditions one can show that no energy is lost in the collision (see Prob. 15–41).

Plastic Impact (e = 0). The impact is said to be *inelastic or plastic* when $e = 0$. In this case there is no restitution impulse given to the particles ($\int \mathbf{R}\,dt = \mathbf{0}$), so that after collision both particles couple or stick *together* and move with a common velocity. In this case the energy lost during collision is a maximum.

---

*The energy loss during collision can be calculated by computing the difference in the total kinetic energy of both particles just before and just after collision.

PROCEDURE FOR ANALYSIS
(Central Impact)

In most cases the *final velocities* of the two smooth colliding particles are to be determined *just after* they are subjected to direct central impact. Provided the coefficient of restitution, the mass of each particle, and each particle's initial velocity *just before* impact are known, the two equations available for solution are:

1. The conservation of momentum applies to the system of particles, $\Sigma m v_1 = \Sigma m v_2$.
2. The coefficient of restitution, $e = ((v_B)_2 - (v_A)_2)/((v_A)_1 - (v_B)_1)$, relates the relative velocities of the particles along the line of impact, from just before to just after collision.

When applying these two equations, the sense of an unknown velocity can be assumed. If the solution yields a negative magnitude, the velocity acts in the opposite sense.

Oblique Impact. When oblique impact occurs between two smooth particles, the particles move away from each other with velocities having unknown directions as well as unknown magnitudes. Provided the initial velocities are known, four unknowns are present in the problem. As shown in Fig. 15–15a, these unknowns may be represented as $(v_A)_2$, $(v_B)_2$, θ_2, and ϕ_2.

PROCEDURE FOR ANALYSIS
(Oblique Impact)

If the y axis is established within the plane of contact and the x axis along the line of impact, the impulsive forces of deformation and restitution act *only in the x direction*, Fig. 15–15b. Resolving the velocity or momentum vectors into components along the x and y axes, Fig. 15–15b, it is possible to write four independent scalar equations in order to determine $(v_{Ax})_2$, $(v_{Ay})_2$, $(v_{Bx})_2$, and $(v_{By})_2$.

1. Momentum of the system is conserved *along the line of impact*, x axis, so that $\Sigma m(v_x)_1 = \Sigma m(v_x)_2$.
2. The coefficient of restitution, $e = ((v_B)_2 - (v_A)_2)/((v_A)_1 - (v_B)_1)$, relates the relative-velocity *components* of the particles *along the line of impact* (x axis).
3. Momentum of particle A is conserved along the y axis, perpendicular to the line of impact, since no impulse acts on the particle in this direction.
4. Momentum of particle B is conserved along the y axis, perpendicular to the line of impact, since no impulse acts on the particle in this direction.

Application of these four equations is illustrated numerically in Example 15–11.

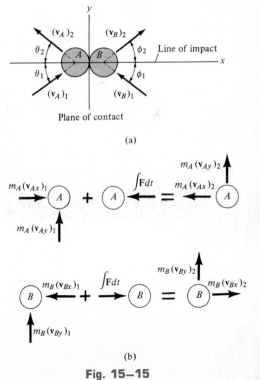

Fig. 15–15

195

Example 15–9

The bag A, having a weight of 6 lb, is released from rest at the position $\theta = 0°$, as shown in Fig. 15–16a. It strikes an 18-lb box B when $\theta = 90°$. If the coefficient of restitution between the bag and box is $e = 0.5$, determine the velocities of the bag and box just after impact.

Solution

This problem involves central impact. Why? Before analyzing the mechanics of the impact, however, it is first necessary to obtain the velocity of the bag *just before* it strikes the box.

(a)

Conservation of Energy. With the datum at $\theta = 0°$, Fig. 15–16b, we have

$$T_0 + V_0 = T_1 + V_1$$

$$0 + 0 = \frac{1}{2}\left(\frac{6}{32.2}\right)(v_A)_1^2 - 6(3)$$

$$(v_A)_1 = 13.90 \text{ ft/s}$$

(b)

Conservation of Momentum. The velocities of A and B just before and just after impact are shown in Fig. 15–16c. Here it is assumed that just after collision both bodies continue to travel to the left. Applying the conservation of momentum to the system, we have

$(\xleftarrow{+})$
$$m_B(v_B)_1 + m_A(v_A)_1 = m_B(v_B)_2 + m_A(v_A)_2$$

$$0 + \frac{6}{32.2}(13.90) = \frac{18}{32.2}(v_B)_2 + \frac{6}{32.2}(v_A)_2$$

$$(v_A)_2 = 13.90 - 3(v_B)_2 \qquad (1)$$

Coefficient of Restitution. Realizing that for separation to occur after collision $(v_B)_2 > (v_A)_2$, Fig. 15–16c, we have

$(\xleftarrow{+})$
$$e = \frac{(v_B)_2 - (v_A)_2}{(v_A)_1 - (v_B)_1}$$

$$0.5 = \frac{(v_B)_2 - (v_A)_2}{13.90 - 0}$$

$$(v_A)_2 = (v_B)_2 - 6.95 \qquad (2)$$

Solving Eqs. (1) and (2) simultaneously yields

$$(v_A)_2 = -1.74 \text{ ft/s} = 1.74 \text{ ft/s} \rightarrow \qquad \text{and} \qquad (v_B)_2 = 5.21 \text{ ft/s} \leftarrow \quad \textit{Ans.}$$

(c)

Fig. 15–16

Example 15–10

The ball B shown in Fig. 15–17a has a mass of 1.5 kg and is suspended from the ceiling by a 1-m-long elastic cord. If the cord is *stretched* downward 0.25 m and the ball is released from rest, determine how far the cord stretches after the ball rebounds from the ceiling. The stiffness of the cord is $k = 800$ N/m and the coefficient of restitution between the ball and ceiling is $e = 0.8$. The ball makes a central impact with the ceiling.

(a)

Solution

It is necessary to first obtain the initial velocity of the ball *just before* it strikes the ceiling.

Conservation of Energy. With the datum located as shown in Fig. 15–17a, realizing that initially $y = y_0 = (1 + 0.25)$ m $= 1.25$ m, we have

$$T_0 + V_0 = T_1 + V_1$$

$$\tfrac{1}{2}m(v_B)_0^2 - W_B y_0 + \tfrac{1}{2}ks^2 = \tfrac{1}{2}m(v_B)_1^2 + 0$$

$$0 - 1.5(9.81)(1.25) + \frac{1}{2}(800)(0.25)^2 = \frac{1}{2}(1.5)(v_B)_1^2$$

$$(v_B)_1 = 2.97 \text{ m/s} \downarrow$$

(b)

Fig. 15–17

The interaction of the ball with the ceiling will now be considered using the principles of impact.[*] Note that the y axis, Fig. 15–17b, represents the line of impact for the ball. Since an unknown portion of the mass of the ceiling is involved in the impact, the conservation of momentum for the ball–ceiling system will not be written. The "velocity" of this portion of ceiling is zero since it remains at rest *both* before and after impact.

Coefficient of Restitution

$$(+\uparrow) \quad e = \frac{(v_B)_2 - (v_A)_2}{(v_A)_1 - (v_B)_1}; \quad 0.8 = \frac{-(v_B)_2 - 0}{0 - 2.97}$$

$$(v_B)_2 = 2.37 \text{ m/s} \downarrow$$

Conservation of Energy. The maximum stretch s in the cord may be determined by again applying the conservation of energy equation to the ball just after collision. Assuming that $y = y_3 = (1 + s_3)$ m, Fig. 15–17a, then

$$T_2 + V_2 = T_3 + V_3$$

$$\tfrac{1}{2}m(v_B)_2^2 + 0 = \tfrac{1}{2}m(v_B)_3^2 - W_B y_3 + \tfrac{1}{2}ks_3^2$$

$$\frac{1}{2}(1.5)(2.37)^2 = 0 - 9.81(1.5)(1 + s_3) + \frac{1}{2}(800)s_3^2$$

$$400s_3^2 - 14.72s_3 - 18.94 = 0$$

Solving this quadratic equation for the positive root yields

$$s_3 = 0.237 \text{ m} = 237 \text{ mm} \qquad \textit{Ans.}$$

[*]The weight of the ball is considered a nonimpulsive force.

Example 15–11

Two smooth disks A and B, having a mass of 1 and 2 kg, respectively, collide with initial velocities as shown in Fig. 15–18a. If the coefficient of restitution for the disks is $e = 0.75$, determine the x and y components of the final velocity of each disk after collision. Neglect friction.

(a)

Solution

The problem involves *oblique impact*. Why? In order to seek a solution, we have established the x and y axes along the line of impact and the plane of contact, respectively, Fig. 15–18a.

Resolving each of the initial velocities into x and y components, we have

$(v_{Ax})_1 = 3 \cos 30° = 2.60$ m/s, $(v_{Ay})_1 = 3 \sin 30° = 1.50$ m/s

$(v_{Bx})_1 = -1 \cos 45° = -0.707$ m/s, $(v_{By})_1 = -1 \sin 45° = -0.707$ m/s

Since the impact occurs only in the x direction (line of impact), Fig. 15–18b, the conservation of momentum for *both* disks can be applied in this direction. Why?

(b)

Conservation of "x" Momentum. In reference to the momentum diagrams, we have

$(\overset{+}{\rightarrow})$ $m_A(v_{Ax})_1 + m_B(v_{Bx})_1 = m_A(v_{Ax})_2 + m_B(v_{Bx})_2$

$1(2.60) + 2(-0.707) = 1(v_{Ax})_2 + 2(v_{Bx})_2$

$(v_{Ax})_2 + 2(v_{Bx})_2 = 1.18$ (1)

(c)

Fig. 15–18

Coefficient of Restitution (x). Both disks are *assumed* to have components of velocity in the $+x$ direction after collision, Fig. 15–18b.

$(\overset{+}{\rightarrow})$ $e = \dfrac{(v_{Bx})_2 - (v_{Ax})_2}{(v_{Ax})_1 - (v_{Bx})_1};$ $0.75 = \dfrac{(v_{Bx})_2 - (v_{Ax})_2}{2.60 - (-0.707)}$

$(v_{Bx})_2 - (v_{Ax})_2 = 2.48$ (2)

Solving Eqs. (1) and (2) for $(v_{Ax})_2$ and $(v_{Bx})_2$ yields

$(v_{Ax})_2 = -1.26$ m/s $(\leftarrow)$ $(v_{Bx})_2 = 1.22$ m/s $(\rightarrow)$ *Ans.*

Conservation of "y" Momentum. The momentum of *each disk* is *conserved* in the y direction (plane of contact), since the disks are smooth and therefore *no impact* occurs in this direction. Hence, in reference to Fig. 15–18b,

$(+ \uparrow)$ $m_A(v_{Ay})_1 = m_A(v_{Ay})_2$ $(v_{Ay})_2 = 1.50$ m/s $(\uparrow)$ *Ans.*

and

$(+ \uparrow)$ $m_B(v_{By})_1 = m_B(v_{By})_2$ $(v_{By})_2 = -0.707$ m/s $(\downarrow)$ *Ans.*

Show that when the velocity components are summed, one obtains the results shown in Fig. 15–18c.

Problems

15–38. Disk A has a mass of 250 g and is sliding on a *smooth* horizontal surface with an initial velocity $(v_A)_1 = 2$ m/s. It makes a direct collision with disk B, which has a mass of 175 g and is originally at rest. If both disks are of the same size and the collision is perfectly elastic ($e = 1$), determine the velocity of each disk just after the collision. Show that the kinetic energy of the disks before and after collision is the same.

15–39. Block B has a mass of 0.75 kg and is sliding forward on the *smooth* surface with a velocity $(v_B)_1 = 4$ m/s when it strikes the 2-kg block A, which is originally at rest. If the coefficient of restitution between the blocks is $e = 0.5$, compute the velocities of A and B just after collision.

Prob. 15–39

*15–40. The 20-lb suitcase A is released from rest at C. After it slides down the smooth ramp it strikes the 10-lb suitcase B, which is originally at rest. If the coefficient of restitution between the suitcases is $e = 0.3$ and the coefficient of kinetic friction between the floor DE and each suitcase is $\mu = 0.4$, determine (a) the velocity of A just before impact, (b) the velocities of A and B just after impact, and (c) the distance B slides before coming to rest.

Prob. 15–40

15–41. If two disks A and B have the same mass and are subjected to direct central impact such that the collision is perfectly elastic ($e = 1$), prove that the kinetic energy before collision equals the kinetic energy after collision. The surface upon which they slide is smooth.

15–42. The cue ball A is given an initial velocity $(v_A)_1 = 5$ m/s. If it makes a direct collision with ball B ($e = 0.8$), determine the velocity of B and the angle θ just after it rebounds from the cushion at C ($e' = 0.6$). Each ball has a mass of 0.4 kg. Neglect the size of each ball.

Prob. 15–42

15–43. The three balls each have the same mass m. If A is released from rest at θ, determine the angle ϕ to which C rises after collision. The coefficient of restitution between each ball is e.

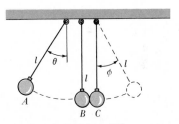

Prob. 15–43

*15–44. The four balls shown each have the same mass m. If A and B are rolling forward with velocity v and strike C, explain why after collision C and D each move off with velocity v. Why doesn't D move off with velocity $2v$? The collision is elastic, $e = 1$. Neglect the size of each ball.

Prob. 15–44

199

15–45. Two smooth billiard balls A and B each have a mass of 200 g. If A strikes B with a velocity $(v_A)_1 = 1.5$ m/s as shown, determine their final velocities just after collision. Ball B is originally at rest and the coefficient of restitution is $e = 0.85$. Neglect the size of each ball.

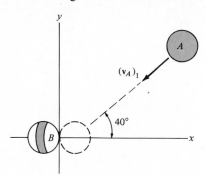

Prob. 15–45

15–46. The "stone" A used in the sport of curling is thrown over the ice track and strikes another "stone" B as shown. If each "stone" has a weight of 47 lb and the coefficient of restitution between the "stones" is $e = 0.8$, determine their speeds just after collision. Initially A has a velocity of 8 ft/s and B is at rest. Neglect friction.

Prob. 15–46

15–47. The 20-lb box slides on the surface for which $\mu_k = 0.3$. The box has a velocity $v_A = 15$ ft/s when it is at A. If it strikes the plate B, which has a weight of 10 lb and is held in position by an unstretched spring of stiffness $k = 400$ lb/ft, determine the maximum compression imparted to the spring. Take $e = 0.8$ between the box and the plate.

Prob. 15–47

*15–48. Determine the horizontal velocity $\mathbf{v}_A$ at which the girl must throw the ball so that it bounces once on the smooth surface and then lands into the cup at C. Take $e = 0.6$ and neglect the size of the cup.

Prob. 15–48

15–49. Two automobile tires A and B, each having a mass of 10 kg, slide on an iced surface and collide at O. After collision, the tires slide in the directions shown 6 m and 4 m before coming to rest at A' and B', respectively. Determine the initial speeds of the tires just before collision. Take $\mu = 0.15$. Neglect the size of the tires. *Hint:* Since the line of impact has not been defined, apply the conservation of momentum along the x' and y' axes, respectively.

Prob. 15–49

15–50. Two smooth disks A and B have the initial velocities shown just before they collide at O. If they have masses $m_A = 8$ kg and $m_B = 6$ kg, respectively, and the surface upon which they slide is smooth, determine their speeds just after impact. The coefficient of restitution is $e = 0.5$.

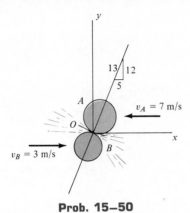

Prob. 15–50

15–51. The two hockey pucks A and B each have a mass of 250 g. If they collide at O and are deflected along the dashed paths, determine their speeds just after impact. Assume that the icy surface over which they slide is smooth. *Hint:* Since the y' axis is *not* along the line of impact, apply the conservation of momentum along the x' and y' axes.

Prob. 15–51

***15–52.** Before a cranberry can make it to your dinner plate, it must pass a bouncing test which rates its quality. If cranberries having an $e \geqslant 0.8$ are to be accepted, determine the dimensions d and h for the barrier so that when a cranberry falls from rest at A it strikes the plate at B and bounces over the barrier at C.

Prob. 15–52

15–53. The girl throws the 0.5-kg ball towards the wall with an initial velocity $v_A = 10$ m/s as shown. Determine (a) the velocity at which it strikes the wall at B, (b) the velocity at which it rebounds from the wall if $e = 0.5$, and (c) the distance s from the wall to where it strikes the ground at C.

Prob. 15–53

15–54. The ball has a mass of 2 kg and falls 1.5 m from an at-rest position onto the inclined surface for which $e = 0.4$. Determine the distance d to where it again strikes the surface. Neglect the size of the ball.

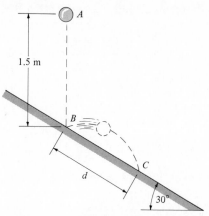

Prob. 15–54

15–55. Ball A strikes ball B with an initial velocity $(v_A)_1$ as shown. If both balls have the same mass and the collision is perfectly elastic, determine the angle θ after collision. Ball B is originally at rest. Neglect the size of each ball.

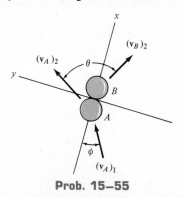

Prob. 15–55

*15–56. The sphere of mass m falls and strikes the triangular block with a vertical velocity $\mathbf{v}$. If the block rests on a smooth surface and has a mass $3m$, determine its velocity just after the collision. The coefficient of restitution is e.

Prob. 15–56

15–57. If disk A is sliding forward along the tangent to disk B and strikes B with a velocity $\mathbf{v}$, determine the velocity of B after the collision and compute the loss of kinetic energy during the collision. Neglect friction. The coefficient of restitution is e, and each disk has the same size and mass m.

Prob. 15–57

Angular Momentum 15.5

The *angular momentum* of a particle about point O is defined as the "moment" of the particle's linear momentum about O. Since this concept is analogous to finding the moment of a force about a point, the angular momentum, $\mathbf{H}_O$, is sometimes referred to as the *moment of momentum*.

Scalar Formulation. If a particle is moving along a curve lying in the *x-y* plane, Fig. 15–19, the angular momentum can be computed about point O (actually the z axis) by using a scalar formulation. The *magnitude* of $\mathbf{H}_O$ is

$$H_O = (d)(mv) \qquad (15\text{–}12)$$

Here d is the moment arm or perpendicular distance from O to the line of action of $m\mathbf{v}$. Common units for this magnitude are $\text{kg} \cdot \text{m}^2/\text{s}$ or $\text{slug} \cdot \text{ft}^2/\text{s}$. The *direction* of $\mathbf{H}_O$ is defined by the right-hand rule. As shown in Fig. 15–19, the curl of the fingers of the right hand indicates the sense of rotation of $m\mathbf{v}$ about O, so that in this case the thumb (or $\mathbf{H}_O$) is directed perpendicular to the *x-y* plane along the $+z$ axis.

Fig. 15–19

Vector Formulation. If the particle is moving along a space curve, Fig. 15–20, the vector cross product can be used to determine the *angular momentum* about O. In this case

$$\mathbf{H}_O = \mathbf{r} \times m\mathbf{v} \qquad (15\text{–}13)$$

Here $\mathbf{r}$ denotes a position vector drawn from the moment point O to the particle P. As shown in the figure, $\mathbf{H}_O$ is *perpendicular* to the shaded plane containing $\mathbf{r}$ and $m\mathbf{v}$.

In order to evaluate the cross product, $\mathbf{r}$ and $m\mathbf{v}$ should be expressed in terms of their Cartesian components, so that the angular momentum is determined by evaluating the determinant:

$$\mathbf{H}_O = \begin{vmatrix} \mathbf{i} & \mathbf{j} & \mathbf{k} \\ r_x & r_y & r_z \\ mv_x & mv_y & mv_z \end{vmatrix} \qquad (15\text{–}14)$$

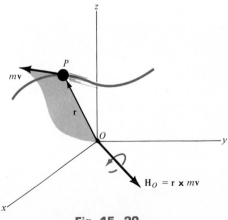

Fig. 15–20

Relation Between Moment of a Force and Angular Momentum. The moments about point O of all the forces acting on the particle may be related to the particle's angular momentum by using the equation of motion. If the mass of the particle is constant, we may write

$$\Sigma \mathbf{F} = m\dot{\mathbf{v}}$$

Performing a cross-product multiplication of each side of this equation by the position vector to obtain the moments of the forces about point O, we have

$$\Sigma \mathbf{M}_O = \mathbf{r} \times \Sigma \mathbf{F} = \mathbf{r} \times m\dot{\mathbf{v}}$$

From Appendix C, the derivative of $\mathbf{r} \times m\mathbf{v}$ can be written as

$$\dot{\mathbf{H}}_O = \frac{d}{dt}(\mathbf{r} \times m\mathbf{v}) = \dot{\mathbf{r}} \times m\mathbf{v} + \mathbf{r} \times m\dot{\mathbf{v}}$$

The first term on the right side, $\dot{\mathbf{r}} \times m\mathbf{v} = m(\dot{\mathbf{r}} \times \dot{\mathbf{r}}) = \mathbf{0}$, since the cross product of a vector with itself is zero. Hence, the above equation becomes

$$\Sigma \mathbf{M}_O = \dot{\mathbf{H}}_O \qquad (15\text{--}15)$$

This equation states that *the resultant moment about point O of all the forces acting on the particle is equal to the time rate of change of the particle's angular momentum about point O*. This result is similar to Eq. 15–1, i.e.,

$$\Sigma \mathbf{F} = \dot{\mathbf{L}} \qquad (15\text{--}16)$$

Here $\mathbf{L} = m\mathbf{v}$, so that *the resultant force acting on the particle is equal to the time rate of change of the particle's linear momentum.*

From the derivations, it is seen that Eqs. 15–15 and 15–16 are actually another way of stating Newton's second law of motion. It will be shown that they have many practical applications when extended and applied to the solution of problems involving a system of particles.

15.6 Angular Momentum of a System of Particles

An equation having the same form as Eq. 15–15 may be derived for the system of n particles shown in Fig. 15–21. The forces acting on the arbitrary ith particle of the system consist of a resultant *external force* $\mathbf{F}_i$ and the internal forces $\displaystyle\sum_{j=1(j \neq i)}^{n} \mathbf{f}_{ij} = \mathbf{f}_i$. Expressing the moments of these forces about point O, using the form of Eq. 15–15, we have

$$(\mathbf{r}_i \times \mathbf{F}_i) + (\mathbf{r}_i \times \mathbf{f}_i) = (\dot{\mathbf{H}}_i)_O$$

Here $\mathbf{r}_i$ represents the position vector drawn from the origin O of an inertial frame of reference to the ith particle, and $(\mathbf{H}_i)_O$ is the angular momentum of

z

f_i F_i

i

r_i

O y

Inertial Coordinate
System

x

Fig. 15–21

the ith particle about O. Similar equations can be written for each of the other particles of the system. When the results are summed vectorially, the result is

$$\Sigma \mathbf{r}_i \times \mathbf{F}_i + \Sigma(\mathbf{r}_i \times \mathbf{f}_i) = \Sigma(\dot{\mathbf{H}}_i)_O$$

The second term is zero since the internal forces occur in equal but opposite collinear pairs, and by the principle of transmissibility the moment of each pair of forces about point O is therefore zero. Hence, dropping the index notation, the above equation can be written in a simplified form as

$$\Sigma \mathbf{M}_O = \dot{\mathbf{H}}_O \qquad (15\text{--}17)$$

which states that *the sum of the moments about point O of all the external forces acting on a system of particles is equal to the time rate of change of the total angular momentum of the system of particles about point O.* Although point O has been chosen as the origin of coordinates, it actually can represent any point *fixed* in the inertial frame of reference.

Angular Impulse and Momentum Principles **15.7**

Principle of Angular Impulse and Momentum. If Eq. 15–15 is rewritten in the form $\Sigma \mathbf{M}_O \, dt = d\mathbf{H}_O$ and integrated, we have, assuming that at time $t = t_1$, $\mathbf{H}_O = (\mathbf{H}_O)_1$ and at time $t = t_2$, $\mathbf{H}_O = (\mathbf{H}_O)_2$,

$$\Sigma \int_{t_1}^{t_2} \mathbf{M}_O \, dt = (\mathbf{H}_O)_2 - (\mathbf{H}_O)_1$$

or

$$(\mathbf{H}_O)_1 + \Sigma \int_{t_1}^{t_2} \mathbf{M}_O \, dt = (\mathbf{H}_O)_2 \qquad (15\text{--}18)$$

This equation is referred to as the *principle of angular impulse and momentum.* The initial and final angular momenta $(\mathbf{H}_O)_1$ and $(\mathbf{H}_O)_2$ are defined as the moment of the linear momentum of the particle ($\mathbf{H}_O = \mathbf{r} \times m\mathbf{v}$) at the instants t_1 and t_2, respectively. The second term on the left side, $\Sigma \int_{t_1}^{t_2} \mathbf{M}_O \, dt$, is called the *angular impulse.* It is computed on the basis of integrating, with respect to time, the moments of all the forces acting on the particle over the time interval t_1 to t_2. Since the moment of a force about point O is defined as $\mathbf{M}_O = \mathbf{r} \times \mathbf{F}$, the angular impulse may be expressed in vector form as

$$\text{angular impulse} = \int_{t_1}^{t_2} \mathbf{M}_O \, dt = \int_{t_1}^{t_2} (\mathbf{r} \times \mathbf{F}) \, dt \qquad (15\text{--}19)$$

Here $\mathbf{r}$ is a position vector which extends from point O to any point on the line of action of $\mathbf{F}$.

In a similar manner, using Eq. 15–17, the principle of angular impulse and momentum for a system of particles may be written as

$$\Sigma(\mathbf{H}_O)_1 + \Sigma \int_{t_1}^{t_2} \mathbf{M}_O \, dt = \Sigma(\mathbf{H}_O)_2 \qquad (15-20)$$

Here the first and third terms represent the angular momenta of the system of particles ($\Sigma \mathbf{H}_O = \Sigma(\mathbf{r}_i \times m\mathbf{v}_i)$) at the instants t_1 and t_2. The second term is the vector sum of the angular impulses given to all the particles during the time period t_1 to t_2. Recall that these impulses are created only by the moments of the external forces acting on the system where, for the ith particle, $\mathbf{M}_O = \mathbf{r}_i \times \mathbf{F}_i$.

Vector Formulation. Using impulse and momentum principles, it is therefore possible to write two vector equations which define the particle motion; namely, Eqs. 15–3 and 15–18, restated as

$$m\mathbf{v}_1 + \Sigma \int_{t_1}^{t_2} \mathbf{F} \, dt = m\mathbf{v}_2$$

$$(\mathbf{H}_O)_1 + \Sigma \int_{t_1}^{t_2} \mathbf{M}_O \, dt = (\mathbf{H}_O)_2 \qquad (15-21)$$

Scalar Formulation. In general, the above equations may be expressed in x, y, z component form, yielding a total of six independent scalar equations. If the particle is confined to move in the x-y plane, three independent scalar equations may be written to express the motion; namely,

$$m(v_x)_1 + \Sigma \int_{t_1}^{t_2} F_x \, dt = m(v_x)_2$$

$$m(v_y)_1 + \Sigma \int_{t_1}^{t_2} F_y \, dt = m(v_y)_2 \qquad (15-22)$$

$$(H_O)_1 + \Sigma \int_{t_1}^{t_2} M_O \, dt = (H_O)_2$$

The first two of these equations represent the principle of linear impulse and momentum in the x and y directions, and the third equation represents the principle of angular impulse and momentum about the z axis.

Conservation of Angular Momentum. When the angular impulses acting on a particle are all zero during the time t_1 to t_2, Eq. 15–18 reduces to the following simplified form,

$$(\mathbf{H}_O)_1 = (\mathbf{H}_O)_2 \qquad (15-23)$$

This equation is known as the *conservation of angular momentum*. It states that from t_1 to t_2 the particle's angular momentum remains constant. Obviously, if no external impulse is applied to the particle, both linear and angular momentum will be conserved. In some cases, however, the particle's angular momentum will be conserved and linear momentum may not. An example of this occurs when the particle is subjected *only* to a *central force* (see Sec. 13.7). As shown in Fig. 15–22, the impulsive central force **F** is always directed toward point O as the particle moves along the path. Hence, the angular impulse (moment) created by **F** about the z axis passing through point O is always zero, and therefore angular momentum of the particle is conserved about this axis.

Using Eq. 15–20, we can also write the conservation of angular momentum for a system of particles, namely,

$$\Sigma (\mathbf{H}_O)_1 = \Sigma (\mathbf{H}_O)_2 \qquad (15\text{–}24)$$

In this case the summation must include the angular momenta of all n particles in the system.

Fig. 15–22

PROCEDURE FOR ANALYSIS

When applying the principles of angular impulse and momentum, or the conservation of angular momentum, it is suggested that the following procedure be used.

Free-Body Diagram. Draw the particle's free-body diagram in order to determine any axis about which angular momentum may be conserved. For this to occur, the moments of the forces (or impulses) about the axis must be zero throughout the time period t_1 to t_2. Apart from the free-body diagram, the direction and sense of the particle's initial and final velocities should also be established. These velocities may be sketched near, *but not on,* the free-body diagram.

An alternative procedure to that mentioned above requires drawing the impulse and momentum diagrams for the particle.

Momentum Equations. Apply the principle of angular impulse and momentum, $(\mathbf{H}_O)_1 + \Sigma \int_{t_1}^{t_2} \mathbf{M}_O \, dt = (\mathbf{H}_O)_2$, or if appropriate, the conservation of angular momentum, $(\mathbf{H}_O)_1 = (\mathbf{H}_O)_2$.

If other equations are needed for the problem solution, when appropriate, use the principle of linear impulse and momentum, the principle of work and energy, the equations of motion, or kinematics.

The following examples numerically illustrate application of the above procedure.

Example 15-12

The 5-kg block of negligible size rests on the smooth horizontal plane, Fig. 15–23a. It is attached at A to a slender rod of negligible mass. The rod is attached to a ball-and-socket joint at B. If a moment $M = (3t)$ N · m, where t is in seconds, is applied to the rod and a horizontal force $P = 10$ N is applied to the block as shown, determine the speed of the block in 4 s starting from rest.

Fig. 15–23

Solution

Free-Body Diagram. If we consider the system of both the rod and block, Fig. 15–23b, then the resultant force reaction at B can be eliminated from the analysis by applying the principle of angular impulse and momentum about the z axis. Why? If this is done, the angular impulses created by the weight and normal reaction N_A are also eliminated, since they act parallel to the z axis and therefore create zero moment about this axis.

Principle of Angular Impulse and Momentum

$$(H_z)_1 + \Sigma \int M_z \, dt = (H_z)_2$$

$$(H_z)_1 + \int M \, dt + r_{BA}P(\Delta t) = (H_z)_2$$

$$0 + \int_0^4 3t \, dt + (0.4)(10)(4) = 5(v_A)_2(0.4)$$

$$24 + 16 = 2(v_A)_2$$

$$(v_A)_2 = 20 \text{ m/s} \qquad\qquad Ans.$$

Example 15–13

The ball B, shown in Fig. 15–24a, has a weight of 0.8 lb and is attached to a cord which passes through a hole at A in a smooth table. When the ball is $r_1 = 1.75$ ft from the hole, it is rotating around in a circle such that its speed is $v_1 = 4$ ft/s. If by applying a force $\mathbf{F}$ the cord is pulled downward through the hole with a constant speed $v_c = 6$ ft/s, determine (a) the speed of the ball at the instant it is $r_2 = 0.6$ ft from the hole, and (b) the amount of work done by the force $\mathbf{F}$ in shortening the cord. Neglect the size of the ball.

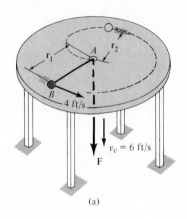

(a)

Solution

Part (a)

Free-Body Diagram. As the ball moves from radial position r_1 to r_2, Fig. 15–24b, the three forces acting on it are all unknown; however, since the cord tension $\mathbf{F}$ passes through the z axis and $\mathbf{W}$ and $\mathbf{N}_B$ are parallel to it, the moments, or angular impulses created by the forces, are all *zero* about this axis. Hence, the conservation of angular momentum applies about the z axis.

Conservation of Angular Momentum. As indicated in Fig. 15–24b, the ball's velocity $\mathbf{v}_2$ is resolved into two components. The radial component, 6 ft/s, is known; however, it produces zero angular momentum about the z axis. Thus,

$$\mathbf{H}_1 = \mathbf{H}_2$$

$$r_1 m_B v_1 = r_2 m_B v_2'$$

$$1.75\left(\frac{0.8}{32.2}\right)4 = 0.6\left(\frac{0.8}{32.2}\right)v_2'$$

$$v_2' = 11.67 \text{ ft/s}$$

(b)

Fig. 15–24

The speed of the ball is thus

$$v_2 = \sqrt{(11.67)^2 + (6)^2}$$

$$= 13.1 \text{ ft/s} \qquad \qquad \textit{Ans.}$$

Part (b). The only force that does work on the ball is $\mathbf{F}$. (The normal force and weight do not move vertically.) The initial and final kinetic energies of the ball can be determined so that from the principle of work and energy we have

$$T_1 + \Sigma U_{1-2} = T_2$$

$$\frac{1}{2}\left(\frac{0.8}{32.2}\right)(4)^2 + U_F = \frac{1}{2}\left(\frac{0.8}{32.2}\right)(13.1)^2$$

$$U_F = 1.94 \text{ ft} \cdot \text{lb} \qquad \qquad \textit{Ans.}$$

Example 15-14

The 2-kg block shown in Fig. 15-25a rests on a smooth horizontal surface and is attached to an elastic cord that has a stiffness $k_c = 20$ N/m and is initially unstretched. If the block is given a velocity $(v_B)_1 = 1.5$ m/s, perpendicular to the cord, determine the rate at which the cord is being stretched and the speed of the block at the instant the cord is stretched 0.2 m.

(a)

Solution

Free-Body Diagram. After the block has been launched, it slides along the dashed path shown in Fig. 15-25b. By inspection, angular momentum about point O (or the z axis) is *conserved*, since the moment of the central force $\mathbf{F}_e$ about O is always zero. Also, when the distance is 0.7 m, only the component of $(\mathbf{v}_B')_2$ is effective in producing angular momentum of the block about O.

(b)

Fig. 15-25

Conservation of Angular Momentum. The component $(\mathbf{v}_B')_2$ can be obtained by applying the conservation of angular momentum about O (the z axis); i.e.,

$$(\mathbf{H}_O)_1 = (\mathbf{H}_O)_2$$

$\curvearrowleft +$

$$r_1 m_B (v_B)_1 = r_2 m_B (v_B')_2$$

$$0.5(2)(1.5) = 0.7(2)(v_B')_2$$

$$(v_B')_2 = 1.07 \text{ m/s}$$

Conservation of Energy. The speed of the block may be obtained by applying the conservation of energy equation at the point where the block was launched and at the point where the cord is stretched 0.2 m.

$$T_1 + V_1 = T_2 + V_2$$

$$\frac{1}{2}(2)(1.5)^2 + 0 = \frac{1}{2}(2)(v_B)_2^2 + \frac{1}{2}(20)(0.2)^2$$

Thus,

$$(v_B)_2 = 1.36 \text{ m/s} \qquad \qquad Ans.$$

Having determined $(v_B)_2$ and its component $(v_B')_2$, the rate of stretch of the cord $(v_B'')_2$ is determined from the Pythagorean theorem,

$$(v_B'')_2 = \sqrt{(v_B)_2^2 - (v_B')_2^2}$$

$$= \sqrt{(1.36)^2 - (1.07)^2}$$

$$= 0.838 \text{ m/s} \qquad \qquad Ans.$$

Problems

15–58. A particle has a linear momentum $\mathbf{L} = \{5\mathbf{i} - 6\mathbf{j} + 10\mathbf{k}\}$ kg · m/s. If it is located at point B (5 m, 2 m, −1 m), determine its angular momentum about point A (−3 m, 6 m, 4 m).

15–59. In each case determine the angular momentum $\mathbf{H}_O$ of each of the particles about point O. Use a scalar solution for part (a) and a Cartesian vector solution for part (b).

(a)

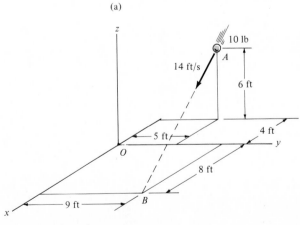

(b)

Prob. 15–59

*15–60. The four spheres each have a mass of 2 kg and are rigidly attached to a crossbar frame of negligible mass. If a couple $M = (5t + 2)$ N · m, where t is in seconds, is applied as shown, determine the speed of each of the spheres in 4 s, starting from rest.

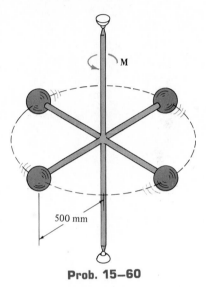

Prob. 15–60

15–61. An amusement-park ride consists of a car which is attached to the cable OA. The car rotates in a horizontal circular path and is brought to a speed $v_1 = 4$ ft/s when $r = 12$ ft. The cable is then pulled in at the constant rate of 0.5 ft/s. Determine the speed of the car in 3 s.

Prob. 15–61

211

15–62. The small cylinder C has a mass of 10 kg and is attached to the end of a rod whose mass may be neglected. If the frame is subjected to a couple $M = (8t^2 + 5) \, \text{N} \cdot \text{m}$, where t is in seconds, and the cylinder is subjected to a force of 60 N, which is always directed as shown, determine the speed of the cylinder when $t = 2$ s. The cylinder has a speed $v_0 = 2$ m/s when $t = 0$.

Prob. 15–62

15–63. A girl having a weight of 80 lb is skating around in a circle of radius $r_A = 20$ ft with a speed of $(v_A)_1 = 5$ ft/s while holding on to the end of a rope. If her partner starts to pull the rope inward with a constant speed of $v_r = 4$ ft/s, determine the girl's speed at the instant $r_B = 10$ ft. How much work is done by her partner after pulling in the rope? Neglect friction and assume the girl remains in a rigid position.

Prob. 15–63

*15–64. A small block having a mass of 0.1 kg is given a horizontal velocity of $v_1 = 0.4$ m/s when $r_1 = 500$ mm. It slides along the smooth conical surface. When it descends to $h = 100$ mm, determine its speed and the angle of descent, that is, the angle measured from the horizontal to the tangent of the path.

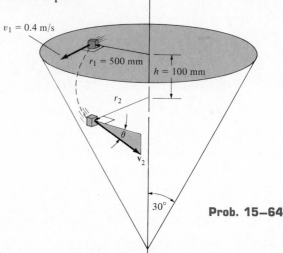

Prob. 15–64

15–65. A gymnast having a mass of 80 kg holds the two rings with his arms down in the position shown as he swings downward. His center of mass is located at point G_1. When he is at the bottom position of his swing, his velocity is $(v_G)_1 = 5$ m/s. At this lower position he *suddenly* lets his arms come up, shifting his center of mass to position G_2. Determine his new velocity in the upswing due to this sudden movement, and the angle θ to which he swings before coming to rest. Treat his body as a particle.

Prob. 15–65

15–66. Two monkeys begin to climb opposite ends of a rope which is draped over a pulley. Monkey A has a mass of 6 kg and climbs at a constant speed of $v_{A/r} = 4$ m/s, measured relative to the rope. Monkey B has a mass of 7 kg and climbs at a constant relative speed $v_{B/r} = 2$ m/s. Determine the speed of each monkey as measured by a fixed observer. Neglect the mass of the pulley and rope and the size of the monkeys.

Prob. 15–66

15–67. The 10-lb block rests on a surface for which $\mu = 0.5$. It is acted upon by a radial force of 2 lb and a horizontal force of 7 lb, always directed at 30° from the tangent to the path as shown. If the block is initially moving in a circular path with a speed $v_1 = 2$ ft/s at the instant the forces are applied, determine the time required before the tension in cord AB becomes 20 lb. Neglect the size of the block for the calculation.

Prob. 15–67

*15–68. The spring has an unstretched length $l_0 = 1.5$ ft and a stiffness $k = 12$ lb/ft. It is attached to a fixed point at A and a block at B, which has a weight of 2 lb. If the block is released from rest from the position shown, determine its speed when it reaches point C after it slides along the smooth guide. After leaving the guide, it is launched onto the smooth horizontal plane. Determine its speed at D at the instant the spring becomes unstretched. Also, how close, d, does the block come within approaching A?

Prob. 15–68

■15–69. The ball B has a weight of 5 lb and is originally rotating in a circle. As shown, the cord AB has a length of 3 ft and passes through the hole at A, which is 2 ft above the plane of motion. If 1.5 ft of cord is pulled through the hole, determine the speed of the ball when it moves in a circular path at C.

Prob. 15–69

213

15–70. An earth satellite of mass 700 kg is launched into a free-flight trajectory about the earth with an initial speed $v_A = 10$ km/s when the distance from the center of the earth is $r_A = 15$ Mm. If the launch angle at this position is $\phi_A = 70°$, determine the speed v_B of the satellite and its closest distance r_B from the center of the earth. The earth has a mass $M_e = 5.976(10^{24})$ kg. *Hint:* Under these conditions, the satellite is subjected only to the earth's gravitational force, $F = GM_e m_s/r^2$, Eq. 13–1. For part of the solution, use the conservation of energy (see Prob. 14–52).

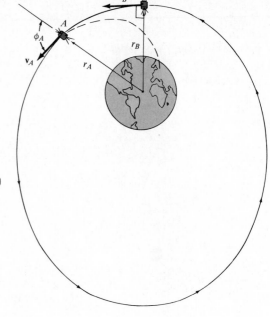

Prob. 15–70

*15.8 Steady Fluid Streams

Knowledge of the forces developed by steadily moving fluid streams is of importance in the design and analysis of turbines, pumps, blades, and fans. To illustrate how the principle of impulse and momentum may be used to determine these forces, consider the diversion of a steady stream of fluid (liquid or gas) by a fixed pipe, Fig. 15–26a. The fluid enters the pipe with a velocity $\mathbf{v}_A$ and exits with a velocity $\mathbf{v}_B$. The impulse and momentum diagrams for the fluid stream are shown in Fig. 15–26b. The force $\Sigma\mathbf{F}$, shown on the impulse diagram, represents the resultant of all the external forces acting on the fluid stream. It is this loading which gives the fluid stream an impulse whereby the original momentum of the fluid is changed in both its magnitude and direction. Since the flow is steady, $\Sigma\mathbf{F}$ will be *constant* during the time interval dt. During this time the fluid stream is in motion, and as a result a small amount of fluid, having a mass dm, is about to enter the pipe with a velocity $\mathbf{v}_A$ at time t. If this element of mass and the mass of fluid in the pipe is considered as a "closed system," then at time $t + dt$ a corresponding element of mass dm must leave the pipe with a velocity $\mathbf{v}_B$. The fluid stream *within* the pipe section has a mass m and an *average velocity* $\mathbf{v}$ which is constant during the time interval dt. Applying the principle of linear impulse and momentum to the fluid stream, we have

$$dm\,\mathbf{v}_A + m\mathbf{v} + \Sigma\mathbf{F}\,dt = dm\,\mathbf{v}_B + m\mathbf{v}$$

(a)

Time t

(b)

(c)

Fig. 15–26

Force Resultant. Solving for the resultant force yields

$$\Sigma\mathbf{F} = \frac{dm}{dt}(\mathbf{v}_B - \mathbf{v}_A) \qquad (15\text{–}25)$$

Provided the motion of the fluid can be represented in the x-y plane, it is usually convenient to express this vector equation in the form of two scalar component equations, i.e.,

$$\Sigma F_x = \frac{dm}{dt}(v_{Bx} - v_{Ax}) \qquad \Sigma F_y = \frac{dm}{dt}(v_{By} - v_{Ay}) \qquad (15\text{–}26)$$

The term dm/dt is called the *mass flow* and indicates the constant amount of fluid which flows either into or out of the pipe per unit of time. If the cross-sectional areas and densities of the fluid at the entrance A and exit B are ρ_A, A_A and ρ_B, A_B, respectively, Fig. 15–26c, then *continuity of mass* requires that $dm = \rho\, dV = \rho_A(ds_A\, A_A) = \rho_B(ds_B\, A_B)$. Hence, during the time dt, since $v_A = ds_A/dt$ and $v_B = ds_B/dt$, we have

$$\frac{dm}{dt} = \rho_A v_A A_A = \rho_B v_B A_B = \rho_A Q_A = \rho_B Q_B \qquad (15\text{–}27)$$

Here $Q = vA$ is the volumetric *flow rate*, which measures the volume of fluid flowing per unit of time.

Moment Resultant. In some cases it is necessary to obtain the support reactions on the fluid-carrying device. If Eq. 15–25 does not provide enough information to do this, the principle of angular impulse and momentum must be used. The formulation of this principle applied to fluid streams can be obtained from Eq. 15–17, $\Sigma \mathbf{M}_O = \dot{\mathbf{H}}_O$, which states that the moment of all the external forces acting on the system about point O is equal to the time rate of change of angular momentum about O. In the case of the pipe shown in Fig. 15–26a, the flow is steady in the x-y plane; hence we have

$$(\curvearrowright+) \qquad \Sigma M_O = \frac{dm}{dt}(d_{OB}\, v_B - d_{OA}\, v_A) \qquad (15\text{–}28)$$

where the moment arms d_{OB} and d_{OA} are directed from O to the *center* of the openings at A and B.

PROCEDURE FOR ANALYSIS

The following procedure provides a method for solving problems involving steady flow.

Kinematic Diagram. In problems where the device is *moving*, a *kinematic diagram* may be helpful for determining the entrance and exit velocities of the fluid flowing onto the device, since a *relative-motion analysis* of velocity will be involved. The *kinematic diagram* in this case is simply a graphical representation of the velocities showing the vector addition of the relative-motion components. Once the velocity of the fluid flowing onto the device is determined, the mass flow is calculated using Eq. 15–27.

Free-Body Diagram. Draw a free-body diagram of the device which is directing the fluid in order to establish the forces $\Sigma \mathbf{F}$ acting on it. These external forces will include the support reactions, the weight of the device and the fluid contained within it, and the static pressure forces of the fluid at the entrance and exit sections of the device.*

Equations of Steady Flow. Apply the equations of steady flow, Eqs. 15–26 and 15–28, using the appropriate components of velocity and force shown on the kinematic and free-body diagrams.

The following examples numerically illustrate application of this procedure.

*In the SI system pressure is measured using the *pascal* (Pa), where $1 \text{ Pa} = 1 \text{ N/m}^2$.

Example 15–15

Determine the reaction components which the pipe joint at A exerts on the elbow in Fig. 15–27a, if water flowing through the pipe is subjected to a static pressure of 100 kPa at A. The discharge at B is $Q_B = 0.2$ m³/s. Water is assumed to have a constant density $\rho_w = 1000$ kg/m³, and the water-filled elbow has a mass of 20 kg and center of gravity at G.

Solution

The absolute velocity of flow at A and B and the mass flow rate can be obtained from Eq. 15–27. Since the density of water is constant, $Q_B = Q_A = Q$. Hence,

$$\frac{dm}{dt} = \rho_w Q = (1000 \text{ kg/m}^3)(0.2 \text{ m}^3/\text{s}) = 200 \text{ kg/s}$$

$$v_B = \frac{Q}{A_B} = \frac{0.2 \text{ m}^3/\text{s}}{\pi(0.05 \text{ m})^2} = 25.46 \text{ m/s}$$

$$v_A = \frac{Q}{A_A} = \frac{0.2 \text{ m}^3/\text{s}}{\pi(0.1 \text{ m})^2} = 6.37 \text{ m/s}$$

Free-Body Diagram. As shown on the free-body diagram, Fig. 15–27b, the *fixed* connection at A exerts a resultant couple $\mathbf{M}_O$ and force components $\mathbf{F}_x$ and $\mathbf{F}_y$ on the elbow. Due to the static pressure of water in the pipe, the pressure force acting on the fluid at A is $F_A = p_A A_A$. Since 1 kPa = 1000 N/m²,

$$F_A = p_A A_A; \quad F_A = (100(10^3) \text{ N/m}^2)[\pi(0.1 \text{ m})^2] = 3141.6 \text{ N}$$

There is no static pressure acting at B, since the water is discharged at atmospheric pressure; i.e., the pressure measured by a gauge at B is equal to zero, $p_B = 0$.

Fig. 15–27

Equations of Steady Flow

$$(\xrightarrow{+}) \quad \Sigma F_x = \frac{dm}{dt}(v_{Bx} - v_{Ax}); \quad -F_x + 3141.6 = 200(0 - 6.37)$$

$$F_x = 4.42 \text{ kN} \leftarrow \qquad Ans.$$

$$(+\uparrow) \quad \Sigma F_y = \frac{dm}{dt}(v_{By} - v_{Ay}); \quad -F_y - 20(9.81) = 200(-25.46 - 0)$$

$$F_y = 4.90 \text{ kN} \downarrow \qquad Ans.$$

If moments are summed about point O, Fig. 15–27b, $\mathbf{F}_x$, $\mathbf{F}_y$, and the static pressure $\mathbf{F}_A$ are eliminated, as well as the moment of momentum of the water entering at A, Fig. 15–27a. Hence,

$$(\curvearrowleft+) \qquad \Sigma M_O = \frac{dm}{dt}(d_{OB}v_B - d_{OA}v_A)$$

$$M_O + 20(9.81)(0.125) = 200[(0.3)(25.46) - 0]$$

$$M_O = 1.50 \text{ kN} \cdot \text{m} \downarrow \qquad Ans.$$

Example 15–16

A 2-in.-diameter water jet having a velocity of 25 ft/s impinges upon a single moving blade, Fig. 15–28a. If the blade is moving at 5 ft/s away from the jet, determine the horizontal and vertical components of force which the blade is exerting on the water, $\rho_w = 62.4$ lb/ft^3. What power does the water generate on the blade?

2 in.

$v_{bl} = 5$ ft/s

$v_w = 25$ ft/s

(a)

Solution

Kinematic Diagram. From a fixed coordinate system, Fig. 15–28b, the rate at which water enters the blade is

$$\mathbf{v}_A = \{20\mathbf{i}\} \text{ ft/s}$$

The *relative-flow velocity* of the water onto the blade is $\mathbf{v}_{w/bl} = \mathbf{v}_w - \mathbf{v}_{bl} = 25\mathbf{i} - 5\mathbf{i} = \{20\mathbf{i}\}$ ft/s. Since the blade is moving with a velocity of $\mathbf{v}_{bl} = \{5\mathbf{i}\}$ ft/s, the velocity of flow at B measured from x, y is the vector sum, shown in Fig. 15–28b. Here,

$$\mathbf{v}_B = \mathbf{v}_{bl} + \mathbf{v}_{w/bl}$$
$$= \{5\mathbf{i} + 20\mathbf{j}\} \text{ ft/s}$$

(b)

The mass flow of water onto the blade is

$$\frac{dm}{dt} = \rho_w(v_{w/bl})A_A = \frac{62.4}{32.2}(20)\left[\pi\left(\frac{1}{12}\right)^2\right] = 0.846 \text{ slug/s}$$

Free-Body Diagram. The free-body diagram of a section of water acting on the blade is shown in Fig. 15–28c. The weight of the water will be neglected in the calculation, since this force is small compared to the reactive components $\mathbf{F}_x$ and $\mathbf{F}_y$.

$-F_x\mathbf{i}$

$F_y\mathbf{j}$

(c)

Fig. 15–28

Equations of Steady Flow

$$\Sigma\mathbf{F} = \frac{dm}{dt}(\mathbf{v}_B - \mathbf{v}_A)$$

$$-F_x\mathbf{i} + F_y\mathbf{j} = 0.846(5\mathbf{i} + 20\mathbf{j} - 25\mathbf{i})$$

Equating the respective $\mathbf{i}$ and $\mathbf{j}$ components gives

$$F_x = 0.846(20) = 16.9 \text{ lb} \leftarrow \qquad \text{Ans.}$$
$$F_y = 0.846(20) = 16.9 \text{ lb} \uparrow \qquad \text{Ans.}$$

The water exerts equal but opposite forces on the blade.

Since the water force which causes the blade to move forward horizontally with a velocity of 5 ft/s is $F_x = 16.9$ lb, then from Eq. 14–11 the power is

$$P = \mathbf{F} \cdot \mathbf{v}; \qquad P = \frac{16.9(5)}{550} = 0.154 \text{ hp} \qquad \text{Ans.}$$

Propulsion with Variable Mass $15.9^{\star}$

In Sec. 15.8 the case in which a *constant* amount of mass *dm* entered and left a "*closed system*" was considered. There are, however, two other important cases involving mass flow, which are represented by a system which is either gaining or losing mass. In this section these cases will be discussed separately.

A System That Loses Mass. Consider a device which at an instant of time has a mass *m* and a forward velocity **v**, Fig. 15–29a. At this same instant the device is expelling an amount of mass m_e such that the mass flow velocity $\mathbf{v}_e$ is *constant* when the measurement is made a small distance away from the device. For the analysis, consider the "*closed system*" at an instant to include *both the mass m and the expelled mass m_e*, as shown by the dashed line in the figure. The impulse and momentum diagrams for the system are shown in Fig. 15–29b. During the time interval *dt*, the velocity of the device is increased from **v** to **v** + *d***v**. This increase in forward velocity, however, does not change the velocity $\mathbf{v}_e$ of the expelled mass, since this mass moves at a constant speed once it has been ejected. To increase the velocity of the device during the time *dt*, an amount of mass dm_e has been ejected and thereby gained in the exhaust. The impulses are created by $\Sigma\mathbf{F}_s$, which represents the resultant of all the external forces which *act on the system* in the direction of motion. This force resultant *does not include* the force which causes the device to move forward, since this force (called a *thrust*) is *internal to the system;* that is, the thrust acts with equal magnitude but opposite direction on the mass *m* of the device and the expelled exhaust mass m_e.* Applying the principle of impulse and momentum to the system, in reference to Fig. 15–29b, we have

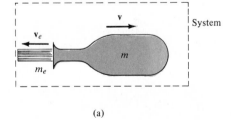

(a)

$$(\xrightarrow{+}) \qquad mv - m_e v_e + \Sigma F_s \, dt = (m - dm_e)(v + dv) - (m_e + dm_e)v_e$$

or

$$\Sigma F_s \, dt = -v \, dm_e + m \, dv - dm_e \, dv - v_e \, dm_e$$

*$\Sigma\mathbf{F}_s$ represents the external resultant force *acting on the system,* which is different from $\Sigma\mathbf{F}$, the resultant force acting only on the device.

(b)

Fig. 15–29

Fig. 15–30

Without loss of accuracy, the third term on the right side of this equation may be neglected since it is a "second-order" differential. Dividing by dt gives

$$\Sigma F_s = m\frac{dv}{dt} - (v + v_e)\frac{dm_e}{dt}$$

Noting that the relative velocity of the device as seen by an observer moving with the particles of the ejected mass is $v_{D/e} = (v + v_e)$, the final result can be written as

$$\Sigma F_s = m\frac{dv}{dt} - v_{D/e}\frac{dm_e}{dt} \tag{15–29}$$

Here the term dm_e/dt represents the rate at which mass is being ejected.

To illustrate an application of Eq. 15–29, consider the rocket shown in Fig. 15–30, which has a weight **W** and is moving upward against an atmospheric drag force **F_D**. The system to be considered consists of the mass of the rocket and the mass of ejected gas m_e. Applying Eq. 15–29 to this system gives

$$(+\uparrow) \qquad -F_D - W = \frac{W}{g}\frac{dv}{dt} - v_{D/e}\frac{dm_e}{dt}$$

The last term of this equation represents the *thrusting force* **T** which the engine exhaust exerts on the rocket, Fig. 15–30. Recognizing that $dv/dt = a$, we may therefore write

$$(+\uparrow) \qquad T - F_D - W = \frac{W}{g}a$$

If a free-body diagram of the rocket is drawn, it becomes obvious that this latter equation represents an application of $\Sigma F = ma$ for the rocket.

A System That Gains Mass. A device such as a scoop or a shovel may gain mass as it moves forward. Consider, for example, the device shown in Fig. 15–31a, which at an instant of time has a mass m and is moving forward with a velocity **v**. At the same instant, the device is collecting a particle stream of mass m_i. The flow velocity $\mathbf{v}_i$ of this injected mass is constant and independent of the velocity **v**. It is required that $v > v_i$. The system to be considered at this instant includes both the mass of the device and the mass of the injected fluid, as shown by the dashed line in the figure. The momentum- and impulse-vector diagrams for this system are shown in Fig. 15–31b. With an increase in mass dm_i gained by the device, there is an assumed increase in velocity dv during the time interval dt. This increase is caused by the impulse created by $\Sigma \mathbf{F}_s$, the resultant of all the external forces *acting on the system* in the direction of motion. The force summation does not include the retarding force of the injected mass acting on the device. Why? Applying the principle of impulse and momentum to the system, we have

$$(\xrightarrow{+}) \qquad mv + m_i v_i + \Sigma F_s\, dt = (m + dm_i)(v + dv) + (m_i - dm_i)v_i$$

(a)

(b)

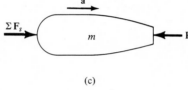

(c)

Fig. 15–31

Using the same procedure as in the previous case, we may write this equation as

$$\Sigma F_s = m\frac{dv}{dt} + (v - v_i)\frac{dm_i}{dt}$$

Since the relative velocity of the device as seen by an observer moving with the particles of the injected mass is $v_{D/i} = (v - v_i)$, the final result can be written as

$$\Sigma F_s = m\frac{dv}{dt} + v_{D/i}\frac{dm_i}{dt} \qquad (15\text{–}30)$$

where dm_i/dt is the rate of mass injected into the device. The last term in this equation represents the magnitude of force **R**, which the injected mass *exerts on the device*. Since $dv/dt = a$, Eq. 15–30 becomes

$$\Sigma F_s - R = ma$$

which represents the application of $\Sigma \mathbf{F} = m\mathbf{a}$, Fig. 15–31$c$.

As in the case of steady flow, problems which are solved using Eqs. 15–29 and 15–30 should be accompanied by the necessary free-body diagram. With this diagram one can then determine ΣF_s *for the system* and isolate the force exerted on the device by the particle stream.

Example 15–17

The initial combined mass of a rocket and its fuel is $m_O = 2.5(10^6)$ kg. If a total of $1.5(10^6)$ kg of fuel is consumed at a constant rate of $7.5(10^3)$ kg/s and expelled at a constant speed of 2500 m/s, determine the velocity of the rocket at the instant the fuel runs out. Neglect the change in gravitational force with altitude and the drag resistance of the air. The rocket is fired vertically from rest.

Solution

Since the rocket is losing mass as it moves upward, Eq. 15–29 can be used for the solution. The only *external force* acting on the *system* consisting of the rocket and a portion of the expelled mass is the weight **W**, Fig. 15–32. Hence,

$$(+\uparrow) \quad \Sigma F_s = m\frac{dv}{dt} - v_{D/e}\frac{dm_e}{dt}; \quad -W = m\frac{dv}{dt} - v_{D/e}\frac{dm_e}{dt} \quad (1)$$

From the problem data, $dm_e/dt = 7.5(10^3)$ kg/s and $v_{D/e} = 2500$ m/s. At any given time t during the flight, the mass of the rocket can be expressed as $m = m_O - (dm_e/dt)t = 2.5(10^6) - 7.5(10^3)t$. Since $W = mg$, Eq. (1) becomes

$$-[2.5(10^6) - 7.5(10^3)t](9.81) = [2.5(10^6) - 7.5(10^3)t]\frac{dv}{dt} - 2500[7.5(10^3)]$$

Fig. 15–32

Rearranging the terms and integrating, realizing that $v_0 = 0$ at $t_0 = 0$, we have

$$\int_0^v dv = \int_0^t \left[\frac{18.75(10^6)}{2.5(10^6) - 7.5(10^3)t} - 9.81\right] dt$$

$$v = -2500 \ln[2.5(10^6) - 7.5(10^3)t] - 9.81t \Big|_0^t$$

$$v = 2500 \ln\frac{2.5(10^6)}{2.5(10^6) - 7.5(10^3)t} - 9.81t \quad (2)$$

This equation gives the speed of the rocket at any instant of time. The time t' needed to consume all the fuel is

$$m_f = (dm_e/dt)t'$$
$$1.5(10^6) = [7.5(10^3)]t'$$

Hence,

$$t' = 200 \text{ s}$$

Substituting into Eq. (2) and solving yields

$$v = 329 \text{ m/s} \uparrow \qquad \qquad \textit{Ans.}$$

Example 15–18

A chain of length l, Fig. 15–33a, has a mass per unit length of γ. Determine the magnitude of force **F** required to (a) raise the chain with a constant speed v_c, starting from rest when $y = 0$; and (b) lower the chain with a constant speed v_c, starting from rest when $y = l$.

Solution

Part (a). As the chain is raised, all the suspended links are given a sudden impulse downward by each added link which is lifted off the ground. Thus, the *suspended portion* of the chain may be considered as a device which is *gaining mass*. A free-body diagram of a portion of the chain which is located at an arbitrary height y above the ground is shown in Fig. 15–33b. The system to be considered is the length of chain y which is suspended by **F** at any instant, including the next link which is about to be added but is still at rest. The forces acting on this system *exclude* the internal force **P** which the added link exerts on the suspended portion of the chain. Hence, $\Sigma F_s = F - \gamma g y$.

It now becomes necessary to find the rate at which mass is being added to the system. The velocity $\mathbf{v}_c$ of the chain at a given instant is equivalent to $\mathbf{v}_{D/i}$. Why? Since v_c is constant, $dv_c/dt = 0$ and $dy/dt = v_c$. Integrating, using the initial condition that $y = 0$ at $t = 0$, gives $y = v_c t$. Thus, the mass of the system at any instant is $m = \gamma y = \gamma v_c t$, and therefore the *rate* at which mass is *added* to the suspended chain is

$$\frac{dm_i}{dt} = \gamma v_c$$

Applying Eq. 15–30 to the system, using this data, we have

$(+\uparrow)$
$$\Sigma F_s = m\frac{dv_c}{dt} + v_{D/i}\frac{dm_i}{dt}$$

$$F - \gamma g y = 0 + v_c(\gamma v_c)$$

Hence,
$$F = \gamma(g y + v_c^2) \qquad \textit{Ans.}$$

Part (b). When the chain is being lowered, the links which are expelled (given zero velocity) *do not* impart an impulse to the *remaining* suspended links. Why? Thus, the system in Part (a) cannot be considered. Instead, the equation of motion will be used to obtain the solution. At time t the portion of chain still off the floor is y. The free-body diagram for a suspended portion of the chain is shown in Fig. 15–33c. Thus,

$+\uparrow \Sigma F = ma;$
$$F - \gamma g y = 0$$
$$F = \gamma g y \qquad \textit{Ans.}$$

(a)

(b)

(c)

Fig. 15–33

Problems

15–71. The fireman sprays a 2-in.-diameter jet of water from a hose at the building. If the water is discharged at 0.80 ft³/s, determine the elevation y to where the center line of the water splashes on the wall.

Prob. 15–71

15–73. The chute is used to divert the flow of water $Q = 0.6$ m³/s. If the water has a cross-sectional area of 0.05 m², determine the force components at the pin A and roller B necessary for equilibrium. Neglect both the weight of the chute and the weight of the water on the chute. $\rho_w = 1$ Mg/m³.

Prob. 15–73

*15–72. The boy at A aims the water hose in the direction shown. If the water is discharged at 30° from the horizontal, and the cross-sectional area of the water stream is approximately 2 in², determine the force the boy must exert on the shield in order to hold the shield perpendicular to the water. $\gamma_w = 62.4$ lb/ft³.

Prob. 15–72

15–74. Water is discharged at 16 m/s against the fixed cone diffuser. If the opening diameter of the nozzle is 40 mm, determine the horizontal force exerted by the water on the diffuser, $\rho_w = 1$ Mg/m³.

Prob. 15–74

15–75. The elbow for a 5-in.-diameter buried pipe is subjected to a static pressure of 10 lb/in^2. The velocity of the water passing through it is $v = 8$ ft/s. Assuming the connections at A and B do not offer any vertical force resistance on the elbow, determine the resultant vertical force **F** that the soil must then exert on the elbow in order to hold it in equilibrium. $\gamma_w = 62.4$ lb/ft^3. Neglect the weight of the elbow and the water within it.

Prob. 15–75

*15–76. The buckets on the Pelton wheel are subjected to a 2-in.-diameter jet of water, which has a velocity of 150 ft/s. If each bucket is traveling at 95 ft/s when the water strikes it, determine the power developed by the wheel. $\gamma_w = 62.4$ lb/ft^3.

Prob. 15–76

15–77. The blade divides the jet of water having a diameter of 3 in. If one fourth of the water flows downward while the other three fourths flow upward, and the flow is $Q = 0.5$ ft^3/s, determine the horizontal and vertical components of force exerted on the blade by the jet. $\gamma_w = 62.4$ lb/ft^3.

Prob. 15–77

15–78. The static pressure of water at C is 40 lb/in^2. If water flows out of the pipe at A and B with velocities $v_A = 25$ ft/s and $v_B = 38$ ft/s, determine the x and y components of force exerted on the elbow at C necessary to hold the pipe assembly in equilibrium. Neglect the weight of water within the pipe and the weight of the pipe. The pipe has a diameter of 0.75 in. at C, and at A and B the diameter is 0.5 in. $\gamma_w = 62.4$ lb/ft^3.

Prob. 15–78

15–79. Water is flowing from the 150-mm-diameter fire hydrant with a velocity $v_B = 15$ m/s. Determine the horizontal and vertical components of force and the moment developed at the base joint A, if the static (gauge) pressure at A is 50 kPa and the density of water is 1 Mg/m^3. The diameter of the fire hydrant at A is 200 mm.

$v_B = 15$ m/s

B

500 mm

A

Prob. 15–79

*15–80. A rocket burns 1000 lb of fuel at a constant rate for $t = 40$ s. If the velocity of the exhaust gas is $v_{e/r} = 7500$ ft/s, measured with respect to the rocket, determine the thrust of the engine.

15–81. The second stage B of the two-stage rocket has a mass of 5 Mg (empty) and is launched from the first stage A with an initial velocity of 600 km/h. The fuel in the second stage has a mass of 0.7 Mg and is consumed at the rate of 4 kg/s. If it is ejected from the rocket at the rate of 3 km/s, measured relative to B, determine the acceleration of B at the instant the engine is fired and just before all the fuel is consumed. Neglect the effects of gravitation and air resistance.

B

$v_B = 600$ km/h

A

Prob. 15–81

15–82. The rocket has a mass of 65 Mg including the fuel. Determine the constant rate at which the fuel must be burned so that its thrust gives the rocket a speed of 200 ft/s in 10 s starting from rest. The fuel is expelled from the rocket at a relative speed of 3000 ft/s. Neglect the effects of air resistance and assume that g is constant.

v

Prob. 15–82

15–83. The 12-Mg jet airplane has a constant speed of 950 km/h when it is flying along a horizontal straight-line path. Air enters the intake scoops S at the rate of 50 m^3/s. If the engine burns fuel at the rate of 0.4 kg/s and the gas is exhausted relative to the plane with a speed of 450 m/s, determine the resultant drag force exerted on the plane by air resistance. Assume that air has a constant density of 1.22 kg/m^3. *Hint:* Since mass both enters and exits the plane, Eqs. 15–29 and 15–30 must be combined to yield

$$\Sigma F_s = m\frac{dv}{dt} - v_{D/e}\frac{dm_e}{dt} + v_{D/i}\frac{dm_i}{dt}$$

$v = 950$ km/h

S

Prob. 15–83

*15–84. The jet is traveling at a speed of 500 mi/h, 30° with the horizontal. If the fuel is being spent at 3 lb/s, and the engine takes in air at 400 lb/s, whereas the exhaust gas has a relative speed of 32,800 ft/s, determine the acceleration of the plane. The drag resistance of the air is $F_D = (0.7v^2)$ lb, where the speed is measured in ft/s. The jet has a weight of 15,000 lb. *Hint:* See Prob. 15–83.

500 mi/h

30°

Prob. 15–84

15–85. The truck has a mass of 50 Mg when empty. When it is unloading 5 m³ of sand at a constant rate of 0.8 m³/s, the sand flows out the back at a speed of 7 m/s, measured relative to the truck, in the direction shown. If the truck is free to roll, determine its initial acceleration just as the load begins to empty. Neglect the mass of the wheels and any frictional resistance to motion. The density of sand is $\rho_s = 1520$ kg/m³.

a

45°

7 m/s

Prob. 15–85

15–86. The earthmover initially carries 10 m³ of sand having a density of 1520 kg/m³. The sand is unloaded horizontally through a 2.5-m² dumping port P at a rate of 900 kg/s measured relative to the port. If the earthmover maintains a constant resultant tractive force of 4 kN at its wheels to provide forward motion, determine its acceleration when half the sand is dumped. When empty, the earthmover has a mass of 30 Mg. Neglect any resistance to forward motion and the mass of the wheels. The rear wheels are free to roll.

a

P

$F = 4$ kN

Prob. 15–86

15–87. The rocket car has a mass of 3 Mg (empty) and carries 150 kg of fuel. If the fuel is consumed at a constant rate of 4 kg/s and ejected from the car with a relative velocity of 250 m/s, determine the maximum speed attained by the car starting from rest. The drag resistance due to the atmosphere is $F_D = (60v^2)$ N, where v is the speed measured in m/s.

Prob. 15–87

*15–88. The car has a mass m_0 and is used to tow the smooth chain having a total length l and a mass per unit of length m'. If the chain is originally piled up, determine the tractive force **F** that must be supplied by the rear wheels of the car necessary to maintain a constant speed v while the chain is being drawn out.

v

F

Prob. 15–88

15–89. If the chain is lowered at a constant velocity $v = 4$ ft/s, determine the normal reaction exerted on the floor as a function of time. The chain has a weight of 5 lb/ft and a total length of 20 ft.

Prob. 15–89

15–90. Determine the magnitude of force **F** as a function of time, which must be applied to the end of the cord at A to raise the hook H with a constant velocity $v = 0.4$ m/s. Initially the chain is at rest on the ground. Neglect the mass of the cord and the hook. The chain has a mass of 2 kg/m.

Prob. 15–90

15–91. The rocket has an initial mass m_0, including the fuel. For practical reasons desired for the passengers, it is required that it maintain a constant upward acceleration a_0. If the fuel is expelled from the rocket at a relative speed $v_{e/r}$, determine the rate at which the fuel should be consumed to maintain the motion. Neglect air resistance, and assume that the gravitational acceleration is constant.

Prob. 15–91

Review 1:
Kinematics and Kinetics of a Particle

The various topics and problems presented in Chapters 12 through 15 have all been *categorized* in order to provide a *clear focus* for learning the various problem-solving principles involved. In engineering practice, however, it is most important to be able to *identify* an appropriate method for the solution of a particular problem. In this regard, one must fully understand the limitations and use of the equations of dynamics, and be able to recognize which equations and principles to use for the problem's solution. For these reasons, we will now summarize the equations and principles of particle dynamics and provide the opportunity for applying them to a variety of problems.

Kinematics. Problems in kinematics require a study only of the geometry of motion, and do not account for the forces causing the motion. When the equations of kinematics are applied, one should clearly establish a fixed origin and select an appropriate coordinate system used to define the position of the particle. Once the positive direction of each coordinate axis is established, then the directions of the components of position, velocity, and acceleration can be determined from the algebraic sign of their numerical quantities.

Rectilinear Motion

Variable Acceleration. If a mathematical (or graphical) relationship is established between *any two* of the *four* variables s, v, a, and t, then a *third* variable can be determined by solving one of the following equations which relates all three variables.

$$v = \frac{ds}{dt} \qquad a = \frac{dv}{dt} \qquad a\,ds = v\,dv$$

Constant Acceleration. Be *absolutely certain* that the acceleration is constant when using the following equations:

$$s = s_0 + v_0 t + \tfrac{1}{2}a_c t^2 \qquad v = v_0 + a_c t \qquad v^2 = v_0{}^2 + 2a_c(s - s_0)$$

Curvilinear Motion

x, y, z Coordinates. Often used when the motion can be resolved into horizontal and vertical components, as in the case of projectile motion.

$$v_x = \dot{x} \qquad a_x = \dot{v}_x$$
$$v_y = \dot{y} \qquad a_y = \dot{v}_y$$
$$v_z = \dot{z} \qquad a_z = \dot{v}_z$$

n, t, b Coordinates. Used when measurements of the motion are made along the path.

$$v = \dot{s}$$
$$a_t = \dot{v} = v\frac{dv}{ds}$$
$$a_n = \frac{v^2}{\rho}$$

where $\rho = \left| \dfrac{[1 + (dy/dx)^2]^{3/2}}{d^2y/dx^2} \right|$ when the path $y = f(x)$ is given.

r, θ, z Coordinates. Can be used for circular types of motion or when the motion is measured radially from a fixed point.

$$v_r = \dot{r} \qquad a_r = \ddot{r} - r\dot{\theta}^2$$
$$v_\theta = r\dot{\theta} \qquad a_\theta = r\ddot{\theta} + 2\dot{r}\dot{\theta}$$
$$v_z = \dot{z} \qquad a_z = \ddot{z}$$

Relative Motion. Consider the origin of the coordinate system to be established at one of the particles (*A*), then for particle *B*

$$\mathbf{r}_B = \mathbf{r}_A + \mathbf{r}_{B/A}$$
$$\mathbf{v}_B = \mathbf{v}_A + \mathbf{v}_{B/A}$$
$$\mathbf{a}_B = \mathbf{a}_A + \mathbf{a}_{B/A}$$

Kinetics. Problems in kinetics involve the analysis of forces which cause the motion. When applying the equations of kinetics, it is absolutely necessary that measurements of the motion be made from an *inertial coordinate system*, i.e., one that does not rotate and is either fixed or translates with constant velocity. If a problem requires *simultaneous solution* of the equations of kinetics and kinematics, then the coordinate systems selected for writing each of the equations should define the *positive directions* of the axes in the *same* manner.

Equations of Motion. These equations are used to solve for the particle's acceleration or the forces causing the motion. If they are used to determine a particle's position, velocity, or time of motion, then kinematics will also have to be considered in the solution. Before applying the equations of motion, *always draw a free-body diagram* to identify all the forces acting on the particle. Also, establish the direction of the particle's acceleration or its components. (A kinetic diagram may accompany the solution in order to graphically account for the $m\mathbf{a}$ vector.)

$$\Sigma F_x = ma_x \qquad \Sigma F_n = ma_n \qquad \Sigma F_r = ma_r$$
$$\Sigma F_y = ma_y \qquad \Sigma F_t = ma_t \qquad \Sigma F_\theta = ma_\theta$$
$$\Sigma F_z = ma_z \qquad \Sigma F_b = 0 \qquad \Sigma F_z = ma_z$$

Work and Energy. The equation of work and energy represents an integrated form of the tangential equation of motion, $\Sigma F_t = ma_t$, combined with kinematics ($a_t\,ds = v\,dv$). *It is used to solve problems involving force, velocity, and displacement.* Before applying this equation, *always draw a free-body diagram* in order to identify the forces which do work on the particle.

$$T_1 + \Sigma U_{1-2} = T_2$$

where

$$T = \tfrac{1}{2}mv^2 \quad \text{(kinetic energy)}$$

$$U_F = \int_{s_1}^{s_2} F \cos\theta\,ds \quad \text{(variable force)}$$

$$U_{F_c} = F_c \cos\theta\,(s_2 - s_1) \quad \text{(constant force)}$$

$$U_W = W\,\Delta y \quad \text{(weight)}$$

$$U_s = -(\tfrac{1}{2}ks_2^2 - \tfrac{1}{2}ks_1^2) \quad \text{(elastic spring)}$$

If the forces acting on the particle are *conservative forces,* i.e., those that *do not* cause a dissipation of energy such as friction, then apply the conservation of energy equation. This equation is easier to use than the equation of work and energy since it applies only at *two points* on the path and *does not* require calculation of the work done by a force as the particle moves along the path.

$$T_1 + V_1 = T_2 + V_2$$

where

$$V_g = Wy \quad \text{(gravitational potential energy)}$$

$$V_e = \tfrac{1}{2}ks^2 \quad \text{(elastic potential energy)}$$

If the *power* developed by a force is to be calculated, use

$$P = \frac{dU}{dt} = \mathbf{F} \cdot \mathbf{v}$$

where $\mathbf{v}$ is the velocity of a particle acted upon by the force $\mathbf{F}$.

Impulse and Momentum. The equation of *linear impulse and momentum* is an integrated form of the equation of motion, $\Sigma \mathbf{F} = m\mathbf{a}$, combined with kinematics ($\mathbf{a} = d\mathbf{v}/dt$). *It is used to solve problems involving force, velocity, and time.* Before applying this equation, one should *always draw the free-body diagram,* in order to identify all the forces that cause impulses on the particle. Once this is done, the impulsive and nonimpulsive forces can be defined. Recall that the nonimpulsive forces can be neglected in the analysis. Also, establish the direction of the particle's velocity just before and just after the impulses are applied. (As an alternative procedure, the impulse and momentum diagrams may accompany the solution in order to graphically account for the terms in the equation.)

$$m\mathbf{v}_1 + \Sigma \int_{t_1}^{t_2} \mathbf{F}\, dt = m\mathbf{v}_2$$

If several particles are involved in the problem, consider applying the *conservation of momentum* to the system. This can be done in a specified direction, provided no external impulses act on the particles in that direction and the internal impulses occur in equal but opposite collinear pairs and thus cancel out.

$$\Sigma m\mathbf{v}_1 = \Sigma m\mathbf{v}_2$$

If the problem involves impact and the coefficient of restitution e is given, then apply the following equation.

$$e = \frac{(v_B)_2 - (v_A)_2}{(v_A)_1 - (v_B)_1} \qquad \text{(along line of impact)}$$

The *principle of angular impulse and momentum* and the *conservation of angular momentum* may be applied about an axis in order to *eliminate* some of the unknown impulses acting on the particle during the time period when its motion is studied. Investigation of the particle's free-body diagram (or the impulse diagram) will aid in choosing the axes for application.

$$(\mathbf{H}_O)_1 + \Sigma \int_{t_1}^{t_2} \mathbf{M}_O\, dt = (\mathbf{H}_O)_2$$

$$(\mathbf{H}_O)_1 = (\mathbf{H}_O)_2$$

The following problems involve application of the above concepts. They are presented in *random order* so as to provide practice in identifying the various types of problems and developing the skills necessary for their solution.

Problems

Except when stated otherwise, for the following problems assume that the coefficients of static and kinetic friction are equal, i.e., $\mu = \mu_s = \mu_k$.

R1–1. Ball A has a velocity of 5 ft/s when it strikes ball B which is initially at rest. After the collision, B moves to the right and strikes ball C which is moving to the left with a velocity of 10 ft/s. If the coefficient of restitution is $e = 0.7$, determine the velocity of ball B just after both collisions. Each ball has a weight of 0.5 lb.

Prob. R1–1

R1–2. The position of a particle may be defined in terms of its rectangular coordinates $x = (3t)$ m, $y = (-4t)$ m, and $z = (2t^2)$ m, where t is in seconds. Express the velocity and acceleration of the particle in terms of its cylindrical coordinates r, θ, z, and determine the magnitudes of the particle's velocity and acceleration when $t = 1.5$ s.

R1–3. The 90-lb force is required to drag the 200-lb block 60 ft up the *rough* inclined plane at constant velocity. If the force is removed when the block reaches point B, and the block is then released from rest, determine the block's velocity when it slides back down the plane and reaches point A.

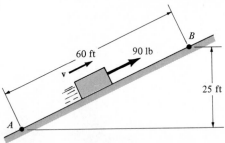

Prob. R1–3

*R1–4. The two particles A and B weigh 2 and 5 lb, respectively. If they are acted upon by the same force system, namely, $\mathbf{F}_1 = \{2\mathbf{i} + 6\mathbf{j} - 2\mathbf{k}\}$ lb, $\mathbf{F}_2 = \{3\mathbf{i} - 1\mathbf{k}\}$ lb, and $\mathbf{F}_3 = \{1\mathbf{i} - t^2\mathbf{j} - 2\mathbf{k}\}$ lb, where t is in seconds, determine the distance between them 3 seconds after they have been released from rest at the position shown.

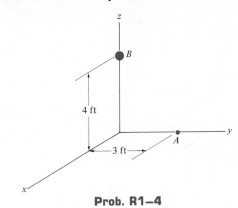

Prob. R1–4

R1–5. An electrically powered train car draws 300 kW of power. If the car has a mass of 18 Mg and travels along a horizontal track, determine the speed it attains in 20 s starting from rest. The mechanical efficiency is $e = 0.8$.

R1–6. Determine the magnitude of the resultant force acting on a 5-kg particle at the instant $t = 2$ s, if the particle is moving along a horizontal path defined by the equations $r = (2t + 10)$ m and $\theta = (1.5t^2 - 6t)$ rad, where t is in seconds.

R1–7. The 3-lb particle P is acted upon by $\mathbf{F}_1$ and $\mathbf{F}_2$. If the particle originally has a velocity $\mathbf{v}_1 = \{3\mathbf{i} + 1\mathbf{j} + 6\mathbf{k}\}$ ft/s, determine its speed after 10 s.

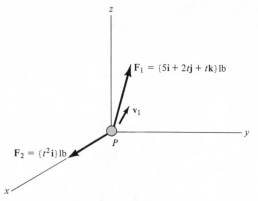

Prob. R1–7

*R1–8. The 6-lb ball is fired from a tube by a spring having a stiffness $k = 20$ lb/in. Determine how far the spring must be compressed to fire the ball from this position to a height of 8 ft, at which point it has a velocity of 6 ft/s.

Prob. R1–8

R1–9. A 50-kg log is given an initial forward velocity $v_1 = 0.25$ m/s in still water. The drag resistance of the water is proportional to the speed of the log at any instant and can be approximated by $F_D = (10v)$ N, where v is measured in m/s. Determine the time needed for the log's speed to become 0.1 m/s. Given an initial velocity $v_1 = 0.25$ m/s, through what distance does the log travel before it stops?

Prob. R1–9

R1–10. The 100-kg crate is subjected to the action of two forces, $F_1 = 800$ N and $F_2 = 1.5$ kN, as shown. If it is originally at rest, determine the distance it slides in order to attain a speed of 6 m/s. The coefficient of friction between the crate and the surface is $\mu = 0.2$.

Prob. R1–10

R1–11. A crate having a mass of 50 kg rests against a stop block S which prevents the crate from moving down the plane. If the coefficient of friction between the plane and the crate is $\mu = 0.3$, determine the time needed for the force $\mathbf{F}$ to give the crate a speed of 2 m/s up the plane. $\mathbf{F}$ always acts parallel to the plane and has a magnitude $F = (300t)$ N, where t is in seconds. *Hint:* First determine the time needed to overcome friction and start the crate moving.

Prob. R1–11

*R1–12. The four 5-lb spheres are rigidly attached to the crossbar frame. The frame has a negligible weight. If a couple moment $M = (5t)$ lb · ft, where t is in seconds, is applied as shown, determine the speed of each of the spheres in 4 s starting from rest.

Prob. R1–12

R1–13. At a given instant the toy cars B and C are located at the positions shown. If car C is traveling at a constant velocity $v_C = 0.2$ m/s, determine the constant acceleration of car B, starting from rest, so that it collides with C at point A. Neglect the size of the cars. Also, what is the relative velocity and relative acceleration of car B with respect to C just before the collision?

Prob. R1–13

■R1–14. Packages having a mass of 2.5 kg ride on the surface of the conveyor belt. If the belt starts from rest and with constant acceleration increases to a speed of 0.75 m/s in 2 s, determine the maximum angle of tilt, θ, so that none of the packages slip on the inclined surface AB of the belt. The coefficient of friction between the belt and each package is $\mu = 0.3$. At what angle ϕ do the packages first begin to slip off the surface of the belt if the belt is moving at a constant speed of 0.75 m/s?

Prob. R1–14

R1–15. The collar of negligible size has a mass of 0.25 kg and is attached to a spring having an unstretched length of 100 mm. If the collar is released from rest at A and travels along the smooth guide, determine its speed just before it strikes B.

Prob. R1–15

*R1–16. The hoisting block and motor M have a total mass of 50 kg. When the motor is turned on, it draws in its attached cable with constant acceleration so that the block moves from rest to a speed of 0.75 m/s when $t = 3$ s. Neglecting the mass of the pulleys and cable, determine the force developed in the cable during this time.

Prob. R1–16

R1–17. The particle P travels with a constant speed of 300 mm/s along the curve. Determine its acceleration when it is located at point (200 mm, 100 mm).

Prob. R1–17

R1–18. The bob of the pendulum has a mass of 0.2 kg and is released from rest when it is in the horizontal position shown. Determine its speed and the tension in the cord at the instant the bob passes through its lowest position.

Prob. R1–18

R1–19. The block rests on a horizontal platform at a distance of 3 ft from its center. If the coefficient of friction between the block and the platform is $\mu = 0.3$, determine the maximum speed which the block can attain before it starts to slip. Assume the angular motion of the platform is slowly increasing.

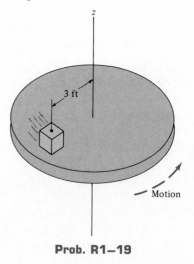

Prob. R1–19

*R1–20. An elastic cord, having an unstretched length of 2 ft and a stiffness $k = 3$ lb/ft, is attached to a fixed point at A and a 6-lb block at B. If the block is given an initial velocity $v_1 = 10$ ft/s outward along the y axis, determine the speed at which the block is moving toward point A the instant the cord becomes slack. How close does the block come to point A after the cord has become slack? Assume that the x-y plane is smooth.

Prob. R1–20

R1–21. The blocks have a coefficient of restitution $e = 0.5$. Determine the velocity of each block after collision and the energy loss during impact. Neglect friction.

Prob. R1–21

R1–22. An ivory ball having a mass of 200 g is released from rest at a height of 400 mm above a very large fixed metal surface. If the ball rebounds to a height of 325 mm above the surface, determine the coefficient of restitution between the ball and the surface.

R1–23. The blocks A and B weigh 10 and 30 lb, respectively. They are connected together by a light cord and ride in the frictionless grooves. Determine the speed of each block after block A moves 6 ft up along the plane. The blocks are released from rest.

Prob. R1–23

*R1–24. A particle moves along a straight-line path with an acceleration $a = (3t^2 - 2)$ mm/s^2, where t is in seconds. When $t = 0$, the particle is located 100 mm to the *left* of the origin; and when $t = 2$ s, it is 500 mm to the *left* of the origin. If the positive position is to the *right* of the origin, determine the particle's position when $t = 4$ s.

R1–25. The motor M pulls in its attached rope such that a particle P on the rope has an acceleration $a_P = 6$ m/s². Determine the towing force exerted by M on the rope in order to move the 50-kg crate C up the inclined plane. The kinetic coefficient of friction between the crate and the plane is $\mu_k = 0.3$. Neglect the mass of the pulleys and rope.

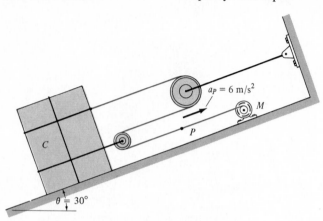

Prob. R1–25

R1–26. Neglecting the effects of friction and the mass of the pulley and cord, determine the acceleration at which the block B will descend. What is the tension in the cord?

Prob. R1–26

R1–27. A car travels along a straight road such that its acceleration varies in accordance with the graph. Construct the v-t and s-t graphs for the motion. When $t = 0$, $v = 0$ and $s = 0$.

Prob. R1–27

*R1–28. The small 2-lb collar starting from rest at A slides down along the smooth wire. During the motion, the collar is acted upon by a force $\mathbf{F} = \{10\mathbf{i} + 6y\mathbf{j} + 2z\mathbf{k}\}$ lb, where x, y, z are in feet. Determine the collar's speed when it strikes the wall at B.

Prob. R1–28

R1–29. Two billiard balls, A and B, are moving with constant velocities $v_A = 0.5$ m/s and $v_B = 1.5$ m/s, respectively. Determine the magnitude of the relative velocity of A with respect to B. If $t = 0$ at the instant shown, determine the distance between the two balls when $t = 1.25$ s.

Prob. R1–29

R1–30. A small metal particle moves downward through a fluid medium while being subjected to the attraction of a magnetic field such that its position is $s = (10t^3 - 2t)$ mm, where t is in seconds. Determine (a) the particle's displacement from $t = 1$ s to $t = 3$ s, and (b) the velocity and acceleration of the particle when $t = 5$ s.

R1–31. Determine the tension developed in the two cords and the acceleration of each block. Neglect the mass of the pulleys and cords.

Prob. R1–31

*R1–32. To test the manufactured properties of 2-lb steel balls, each ball is released from rest as shown and strikes a 45° inclined surface. If the coefficient of restitution is $e = 0.8$, determine the distance s to where the ball strikes the horizontal plane at A. At what speed does the ball strike point A?

Prob. R1–32

R1–33. Two planes A and B are flying side by side at a constant speed of 900 km/h. Maintaining this speed, plane A begins to travel along the spiral path $r = (1500\theta)$ km, where θ is in radians, whereas plane B continues to fly in a straight line. Determine the relative speed of plane A with respect to plane B when $r = 750$ km.

Prob. R1–33

R1–34. Determine the velocity of each block 2 s after the blocks are released from rest. Neglect the mass of the pulleys and cord.

Prob. R1–34

R1–35. The chain has a mass of 3 kg/m. If the coefficient of friction between the chain and the plane is $\mu = 0.2$, determine the velocity at which the end A will pass point B when the chain is released from rest. Neglect friction.

Prob. R1–35

*R1–36. A projectile having a mass of 15 kg is fired horizontally from a cannon located 20 m from the ground. If an average force of 600 kN is exerted on the projectile for 0.02 s while it is being fired through the cannon barrel, determine the range R, measured from the end of the cannon A to where the projectile strikes the ground at B. Neglect friction.

Prob. R1–36

R1–37. A 20-kg crate is originally at rest on a horizontal surface for which the coefficient of friction is $\mu = 0.6$. If a horizontal force F is applied such that it varies with time as shown, determine the speed of the crate in 10 s. *Hint:* First determine the time needed to overcome friction and start the crate moving.

Prob. R1–37

R1–38. A 2-lb pendulum bob is released at $\theta = 0°$ with a velocity $\mathbf{v}_A$. When it reaches the lower position $\theta = 90°$, a 0.1-lb bullet strikes the bob from below with a velocity of 1600 ft/s and becomes embedded in the bob in 0.2 s after the initial contact. Determine the minimum velocity $\mathbf{v}_A$ of the bob so that when the impact occurs the cord OC does not become slack. Assume the bullet creates a constant force on the bob during the impact.

Prob. R1–38

R1–39. The 1.5-kg block slides along a smooth plane and strikes a *nonlinear spring* with a speed $v = 4$ m/s. The spring is termed "nonlinear," since it has a resistance $F_s = ks^2$, where $k = 900$ N/m^2. Determine the speed of the block after it has compressed the spring $s = 0.2$ m.

Prob. R1–39

*R1–40. A 3-lb block, initially at rest at point A, slides along the smooth parabolic surface. Determine the normal force acting on the block when it reaches point B. Neglect the size of the block.

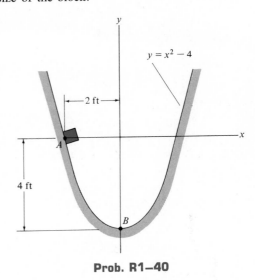

Prob. R1–40

R1–41. Assuming that the impulse acting on a 2-g bullet as it passes horizontally through the barrel of a rifle varies with time in the manner shown, determine the maximum net force $\mathbf{F}_O$ applied to the bullet when it is fired. The muzzle velocity is 500 m/s when $t = 0.75$ ms. Neglect friction between the bullet and the rifle barrel.

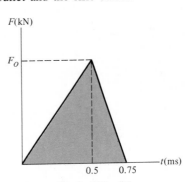

Prob. R1–41

R1–42. Block B rests on a smooth surface. If the coefficient of friction between A and B is $\mu = 0.4$, determine the acceleration of each block if (a) $F = 6$ lb, and (b) $F = 50$ lb.

Prob. R1–42

R1–43. The 2-lb collar is given an initial velocity of 20 ft/s when it is at A. If the spring has an unstretched length of 2 ft and a stiffness of $k = 100$ lb/ft, determine the velocity of the collar when $s = 1$ ft. Neglect friction.

Prob. R1–43

*R1–44. A 0.2-kg spool slides down along a smooth rod. If the rod has a constant angular rate of rotation $\dot{\theta} = 2$ rad/s in the vertical plane, show that the equations of motion for the spool are $\ddot{r} - 4r - 9.81 \sin\theta = 0$ and $0.8\dot{r} + N_s - 1.962 \cos\theta = 0$, where N_s is the magnitude of the normal force of the rod on the spool. Using the methods of differential equations, it can be shown that the solution of the first of these equations is $r = C_1 e^{-2t} + C_2 e^{2t} - (9.81/8) \sin 2t$. If r, $\dot{r}$, and θ are zero when $t = 0$, evaluate the constants C_1 and C_2 and determine r at the instant $\theta = \pi/4$ rad.

Prob. R1–44

R1–45. The 2-lb box moves along the smooth curved ramp. If the box has a velocity $v_A = 30$ ft/s when it is at A, determine its velocity and the normal force acting on the ramp when the box is located at B and C. The radius of curvature of the path at B and C is 5 ft.

Prob. R1–45

R1–46. The 100-lb crate is lifted up the 30° incline by using the pulley system and motor M. If the crate starts from rest and by constant acceleration attains a speed of 12 ft/s in 30 ft, determine (a) the average horsepower developed by the motor, and (b) the instantaneous power when the crate has moved 30 ft. Neglect friction along the plane and the mass of the pulleys and cord.

Prob. R1–46

■**R1–47.** The man throws a ball with an initial speed of 20 m/s. Determine the two possible angles θ_1 and θ_2 of release so that the ball strikes point B.

Prob. R1–47

*****R1–48.** The 20-lb block B rests on the surface of a table for which the coefficient of friction is $\mu = 0.1$. Determine the speed of the 10-lb block A after it has moved downward 2 ft from rest. Neglect the mass of the pulleys and cords.

Prob. R1–48

R1–49. An electric train having a mass of 30 Mg travels up a 10° incline with a constant speed of 60 km/h. Determine the power required to overcome the force of gravity.

R1–50. The spool, which has a mass of 4 kg, slides along the rotating rod. At the instant shown, the angular rate of rotation of the rod is $\dot{\theta} = 6$ rad/s and the rotation is increasing at $\ddot{\theta} = 2$ rad/s². At this same instant, the spool has a velocity of 3 m/s and an acceleration of 1 m/s², both measured relative to the rod and directed away from the center O when $r = 0.5$ m. Determine the radial frictional force and the normal force, both exerted by the rod on the spool at this instant.

Prob. R1–50

R1–51. Disk A weighs 2 lb and is sliding on a smooth horizontal plane with a velocity of 3 ft/s. Disk B weighs 11 lb and is initially at rest. If after the impact A has a velocity of 1 ft/s directed along the positive x axis, determine the velocity of B after impact. How much kinetic energy is lost in the collision?

Prob. R1–51

Planar Kinematics of a Rigid Body

16

In this chapter the kinematics or the geometry of the planar motion for a rigid body will be discussed. This study is important since the design of gears, cams, and mechanisms used for machine operations depends upon the geometry of their motions. Furthermore, once the kinematics of a rigid body is thoroughly understood, it will be possible to apply the equations of motion, which relate the forces on the body to the body's motion.

A rigid body can be subjected to three types of planar motion; namely, translation, rotation about a fixed axis, and general plane motion. We will first define each of these motions, then discuss the analysis of each separately. In all cases it will be shown that rigid-body planar motion is completely specified provided the motions of any two points on the body are known. The more complex study of three-dimensional kinematics of a rigid body is presented in Chapter 20.

Rigid-Body Motion 16.1

When each of the particles of a rigid body moves along a path which is equidistant from a fixed plane, the body is said to undergo *planar motion*. As previously stated, there are three types of planar motion; in order of increasing complexity, they are:

1. *Translation.* This type of motion occurs if any line segment on the body remains parallel to its original direction during the motion. When the paths of motion for all the particles of the body are along parallel straight lines,

Fig. 16—1

Fig. 16–1a, the motion is called *rectilinear translation*. However, if the paths of motion are along curved lines which are all parallel, Fig. 16–1b, the motion is called *curvilinear translation*.

2. *Rotation about a fixed axis*. When a rigid body rotates about a fixed axis, all the particles of the body, except those which lie on the axis of rotation, move along circular paths, Fig. 16–1c.

3. *General plane motion*. When a body is subjected to general plane motion, it undergoes a combination of translation *and* rotation, Fig. 16–1d. The translation occurs within a reference plane, and the rotation occurs about an axis perpendicular to the reference plane.

The above planar motions are exemplified by the moving parts of the crank mechanism shown in Fig. 16–2.

Fig. 16—2

Translation 16.2

Consider a rigid body which is subjected to either rectilinear or curvilinear translation in the *x-y* plane, Fig. 16–3.

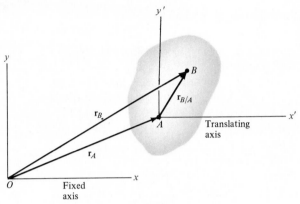

Fig. 16–3

Position. The locations of points A and B in the body are defined from the fixed x, y reference frame by using *position vectors* $\mathbf{r}_A$ and $\mathbf{r}_B$. The translating x', y' coordinate system is *fixed in the body* and has its origin located at A, hereafter referred to as the *base point*. The position of B with respect to A is denoted by the *relative-position vector* $\mathbf{r}_{B/A}$ ("$\mathbf{r}$ of B with respect to A"). By vector addition,

$$\mathbf{r}_B = \mathbf{r}_A + \mathbf{r}_{B/A}$$

Velocity. A relationship between the instantaneous velocities of A and B is obtained by taking the time derivative of the position equation, which yields $\mathbf{v}_B = \mathbf{v}_A + d\mathbf{r}_{B/A}/dt$. Here $\mathbf{v}_A$ and $\mathbf{v}_B$ denote *absolute velocities* since these vectors are measured from the x, y axes. The term $d\mathbf{r}_{B/A}/dt = \mathbf{0}$, since the *magnitude* of $\mathbf{r}_{B/A}$ is constant by definition of a rigid body; and because the body is translating, the *direction* of $\mathbf{r}_{B/A}$ is constant. Therefore,

$$\mathbf{v}_B = \mathbf{v}_A$$

Acceleration. Taking the time derivative of the velocity equation yields a similar relationship between the instantaneous accelerations of A and B,

$$\mathbf{a}_B = \mathbf{a}_A$$

The above two equations indicate that *all points in a rigid body subjected to either curvilinear or rectilinear translation move with the same velocity and acceleration.* As a result, the kinematics of particle motion, discussed in Chapter 12, may also be applied to specify the kinematics of the particles located in a translating rigid body.

16.3 Rotation About a Fixed Axis

The motion of any point P located in a body which is rotating about a fixed axis depends upon the angular motion of the body about the axis. For this reason, we will first study the angular-motion properties of the body before defining the motion of P.

(a)

Fig. 16-4a

Angular Motion. Since a point is without dimension, it has no angular motion; instead, this property is descriptive of the motion of lines or bodies. To study its effects, we will consider the body shown in Fig. 16-4a.

Angular Position. At the instant shown, the *angular position* of the radial line r is defined by the angle θ, measured between a *fixed* reference line and the line r. Here r extends perpendicular to the axis of rotation from point O to a point P in the body.

Angular Displacement. The change in the angular position, often measured as a differential $d\theta$, is called the *angular displacement*.* This vector has a *magnitude of $d\theta$,* which is measured in degrees, radians, or revolutions, where 1 rev = 2π rad. The *direction* of $d\boldsymbol{\theta}$ is determined by the right-hand rule; that is, the fingers of the right hand are curled with the sense of rotation, so that in this case the thumb, or $d\boldsymbol{\theta}$, points upward, Fig. 16-4a.

Angular Velocity. The time rate of change in the angular position is called the *angular velocity* $\boldsymbol{\omega}$ (omega). The *magnitude* of this vector is

$$\omega = \frac{d\theta}{dt} \tag{16-1}$$

which is often measured in rad/s. The *direction* of $\boldsymbol{\omega}$ is the same as that of the angular displacement, Fig. 16-4a.

Angular Acceleration. The *angular acceleration* $\boldsymbol{\alpha}$ (alpha) measures the time rate of change of the angular velocity. Hence, the *magnitude* of this vector may be written as

$$\alpha = \frac{d\omega}{dt} \tag{16-2}$$

Or taking the second time derivative of Eq. 16-1, it is possible to express α as

$$\alpha = \frac{d^2\theta}{dt^2} \tag{16-3}$$

The line of action of $\boldsymbol{\alpha}$ is the same as that for $\boldsymbol{\omega}$, Fig. 16-4a; however, its sense of *direction* depends upon whether $\boldsymbol{\omega}$ is increasing or decreasing with time. In particular, if $\boldsymbol{\omega}$ is decreasing $\boldsymbol{\alpha}$ is called an *angular deceleration* and therefore has a sense of direction which is opposite to $\boldsymbol{\omega}$.

*It is shown in Sec. 20.1 that finite rotations or finite angular displacements are *not* vector quantities, although differential rotations $d\boldsymbol{\theta}$ are vectors.

By eliminating dt from Eqs. 16–1 and 16–2, it is possible to obtain a differential relation between the angular acceleration, angular velocity, and angular displacement; namely,

$$\alpha \, d\theta = \omega \, d\omega \qquad (16–4)$$

The similarity between the differential relations for angular motion and those developed for rectilinear motion of a particle ($v = ds/dt$, $a = dv/dt$, and $a \, ds = v \, dv$) should be apparent.

Constant Angular Acceleration. If the angular acceleration of the body is constant, $\boldsymbol{\alpha} = \boldsymbol{\alpha}_c$, Eqs. 16–1, 16–2, and 16–4, when integrated, yield a set of formulas which relate the body's angular velocity, its angular position, and time. These equations are similar to Eqs. 12–4 to 12–6 used for rectilinear motion. The results are

$$\omega = \omega_0 + \alpha_c t \qquad (16–5)$$
$$\theta = \theta_0 + \omega_0 t + \tfrac{1}{2}\alpha_c t^2 \qquad (16–6)$$
$$\omega^2 = \omega_0^2 + 2\alpha_c(\theta - \theta_0) \qquad (16–7)$$

Here θ_0 and ω_0 are the initial values of the body's angular position and angular velocity, respectively.

Motion of Point P. As the rigid body in Fig. 16–4b rotates, point P travels along a *circular path* of radius r and with center at point O.

Position. The position of P is defined by the position vector $\mathbf{r}$, which extends from O to P.

Velocity. The velocity of P has a magnitude which can be determined from the circular motion of P using its polar coordinate components $v_r = \dot{r}$ and $v_\theta = r\dot{\theta}$, Eqs. 12–25. Since r is constant, the radial component $v_r = \dot{r} = 0$, so that $v = v_\theta = r\dot{\theta}$. Because $\omega = \dot{\theta}$, Eq. 16–1, the result is

$$v = \omega r \qquad (16–8)$$

As shown in Fig. 16–4b, the *direction* of $\mathbf{v}$ is *tangent* to the circular path.

Both the magnitude and direction of $\mathbf{v}$ can be accounted for by using the cross product of $\boldsymbol{\omega}$ and $\mathbf{r}$ (see Appendix C). We have

$$\mathbf{v} = \boldsymbol{\omega} \times \mathbf{r} \qquad (16–9)$$

(b)

Fig. 16–4b

The order of the vectors in this formulation is important, since the cross product is not commutative, i.e., $\boldsymbol{\omega} \times \mathbf{r} \neq \mathbf{r} \times \boldsymbol{\omega}$. In this regard, notice in Fig. 16–4b how the correct direction of $\mathbf{v}$ is established by the right-hand rule. The fingers of the right hand are curled from $\boldsymbol{\omega}$ to $\mathbf{r}$ ($\boldsymbol{\omega}$ "cross" $\mathbf{r}$). The thumb indicates the correct direction of $\mathbf{v}$, that is, tangent to the path in the direction of motion.

Acceleration. For convenience, the acceleration of P will be expressed in terms of its normal and tangential components.* Using $a_t = dv/dt$ and $a_n = v^2/\rho$, noting that $\rho = r$, $v = \omega r$, and $\alpha = d\omega/dt$, we have

$$a_t = \alpha r \qquad (16\text{--}10)$$

$$a_n = \omega^2 r \qquad (16\text{--}11)$$

(c)

Fig. 16–4c

The *tangential component of acceleration,* Fig. 16–4c, represents the time rate of change in the magnitude of velocity. If the speed of P is increasing, then $\mathbf{a}_t$ acts in the same *direction* as $\mathbf{v}$; however, if the speed is decreasing, $\mathbf{a}_t$ acts in the opposite direction of $\mathbf{v}$.

The *normal component of acceleration* represents the time rate of change in the direction of velocity. The *direction* of $\mathbf{a}_n$ is always toward O, the center of the circular path, Fig. 16–4c.

Like the velocity, the acceleration of point P may be expressed in terms of the vector cross product. Taking the time derivative of Eq. 16–9 yields

$$\mathbf{a} = \frac{d\mathbf{v}}{dt} = \frac{d\boldsymbol{\omega}}{dt} \times \mathbf{r} + \boldsymbol{\omega} \times \frac{d\mathbf{r}}{dt}$$

Recalling that $\boldsymbol{\alpha} = d\boldsymbol{\omega}/dt$, and using Eq. 16–9 ($d\mathbf{r}/dt = \mathbf{v} = \boldsymbol{\omega} \times \mathbf{r}$), we have

$$\mathbf{a} = \boldsymbol{\alpha} \times \mathbf{r} + \boldsymbol{\omega} \times (\boldsymbol{\omega} \times \mathbf{r})$$

By definition of the cross product, the first term on the right has a magnitude $a_t = \alpha r$ and a direction defined by $\mathbf{a}_t$. Likewise, the second term has a magnitude $a_n = \omega^2 r$ and a direction defined by $\mathbf{a}_n$. Thus,

$$\begin{aligned} \mathbf{a} &= \mathbf{a}_t + \mathbf{a}_n \\ &= \boldsymbol{\alpha} \times \mathbf{r} + \boldsymbol{\omega} \times (\boldsymbol{\omega} \times \mathbf{r}) \end{aligned} \qquad (16\text{--}12)$$

Since $\mathbf{a}_t$ and $\mathbf{a}_n$ are perpendicular, Fig. 16–4c, if needed the magnitude of acceleration can be determined from the Pythagorean theorem; namely, $a = \sqrt{a_n^2 + a_t^2}$.

*Polar coordinates can also be used. Since $a_r = \ddot{r} - r(\dot{\theta})^2$ and $a_\theta = r\ddot{\theta} + 2\dot{r}\dot{\theta}$, substituting $\dot{r} = \ddot{r} = 0$, $\dot{\theta} = \omega$, $\ddot{\theta} = \alpha$, we obtain Eqs. 16–10 and 16–11.

PROCEDURE FOR ANALYSIS

In order to determine the velocity and acceleration of a point located in a rigid body that is rotating about a fixed axis, it is first necessary to know the body's angular velocity and angular acceleration.

Angular Motion. If α or ω are unknown, then the relationships between the angular motions are defined by the differential equations:

$$\omega = \frac{d\theta}{dt} \qquad \alpha = \frac{d\omega}{dt} \qquad \alpha\,d\theta = \omega\,d\omega$$

If one is *absolutely certain* that the body's angular acceleration is *constant*, the following equations can be used:

$$\theta = \theta_0 + \omega_0 t + \tfrac{1}{2}\alpha_c t^2$$
$$\omega = \omega_0 + \alpha_c t$$
$$\omega^2 = \omega_0^2 + 2\alpha_c(\theta - \theta_0)$$

Motion of P. When the motion of a point P in the body is to be determined, it is suggested that a *kinematic diagram* accompany the problem solution. This diagram is simply a graphical representation showing the motion of the point (Fig. 16–4b or c).

In most cases the velocity and the two components of acceleration can be determined from the scalar equations

$$v = \omega r$$
$$a_t = \alpha r$$
$$a_n = \omega^2 r$$

However, if the geometry of the problem is difficult to visualize, the following vector equations should be used:

$$\mathbf{v} = \boldsymbol{\omega} \times \mathbf{r}$$
$$\mathbf{a}_t = \boldsymbol{\alpha} \times \mathbf{r}$$
$$\mathbf{a}_n = \boldsymbol{\omega} \times (\boldsymbol{\omega} \times \mathbf{r})$$

For application, $\mathbf{r}$, $\boldsymbol{\omega}$, and $\boldsymbol{\alpha}$ are expressed in terms of their $\mathbf{i}, \mathbf{j}, \mathbf{k}$ components and, if necessary, the cross products are computed by using a determinant expansion (see Eq. C–12).

The following examples numerically illustrate application of these equations.

Example 16–1

A cord is wrapped around a wheel which is initially at rest as shown in Fig. 16–5. If a force is applied to the cord and gives it an acceleration $a = (4t)$m/s^2, where t is in seconds, determine as a function of time (a) the angular velocity of the wheel, and (b) the angular position of line OP in radians.

Fig. 16–5

Solution

Part (a). The wheel is subjected to rotation about a fixed axis passing through point O, Fig. 16–5. Thus, point P on the wheel has motion about a circular path, and therefore the acceleration of this point has both tangential and normal components. In particular, $(a_P)_t = (4t)$ m/s^2, since the cord is connected to the wheel and *tangent* to it at P. Hence, the angular acceleration of the wheel is

$(\curvearrowright +)$
$$(a_P)_t = \alpha r$$
$$4t = \alpha(0.2)$$
$$\alpha = 20t \text{ rad/s}^2$$

The wheel's angular velocity $\boldsymbol{\omega}$ can be determined by using $\alpha = d\omega/dt$, since this equation relates α, t, and ω. Integrating, with the initial condition that $\omega_0 = 0$ at $t_0 = 0$, yields

$(\curvearrowright +)$
$$\alpha = \frac{d\omega}{dt} = 20t$$
$$\int_0^\omega d\omega = \int_0^t 20t \, dt$$
$$\omega = 10t^2 \text{ rad/s} \downarrow \qquad\qquad \textit{Ans.}$$

Why is it not possible to use Eq. 16–5 ($\omega = \omega_0 + \alpha_c t$) to obtain this result?

Part (b). The angular position of the radial line OP can be computed using $\omega = d\theta/dt$, since this equation relates θ, ω, and t. Using the initial condition $\theta_0 = 0$ at $t_0 = 0$, we have

$(\curvearrowright +)$
$$\frac{d\theta}{dt} = \omega = 10t^2$$
$$\int_0^\theta d\theta = \int_0^t 10t^2 \, dt$$
$$\theta = \tfrac{10}{3} t^3 \text{ rad} \qquad\qquad \textit{Ans.}$$

Example 16–2

Disk A, shown in Fig. 16–6a, starts from rest and rotates with a constant angular acceleration of $\alpha_A = 2$ rad/s^2. If disk A is in contact with disk B and no slipping occurs between the disks, determine the angular velocity and angular acceleration of B just after A turns 10 revolutions.

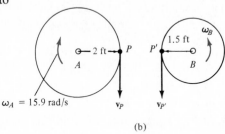

(a)

Solution

The angular velocity of A can be determined using the equations of *constant* angular acceleration since $\alpha_A = 2$ rad/s^2. There are 2π radians to one revolution, so that

$$\theta_A = 10 \text{ rev} \left(\frac{2\pi \text{ rad}}{1 \text{ rev}} \right) = 62.83 \text{ rad}$$

Thus,

$(\curvearrowright+)$
$$\omega^2 = \omega_0^2 + 2\alpha_c(\theta - \theta_0)$$
$$\omega_A^2 = 0 + 2(2)(62.83 - 0)$$
$$\omega_A = 15.9 \text{ rad/s} \downarrow$$

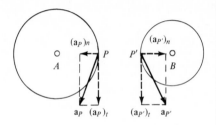

(b)

As shown in Fig. 16–6b, the speed of the contacting point P on the rim of A is

$(+\downarrow)$
$$v_P = \omega_A r_A = (15.9)(2) = 31.8 \text{ ft/s} \downarrow$$

The velocity is always tangent to the path of motion, so that the speed of point P' on B is the *same* as the speed of P on A. The angular velocity of B is therefore

$(\curvearrowleft+)$
$$\omega_B = \frac{v_{P'}}{r_B} = \frac{31.8}{1.5} = 21.2 \text{ rad/s} \uparrow \qquad \textit{Ans.}$$

As shown in Fig. 16–6c, the *tangential components* of acceleration of both disks are equal, since the disks are in contact with one another. Hence,

$$(a_P)_t = (a_{P'})_t \quad \text{or} \quad \alpha_A r_A = \alpha_B r_B$$

$$\alpha_B = \alpha_A \left(\frac{r_A}{r_B} \right) = 2 \left(\frac{2}{1.5} \right) = 2.67 \text{ rad/s}^2 \uparrow \qquad \textit{Ans.}$$

(c)

Fig. 16–6

Notice that the normal components of acceleration $(a_P)_n$ and $(a_{P'})_n$ act in *opposite directions*, since the paths of motion for both points are *different*. Furthermore, $(a_P)_n \neq (a_{P'})_n$, since the *magnitudes* of these components depend upon the radius and angular velocity of each disk, i.e., $(a_P)_n = \omega_A^2 r_A$ and $(a_{P'})_n = \omega_B^2 r_B$.

Problems

16–1. A wheel has an initial clockwise angular velocity of 10 rad/s and a constant angular acceleration of 3 rad/s². Determine the number of revolutions it must undergo to acquire a clockwise angular velocity of 15 rad/s. What time is required?

16–2. A flywheel has its angular speed increased uniformly from 15 rad/s to 60 rad/s in 80 s. If the diameter of the wheel is 2 ft, determine the magnitudes of the normal and tangential components of acceleration of a point on the rim of the wheel when $t = 80$ s, and the total distance the point travels during the time period.

16–3. The tub of a washing machine is rotating at 50 rad/s when the power is turned off. If it takes 15 s for the tub to come to rest, determine (a) the constant angular deceleration, and (b) the total number of revolutions the tub makes.

*16–4. A belt is wrapped around the inner hubs of pulleys A and B as shown. Each hub is fixed to its respective pulley and turns with it. If A has a constant angular acceleration $\alpha_A = 6$ rad/s², determine the velocity of block C in 3 s if it starts from rest. How high does it rise in 3 s?

Prob. 16–4

16–5. The operation of "reverse" for a three-speed automotive transmission is illustrated schematically in the figure. If the crank shaft G is turning with an angular speed of 60 rad/s, determine the angular speed of the drive shaft H. Each of the gears rotates about a fixed axis. Note that gears A and B, C and D, E and F are in mesh. The radius of each of these gears is reported in the figure.

$r_A = 90$ mm
$r_B = r_C = 30$ mm
$r_D = 50$ mm
$r_E = 70$ mm
$r_F = 60$ mm

Prob. 16–5

16–6. If the angular velocity of the drum is increased uniformly from 6 rad/s when $t = 0$ to 12 rad/s when $t = 5$ s, determine the magnitudes of the velocity and acceleration of points A and B on the conveyor belt when $t = 1$ s. At this instant the points are located as shown.

Prob. 16–6

16–7. The phonograph turntable T is driven by the frictional idler wheel A, which simultaneously bears against the inner rim of the turntable and the motor-shaft spindle B. Determine the required diameter d of the spindle if the motor turns it at 25 rad/s and it is required that the turntable rotate at 33 revolutions per minute.

Prob. 16–7

*16–8.** The angular position of a disk is defined by $\theta = (t + 4t^2)$ rad, where t is in minutes. Determine the number of revolutions, the angular velocity, and angular acceleration of the disk when $t = 90$ s.

16–9. The figure shows the internal gearing of a "spinner" used for drilling wells. With constant angular acceleration, the motor M rotates the shaft S to 100 rev/min in $t = 2$ s starting from rest. Determine the angular acceleration of the drill-pipe connection D and the number of revolutions it makes during the 2-s start up.

Prob. 16–9

16–10. The power of a bus engine is transmitted using the belt-and-pulley arrangement shown. If the engine turns pulley A at 60 rad/s, compute the angular velocities of the generator pulley B and the air-conditioning pulley C. The hub at D is rigidly *connected* to B and turns with it.

Prob. 16–10

16–11. The mechanics for the transmission of power in an outboard motor are shown. The gear A on the drive shaft has a radius $r_A = 0.5$ in. and the meshed pinion gear B on the propeller shaft has a radius $r_B = 1.2$ in. Determine the angular velocity of the propeller in $t = 1.5$ s, if the drive shaft rotates with an angular acceleration $\alpha = (400t^3)$ rad/s^2, where t is in seconds. The propeller is originally at rest and the motor frame does not move.

Probs. 16–11 / 16–12

*16–12.** For the outboard motor in Prob. 16–11, determine the magnitudes of the velocity and acceleration of a point P located on the tip of the propeller at the instant $t = 0.75$ s.

16–13. Starting from rest, pulley A is given an angular acceleration $\alpha = (6\theta)$ rad/s^2, where θ is in radians. Determine the speed of block B when it has risen $s = 6$ m. The pulley has an inner hub D which is fixed to C and turns with it.

150 mm

50 mm

A

D

C

75 mm

B

s

Prob. 16–13

16–14. If the hoisting gear A has an initial angular velocity $\omega_A = 8$ rad/s and an angular deceleration $\alpha_A = -1.5$ rad/s^2, determine the velocity and acceleration of block C in 2 s.

Motion

100 mm

A

200 mm

50 mm

B

C

Probs. 16–14 / 16–15

■16–15. Solve Prob. 16–14 assuming that the angular deceleration of gear A is defined by the relation $\alpha_A = (-0.5e^{-0.4t^2})$ rad/s^2, where t is in seconds. Use Simpson's rule with $n = 50$ to evaluate the integral.

*16–16. Determine the distance the load W is lifted in $t = 5$ s using the hoist. The shaft of the motor M turns with an angular velocity $\omega = 100(4 + t)$ rad/s, where t is in seconds.

300 mm

30 mm

B

225 mm

40 mm

M

50 mm

A

W

Prob. 16–16

16–17. Due to an increase in power, the motor M rotates the shaft A with an angular acceleration $\alpha = (0.06\theta^2)$ rad/s^2, where θ is in radians. If the shaft is initially turning at $\omega_1 = 50$ rad/s, determine the angular velocity of gear B after the shaft undergoes an angular displacement $\Delta\theta = 10$ rev.

M

Shaft Diameter is 24 mm

A

B

60 mm

Prob. 16–17

16–18. The angular acceleration of a wheel having a diameter of 2 ft is defined by $\alpha = (3 + 0.1\theta)$ rad/s^2, where θ is in radians. Determine the magnitudes of the velocity and acceleration of a point P on its rim when $\theta = 20$ rad. How much time is needed to reach this angular position? Originally, $\theta = 0$, and $\omega = 0$ when $t = 0$.

Prob. 16–18

16–19. The construction boom AB is used to place the core C of a nuclear reactor into its pressure vessel P. When the top of the boom B is positioned directly over the top of the pressure vessel and held in that position, the core is observed to have a swinging motion such that $\alpha = (-0.50 \sin \theta)$ rad/s^2. If $\omega = 0$ at $\theta = 10°$, determine the magnitudes of the velocity and acceleration of the center of mass G of the core when $\theta = 0°$.

Prob. 16–19

*16–20. Just after the fan is turned on, the motor gives the blade an angular acceleration $\alpha = (20e^{-0.6t})$ rad/s^2, where t is in seconds. Determine the speed of the tip P of one of the blades when $t = 3$ s. How many revolutions has the blade turned in 3 s? When $t = 0$ the blade is at rest.

Prob. 16–20

16–21. A tape having a thickness s wraps around the wheel which is turning at a constant rate $\boldsymbol{\omega}$. Assuming the unwrapped portion of tape remains horizontal, determine the acceleration of point P of the unwrapped tape when the radius of the wrapped tape is r. *Hint:* Since $v_P = \omega r$, take the time derivative and note that $dr/dt = \omega(s/2\pi)$.

Prob. 16–21

16.4 Absolute General Plane Motion Analysis

A body subjected to *general plane motion* undergoes a *simultaneous* translation and rotation. If the body is represented by a thin slab, the slab translates in the plane and rotates about an axis perpendicular to the plane. To analyze this motion, the *position* of a *point* located in a body should be related to the *angular position* of a line contained in the body. Then, by *direct application* of the time-differential equations $v = ds/dt$, $a = dv/dt$, $\omega = d\theta/dt$, and $\alpha = d\omega/dt$, the *motion* of the point and the angular motion of the line can be related. In some cases, this procedure may also be used to relate the motions of one body to those of an adjacent connected body.

PROCEDURE FOR ANALYSIS

The following procedure provides a method for relating the velocity and acceleration of a point P in a body to the angular velocity and angular acceleration of a line contained in the same body.

Position Coordinate Equation. Locate point P lying in the body using a rectilinear position coordinate s, which is measured from a *fixed origin* and is directed along the path of motion of point P. Also measure from a fixed reference line the angular position θ of a line lying in the body. From the dimensions of the body, relate s to θ, $s = f(\theta)$, using geometry and/or trigonometry.

Time Derivatives. Take the first derivative of $s = f(\theta)$ with respect to time to get a relationship between v and ω. Take the second time derivative to get a relationship between a and α.

This procedure is illustrated in the following example problems.

Example 16–3

At a given instant, the cylinder of radius r, shown in Fig. 16–7, has an angular velocity $\boldsymbol{\omega}$ and angular acceleration $\boldsymbol{\alpha}$. Determine the velocity and acceleration of its center G if it rolls without slipping.

Solution

Position-Coordinate Equation. By inspection, point G moves *horizontally* to the right from G to G' as the cylinder rolls, Fig. 16–7. Consequently its new location G' will be specified by the *horizontal* position coordinate s_G, which is measured from the original position (G) of the cylinder's center. Notice also, that as the cylinder rolls (without slipping), points on its surface contact the ground such that the arc length AB of contact must be equal to the distance s_G. Consequently, motion from the colored to the dashed position requires the radial line GB to rotate θ to the position $G'C$. Since the arc $AB = r\theta$, then G travels a distance $s_G = r\theta$.

Time Derivatives. Taking successive time derivatives of this equation, realizing that r is constant, $\omega = d\theta/dt$, and $\alpha = d\omega/dt$, gives the necessary relationships

$$s_G = r\theta$$

$$v_G = r\omega \qquad\qquad Ans.$$

$$a_G = r\alpha \qquad\qquad Ans.$$

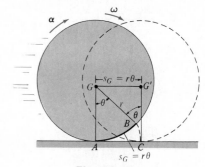

Fig. 16–7

Example 16–4

The end of rod R shown in Fig. 16–8 maintains contact with the cam by means of a spring. If the cam rotates about an axis through point O with an angular acceleration $\boldsymbol{\alpha}$ and angular velocity $\boldsymbol{\omega}$, compute the velocity and acceleration of R when the cam is in the arbitrary position θ.

Solution

Position Coordinate Equation. Coordinates x and θ are chosen for the analysis in order to express the motion of the cam, which is defined by the angular motion of line OA, i.e., $\omega = d\theta/dt$ and the rectilinear motion of the rod (or horizontal component of the motion of point B), i.e., $v = dx/dt$. These coordinates are measured from the fixed point O and may be related to each other using trigonometry. Since $OC = CB = r\cos\theta$, Fig. 16–8, then

$$x = 2r\cos\theta$$

Time Derivatives. Using the chain rule of calculus, we have

$$\frac{dx}{dt} = -2r\sin\theta\,\frac{d\theta}{dt}$$

$$v = -2r\,\omega\sin\theta \qquad\qquad Ans.$$

$$\frac{dv}{dt} = -2r\left(\frac{d\omega}{dt}\right)\sin\theta - 2r\omega\left(\cos\theta\,\frac{d\theta}{dt}\right)$$

$$a = -2r\,\alpha\sin\theta - 2r\omega^2\cos\theta \qquad\qquad Ans.$$

Fig. 16–8

Problems

16–22. Compute the angular velocity of rod *AB* when $\theta = 30°$. The shaft and the center of the roller *C* move forward at a constant rate $v = 5$ m/s.

Prob. 16–22

16–23. The bar *DC* rotates uniformly about the fixed pin *D* with a constant angular velocity **ω**. Determine the velocity and acceleration of the bar *AB*, which is confined to move vertically.

Prob. 16–23

*16–24. At the instant shown, $\theta = 60°$, and rod *AB* is subjected to a deceleration of 16 m/s² when the velocity is 10 m/s. Determine the angular velocity and angular acceleration of link *CD* at this instant.

Prob. 16–24

16–25. The *Scotch yoke* is used to convert the constant circular motion of crank *OA* into translating motion of rod *BC*. If *OA* is rotating with a constant angular velocity $\omega = 5$ rad/s, determine the velocity and acceleration of *BC* for any angle θ of the crank.

Probs. 16–25 / 16–26

16–26. The *Scotch yoke* is used to convert the constant circular motion of crank *OA* into translating motion of rod *BC*. When $\theta = 30°$, *OA* is rotating with an angular velocity $\omega = 5$ rad/s and angular acceleration $\alpha = 2$ rad/s². Determine the velocity and acceleration of *BC* at this instant.

16–27. Determine the velocity and acceleration of the plate C as a function of θ, if the circular cam is rotating about the fixed point O with a constant angular velocity $\omega = 4$ rad/s.

Probs. 16–27 / 16–28

*16–28. Determine the velocity and acceleration of the plate at the instant $\theta = 30°$, if at this instant the circular cam is rotating about the fixed point O with an angular velocity $\omega = 4$ rad/s and an angular acceleration $\alpha = 2$ rad/s^2.

16–29. The 2-m-long bar is confined to move in the horizontal and vertical slots A and B. If the velocity of the slider block at A is 8 m/s, determine the bar's angular velocity and the velocity of block B at the instant $\theta = 60°$.

Prob. 16–29

16–30. At the instant $\theta = 30°$, the angular velocity of the crankshaft AB is $\omega = 150$ rad/s. Determine the velocity of the piston C at this instant.

Prob. 16–30

16–31. The circular cam rotates about the fixed point O with a constant angular velocity ω. Determine the velocity v of the follower rod AB as a function of θ.

Prob. 16–31

*16–32. Wheel A rolls without slipping over the surface of the *fixed* cylinder B. Determine the angular velocity of A if its center C has a speed $v_C = 5$ m/s. How many revolutions will A have made about its center just after link DC completes one revolution?

Prob. 16–32

261

16–33. The end A of the bar is moving downward along the slotted guide with a constant velocity $\mathbf{v}_A$. Determine the angular velocity $\boldsymbol{\omega}$ and angular acceleration $\boldsymbol{\alpha}$ of the bar as a function of its position y.

Prob. 16–33

16–34. The slotted yoke is pinned at A while end B is used to move the ram R horizontally. If the disk rotates with a constant angular velocity $\boldsymbol{\omega}$, determine the velocity and acceleration of the ram. The crank pin C is fixed to the disk and turns with it.

Prob. 16–34

16–35. The Geneva wheel A provides intermittent rotary motion $\boldsymbol{\omega}_A$ for continuous motion $\omega_D = 2$ rad/s of disk D. By choosing $d = (100\sqrt{2})$ mm, the wheel has zero angular velocity at the instant peg B enters or leaves one of the four slots. Determine the magnitude of the angular velocity $\boldsymbol{\omega}_A$ of the Geneva wheel for any angle θ at which peg B is in contact with the slot.

Prob. 16–35

Relative-Motion Analysis: Velocity 16.5

It was pointed out in Sec. 16.1 that general plane motion of a rigid body consists of a *combination* of translation and rotation. Oftentimes it is convenient to view these "component" motions *separately* using a *relative-motion analysis*. Two sets of coordinate axes will be used to do this. The x, y coordinate system is fixed and will be used to measure the *absolute* positions, velocities, and accelerations of two points A and B in the body, Fig. 16–9a. The origin of the x', y' coordinate system will be attached to the selected "base point" A, which generally has a *known* motion. The axes of the x', y' coordinate system are *not fixed to the body;* rather they will be allowed to *translate* with respect to the fixed frame.

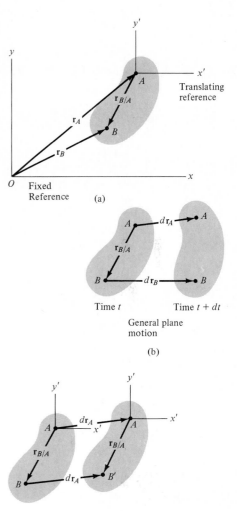

(a)

Position. The position vector $\mathbf{r}_A$ in Fig. 16–9a specifies the location of the "base point" A, whereas the relative-position vector $\mathbf{r}_{B/A}$ locates point B in the body with respect to A. By vector addition, the *position* of B can be determined from the equation

$$\mathbf{r}_B = \mathbf{r}_A + \mathbf{r}_{B/A}$$

Displacement. During an instant of time dt, points A and B undergo displacements $d\mathbf{r}_A$ and $d\mathbf{r}_B$ as shown in Fig. 16–9b. As stated above, here we will consider the motion by its component parts. For example, the *entire body* first *translates* by an amount $d\mathbf{r}_A$ so that A, the base point, moves to its *final position* and B moves to B', Fig. 16–9c. The body is then rotated by an amount $d\theta$ about A, so that B' undergoes a *relative displacement* $d\mathbf{r}_{B/A}$ and thus moves to its final position B, Fig. 16–9d. Note that since the body is rigid $\mathbf{r}_{B/A}$ has a fixed magnitude and thus $d\mathbf{r}_{B/A}$ accounts only for a change in the *direction* of $\mathbf{r}_{B/A}$. Due to the rotation about A the magnitude of the relative displacement is $d\mathbf{r}_{B/A} = \mathbf{r}_{B/A}d\theta$ and the displacement of B in Fig. 16–9b is thus

$$d\mathbf{r}_B = d\mathbf{r}_A + d\mathbf{r}_{B/A}$$

Velocity. To determine the relationship between the velocities of points A and B, it is necessary to take the time derivative of the position equation, or simply divide the displacement equation by dt. This yields

$$\frac{d\mathbf{r}_B}{dt} = \frac{d\mathbf{r}_A}{dt} + \frac{d\mathbf{r}_{B/A}}{dt}$$

The terms $d\mathbf{r}_B/dt = \mathbf{v}_B$ and $d\mathbf{r}_A/dt = \mathbf{v}_A$ are measured from the fixed x, y axes and represent the *absolute velocities* of points A and B, respectively. The magnitude of the third term is $d(r_{B/A}\theta)/dt = r_{B/A}\dot{\theta} = r_{B/A}\omega$, where ω is the angular velocity of the body at the instant considered. Note that this rotation is *absolute* since it is the *same* when measured from either the fixed x, y or translating x', y' axes. We will denote this third term as the *relative velocity* $\mathbf{v}_{B/A}$, since it represents the velocity of B with respect to A as measured by an

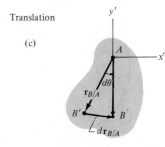

Rotation about the base point A

Fig. 16–9a–d   (d)

Fig. 16–9e–h

observer fixed to the translating x', y' axes. Since the body is rigid, it is important to realize that this observer sees point B move along a *circular arc* that has a radius of curvature $r_{B/A}$. In other words, *the body appears to move as if it were rotating with an angular velocity* $\boldsymbol{\omega}$ *about a fixed axis passing through* A. Consequently, $\mathbf{v}_{B/A}$ has a magnitude of $v_{B/A} = \omega r_{B/A}$ and a *direction* which is perpendicular to $r_{B/A}$. We therefore have

$$\mathbf{v}_B = \mathbf{v}_A + \mathbf{v}_{B/A} \qquad (16\text{–}13)$$

where

$\mathbf{v}_B$ = absolute velocity of point B

$\mathbf{v}_A$ = absolute velocity of the base point A

$\mathbf{v}_{B/A}$ = relative velocity of "B with respect to A" as measured by a *translating* observer; the *magnitude* is $v_{B/A} = \omega r_{B/A}$ and the *direction* is perpendicular to the line of action of $\mathbf{r}_{B/A}$

$\boldsymbol{\omega}$ = absolute angular velocity of the body

$\mathbf{r}_{B/A}$ = relative-position vector drawn from A to B

Each of the three terms in Eq. 16–13 is represented graphically on the *kinematic diagrams* in Fig. 16–9e, f, and g.* Here it is seen that at a given instant the velocity of B, Fig. 16–9e, is determined by considering the entire body to translate with a velocity of $\mathbf{v}_A$, Fig. 16–9f, and then rotate about the base point A with an angular velocity $\boldsymbol{\omega}$, Fig. 16–9g. Vector addition of these two effects, applied to B, yields $\mathbf{v}_B$, as shown in Fig. 16–9h.

The velocity equation 16–13 may be used in a practical manner to study the motion of a rigid body which is either pin-connected to or in contact with other moving bodies. To obtain the necessary data when applying this equation, points A and B should generally be selected at joints which are pin-connected, or at points in contact with adjacent bodies which have a *known motion*. For example, both points A and B on link AB, Fig. 16–10a, have circular paths of motion. Hence, at the instant shown, the magnitudes of their velocities are determined by the angular motion of the wheel and link BC so

---

*These diagrams are similar to those used for displacement, Fig. 16–9b, c, and d.

(a)   **Fig. 16–10**

(b)

that $v_A = \omega_A r$ and $v_B = \omega_{BC} l$. The *directions* of $\mathbf{v}_A$ and $\mathbf{v}_B$ are always *tangent* to their paths of motion, Fig. 16–10b. In the case of the wheel in Fig. 16–11, which rolls without slipping, point A can be selected at the ground. Here A momentarily has zero velocity since it is in contact with the ground—which, of course, is at rest. Furthermore, the center of the wheel, B, moves along a horizontal path so that $\mathbf{v}_B$ is horizontal.

Fig. 16–11

PROCEDURE FOR ANALYSIS

The velocity equation $\mathbf{v}_B = \mathbf{v}_A + \mathbf{v}_{B/A}$ can be applied to any two points A and B located in the *same* rigid body. For application, it is suggested that the following procedure be used.

Kinematic Diagrams. Draw a kinematic diagram of the body such as shown in Fig. 16–9e, and indicate on it the *absolute velocities* $\mathbf{v}_A$ and $\mathbf{v}_B$ of points A and B and the angular velocity $\boldsymbol{\omega}$. Usually the base point A is selected as a point having a known velocity.

Also, draw a kinematic diagram of the body such as shown in Fig. 16–9g and indicate on it the *relative velocity* $\mathbf{v}_{B/A}$. Since the body is considered to be "pinned" at the base point A, the *magnitude* of $\mathbf{v}_{B/A}$ is $v_{B/A} = \omega r_{B/A}$. The *direction* is established from the diagram, such that $\mathbf{v}_{B/A}$ acts perpendicular to $\mathbf{r}_{B/A}$ in accordance with the rotational motion $\boldsymbol{\omega}$ of the body.*

Velocity Equation. Write the velocity equation in symbolic form: $\mathbf{v}_B = \mathbf{v}_A + \mathbf{v}_{B/A}$, and underneath each of the terms represent the respective vectors by their magnitudes and directions. To do this, use the data tabulated on the kinematic diagrams. Since motion occurs in the plane, this "vector" equation can be expressed in terms of two scalar component equations which provide a solution for at most two unknowns. If the solution yields a *negative* answer for an *unknown,* it indicates its *direction* is *opposite* to that shown on the kinematic diagram.

The following example problems numerically illustrate this scalar method of application.

*The notation $\mathbf{v}_B = \mathbf{v}_A + \mathbf{v}_{B/A\text{(pin)}}$ may be helpful in recalling that A is "pinned."

265

Example 16–5

The link shown in Fig. 16–12a is guided by two blocks at A and B, which move in the fixed slots. If the velocity of A is 2 m/s downward, determine the velocity of B at the instant $\theta = 45°$.

0.2 m

$v_A = 2$ m/s

$\theta = 45°$

0.1 m

(a)

Solution

Kinematic Diagrams. Since points A and B are restricted to move along the fixed slots and $\mathbf{v}_A$ is directed downward, the velocity $\mathbf{v}_B$ must be directed horizontally to the right, Fig. 16–12b. This motion causes the link to rotate counterclockwise; that is, by the right-hand rule the angular velocity $\boldsymbol{\omega}$ is directed outward, perpendicular to the page (the plane of motion). Knowing the magnitude and direction of $\mathbf{v}_A$ and the lines of action of $\mathbf{v}_B$ and $\boldsymbol{\omega}$, it is possible to apply the velocity equation $\mathbf{v}_B = \mathbf{v}_A + \mathbf{v}_{B/A}$ to points A and B in order to solve for the two unknown magnitudes v_B and ω.

As shown in Fig. 16–12c, the relative velocity $\mathbf{v}_{B/A}$ acts up to the right, *perpendicular* to $\mathbf{r}_{B/A}$, in accordance with the angular velocity $\boldsymbol{\omega}$. The magnitude of $\mathbf{v}_{B/A}$ is

$$v_{B/A} = \omega r_{B/A} = \omega(0.2 \text{ m})$$

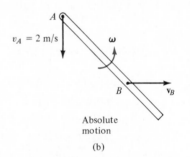

$v_A = 2$ m/s

ω

v_B

Absolute motion

(b)

Velocity Equation

$$\mathbf{v}_B = \mathbf{v}_A + \mathbf{v}_{B/A}$$

$$v_B = 2 \text{ m/s} + \omega(0.2 \text{ m})$$
$$\rightarrow \qquad \downarrow \qquad \nearrow^{45°}$$

Equating the horizontal and vertical components, with the assumed positive directions to the right and upward, we have

$(\xrightarrow{+})$ $\qquad\qquad v_B = 0 + 0.2\omega \cos 45°$

$(+\uparrow)$ $\qquad\qquad 0 = -2 + 0.2\omega \sin 45°$

Thus,

$$\omega = 14.14 \text{ rad/s} \; \curvearrowleft$$
$$v_B = 2 \text{ m/s} \rightarrow \qquad\qquad\qquad \textit{Ans.}$$

ω

$45°$

$\mathbf{r}_{B/A}$

$v_{B/A}$

0.2 m

$45°$

Relative motion

(c)

Fig. 16–12

Since both results are *positive*, the *directions* of $\mathbf{v}_B$ and $\boldsymbol{\omega}$ are indeed correct as shown in Fig. 16–12b. It should be emphasized that these results are *valid only* at the instant $\theta = 45°$. A recalculation for $\theta = 44°$ yields $v_B = 2.07$ m/s and $\omega = 14.40$ rad/s; whereas when $\theta = 46°$, $v_B = 1.93$ m/s and $\omega = 13.90$ rad/s, etc.

Having determined $\boldsymbol{\omega}$, as an exercise apply the velocity equation $\mathbf{v}_C = \mathbf{v}_A + \mathbf{v}_{C/A}$ to points A and C and show that $v_C = 3.16$ m/s, directed at $\theta = 18.4°$ up from the horizontal.

Example 16–6

The cylinder shown in Fig. 16–13*a* rolls freely on the surface of a conveyor belt which is moving at 2 ft/s. Assuming that no slipping occurs between the cylinder and the belt, determine the velocity of point *A*. The cylinder has a clockwise angular velocity $\omega = 15$ rad/s at the instant shown.

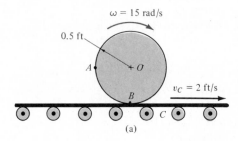

(a)

Fig. 16–13

Solution

Kinematic Diagrams. Since no slipping occurs, point *B* on the cylinder has the same velocity as the conveyor, Fig. 16–13*b*. Knowing the angular velocity of the cylinder, we will apply the velocity equation to *B*, the base point, and *A* to determine the *x* and *y* components of $\mathbf{v}_A$, i.e., $\mathbf{v}_A = \mathbf{v}_B + \mathbf{v}_{A/B}$.

The relative velocity $\mathbf{v}_{A/B}$ has a direction as shown in Fig. 16–13*c* and a magnitude of

$$v_{A/B} = \omega r_{A/B} = (15 \text{ rad/s})\left(\frac{0.5 \text{ ft}}{\cos 45°}\right) = 10.6 \text{ ft/s} \quad \measuredangle^{45°}$$

Absolute motion

(b)

Velocity Equation

$$\mathbf{v}_A = \mathbf{v}_B + \mathbf{v}_{A/B}$$

$$(v_A)_x + (v_A)_y = 2 \text{ ft/s} + 10.6 \text{ ft/s}$$
$$\rightarrow \qquad \uparrow \qquad \rightarrow \qquad \measuredangle^{45°}$$

Equating the horizontal and vertical components gives

$(\xrightarrow{+})$ $\qquad (v_A)_x = 2 + 10.6 \cos 45° = 9.50 \text{ ft/s}$

$(+\uparrow)$ $\qquad (v_A)_y = 0 + 10.6 \sin 45° = 7.50 \text{ ft/s}$

Relative motion

(c)

Thus,

$$v_A = \sqrt{(9.50)^2 + (7.50)^2} = 12.1 \text{ ft/s} \qquad \textit{Ans.}$$

$$\theta = \tan^{-1}\left(\frac{7.50}{9.50}\right) = 38.3° \quad \measuredangle^{\mathbf{v}_A}_{\theta}$$

Example 16–7

The bar AB of the linkage shown in Fig. 16–14a has a clockwise angular velocity of 30 rad/s when $\theta = 60°$. Compute the angular velocity of members BC and DC at this instant.

Solution

Kinematic Diagrams. From the figure, AB and DC are subjected to *rotation about a fixed axis* and BC is subjected to *general plane motion*. The magnitude and direction of $\mathbf{v}_B$ may be determined from the kinematic diagram shown in Fig. 16–14b. Because of the rotation, $\mathbf{v}_B$ *always* acts tangent to the dashed circular path. The magnitude of $\mathbf{v}_B$ is

$$v_B = 30 \text{ rad/s } (0.2 \text{ m}) = 6 \text{ m/s} \quad \text{\searrow}^{30°}$$

Likewise, from the kinematic diagram of bar DC, Fig. 16–4c, $\mathbf{v}_C$ acts tangent to its circular path.

Knowing $\mathbf{v}_B$ and the direction of $\mathbf{v}_C$, Fig. 16–14d, we can determine the two unknown magnitudes v_C and ω_{BC} by applying the velocity equation to B, the base point, and C lying on rod BC, i.e., $\mathbf{v}_C = \mathbf{v}_B + \mathbf{v}_{C/B}$.

As shown in Fig. 16–14e, the relative velocity $\mathbf{v}_{C/B}$ acts upward and has a magnitude of

$$v_{C/B} = \omega_{BC}(r_{C/B})$$
$$= \omega_{BC}(0.2 \text{ m})\uparrow$$

Velocity Equation

$$\mathbf{v}_C = \mathbf{v}_B + \mathbf{v}_{C/B}$$
$$v_C = 6 \text{ m/s} + \omega_{BC}(0.2 \text{ m})$$
$$\rightarrow \qquad \searrow^{30°} \qquad \uparrow$$

Equating the horizontal and vertical components yields

$(\xrightarrow{+}) \qquad\qquad v_C = 6 \cos 30° + 0 = 5.20 \text{ m/s}$

$(+\uparrow) \qquad\qquad 0 = -6 \sin 30° + 0.2\omega_{BC}$

$$\omega_{BC} = 15 \text{ rad/s} \uparrow \qquad\qquad\qquad \textit{Ans.}$$

Due to the circular motion of DC, Fig. 16–14c,

$$5.20 \text{ m/s} = \omega_{DC}(0.1 \text{ m})$$
$$\omega_{DC} = 52 \text{ rad/s} \uparrow \qquad\qquad\qquad \textit{Ans.}$$

(a)

(b)

(c)

Absolute motion

(d)

Relative motion

(e)

Fig. 16–14

Velocity: Cartesian Vector Analysis 16.6

In Section 16.5 the velocity equation $\mathbf{v}_B = \mathbf{v}_A + \mathbf{v}_{B/A}$ was applied using a scalar method for solution. It is also possible to apply this equation using a Cartesian vector analysis.

Since the relative velocity $\mathbf{v}_{B/A}$ represents the effect of *circular motion,* observed from translating axes having their origin at the base point A, this term can be expressed by the cross product $\mathbf{v}_{B/A} = \boldsymbol{\omega} \times \mathbf{r}_{B/A}$, Eq. 16–9. Hence, for application,

$$\mathbf{v}_B = \mathbf{v}_A + (\boldsymbol{\omega} \times \mathbf{r}_{B/A}) \qquad (16\text{--}14)$$

where

       $\mathbf{v}_B$ = absolute velocity of B

       $\mathbf{v}_A$ = absolute velocity of the base point A

       $\boldsymbol{\omega}$ = absolute angular velocity of the body

       $\mathbf{r}_{B/A}$ = relative-position vector drawn from A to B

Applying Eq. 16–14 in Cartesian vector form is very methodical and therefore has an advantage for solving problems involving mechanisms consisting of *several connected members*.

PROCEDURE FOR ANALYSIS

The following procedure provides a method for applying Eq. 16–14 to two points A and B located in the *same* rigid body.

Kinematic Diagram. Draw a kinematic diagram of the body, which is used to indicate the absolute velocities $\mathbf{v}_A$ and $\mathbf{v}_B$ of points A and B, the angular velocity $\boldsymbol{\omega}$, and the relative-position vector $\mathbf{r}_{B/A}$.

Velocity Equation. Establish the fixed x, y, z axes and express the vectors shown on the kinematic diagram in terms of their $\mathbf{i}$, $\mathbf{j}$, $\mathbf{k}$ components. Substitute these vectors into Eq. 16–14 and evaluate the cross product. Then equate the respective $\mathbf{i}$ and $\mathbf{j}$ components to obtain two scalar equations which may be solved for at most two unknowns.

The following two examples numerically illustrate this procedure.

Example 16–8

The link shown in Fig. 16–15a is guided by two blocks at A and B which move in the fixed slots. If the velocity of A is 2 m/s downward, determine the velocity of B at the instant $\theta = 45°$.

(a)   **Fig. 16–15**

Solution

Kinematic Diagram. A kinematic diagram showing the velocities of points A and B and the angular velocity of the link is given in Fig. 16–15b. The two unknowns are represented by the magnitudes of $\boldsymbol{\omega}$ and $\mathbf{v}_B$.

Velocity Equation. Expressing each of the vectors in Fig. 16–15b in terms of their **i**, **j**, and **k** components, we have

$$\mathbf{v}_A = \{-2\mathbf{j}\} \text{ m/s}, \qquad \mathbf{v}_B = v_B\mathbf{i}, \qquad \boldsymbol{\omega} = \omega\mathbf{k}$$
$$\mathbf{r}_{B/A} = \{0.2 \sin 45°\mathbf{i} - 0.2 \cos 45°\mathbf{j}\} \text{ m}$$

(b)

Applying Eq. 16–14 to A, the base point, and B, we have

$$\mathbf{v}_B = \mathbf{v}_A + (\boldsymbol{\omega} \times \mathbf{r}_{B/A})$$
$$v_B\mathbf{i} = -2\mathbf{j} + [\omega\mathbf{k} \times (0.2 \sin 45°\mathbf{i} - 0.2 \cos 45°\mathbf{j})]$$

or

$$v_B\mathbf{i} = -2\mathbf{j} + 0.2\omega \sin 45°\mathbf{j} + 0.2\omega \cos 45°\mathbf{i}$$

Equating the **i** and **j** components gives

$$v_B = 0.2\omega \cos 45°, \qquad 0 = -2 + 0.2\omega \sin 45°$$

Thus,

$$\omega = 14.14 \text{ rad/s} \curvearrowleft$$
$$v_B = 2 \text{ m/s} \rightarrow \qquad\qquad\qquad Ans.$$

Now that $\boldsymbol{\omega}$ has been determined, as an exercise apply Eq. 16–14 to points A and C and show that $\mathbf{v}_C = \{3.0\mathbf{i} + 1.0\mathbf{j}\}$ m/s.

Example 16–9

The bar AB of the linkage shown in Fig. 16–16a has a clockwise angular velocity of 30 rad/s when $\theta = 60°$. Compute the angular velocities of members BC and DC at this instant.

(a)

Fig. 16–16

Solution

Kinematic Diagram. The velocities of points B and C are defined by the rotation of links AB and DC about their fixed axes. The relative-position vectors and the angular velocity of each link are shown on the kinematic diagram in Fig. 16–16b.

(b)

Velocity Equation. Expressing each vector in Cartesian vector form, we have

$$\boldsymbol{\omega}_{AB} = \{-30\mathbf{k}\} \text{ rad/s}, \qquad \boldsymbol{\omega}_{BC} = \omega_{BC}\mathbf{k}, \qquad \boldsymbol{\omega}_{DC} = \omega_{DC}\mathbf{k}$$

$$\mathbf{r}_{B/A} = \{0.2 \cos 60°\mathbf{i} + 0.2 \sin 60°\mathbf{j}\} \text{ m}, \qquad \mathbf{r}_{C/B} = \{0.2\mathbf{i}\} \text{ m},$$

$$\mathbf{r}_{C/D} = \{-0.1\mathbf{j}\} \text{ m}$$

For link AB:

$$\mathbf{v}_B = \boldsymbol{\omega}_{AB} \times \mathbf{r}_{B/A}$$
$$= (-30\mathbf{k}) \times (0.2 \cos 60°\mathbf{i} + 0.2 \sin 60°\mathbf{j})$$
$$= \{5.20\mathbf{i} - 3.0\mathbf{j}\} \text{ m/s}$$

For link BC:

$$\mathbf{v}_C = \mathbf{v}_B + (\boldsymbol{\omega}_{BC} \times \mathbf{r}_{C/B})$$
$$= 5.20\mathbf{i} - 3.0\mathbf{j} + [(\omega_{BC}\mathbf{k}) \times (0.2\mathbf{i})]$$
$$= 5.20\mathbf{i} + (0.2\omega_{BC} - 3.0)\mathbf{j}$$

For link DC:

$$\mathbf{v}_C = \boldsymbol{\omega}_{DC} \times \mathbf{r}_{C/D}$$
$$5.20\mathbf{i} + (0.2\omega_{BC} - 3.0)\mathbf{j} = (\omega_{DC}\mathbf{k}) \times (-0.1\mathbf{j})$$
$$5.20\mathbf{i} + (0.2\omega_{BC} - 3.0)\mathbf{j} = 0.1\omega_{DC}\mathbf{i}$$

Equating the respective $\mathbf{i}$ and $\mathbf{j}$ components and solving, we have

$$\omega_{DC} = 52 \text{ rad/s} \;\text{\Large\curvearrowleft} \qquad\qquad \textit{Ans.}$$
$$\omega_{BC} = 15 \text{ rad/s} \;\text{\Large\curvearrowleft} \qquad\qquad \textit{Ans.}$$

Problems

*16–36. If bar AB has an angular velocity $\omega_{AB} = 6$ rad/s, determine the velocity of the slider block C at the instant shown.

Prob. 16–36

16–37. Knowing that the angular velocity of link AB is $\omega_{AB} = 4$ rad/s, determine the velocity of the collar at C and the angular velocity of link CB at the instant shown. Link CB is horizontal at this instant.

Prob. 16–37

16–38. Rod AB is rotating with an angular velocity $\omega_{AB} = 5$ rad/s. Determine the velocity of the collar C at the instant shown.

Prob. 16–38

16–39. The pinion gear A rolls on the fixed gear rack B with an angular velocity $\omega = 4$ rad/s. Determine the velocity of the gear rack C.

Prob. 16–39

*16–40. The wheel is rotating with an angular velocity $\omega = 8$ rad/s. Determine the velocity of the collar A at this instant.

Prob. 16–40

16–43. If link AB is rotating at 3 rad/s, determine the angular velocities of links BC and CD at the instant shown.

Prob. 16–43

16–41. At the instant shown, the truck is moving to the right at 3 m/s, while the spool is rolling counterclockwise at $\omega = 8$ rad/s without slipping at B. Determine the velocity of the spool's center G.

Prob. 16–41

16–42. The bicycle has a velocity $v = 4$ ft/s, and at the same instant the rear wheel has a clockwise angular velocity $\omega = 3$ rad/s, which causes it to slip at its contact point A. Determine the velocity of point A.

Prob. 16–42

*16–44. If the end of the cord is pulled downward with a speed $v_C = 120$ mm/s, determine the angular velocities of pulleys A and B and the speed of block D. Assume that the cord does not slip on the pulleys.

Prob. 16–44

273

16–45. Part of an automatic transmission consists of a *fixed* ring gear R, three equal planet gears P, the sun gear S, and the planet carrier C, which is shaded. If the sun gear is rotating at $\omega_S = 6$ rad/s, determine the angular velocity of the *planet carrier*. Note that C is pin-connected to the center of each of the planet gears.

Prob. 16–45

16–46. The center of the wheel has a velocity of 2 ft/s up the inclined plane, and at the same instant the wheel has a clockwise angular velocity $\omega = 4$ rad/s, which causes it to slip at its contact point A. Determine the velocities of points A and B at this instant. Note that B is located on the *horizontal* line CB.

Prob. 16–46

16–47. The crank and counterweight of the oil pumping unit rotate about the fixed point O with a constant angular velocity $\omega_{OA} = 8$ rad/s. Determine the angular velocity of the walking beam BCD at the instant shown.

Prob. 16–47

*16–48. When the crank on the Chinese windlass is turning, the rope on shaft A unwinds while that on shaft B winds up. Determine the speed at which the block lowers if the crank is turning with an angular velocity $\omega = 4$ rad/s. What is the angular velocity of the pulley at C? The rope segments on each side of the pulley are both parallel and vertical, and the rope does not slip on the pulley.

Prob. 16–48

16–49. Mechanical toy animals often use a walking mechanism as shown idealized in the figure. If the driving crank AB is propelled by a spring motor such that $\omega_{AB} = 5$ rad/s, determine the velocity of the rear foot E at the instant shown. Although not part of this problem, the upper end of the foreleg has a slotted guide which is constrained by the fixed pin at G.

Prob. 16–49

16–50. If rod AB is rotating with an angular velocity $\omega_{AB} = 3$ rad/s, determine the angular velocity of rods BC and CD at the instant shown.

Prob. 16–50

16–51. If the hub gear H and ring gear R have angular velocities $\omega_H = 5$ rad/s and $\omega_R = 20$ rad/s, respectively, determine the angular velocity ω_S of the spur gear S and the angular velocity of its attached arm OA.

Probs. 16–51 / 16–52

*16–52. If the hub gear H has an angular velocity $\omega_H = 5$ rad/s, determine the angular velocity of the ring gear R so that the arm OA attached to the spur gear S remains stationary ($\omega_{OA} = 0$). What is the angular velocity of the spur gear?

16–53. If gear A has an angular velocity $\omega_A = 6$ rad/s, determine the angular velocities of gears B and C at the instant shown.

Prob. 16–53

275

16–54. The epicyclic gear train consists of the sun gear A which is in mesh with the planet gear B. This gear has an inner hub C which is fixed to B and in mesh with the fixed ring gear R. If the connecting rod DE attached to B and C is rotating at $\omega_{DE} = 18$ rad/s, determine the angular velocities of the planet and sun gears.

Prob. 16–54

16–55. The two-cylinder engine is designed so that the pistons are connected to the crankshaft BE using a master rod ABC and articulated rod AD. If the crankshaft is rotating at $\omega = 30$ rad/s, determine the velocities of the pistons C and D at the instant shown.

Prob. 16–55

*16–56. The mechanism is used on a machine for the manufacturing of a wire product. Because of the rotational motion of link AB and the sliding of block F, the segmental gear lever DE undergoes general plane motion. If AB is rotating at $\omega_{AB} = 5$ rad/s, determine the velocity of point E at the instant shown.

Prob. 16–56

Instantaneous Center of Zero Velocity **16.7**

The velocity of any point B located in a rigid body can be obtained in a simple way if one chooses the base point IC to be a point that has *zero velocity* at the instant considered. In this case, $\mathbf{v}_{IC} = \mathbf{0}$, and therefore the velocity equation, Eq. 16–13, becomes $\mathbf{v}_B = \mathbf{v}_{IC} + \mathbf{v}_{B/IC} = \mathbf{v}_{B/IC}$. For a body having general plane motion, point IC so chosen is called the *instantaneous center of zero velocity* and it lies on the *instantaneous axis of zero velocity* (IA). Hence, the IA is always perpendicular to the plane used to represent the motion, and the intersection of the IA with this plane defines the IC. Since $\mathbf{v}_B = \mathbf{v}_{B/IC}$, point B moves momentarily about the IC in a circular path; in other words, the body appears to rotate about the IA, Fig. 16–17a. If the relative-position vector $\mathbf{r}_{B/IC}$ is established, then the *magnitude* of $\mathbf{v}_B$ is simply $v_B = \omega r_{B/IC}$, where ω is the angular velocity of the body. The *direction* of $\mathbf{v}_B$ must always be *perpendicular* to $\mathbf{r}_{B/IC}$, as shown in Fig. 16–17a.

When a body is subjected to general plane motion, the point determined as the instantaneous center of zero velocity for the body can only be used for an *instant of time*. Since the body changes its position from one instant to the next, then for each position of the body a unique instantaneous center must be determined. The locus of points which define the IC during the body's motion is called a *centrode*, Fig. 16–17b. Thus, each point on the centrode acts as the IC for the body only for an instant of time.

Although the IC may be conveniently used to determine the velocity of any point in a body, it generally *does not have zero acceleration* and therefore it *should not* be used for finding the accelerations of points in a body.

Location of the IC. If the location of the IC is unknown, it may be determined by using the fact that the relative-position vector extending from the IC to a point is always perpendicular to the velocity of the point. Several possibilities exist:

(a) (b) (c) (d)

Fig. 16–17

1. *Given the velocity of a point in the body and the angular velocity of the body*. In this case if $\mathbf{v}_B$ and $\boldsymbol{\omega}$ are known, the *IC* is located along the line drawn perpendicular to $\mathbf{v}_B$ at B, such that the distance from B to the *IC* is $r_{B/IC} = v_B/\omega$, Fig. 16–17a. Note that the *IC* lies on that side of B which causes rotation about the *IC* which is consistent with the direction of motion caused by $\boldsymbol{\omega}$ and $\mathbf{v}_B$.

2. *Given the lines of action of two nonparallel velocities*. Consider the body in Fig. 16–17b, where the lines of action of the velocities $\mathbf{v}_A$ and $\mathbf{v}_B$ are known. From each of these lines construct at points A and B perpendicular line segments which then define the lines of action of $\mathbf{r}_{A/IC}$ and $\mathbf{r}_{B/IC}$, respectively. Extending these perpendiculars to their *point of intersection* as shown locates the *IC* at the instant considered. The magnitudes of $\mathbf{r}_{A/IC}$ and $\mathbf{r}_{B/IC}$ are generally determined from the geometry of the body and trigonometry. Furthermore, if the magnitude and sense of $\mathbf{v}_A$ are known, then the angular velocity of the body is determined from $\omega = v_A/r_{A/IC}$. Once computed, $\boldsymbol{\omega}$ can then be used to determine $v_B = \omega r_{B/IC}$.

3. *Given the magnitude and direction of two parallel velocities*. When the velocities of points A and B are parallel and have known magnitudes v_A and v_B, then the location of the *IC* is determined by proportional triangles. Examples are shown in Fig. 16–17c and d. In both cases $v_A/r_{A/IC} = v_B/r_{B/IC}$. If d is the known distance between points A and B, then in Fig. 16–17c, $r_{A/IC} = d - r_{B/IC}$ and in Fig. 16–17d, $r_{A/IC} = r_{B/IC} - d$. Substituting each of these equations into the previous equation yields a solution for $r_{B/IC}$.

PROCEDURE FOR ANALYSIS

The velocity of a point in a body which is subjected to general plane motion can be determined with reference to its instantaneous center of zero velocity, provided the location of the *IC* is first established. This is done by using one of the methods described above. As shown on the kinematic diagram in Fig. 16–18, the body is imagined as "extended and pinned" at the *IC* such that it rotates about this pin with its instantaneous angular velocity $\boldsymbol{\omega}$. The *magnitude* of velocity for the arbitrary points A, B, and C in the body can then be determined by using the equation $v = \omega r$, where r is the radial line drawn from the *IC* to the point. The line of action of each velocity vector is *perpendicular* to its associated radial line, and the velocity has a *direction* which tends to move the point in a manner consistent with the angular rotation of the radial line, Fig. 16–18.

The following examples illustrate the method of determining the *IC* and computing the velocities of points in a rigid body subjected to general plane motion. See also the "velocity analysis" of Examples 16–13 to 16–15, which further illustrates application of this method.

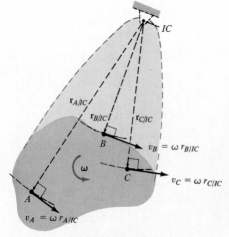

Fig. 16–18

Example 16–10

Determine the location of the instantaneous center of zero velocity for (a) the wheel shown in Fig. 16–19a, which is rolling without slipping along the ground; (b) the crankshaft BC shown in Fig. 16–19b; and (c) the link CB shown in Fig. 16–19c.

Fig. 16–19

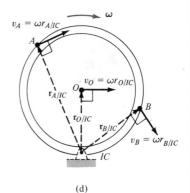

(d)

Solution

Part (a). The wheel rolls without slipping, and therefore the point of *contact* of the wheel with the ground has *zero velocity*. Hence, this point represents the *IC* for the wheel, Fig. 16–19d. If it is imagined that the wheel is momentarily pinned at this point, the velocities of points A, B, O, and so on, can be found using $v = \omega r$. Here the distances $r_{A/IC}$, $r_{B/IC}$, and $r_{O/IC}$ must be determined from the geometry of the wheel.

Part(b). As shown in Fig. 16–19b, link AB rotates about the fixed pin A. Thus, point B has a speed of $v_B = \omega d$, caused by the clockwise rotation of AB. Also, $\mathbf{v}_B$ is perpendicular to AB, so that it acts at an angle θ from the horizontal as shown in Fig. 16–19e. The motion of point B causes the piston to move forward *horizontally* with a velocity $\mathbf{v}_C$. Consequently, point C on the rod moves horizontally with this same velocity. When lines are drawn perpendicular to $\mathbf{v}_B$ and $\mathbf{v}_C$, Fig. 16–19e, they intersect at the *IC*. The magnitudes of $\mathbf{r}_{B/IC}$ and $\mathbf{r}_{C/IC}$ are determined strictly from the geometry of construction.

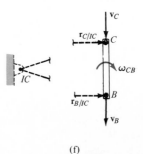

(e)

Part(c). Points B and C follow circular paths of motion since rods AB and DC are each subjected to rotation about a fixed axis, Fig. 16–19c. In particular, $v_C = \omega_{DC}b$. Since the velocity is always tangent to the path, at the instant considered, $\mathbf{v}_C$ on rod DC and $\mathbf{v}_B$ on rod AB are both directed vertically downward, along the axis of link CB, Fig. 16–19f. Furthermore, since CB is *rigid,* no relative displacement occurs between points B and C, so that $\mathbf{v}_B = \mathbf{v}_C$. Radial lines drawn perpendicular to these two velocities form parallel lines which intersect at "infinity;" i.e., $r_{C/IC} \to \infty$ and $r_{B/IC} \to \infty$. Since $v_C = r_{C/IC}\omega_{CB}$, then $\omega_{CB} = v_C/r_{C/IC} = (\omega_{DC}b)/\infty = 0$. As a result, rod CB momentarily *translates* with a speed of $v_C = \omega_{DC}b$. An instant later, however, CB will move to a new position, causing the instantaneous center to *change* to some finite location.

(f)

Example 16–11

Block D shown in Fig. 16–20a moves with a speed of 3 m/s. Determine the angular velocities of links BD and AB, and the velocity of point B at the instant shown.

(a)

Solution

Since D moves to the right at 3 m/s, it causes arm AB to rotate about point A in a clockwise direction. Hence, $\mathbf{v}_B$ is directed perpendicular to AB as shown in Fig. 16–20b. The instantaneous center of zero velocity for BD is located at the intersection of the line segments drawn perpendicular to $\mathbf{v}_B$ and $\mathbf{v}_D$, Fig. 16–20b. From the geometry,

$$r_{B/IC} = 0.4 \tan 45° \text{ m} = 0.4 \text{ m}$$

$$r_{D/IC} = \frac{0.4 \text{ m}}{\cos 45°} = 0.566 \text{ m}$$

Since the magnitude of $\mathbf{v}_D$ is known, the angular velocity of the link is

$$\omega_{BD} = \frac{v_D}{r_{D/IC}} = \frac{3 \text{ m/s}}{0.566 \text{ m}} = 5.30 \text{ rad/s} \; \curvearrowleft \qquad Ans.$$

(b)

The velocity of B is therefore

$$v_B = \omega_{BD}(r_{B/IC}) = 5.30 \text{ rad/s}(0.4 \text{ m}) = 2.12 \text{ m/s} \quad \searrow^{45°} \qquad Ans.$$

Note that link AB is subjected to rotation about a fixed axis passing through A, Fig. 16–20c, and since v_B is known, the angular velocity of AB is

$$\omega_{AB} = \frac{v_B}{r_{B/A}} = \frac{2.12 \text{ m/s}}{0.4 \text{ m}} = 5.30 \text{ rad/s} \; \downarrow \qquad Ans.$$

(c)

Fig. 16–20

Example 16–12

The cylinder shown in Fig. 16–21a rolls without slipping between the two moving plates E and D. Determine the angular velocity of the cylinder and the velocity of its center C at the instant shown.

(a)

Fig. 16–21

(b)

Solution

Since no slipping occurs, the contact points A and B on the cylinder have the same velocity as the plates E and D. Furthermore, the velocities $\mathbf{v}_A$ and $\mathbf{v}_B$ are *parallel,* so that by the proportionality of right triangles the *IC* is located at a point on line AB, Fig. 16–21b. Assuming this point to be a distance x from B, we have

$$v_B = \omega x; \qquad\qquad 0.4 = \omega x$$
$$v_A = \omega(0.25 - x); \qquad 0.25 = \omega(0.25 - x)$$

Dividing one of these equations into the other eliminates ω and yields

$$0.4(0.25 - x) = 0.25x$$
$$x = \frac{0.1}{0.65} = 0.154 \text{ m}$$

Hence, the angular velocity is

$$\omega = \frac{v_B}{x} = \frac{0.4}{0.154} = 2.60 \text{ rad/s} \downdownarrows \qquad\qquad \textit{Ans.}$$

The velocity of point C is therefore

$$v_C = \omega r_{C/IC} = 2.60(0.154 - 0.125)$$
$$= 0.0754 \text{ m/s} \leftarrow \qquad\qquad \textit{Ans.}$$

Problems

16–57. Solve Prob. 16–36 using the method of instantaneous center of zero velocity.

16–58. Solve Prob. 16–37 using the method of instantaneous center of zero velocity.

16–59. Solve Prob. 16–42 using the method of instantaneous center of zero velocity.

*16–60. Solve Prob. 16–40 using the method of instantaneous center of zero velocity.

16–61. Solve Prob. 16–43 using the method of instantaneous center of zero velocity.

16–62. An acrobat is executing a somersault. At the instant shown he has an angular velocity of 1.5 rad/s, and his mass center G has a velocity of 0.75 m/s vertically upward. Determine the velocities of his head H and foot F at this instant.

Prob. 16–62

16–63. The crankshaft AB is rotating at 500 rad/s about a fixed axis passing through A. Determine the speed of the piston P at the instant it is in the position shown.

Prob. 16–63

*16–64. As the car travels forward at 80 ft/s on a wet road, due to slipping the rear wheels have an angular velocity $\omega = 100$ rad/s. Determine the speeds of points A, B, and C caused by the motion.

Prob. 16–64

16–65. In an automobile transmission the planet pinions A and B rotate on shafts that are mounted on the planet-pinion carrier CD. As shown, CD is attached to a shaft at E which is aligned with the center of the *fixed* sun-gear S. This shaft is not attached to the sun gear. If CD is rotating at $\omega_{CD} =$ 8 rad/s, determine the angular velocity of the ring gear R.

Prob. 16–65

16–66. The similar links AB and CD rotate about the fixed pins at A and C. If AB has an angular velocity $\omega_{AB} =$ 8 rad/s, determine the angular velocity of BDP and the velocity of point P.

Prob. 16–66

16–67. The shaper mechanism is designed to give a slow cutting stroke and a quick return to a blade attached to the slider at C. Determine the velocity of the slider at C at the instant shown, if the link AB is rotating at 4 rad/s.

Prob. 16–67

*16–68. At the instant shown, the disk is rotating at $\omega =$ 4 rad/s. If the end of the cord wrapped around the disk is fixed at D, determine the velocities of points A, B, and C.

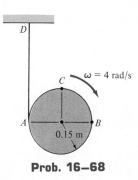

Prob. 16–68

16–69. The epicyclic gear train is driven by the rotating link *DE*, which has an angular velocity $\omega_{DE} = 5$ rad/s. If the ring gear *F* is fixed, determine the angular velocities of gears *A*, *B*, and *C*.

Prob. 16–69

16–70. If the slider block *A* is moving to the right at $v_A = 8$ ft/s, determine the velocity of blocks *B* and *C* at the instant shown.

Prob. 16–70

16–71. As the crankshaft *AB* rotates at $\omega_{AB} = 50$ rad/s about the fixed axis through point *A*, the wrist pins *C* and *D* rock between their bearing supports *E* and *F*. If *C* and *D* are allowed to turn freely about their centers, while they are held from slipping at *E* or *F*, determine the angular velocity of the connecting rod *CBD* at the instant shown.

Prob. 16–71

Relative-Motion Analysis: Acceleration 16.8

An equation that relates the accelerations of two points in a rigid body subjected to general plane motion may be determined by differentiating the velocity equation $\mathbf{v}_B = \mathbf{v}_A + \mathbf{v}_{B/A}$ with respect to time. Thus,

$$\frac{d\mathbf{v}_B}{dt} = \frac{d\mathbf{v}_A}{dt} + \frac{d\mathbf{v}_{B/A}}{dt}$$

The terms $d\mathbf{v}_B/dt = \mathbf{a}_B$ and $d\mathbf{v}_A/dt = \mathbf{a}_A$ are measured from a set of *fixed x, y axes* and represent the *absolute accelerations* of points B and A. The last term represents the acceleration of B with respect to A as measured by an observer fixed to the translating x', y' axes having their origin at point A. In Sec. 16.5, it was shown that to this observer point B appears to move along a *circular arc* that has a radius of curvature $r_{B/A}$. In other words, the body appears to move as if it were rotating about a fixed axis passing through point A. Consequently, $\mathbf{a}_{B/A}$ can be expressed in terms of its tangential and normal components of motion; i.e., $\mathbf{a}_{B/A} = (\mathbf{a}_{B/A})_t + (\mathbf{a}_{B/A})_n$, where $(a_{B/A})_t = \alpha r_{B/A}$ and $(a_{B/A})_n = \omega^2 r_{B/A}$. Hence, the above equation can be written in the form

$$\mathbf{a}_B = \mathbf{a}_A + (\mathbf{a}_{B/A})_t + (\mathbf{a}_{B/A})_n \qquad (16\text{--}15)$$

where

$\mathbf{a}_B$ = absolute acceleration of point B

$\mathbf{a}_A$ = absolute acceleration of point A

$(\mathbf{a}_{B/A})_t$ = relative tangential acceleration component of "B with respect to A;" the *magnitude* is $(a_{B/A})_t = \alpha r_{B/A}$ and the *direction* is perpendicular to the line of action of $\mathbf{r}_{B/A}$

$(\mathbf{a}_{B/A})_n$ = relative normal acceleration component of "B with respect to A;" the *magnitude* is $(a_{B/A})_n = \omega^2 r_{B/A}$ and the *direction* is always from B towards A

α = absolute angular acceleration of the body

ω = absolute angular velocity of the body

$\mathbf{r}_{B/A}$ = relative-position vector drawn from A to B

Each of the four terms in Eq. 16–15 is represented graphically on the *kinematic diagrams* shown in Fig. 16–22.* Here it is seen that at a given instant the acceleration of B, Fig. 16–22a, is determined by considering the body to translate with an acceleration $\mathbf{a}_A$, Fig. 16–22b, and simultaneously rotate about the base point A with an instantaneous angular velocity $\boldsymbol{\omega}$ and angular acceleration $\boldsymbol{\alpha}$, Fig. 16–22c. Vector addition of these two effects, applied to B, yields $\mathbf{a}_B$, as shown in Fig. 16–22d. It should be noted from Fig. 16–22a that points A and B move along *curved paths,* and as a result the accelerations of these points have *both tangential and normal components*. (Recall that the acceleration of a point is *tangent to the path only* when the path is *rectilinear*.)

*These diagrams are similar to those used for displacement and velocity, Fig. 16–9.

Path of point A

Path of point B

General plane motion

(a)

Translation

(b)

Rotation about the base point A

(c)

$$\mathbf{a}_B = \mathbf{a}_A + (\mathbf{a}_{B/A})_t + (\mathbf{a}_{B/A})_n$$

(d)

Fig. 16–22

Path of B

(a)

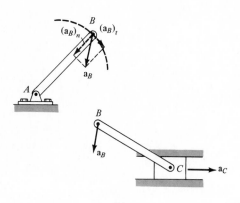

(b)

Fig. 16—23

If Eq. 16–15 is applied in a practical manner to study the accelerated motion of a rigid body which is pin-connected to two other bodies, it should be realized that points which are *coincident at the pin* move with the *same acceleration,* since the path of motion over which they travel is the *same.* For example, point B lying on either rod AB or BC of the crank mechanism shown in Fig. 16–23a has the same acceleration, since the rods are pin-connected at B. Here the motion of B is along a *curved path,* so that $\mathbf{a}_B$ is calculated on the basis of its tangential and normal components $(a_B)_t = \alpha r_{B/A}$ and $(a_B)_n = \omega^2 r_{B/A}$, which are defined by the angular motion of AB. At the other end of rod BC, however, point C moves along a *rectilinear path,* which is defined by the piston. Hence, in this case, the acceleration $\mathbf{a}_C$ is directed along the path, Fig. 16–23b.

If two bodies contact one another *without slipping,* and the *points in contact* move along *different paths,* the *tangential components* of acceleration of the points will be the *same;* however, the *normal components* will *not* be the same. For example, consider the two meshed gears in Fig. 16–24a. Point A is located on gear B and A' is located on gear C. Due to the rotational motion, $(\mathbf{a}_A)_t = (\mathbf{a}_{A'})_t$; however, since both points follow different curved paths, $(\mathbf{a}_A)_n \neq (\mathbf{a}_{A'})_n$ and therefore $\mathbf{a}_A \neq \mathbf{a}_{A'}$, Fig. 16–24b.

(a)

(b)

Fig. 16—24

PROCEDURE FOR ANALYSIS

The acceleration equation $\mathbf{a}_B = \mathbf{a}_A + (\mathbf{a}_{B/A})_t + (\mathbf{a}_{B/A})_n$ applies between any two points located in the *same* rigid body. For application, it is suggested that the following procedure be used.

Velocity Analysis. If the angular velocity $\boldsymbol{\omega}$ of the body is unknown, determine it by using a velocity analysis as discussed in Secs. 16.5, 16.6, or 16.7. Determine also the velocities $\mathbf{v}_A$ and $\mathbf{v}_B$ of points A and B *if these points move along curved paths.*

Kinematic Diagrams. Draw a kinematic diagram of the body such as shown in Fig. 16–22a and indicate on it the absolute accelerations of points A and B, $\mathbf{a}_A$ and $\mathbf{a}_B$, the angular velocity $\boldsymbol{\omega}$, and the angular acceleration $\boldsymbol{\alpha}$. Usually the base point A is selected as a point having a known acceleration. In particular, if points A and B move along *curved paths,* their accelerations should be expressed in terms of their tangential and normal components; i.e., $\mathbf{a}_A = (\mathbf{a}_A)_t + (\mathbf{a}_A)_n$ and $\mathbf{a}_B = (\mathbf{a}_B)_t + (\mathbf{a}_B)_n$. Here $(a_A)_n = (v_A)^2/\rho_A$ and $(a_B)_n = (v_B)^2/\rho_B$, where ρ_A and ρ_B define the radii of curvature of the paths of points A and B, respectively.

Also, draw a kinematic diagram of the body such as shown in Fig. 16–22c and indicate on it the two components of the relative acceleration, $(\mathbf{a}_{B/A})_t$ and $(\mathbf{a}_{B/A})_n$. Since the body is considered to be pinned at the base point A, these components have *magnitudes* of $(a_{B/A})_t = \alpha r_{B/A}$ and $(a_{B/A})_n = \omega^2 r_{B/A}$ and their *directions* are established from the diagram such that $(\mathbf{a}_{B/A})_t$ acts perpendicular to $\mathbf{r}_{B/A}$, in accordance with the rotational motion $\boldsymbol{\alpha}$ of the body, and $(\mathbf{a}_{B/A})_n$ is directed from B toward A.*

Acceleration Equation. Write the acceleration equation in symbolic form: $\mathbf{a}_B = \mathbf{a}_A + (\mathbf{a}_{B/A})_t + (\mathbf{a}_{B/A})_n$, and underneath each of the terms represent the respective vectors by their magnitudes and directions. To do this use the data tabulated on the kinematic diagrams. Since motion occurs in the plane, this "vector" equation can be expressed in terms of two scalar component equations which provide a solution for at most two unknowns. If a solution yields a *negative* answer for an *unknown,* it indicates that its *direction* is *opposite* to that shown on the kinematic diagram.

The following example problems numerically illustrate this scalar method of application.

*Perhaps the notation $\mathbf{a}_B = \mathbf{a}_A + (\mathbf{a}_{B/A(\text{pin})})_t + (\mathbf{a}_{B/A(\text{pin})})_n$ may be helpful in recalling that A is pinned.

Example 16–13

The rod AB shown in Fig. 16–25a is confined to move along the inclined planes at A and B. If point A has an acceleration of 3 m/s^2 and a velocity of 2 m/s, both directed down the plane at the instant the rod becomes horizontal, determine the angular acceleration of the rod at this instant.

(a)

Solution

We will apply the acceleration equation to points A and B on the rod.

Velocity Analysis. Since the velocity of A is known and the velocity of B is tangent to its path, the instantaneous center of zero velocity for the rod is located as shown in Fig. 16–25b. From the geometry,

$$r_{A/IC} = r_{B/IC} = 10 \cos 45° = 7.07 \text{ m}$$

Thus,

$$\omega = \frac{v_A}{r_{A/IC}} = \frac{2 \text{ m/s}}{7.07 \text{ m}} = 0.283 \text{ rad/s} \; \curvearrowleft$$

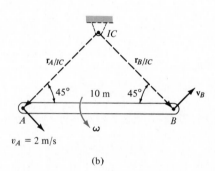

(b)

Kinematic Diagrams. The accelerated motions of points A and B are shown on the kinematic diagram in Fig. 16–25c. Since B moves along a straight-line path, it has an acceleration along the path.* There are two unknowns in Fig. 16–25c, i.e., a_B and α. Point A will be chosen as the base point, so that $\mathbf{a}_B = \mathbf{a}_A + \mathbf{a}_{B/A}$.

The relative accelerated motion of B with respect to A is shown on the kinematic diagram in Fig. 16–25d.

(c)

Acceleration Equation

$$\mathbf{a}_B = \mathbf{a}_A + (\mathbf{a}_{B/A})_t + (\mathbf{a}_{B/A})_n$$

$$a_B = 3 \text{ m/s}^2 + \alpha(10 \text{ m}) + (0.283 \text{ rad/s})^2(10 \text{ m})$$

Equating the horizontal and vertical components yields

$(\overset{+}{\rightarrow})$ $a_B \cos 45° = 3 \cos 45° - (0.283)^2(10)$

$(+\uparrow)$ $a_B \sin 45° = -3 \sin 45° + \alpha(10)$

Solving, we have

$$a_B = 1.87 \text{ m/s}^2 \quad \nearrow^{45°}$$

$$\alpha = 0.344 \text{ rad/s}^2 \; \curvearrowleft \qquad \textit{Ans.}$$

Relative
motion

(d)

*If the path were *curved*, B would have both normal and tangential components of acceleration.

Fig. 16–25

Example 16–14

The collar C in Fig. 16–26a is moving downward with an acceleration of 1 m/s^2, and at the instant shown it has a speed of 2 m/s. Determine the angular acceleration of link CB at this instant.

Solution

We will apply the acceleration equation to points C and B on link CB.

Velocity Analysis. Since $\mathbf{v}_C$ is downward, $\mathbf{v}_B$ is directed horizontally (to the right). The instantaneous center of zero velocity for CB is located as shown in Fig. 16–26b. From the figure, $r_{C/IC} = r_{B/IC} = 0.2$ m and $\boldsymbol{\omega}_{CB}$ is assumed to act counterclockwise. Show that

$$\omega_{CB} = 10 \text{ rad/s} \curvearrowleft \qquad v_B = 2 \text{ m/s} \rightarrow$$

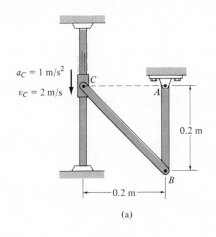

$a_C = 1$ m/s^2

$v_C = 2$ m/s

0.2 m

0.2 m

(a)

Kinematic Diagrams. As shown on the kinematic diagram in Fig. 16–26c, the acceleration of point C is directed downward since it moves along a straight-line path. This point will be chosen as the base point since $\mathbf{a}_C$ is known. Point B moves along a *curved path* and therefore $\mathbf{a}_B$ is represented by its normal and tangential components. From bar AB, the normal component is

$$(a_B)_n = \frac{(v_B)^2}{r_{B/A}} = \frac{(2)^2}{0.2} = 20 \text{ m/s}^2$$

There are two unknowns shown on the kinematic diagram in Fig. 16–26c, α_{CB} and $(a_B)_t$.

The relative accelerated motion of B with respect to C is shown on the kinematic diagram in Fig. 16–26d.

0.2 m

$r_{C/IC}$

IC

$v_C = 2$ m/s

ω_{CB}

0.2 m

$r_{B/IC}$

B

v_B

(b)

Acceleration Equation

$$(\mathbf{a}_B)_t + (\mathbf{a}_B)_n = \mathbf{a}_C + (\mathbf{a}_{B/C})_t + (\mathbf{a}_{B/C})_n$$

$$\underset{\rightarrow}{(a_B)_t} + \underset{\uparrow}{20 \text{ m/s}^2} = \underset{\downarrow}{1 \text{ m/s}^2} + \underset{\angle 45°}{\alpha_{CB}(0.2 \sqrt{2} \text{ m})} + \underset{45° \searrow}{(10 \text{ rad/s})^2(0.2 \sqrt{2} \text{ m})}$$

Equating components in the horizontal and vertical directions gives

$$(\overset{+}{\rightarrow}) \quad (a_B)_t = \alpha_{CB}(0.2 \sqrt{2}) \cos 45° - (10)^2(0.2 \sqrt{2}) \cos 45°$$

$$(+\uparrow) \quad 20 = -1 + \alpha_{CB}(0.2 \sqrt{2}) \sin 45° + (10)^2(0.2 \sqrt{2}) \sin 45°$$

Solving, we get

$$(a_B)_t = -19 \text{ m/s}^2 = 19 \text{ m/s}^2 \leftarrow$$

$$\alpha_{CB} = 5 \text{ rad/s}^2 \curvearrowleft \qquad \qquad \textit{Ans.}$$

$\omega_{CB} = 10$ rad/s

α_{CB}

$a_C = 1$ m/s^2

$(a_B)_n = 20$ m/s^2

$\mathbf{a}_B$

B

$(a_B)_t$

Absolute motion

(c)

C

$45°$

$\omega_{CB} = 10$ rad/s

α_{CB}

$r_{B/C}$

$(a_{B/C})_n = (\omega_{CB})^2 r_{B/C}$

$0.2\sqrt{2}$ m

$(a_{B/C})_t = \alpha_{CB} \, r_{B/C}$

B

Relative motion

(d)

Fig. 16–26

Example 16–15

The cylinder shown in Fig. 16–27a rolls without slipping, such that at the instant shown its center has a velocity of 1.5 ft/s and an acceleration of 4 ft/s². Determine the acceleration of point B.

(a)

Solution

We will apply the acceleration equation to points B and O on the cylinder.

Velocity Analysis. Since no slipping occurs, $\mathbf{v}_A = \mathbf{0}$, so that point A represents the instantaneous center of zero velocity. Hence,

$$\omega = \frac{v_O}{r} = \frac{1.5}{0.75} = 2 \text{ rad/s} \downarrow$$

This result can also be obtained *directly* from the formulation in Example 16–3, i.e., $v_O = r\omega$.

Kinematic Diagrams. Using the result of Example 16–3, $a_O = r\alpha$, the angular acceleration is

$$a_O = r\alpha; \qquad\qquad 4 = 0.75\alpha$$
$$\alpha = 5.33 \text{ rad/s}^2 \downarrow$$

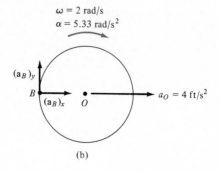

(b)

Since $\mathbf{a}_O$ is known, O will be chosen as the base point. Point B moves along a *curved path* having an unknown radius of curvature.* Its acceleration will be represented by its unknown x and y components, Fig. 16–27b.

The relative accelerated motion of B with respect to O is shown on the kinematic diagram in Fig. 16–27c.

(c)

Fig. 16–27

Acceleration Equation

$$(\mathbf{a}_B)_x + (\mathbf{a}_B)_y = \mathbf{a}_O + (\mathbf{a}_{B/O})_t + (\mathbf{a}_{B/O})_n$$
$$(\mathbf{a}_B)_x + (\mathbf{a}_B)_y = 4 \text{ ft/s}^2 + 5.33 \text{ rad/s}^2(0.75 \text{ ft}) + (2 \text{ rad/s})^2(0.75 \text{ ft})$$
$$\rightarrow \qquad \uparrow \qquad\quad \rightarrow \qquad\qquad\quad \uparrow \qquad\qquad\qquad \rightarrow$$

Equating the horizontal and vertical components yields

$$\xrightarrow{+} \qquad\qquad (a_B)_x = 4 + (2)^2(0.75) = 7.0$$
$$+\uparrow \qquad\qquad (a_B)_y = 5.33(0.75) = 4.0$$

Thus,

$$a_B = \sqrt{(7.0)^2 + (4.0)^2} = 8.06 \text{ ft/s}^2 \qquad\qquad \textit{Ans.}$$

$$\theta = \tan^{-1}\left(\frac{4.0}{7.0}\right) = 29.7° \; \measuredangle^\theta \qquad\qquad \textit{Ans.}$$

*The radius of curvature ρ is *not* equal to the radius of the cylinder since the cylinder is *not* rotating about point O. Furthermore, ρ is *not* defined as the distance from A(IC) to B, since the location of the IC depends only on the velocity of a point and *not* the geometry of its path.

Acceleration: Cartesian Vector Analysis 16.9

In Sec. 16.8 the acceleration equation $\mathbf{a}_B = \mathbf{a}_A + (\mathbf{a}_{B/A})_t + (\mathbf{a}_{B/A})_n$ was applied using a scalar method for solution. It is also possible to apply this equation using a Cartesian vector analysis.

Since the relative-acceleration components represent the effect of *circular motion* observed from translating axes having their origin at the base point A, these terms can be expressed by the cross products $(\mathbf{a}_{B/A})_t = \boldsymbol{\alpha} \times \mathbf{r}_{B/A}$ and $(\mathbf{a}_{B/A})_n = \boldsymbol{\omega} \times (\boldsymbol{\omega} \times \mathbf{r}_{B/A})$, Eq. 16–12. Since *planar motion* occurs $\boldsymbol{\omega}$ is perpendicular to $\mathbf{r}_{B/A}$ and $(\mathbf{a}_{B/A})_n$ *always* acts from B toward A, Fig. 16–22c. Noting that this *direction* is the same as $-\mathbf{r}_{B/A}$ and the *magnitude* is $(a_{B/A})_n = \omega^2 r_{B/A}$, we can express $(\mathbf{a}_{B/A})_n$ in a much simpler vector form as $(\mathbf{a}_{B/A})_n = -\omega^2 \mathbf{r}_{B/A}$. Hence, Eq. 16–15 becomes

$$\mathbf{a}_B = \mathbf{a}_A + \boldsymbol{\alpha} \times \mathbf{r}_{B/A} - \omega^2 \mathbf{r}_{B/A} \qquad (16\text{–}16)$$

where

$\mathbf{a}_B$ = absolute acceleration of point B

$\mathbf{a}_A$ = absolute acceleration of the base point A

$\boldsymbol{\alpha}$ = absolute angular acceleration of the body

$\boldsymbol{\omega}$ = absolute angular velocity of the body

$\mathbf{r}_{B/A}$ = relative-position vector drawn from A to B

Applying Eq. 16–16 in Cartesian vector form is very methodical and thus it has an advantage for solving problems involving mechanisms having *several connected members*.

PROCEDURE FOR ANALYSIS

The following procedure provides a method for applying Eq. 16–16 to two points A and B located in the *same* rigid body.

Kinematic Diagram. Draw a kinematic diagram of the body such as shown in Fig. 16–22a and indicate on it $\mathbf{a}_A$ and $\mathbf{a}_B$, $\boldsymbol{\alpha}$, $\boldsymbol{\omega}$, and $\mathbf{r}_{B/A}$. If $\boldsymbol{\omega}$ is to be determined, use the analysis outlined in Sec. 16.6.

Acceleration Equation. Establish the fixed x, y, z axes and express each of the vectors on the kinematic diagram in terms of their $\mathbf{i}$, $\mathbf{j}$, $\mathbf{k}$ components. Substitute these vectors into Eq. 16–16 and evaluate the cross product. Then equate the respective $\mathbf{i}$ and $\mathbf{j}$ components to obtain two scalar equations which may be solved for at most two unknowns.

The following two examples numerically illustrate this procedure.

Example 16–16

The collar C in Fig. 16–28a is moving downward with an acceleration of 1 m/s^2. At the instant shown, it has a speed of 2 m/s which gives links CB and AB an angular velocity $\omega_{AB} = \omega_{CB} = 10$ rad/s. (Can you show this?) Determine the angular accelerations of CB and AB at this instant.

$a_C = 1$ m/s^2

$v_C = 2$ m/s

$\omega_{AB} = 10$ rad/s

0.2 m

$\omega_{CB} = 10$ rad/s

0.2 m

(a)

Solution

Kinematic Diagrams. The kinematic diagrams of both links AB and CB are shown in Fig. 16–28b.

Expressing each of the vectors in Cartesian vector form, we have

$$\mathbf{r}_{B/A} = \{-0.2\mathbf{j}\} \text{ m} \qquad \mathbf{r}_{B/C} = \{0.2\mathbf{i} - 0.2\mathbf{j}\} \text{ m}$$
$$\boldsymbol{\omega}_{AB} = \{10\mathbf{k}\} \text{ rad/s} \qquad \boldsymbol{\omega}_{CB} = \{10\mathbf{k}\} \text{ rad/s}$$
$$\boldsymbol{\alpha}_{AB} = \alpha_{AB}\mathbf{k} \qquad \boldsymbol{\alpha}_{CB} = \alpha_{CB}\mathbf{k}$$
$$\mathbf{a}_C = \{-1\mathbf{j}\} \text{ m/s}^2$$

Acceleration Equations. Applying Eq. 16–12 to point B on link AB, we have

$$\mathbf{a}_B = \boldsymbol{\alpha}_{AB} \times \mathbf{r}_{B/A} - \omega_{AB}^2 \mathbf{r}_{B/A}$$
$$\mathbf{a}_B = (\alpha_{AB}\mathbf{k}) \times (-0.2\mathbf{j}) - (10)^2(-0.2\mathbf{j})$$
$$\mathbf{a}_B = 0.2\alpha_{AB}\mathbf{i} + 20\mathbf{j}$$

Applying Eq. 16–16 to points B and C on rod BC, using the above value for $\mathbf{a}_B$, yields

$$\mathbf{a}_B = \mathbf{a}_C + \boldsymbol{\alpha}_{CB} \times \mathbf{r}_{B/C} - \omega_{CB}^2 \mathbf{r}_{B/C}$$
$$0.2\alpha_{AB}\mathbf{i} + 20\mathbf{j} = -1\mathbf{j} + (\alpha_{CB}\mathbf{k}) \times (0.2\mathbf{i} - 0.2\mathbf{j}) - (10)^2(0.2\mathbf{i} - 0.2\mathbf{j})$$
$$0.2\alpha_{AB}\mathbf{i} + 20\mathbf{j} = -1\mathbf{j} + 0.2\alpha_{CB}\mathbf{j} + 0.2\alpha_{CB}\mathbf{i} - 20\mathbf{i} + 20\mathbf{j}$$

Equating the respective $\mathbf{i}$ and $\mathbf{j}$ components, we obtain

$$0.2\alpha_{AB} = 0.2\alpha_{CB} - 20 \qquad (1)$$
$$20 = -1 + 0.2\alpha_{CB} + 20 \qquad (2)$$

Hence,

$$\alpha_{CB} = 5 \text{ rad/s}^2$$
$$\alpha_{AB} = -95 \text{ rad/s}^2$$

Thus,

$$\boldsymbol{\alpha}_{CB} = \{5\mathbf{k}\} \text{ rad/s}^2 \qquad \qquad Ans.$$
$$\boldsymbol{\alpha}_{AB} = \{-95\mathbf{k}\} \text{ rad/s}^2 \qquad \qquad Ans.$$

y

x

C

A

a_C

$\mathbf{r}_{B/C}$

$\omega_{AB}, \boldsymbol{\alpha}_{AB}$

0.2 m

$\mathbf{r}_{B/A}$

$\omega_{CB}, \boldsymbol{\alpha}_{CB}$

B

0.2 m

(b)

Fig. 16–28

Example 16–17

The crankshaft AB of an engine turns with a clockwise angular accelera-tion of 20 rad/s^2, Fig. 16–29a. Determine the acceleration of the piston at the instant AB is in the position shown. At this instant ω_{AB} = 10 rad/s and ω_{BC} = 2.43 rad/s.

Solution

Kinematic Diagrams. The kinematic diagrams of AB and BC are shown in Fig. 16–29b. Expressing each of the vectors in Cartesian vector form yields

$$\mathbf{r}_{B/A} = \{-0.25 \sin 45°\mathbf{i} + 0.25 \cos 45°\mathbf{j}\} \text{ ft} = \{-0.177\mathbf{i} + 0.177\mathbf{j}\} \text{ ft}$$

$$\mathbf{r}_{C/B} = \{0.75 \sin 13.6°\mathbf{i} + 0.75 \cos 13.6°\mathbf{j}\} \text{ ft} = \{0.176\mathbf{i} + 0.729\mathbf{j}\} \text{ ft}$$

$$\boldsymbol{\omega}_{AB} = \{-10\mathbf{k}\} \text{ rad/s} \qquad \boldsymbol{\alpha}_{AB} = \{-20\mathbf{k}\} \text{ rad/s}^2$$

$$\boldsymbol{\omega}_{BC} = \{2.43\mathbf{k}\} \text{ rad/s} \qquad \boldsymbol{\alpha}_{BC} = \alpha_{BC}\mathbf{k}$$

$$\mathbf{a}_C = a_C\mathbf{j}$$

(a)

Acceleration Equations. Applying Eq. 16–12 to point B on link AB, we have

$$\mathbf{a}_B = \boldsymbol{\alpha}_{AB} \times \mathbf{r}_{B/A} - \omega_{AB}^2\mathbf{r}_{B/A}$$

$$= (-20\mathbf{k}) \times (-0.177\mathbf{i} + 0.177\mathbf{j}) - (10)^2(-0.177\mathbf{i} + 0.177\mathbf{j})$$

$$= \{21.24\mathbf{i} - 14.16\mathbf{j}\} \text{ ft/s}^2$$

Applying Eq. 16–16 to points B and C on rod BC yields

$$\mathbf{a}_C = \mathbf{a}_B + \boldsymbol{\alpha}_{BC} \times \mathbf{r}_{C/B} - \omega_{BC}^2\mathbf{r}_{C/B}$$

$$a_C\mathbf{j} = 21.24\mathbf{i} - 14.16\mathbf{j} + (\alpha_{BC}\mathbf{k}) \times (0.176\mathbf{i} + 0.729\mathbf{j})$$
$$- (2.43)^2(0.176\mathbf{i} + 0.729\mathbf{j})$$

$$a_C\mathbf{j} = 21.24\mathbf{i} - 14.16\mathbf{j} + 0.176\alpha_{BC}\mathbf{j} - 0.729\alpha_{BC}\mathbf{i} - 1.04\mathbf{i} - 4.30\mathbf{j}$$

Equating the respective $\mathbf{i}$ and $\mathbf{j}$ components, we have

$$0 = 20.20 - 0.729\alpha_{BC}$$

$$a_C = 0.176\alpha_{BC} - 18.46$$

(b)

Fig. 16–29

Solving yields

$$\alpha_{BC} = 27.7 \text{ rad/s}^2 \curvearrowleft$$

$$a_C = -13.6 \text{ ft/s}^2 \qquad\qquad Ans.$$

The negative sign indicates that the piston is decelerating, i.e., $\mathbf{a}_C$ = $\{-13.6\mathbf{j}\}$ ft/s^2. This causes the speed of the piston to decrease until the crankshaft AB becomes vertical, at which time the piston is momentarily at rest.

Problems

*16–72. At a given instant the bottom A of the ladder has the velocity and acceleration shown. Determine the acceleration of the top of the ladder, B, and the ladder's angular acceleration at this same instant.

$v_A = 6$ ft/s
$a_A = 4$ ft/s²

Prob. 16–72

16–73. Determine the acceleration of piston P in Prob. 16–63 at the instant shown. The crankshaft is rotating at a constant rate of $\omega = 500$ rad/s as shown.

16–74. The ends of the bar AB are confined to move along the paths shown. At a given instant, A has a velocity of 8 ft/s and an acceleration of 3 ft/s². Determine the angular velocity and angular acceleration of AB at this instant.

30°
4 ft

30°
4 ft

$v_A = 8$ ft/s
A
$a_A = 3$ ft/s²

Prob. 16–74

16–75. If rod AB has the angular motion at the instant shown, determine the acceleration of the collar C at this instant.

$\omega_{AB} = 5$ rad/s
$\alpha_{AB} = 3$ rad/s² 2 ft

B

A 45° 2.5 ft

60°

C

Prob. 16–75

*16–76. At a given instant, link AB has an angular velocity $\omega_{AB} = 4$ rad/s and an angular acceleration $\alpha_{AB} = 12$ rad/s². Determine the angular motions of links BC and CD at this instant.

2 ft
B C

1.5 ft

$\omega_{AB} = 4$ rad/s
$\alpha_{AB} = 12$ rad/s² 2.5 ft

60°

45°
A D

Prob. 16–76

16–77. The wheel is moving to the right such that it has an angular acceleration $\alpha = 4$ rad/s^2 and angular velocity $\omega = 2$ rad/s at the instant shown. If it does not slip at A, determine the acceleration of point B.

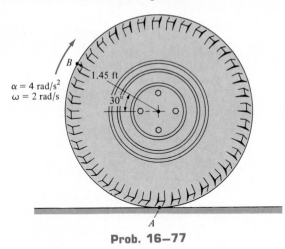

$\alpha = 4$ rad/s^2
$\omega = 2$ rad/s

Prob. 16–77

16–78. At a given instant the wheel is rotating with the angular motions shown. Determine the acceleration of the collar at A at this instant.

$\omega = 8$ rad/s
$\alpha = 16$ rad/s^2

Prob. 16–78

16–79. At a given instant the gear has the angular motions shown. Determine the accelerations of points A and B on the link and the link's angular acceleration at this instant.

$\omega = 6$ rad/s
$\alpha = 12$ rad/s^2

Prob. 16–79

*16–80. The flywheel rotates with an angular velocity $\omega = 2$ rad/s and an angular acceleration $\alpha = 6$ rad/s^2. Determine the angular acceleration of links AB and BC at this instant.

$\omega = 2$ rad/s
$\alpha = 6$ rad/s^2

Prob. 16–80

16–81. At the instant shown, link AB has an angular velocity $\omega_{AB} = 2$ rad/s and an angular acceleration $\alpha_{AB} = 6$ rad/s^2. Determine the acceleration of the pin at C and the angular acceleration of link CB at this instant. $\theta = 60°$.

Prob. 16–81

16–82. At a given instant, the cables supporting the pipe have the motions shown. Determine the angular velocity and angular acceleration of the pipe and the velocity and acceleration of point B located on the pipe.

Prob. 16–82

16–83. At a given instant gears A and B have the angular motions shown. Determine the angular acceleration of gear C and the acceleration of its center point D at this instant. Note that the inner hub of gear C is in mesh with gear A and its outer rim is in mesh with gear B.

Prob. 16–83

*16–84. The tied crank and gear mechanism gives rocking motion to crank AC, necessary for the operation of a printing press. If link DE has the angular motion shown, determine the respective angular velocities and angular accelerations of gear F and crank AC at this instant.

Prob. 16–84

16–85. The mechanism produces intermittent motion of link AB. If the sprocket S is turning with an angular acceleration $\alpha_S = 2$ rad/s^2 and has an angular velocity $\omega_S = 6$ rad/s at the instant shown, determine the angular velocity and angular acceleration of link AB at this instant. The sprocket S is mounted on a shaft which is *separate* from a collinear shaft attached to AB at A. The pin at C is attached to one of the chain links such that it moves vertically downward.

Prob. 16–85

Relative-Motion Analysis Using Rotating Axes 16.10

In the previous sections, the relative-motion analysis for velocity and acceleration was described using a translating coordinate system. This type of analysis is useful for determining the motion of points in the *same* rigid body, or the motion of points located in several pin-connected rigid bodies. In some problems, however, rigid bodies (mechanisms) are constructed such that *sliding* will occur at their connections. The kinematic analysis for such cases is best performed if the motion is analyzed using a coordinate system which both *translates* and *rotates*. Furthermore, this frame of reference is useful for analyzing the motions of two points on a mechanism which are *not* located in the *same* rigid body and for specifying the kinematics of particle motion when one of the particles is moving along a rotating path.

In the following analysis two equations are developed which relate the velocity and acceleration of a point to the origin of a moving frame of reference subjected to both a translation and a rotation in the plane.* Due to the generality in the derivation which follows, these two points may represent either two particles moving independently of one another or two points located in the same (or different) rigid bodies.

Position. Consider the two points A and B shown in Fig. 16–30a. Their location is specified by the position vectors $\mathbf{r}_A$ and $\mathbf{r}_B$, which are measured from the fixed X, Y, Z coordinate system. As shown in the figure, the "base point" A represents the origin of the x, y, z coordinate system, which is as-

---

*The more general, three-dimensional motion of the points is developed in Sec. 20.4.

(a)

(b)

(c)

Fig. 16–30

sumed to be both translating and rotating with respect to the X, Y, Z system. The position of B with respect to A is specified by the relative-position vector $\mathbf{r}_{B/A}$. The components of this vector may be expressed either in terms of unit vectors along the X, Y axes, i.e., $\mathbf{I}$ and $\mathbf{J}$, or by unit vectors along the x, y axes, i.e., $\mathbf{i}$ and $\mathbf{j}$. Although the magnitude of $\mathbf{r}_{B/A}$ is the same when measured in both coordinate systems, the direction of this vector will be measured differently if the x, y axes are not parallel to the X, Y axes. For the proof which follows, $\mathbf{r}_{B/A}$ will be measured relative to the moving x, y frame of reference. Thus, if B has coordinates (x_B, y_B), Fig. 16–30a, then

$$\mathbf{r}_{B/A} = x_B \mathbf{i} + y_B \mathbf{j}$$

Using vector addition, the three position vectors in Fig. 16–30a may be related by the equation

$$\mathbf{r}_B = \mathbf{r}_A + \mathbf{r}_{B/A} \qquad (16\text{–}17)$$

At the instant considered, point A has a velocity $\mathbf{v}_A$ and an acceleration $\mathbf{a}_A$, while the angular velocity and angular acceleration of the x, y axes are $\mathbf{\Omega}$ (omega) and $\dot{\mathbf{\Omega}} = d\mathbf{\Omega}/dt$, respectively. All these vectors are measured from the X, Y, Z frame of reference, although they may be expressed either in terms of $\mathbf{I}, \mathbf{J}, \mathbf{K}$ or $\mathbf{i}, \mathbf{j}, \mathbf{k}$ components. Since planar motion is specified, then by the right-hand rule $\mathbf{\Omega}$ and $\dot{\mathbf{\Omega}}$ are always directed *perpendicular* to the reference plane of motion, whereas $\mathbf{v}_A$ and $\mathbf{a}_A$ lie in this plane.

Velocity. The velocity of point B is determined by taking the time derivative of Eq. 16–17, which yields

$$\mathbf{v}_B = \mathbf{v}_A + \frac{d\mathbf{r}_{B/A}}{dt} \qquad (16\text{–}18)$$

The last term in this equation is evaluated as follows:

$$\frac{d\mathbf{r}_{B/A}}{dt} = \frac{d}{dt}(x_B \mathbf{i} + y_B \mathbf{j})$$

$$= \frac{dx_B}{dt}\mathbf{i} + x_B \frac{d\mathbf{i}}{dt} + \frac{dy_B}{dt}\mathbf{j} + y_B \frac{d\mathbf{j}}{dt}$$

$$= \left(\frac{dx_B}{dt}\mathbf{i} + \frac{dy_B}{dt}\mathbf{j}\right) + \left(x_B \frac{d\mathbf{i}}{dt} + y_B \frac{d\mathbf{j}}{dt}\right) \qquad (16\text{–}19)$$

The two terms in the first set of parentheses represent the components of velocity of point B as measured by an observer attached to the moving coordinate system. These terms will be denoted by vector $(\mathbf{v}_{B/A})_{\text{rel}}$. In the second set of parentheses the instantaneous time rate of change of the unit vectors $\mathbf{i}$ and $\mathbf{j}$ is measured by an observer located in the fixed X, Y, Z coordinate system. These changes, $d\mathbf{i}$ and $d\mathbf{j}$, are due *only* to the instantaneous *rotation* $d\theta$ of the x, y, z axes, Fig. 16–30b. As shown, the *magnitudes* of both $d\mathbf{i}$ and $d\mathbf{j}$ equal 1 $(d\theta)$, since $i = j = 1$. The *direction* of $d\mathbf{i}$ is defined by $+\mathbf{j}$, since $d\mathbf{i}$

is tangent to the path described by the tip of $\mathbf{i}$ in the limit as $\Delta t \to dt$. Likewise, $d\mathbf{j}$ acts in the $-\mathbf{i}$ direction. Hence,

$$\frac{d\mathbf{i}}{dt} = \frac{d\theta}{dt}(\mathbf{j}) = \Omega\mathbf{j} \qquad \frac{d\mathbf{j}}{dt} = \frac{d\theta}{dt}(-\mathbf{i}) = -\Omega\mathbf{i}$$

Viewing the axes in three dimensions, Fig. 16–30c, and noting that $\Omega = \Omega\mathbf{k}$, we can express the above derivatives in terms of the cross product as

$$\frac{d\mathbf{i}}{dt} = \Omega \times \mathbf{i} \qquad \frac{d\mathbf{j}}{dt} = \Omega \times \mathbf{j} \qquad (16\text{--}20)$$

Substituting these results into Eq. 16–19 and using the distributive property of the vector cross product, we obtain

$$\frac{d\mathbf{r}_{B/A}}{dt} = \mathbf{v}_{B/A} + \Omega \times (x_B\mathbf{i} + y_B\mathbf{j}) = (\mathbf{v}_{B/A})_{\text{rel}} + \Omega \times \mathbf{r}_{B/A} \quad (16\text{--}21)$$

Hence, Eq. 16–18 becomes

$$\mathbf{v}_B = \mathbf{v}_A + \Omega \times \mathbf{r}_{B/A} + (\mathbf{v}_{B/A})_{\text{rel}} \qquad (16\text{--}22)$$

where

$\mathbf{v}_B$ = velocity of B, measured from the X, Y, Z reference

$\mathbf{v}_A$ = velocity of the origin A of the x, y, z reference, measured from the X, Y, Z reference

$(\mathbf{v}_{B/A})_{\text{rel}}$ = relative velocity of "B with respect to A," as measured by an observer attached to the *rotating* x, y, z reference

Ω = angular velocity of the x, y, z reference, measured from the X, Y, Z reference

$\mathbf{r}_{B/A}$ = relative position of "B with respect to A"

Comparing Eq. 16–22 with Eq. 16–14 ($\mathbf{v}_B = \mathbf{v}_A + \Omega \times \mathbf{r}_{B/A}$), which is valid for a translating frame of reference, it can be seen that the only difference between the equations is represented by the term $(\mathbf{v}_{B/A})_{\text{rel}}$.

When applying Eq. 16–22 it is often useful to understand what each of the terms represents. In order of appearance, they are as follows:

$\mathbf{v}_B$ {absolute velocity of B

                (equals)

$\mathbf{v}_A$ {absolute velocity of origin of x, y, z frame

                (plus)   } motion of x, y, z frame

$\Omega \times \mathbf{r}_{B/A}$ {angular velocity effect caused by rotation of x, y, z frame

                (plus)

$(\mathbf{v}_{B/A})_{\text{rel}}$ {relative velocity of B with respect to A } motion of B within x, y, z frame

Acceleration. The acceleration of point B, observed from the X, Y, Z coordinate system, may be expressed in terms of its motion measured with respect to the rotating or moving system of coordinates by taking the time derivative of Eq. 16–22, i.e.,

$$\frac{d\mathbf{v}_B}{dt} = \frac{d\mathbf{v}_A}{dt} + \frac{d\mathbf{\Omega}}{dt} \times \mathbf{r}_{B/A} + \mathbf{\Omega} \times \frac{d\mathbf{r}_{B/A}}{dt} + \frac{d(\mathbf{v}_{B/A})_{\text{rel}}}{dt}$$

$$\mathbf{a}_B = \mathbf{a}_A + \dot{\mathbf{\Omega}} \times \mathbf{r}_{B/A} + \mathbf{\Omega} \times \frac{d\mathbf{r}_{B/A}}{dt} + \frac{d(\mathbf{v}_{B/A})_{\text{rel}}}{dt} \qquad (16\text{–}23)$$

Here $\dot{\mathbf{\Omega}} = d\mathbf{\Omega}/dt$ is the angular acceleration of the x, y, z coordinate system. For planar motion $\mathbf{\Omega}$ is always perpendicular to the plane of motion, and therefore $\dot{\mathbf{\Omega}}$ measures *only the change in magnitude* of $\mathbf{\Omega}$. The derivative $d\mathbf{r}_{B/A}/dt$ in Eq. 16–23 is defined by Eq. 16–21, so that

$$\mathbf{\Omega} \times \frac{d\mathbf{r}_{B/A}}{dt} = \mathbf{\Omega} \times (\mathbf{v}_{B/A})_{\text{rel}} + \mathbf{\Omega} \times (\mathbf{\Omega} \times \mathbf{r}_{B/A}) \qquad (16\text{–}24)$$

Computing the time derivative of $(\mathbf{v}_{B/A})_{\text{rel}} = (v_{B/A})_x \mathbf{i} + (v_{B/A})_y \mathbf{j}$, we have

$$\frac{d(\mathbf{v}_{B/A})_{\text{rel}}}{dt} = \left[\frac{d(v_{B/A})_x}{dt} \mathbf{i} + \frac{d(v_{B/A})_y}{dt} \mathbf{j} \right] + \left[(v_{B/A})_x \frac{d\mathbf{i}}{dt} + (v_{B/A})_y \frac{d\mathbf{j}}{dt} \right]$$

The two terms in the first set of brackets represent the components of acceleration of point B as measured by an observer attached to the moving coordinate system. These terms will be denoted by vector $(\mathbf{a}_{B/A})_{\text{rel}}$. The terms in the second set of brackets can be simplified using Eqs. 16–20. Hence,

$$\frac{d(\mathbf{v}_{B/A})_{\text{rel}}}{dt} = (\mathbf{a}_{B/A})_{\text{rel}} + \mathbf{\Omega} \times (\mathbf{v}_{B/A})_{\text{rel}}$$

Substituting this equation and Eq. 16–24 into Eq. 16–23 and rearranging terms yields

$$\mathbf{a}_B = \mathbf{a}_A + \dot{\mathbf{\Omega}} \times \mathbf{r}_{B/A} + \mathbf{\Omega} \times (\mathbf{\Omega} \times \mathbf{r}_{B/A}) + 2\mathbf{\Omega} \times (\mathbf{v}_{B/A})_{\text{rel}} + (\mathbf{a}_{B/A})_{\text{rel}} \qquad (16\text{–}25)$$

where

$\mathbf{a}_B$ = acceleration of B, measured from the X, Y, Z reference

$\mathbf{a}_A$ = acceleration of the origin A of the x, y, z reference, measured from the X, Y, Z reference

$(\mathbf{a}_{B/A})_{\text{rel}}, (\mathbf{v}_{B/A})_{\text{rel}}$ = relative acceleration and relative velocity of "B with respect to A," as measured by an observer attached to the *rotating* x, y, z reference

$\dot{\mathbf{\Omega}}, \mathbf{\Omega}$ = angular acceleration and angular velocity of the x, y, z reference, measured from the X, Y, Z reference

$\mathbf{r}_{B/A}$ = relative position of "B with respect to A"

If the motions of points A and B are along *curved paths*, it is often convenient to express the accelerations $\mathbf{a}_B$, $\mathbf{a}_A$, and $(\mathbf{a}_{B/A})_{\text{rel}}$ in Eq. 16–25 in terms of their normal and tangential components. If Eq. 16–25 is compared with Eq. 16–16 $(\mathbf{a}_B = \mathbf{a}_A + \dot{\boldsymbol{\Omega}} \times \mathbf{r}_{B/A} + \boldsymbol{\Omega} \times (\boldsymbol{\Omega} \times \mathbf{r}_{B/A})$, which is valid for a translating frame of reference, it can be seen that the difference between the equations is represented by the terms $2\boldsymbol{\Omega} \times (\mathbf{v}_{B/A})_{\text{rel}}$ and $(\mathbf{a}_{B/A})_{\text{rel}}$. In particular, $2\boldsymbol{\Omega} \times (\mathbf{v}_{B/A})_{\text{rel}}$ is called the *Coriolis acceleration*, named after the French engineer G. C. Coriolis (1792–1843), who was the first to determine it. This term represents the difference in the acceleration of B as measured from nonrotating and rotating x, y, z axes. As indicated by the vector cross product, the Coriolis acceleration will *always* be perpendicular to both $\boldsymbol{\Omega}$ and $(\mathbf{v}_{B/A})_{\text{rel}}$. It is an important component of the acceleration which must be considered whenever rotating reference frames are used. This often occurs, for example, when studying the accelerations and forces which act on rockets, long-range projectiles, or other bodies having motions which are largely affected by the rotation of the earth.

The following interpretation of the terms in Eq. 16–25 may be useful when applying this equation to the solution of problems.

$\mathbf{a}_B$ {absolute acceleration of B

(equals)

$\mathbf{a}_A$ {absolute acceleration of origin of x, y, z frame

(plus)

$\dot{\boldsymbol{\Omega}} \times \mathbf{r}_{B/A}$ {angular acceleration effect caused by rotation of x, y, z frame | motion of x, y, z frame

(plus)

$\boldsymbol{\Omega} \times (\boldsymbol{\Omega} \times \mathbf{r}_{B/A})$ {angular velocity effect caused by rotation of x, y, z frame

(plus)

$2\boldsymbol{\Omega} \times (\mathbf{v}_{B/A})_{\text{rel}}$ {combined effect of B moving relative to x, y, z coordinates and rotation of x, y, z frame}   interacting motion

(plus)

$(\mathbf{a}_{B/A})_{\text{rel}}$ {relative acceleration of B with respect to A}   motion of B within x, y, z frame

PROCEDURE FOR ANALYSIS

The following procedure provides a method for applying Eqs. 16–22 and 16–25 to the solution of problems involving the planar motion of particles or rigid bodies.

Coordinate Axes.   Choose an appropriate location for the origin and proper orientation of the axes for both the X, Y, Z and moving x, y, z reference frames. Most often solutions are easily obtained if at the instant considered: (1) the origins are coincident, (2) the axes are collinear, and/or (3) the axes are parallel. The moving frame should be selected fixed to the body or device where the relative motion occurs.

Kinematic Equations.   After defining the origin A of the moving reference and specifying the moving point B, Eqs. 16–22 and 16–25 should be written in symbolic form

$$\mathbf{v}_B = \mathbf{v}_A + \boldsymbol{\Omega} \times \mathbf{r}_{B/A} + (\mathbf{v}_{B/A})_{\text{rel}}$$

$$\mathbf{a}_B = \mathbf{a}_A + \dot{\boldsymbol{\Omega}} \times \mathbf{r}_{B/A} + \boldsymbol{\Omega} \times (\boldsymbol{\Omega} \times \mathbf{r}_{B/A}) + 2\boldsymbol{\Omega} \times (\mathbf{v}_{B/A})_{\text{rel}} + (\mathbf{a}_{B/A})_{\text{rel}}$$

Each of the vectors in these equations should be defined from the problem data and expressed in Cartesian vector form. This essentially requires a determination of (1) the motion of the moving reference, i.e., $\mathbf{v}_A$, $\mathbf{a}_A$, $\boldsymbol{\Omega}$, and $\dot{\boldsymbol{\Omega}}$; and (2) the motion of B measured with respect to the moving reference, i.e., $\mathbf{r}_{B/A}$, $(\mathbf{v}_{B/A})_{\text{rel}}$, and $(\mathbf{a}_{B/A})_{\text{rel}}$. The components of all these vectors may be selected along either the X, Y, Z axes or the x, y, z axes. The choice is arbitrary provided a consistent set of unit vectors is used.

Finally, substitute the data into the kinematic equations and perform the vector operations.

The following examples numerically illustrate this procedure.

Example 16–18

At the instant $\theta = 60°$, the rod in Fig. 16–31 has an angular velocity of 3 rad/s and an angular acceleration of 2 rad/s². At this same instant, the collar C is traveling outward along the rod such that when $x = 0.2$ m the velocity is 2 m/s and the acceleration is 3 m/s², both measured relative to the rod. Determine the Coriolis acceleration and the velocity and acceleration of the collar at this instant.

Fig. 16–31

Solution

Coordinate Axes. The origin of both coordinate systems is located at point O, Fig. 16–31. Since motion of the collar is reported relative to the rod, the moving x, y, z frame of reference is *attached* to the rod.

Kinematic Equations

$$\mathbf{v}_C = \mathbf{v}_O + \mathbf{\Omega} \times \mathbf{r}_{C/O} + (\mathbf{v}_{C/O})_{\text{rel}} \qquad (1)$$

$$\mathbf{a}_C = \mathbf{a}_O + \dot{\mathbf{\Omega}} \times \mathbf{r}_{C/O} + \mathbf{\Omega} \times (\mathbf{\Omega} \times \mathbf{r}_{C/O}) + 2\mathbf{\Omega} \times (\mathbf{v}_{C/O})_{\text{rel}} + (\mathbf{a}_{C/O})_{\text{rel}} \qquad (2)$$

It will be simpler to express the data in terms of $\mathbf{i}, \mathbf{j}, \mathbf{k}$ component vectors rather than $\mathbf{I}, \mathbf{J}, \mathbf{K}$ components. Hence,

Motion *of moving reference*	*Motion of C with respect* *to moving reference*
$\mathbf{v}_O = \mathbf{0}$	$\mathbf{r}_{C/O} = \{0.2\mathbf{i}\}$ m
$\mathbf{a}_O = \mathbf{0}$	$(\mathbf{v}_{C/O})_{\text{rel}} = \{2\mathbf{i}\}$ m/s
$\mathbf{\Omega} = \{-3\mathbf{k}\}$ rad/s	$(\mathbf{a}_{C/O})_{\text{rel}} = \{3\mathbf{i}\}$ m/s²
$\dot{\mathbf{\Omega}} = \{-2\mathbf{k}\}$ rad/s²	

From Eq. (2) the Coriolis acceleration is defined as

$$\mathbf{a}_{\text{Cor}} = 2\mathbf{\Omega} \times (\mathbf{v}_{C/O})_{\text{rel}} = 2(-3\mathbf{k}) \times (2\mathbf{i}) = \{-12\mathbf{j}\} \text{ m/s}^2 \qquad \textit{Ans.}$$

This vector is shown dashed in Fig. 16–31. If desired, it may be resolved into $\mathbf{I}, \mathbf{J}$ components acting along the X and Y axes, respectively.

The velocity and acceleration of the collar are determined by substituting the data into Eqs. (1) and (2) and evaluating the cross products, which yields

$$\mathbf{v}_C = \mathbf{v}_O + \mathbf{\Omega} \times \mathbf{r}_{C/O} + (\mathbf{v}_{C/O})_{\text{rel}}$$
$$= \mathbf{0} + (-3\mathbf{k}) \times (0.2\mathbf{i}) + 2\mathbf{i}$$
$$= \{2\mathbf{i} - 0.6\mathbf{j}\} \text{ m/s} \qquad \textit{Ans.}$$

$$\mathbf{a}_C = \mathbf{a}_O + \dot{\mathbf{\Omega}} \times \mathbf{r}_{C/O} + \mathbf{\Omega} \times (\mathbf{\Omega} \times \mathbf{r}_{C/O}) + 2\mathbf{\Omega} \times (\mathbf{v}_{C/O})_{\text{rel}} + (\mathbf{a}_{C/O})_{\text{rel}}$$
$$= \mathbf{0} + (-2\mathbf{k}) \times (0.2\mathbf{i}) + (-3\mathbf{k}) \times [(-3\mathbf{k}) \times (0.2\mathbf{i})]$$
$$\qquad\qquad\qquad + 2(-3\mathbf{k}) \times (2\mathbf{i}) + 3\mathbf{i}$$
$$= \mathbf{0} - 0.4\mathbf{j} - 1.8\mathbf{i} - 12\mathbf{j} + 3\mathbf{i}$$
$$= \{1.2\mathbf{i} - 12.4\mathbf{j}\} \text{ m/s}^2 \qquad \textit{Ans.}$$

Example 16–19

The rod AB, shown in Fig. 16–32a, rotates clockwise such that it has an angular velocity $\omega_{AB} = 2$ rad/s when $\theta = 45°$. Determine the angular velocity of rod DE at this instant. The collar at C is pin-connected to AB and slides over rod DE.

Solution

Coordinate Axes. The origins of both the fixed and moving frames of reference are located at D, Fig. 16–32a. Furthermore, the x, y, z reference is attached to and rotates with rod DE.

Velocity Equation

$$\mathbf{v}_C = \mathbf{v}_D + \mathbf{\Omega} \times \mathbf{r}_{C/D} + (\mathbf{v}_{C/D})_{rel} \qquad (1)$$

All vectors will be expressed in terms of $\mathbf{i}$, $\mathbf{j}$, $\mathbf{k}$ components.

Motion of moving reference	*Motion of C with respect to moving reference*
$\mathbf{v}_D = 0$	$\mathbf{r}_{C/D} = \{0.4\mathbf{i}\}$ m
$\mathbf{\Omega} = -\omega_{DE}\mathbf{k}$	$(\mathbf{v}_{C/D})_{rel} = (v_{C/D})_{rel}\mathbf{i}$

Since the collar moves along a *circular path*, it has the velocity shown in Fig. 16–32b.

Motion of Point C:

$$v_C = \omega_{AB}r_{C/A} = 2(0.4\sqrt{2}) = 1.131 \text{ m/s}$$
$$\mathbf{v}_C = 1.131 \cos 45°\mathbf{i} - 1.131 \sin 45°\mathbf{j}$$
$$= \{0.8\mathbf{i} - 0.8\mathbf{j}\} \text{ m/s}$$

Substituting the data into Eq. (1), we have

$$\mathbf{v}_C = \mathbf{v}_D + \mathbf{\Omega} \times \mathbf{r}_{C/D} + (\mathbf{v}_{C/D})_{rel}$$
$$0.8\mathbf{i} - 0.8\mathbf{j} = 0 + (-\omega_{DE}\mathbf{k}) \times (0.4\mathbf{i}) + (v_{C/D})_{rel}\mathbf{i}$$
$$0.8\mathbf{i} - 0.8\mathbf{j} = 0 - 0.4\omega_{DE}\mathbf{j} + (v_{C/D})_{rel}\mathbf{i}$$

Equating the respective $\mathbf{i}$ and $\mathbf{j}$ components yields

$$(v_{C/D})_{rel} = 0.8 \text{ m/s}$$
$$\omega_{DE} = \frac{0.8}{0.4} = 2 \text{ rad/s}$$

Hence,

$$\boldsymbol{\omega}_{DE} = \mathbf{\Omega} = \{-2\mathbf{k}\} \text{ rad/s} \qquad \qquad \textit{Ans.}$$

Example 16–20

Two planes are flying at the same elevation and have the motions shown in Fig. 16–33. Determine the velocity and acceleration of A as measured by the pilot of B.

Fig. 16–33

Solution

Coordinate Axes. Since the relative motion of A with respect to B is being sought, the x, y, z axes are attached to the pilot at B, Fig. 16–33. At the *instant* considered, the origin B coincides with the origin of the fixed X, Y, Z frame.

Kinematic Equations

$$\mathbf{v}_A = \mathbf{v}_B + \mathbf{\Omega} \times \mathbf{r}_{A/B} + (\mathbf{v}_{A/B})_{\text{rel}} \qquad (1)$$

$$\mathbf{a}_A = \mathbf{a}_B + \dot{\mathbf{\Omega}} \times \mathbf{r}_{A/B} + \mathbf{\Omega} \times (\mathbf{\Omega} \times \mathbf{r}_{A/B}) + 2\mathbf{\Omega} \times (\mathbf{v}_{A/B})_{\text{rel}} + (\mathbf{a}_{A/B})_{\text{rel}} \qquad (2)$$

Motion of moving reference:

$$\mathbf{v}_B = \{600\mathbf{j}\}\ \text{km/h}$$

$$\mathbf{a}_B = (\mathbf{a}_B)_n + (\mathbf{a}_B)_t = \{900\mathbf{i} - 100\mathbf{j}\}\ \text{km/h}^2$$

$$(a_B)_n = \frac{v_B^2}{\rho} = \frac{(600)^2}{400} = 900\ \text{km/h}^2$$

$$\mathbf{\Omega} = \{-1.5\mathbf{k}\}\ \text{rad/h} \qquad \Omega = \frac{v_B}{\rho} = \frac{600\ \text{km/h}}{400\ \text{km}} = 1.5\ \text{rad/h}\ \curvearrowright$$

$$\dot{\mathbf{\Omega}} = \{0.25\mathbf{k}\}\ \text{rad/h}^2 \qquad \dot{\Omega} = \frac{(a_B)_t}{\rho} = \frac{100\ \text{km/h}^2}{400\ \text{km}} = 0.25\ \text{rad/h}^2\ \curvearrowleft$$

Motion of A with respect to moving reference:

$$\mathbf{r}_{A/B} = \{-4\mathbf{i}\}\ \text{km} \qquad (\mathbf{v}_{A/B})_{\text{rel}} = ? \qquad (\mathbf{a}_{A/B})_{\text{rel}} = ?$$

Substituting this data into Eqs. (1) and (2), realizing that $\mathbf{v}_A = \{700\mathbf{j}\}\ \text{km/h}$ and $\mathbf{a}_A = \{50\mathbf{j}\}\ \text{km/h}^2$, we have

$$\mathbf{v}_A = \mathbf{v}_B + \mathbf{\Omega} \times \mathbf{r}_{A/B} + (\mathbf{v}_{A/B})_{\text{rel}}$$

$$700\mathbf{j} = 600\mathbf{j} + (-1.5\mathbf{k}) \times (-4\mathbf{i}) + (\mathbf{v}_{A/B})_{\text{rel}}$$

$$(\mathbf{v}_{A/B})_{\text{rel}} = \{94\mathbf{j}\}\ \text{km/h} \qquad\qquad\qquad\qquad\qquad\qquad Ans.$$

$$\mathbf{a}_A = \mathbf{a}_B + \dot{\mathbf{\Omega}} \times \mathbf{r}_{A/B} + \mathbf{\Omega} \times (\mathbf{\Omega} \times \mathbf{r}_{A/B}) + 2\mathbf{\Omega} \times (\mathbf{v}_{A/B})_{\text{rel}} + (\mathbf{a}_{A/B})_{\text{rel}}$$

$$50\mathbf{j} = (900\mathbf{i} - 100\mathbf{j}) + (0.25\mathbf{k}) \times (-4\mathbf{i})$$

$$\qquad + (-1.5\mathbf{k}) \times [(-1.5\mathbf{k}) \times (-4\mathbf{i})] + 2(-1.5\mathbf{k}) \times (94\mathbf{j}) + (\mathbf{a}_{A/B})_{\text{rel}}$$

$$(\mathbf{a}_{A/B})_{\text{rel}} = \{-1191\mathbf{i} + 151\mathbf{j}\}\ \text{km/h}^2 \qquad\qquad\qquad\qquad Ans.$$

Problems

16–86. Block A, which is attached to a cord, moves along the slot of a horizontal forked rod. If the cord is pulled down through the hole at O at a constant rate of 2 m/s, determine the acceleration of the block at the instant shown. The rod rotates about O with a constant angular velocity ω = 4 rad/s.

Probs. 16–86 / 16–87

16–87. Block A, which is attached to a cord, moves along the slot of a horizontal forked rod. If the cord is pulled down through the hole at O with a velocity of 2 m/s and an acceleration of 3 m/s², determine the acceleration of the block at the instant shown. At this instant, the rod is rotating about O with an angular velocity ω = 4 rad/s and has an *angular deceleration* α = 2 rad/s².

*16–88. The man stands on the platform at O and runs out toward the edge such that when he is at A, y = 5 ft, he has a velocity of 2 ft/s and an acceleration of 3 ft/s², both measured with respect to the platform. If the platform has the angular motions shown, determine the man's velocity and acceleration at this instant.

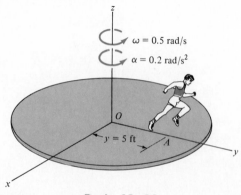

Prob. 16–88

16–89. A girl stands at A on a platform which is rotating with a constant angular velocity ω = 0.5 rad/s. If she walks at a constant speed v = 2 ft/s on the platform, determine her acceleration (a) when she reaches point D in going along the path ADC, and (b) when she reaches point B if she follows the path ABC.

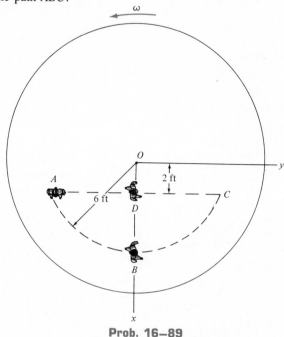

Prob. 16–89

16–90. The rod AB rotates counterclockwise with a constant angular velocity of 2 rad/s. Determine the velocity and acceleration of point C located on the double collar when θ = 45°. The collar consists of two slider blocks which are constrained to move along the circular shaft and the rod AB.

Prob. 16–90

16–91. At the instant $\theta = 60°$, link CD has an angular velocity $\omega_{CD} = 4$ rad/s and an angular acceleration $\alpha_{CD} = 2$ rad/s². Determine the angular velocity and angular acceleration of rod AB at this instant. The collar at C is pin-connected to DC and slides over AB.

Prob. 16–91

16–93. The "quick return" mechanism consists of a crank AB, slider block B, and slotted link CD. If the crank has the angular motions shown, determine the angular motions of the slotted link at this instant.

Prob. 16–93

*16–92. At a given instant rod AB has the angular motions shown. Determine the angular velocity and angular acceleration of rod CD at this instant. There is a collar at D.

Prob. 16–92

16–94. The ring pivots about point O and has the angular motion shown. Also at this instant, the collar is traveling down along the ring with a velocity of 5 m/s and acceleration of 8 m/s², both measured with respect to the ring. Determine the velocity and acceleration of the collar.

Prob. 16–94

16–95. A ride in an amusement park consists of a rotating platform P, having a constant angular velocity $\omega_P = 1.5$ rad/s, and four cars, C, mounted on the platform, which have constant angular velocities $\omega_{C/P} = 2$ rad/s measured relative to the platform. At the instant shown, determine the velocity and acceleration of the passenger at A.

Prob. 16–95

*16–96. The two-link mechanism serves to amplify angular motion. Link AB has a pin at B which is confined to move within the slot of link CD. If at the instant shown AB (input) has an angular velocity $\omega_{AB} = 3$ rad/s and an angular acceleration $\alpha_{AB} = 2$ rad/s^2, determine the angular velocity and angular acceleration of CD (output) at this instant.

Prob. 16–96

16–97. Particles B and A move along the parabolic and circular paths, respectively. If B has a velocity of 7 m/s in the direction shown and its speed is increasing at 4 m/s^2, while A has a velocity of 8 m/s in the direction shown and its speed is decreasing at 6 m/s^2, determine the relative velocity and relative acceleration of B with respect to A.

Prob. 16–97

16–98. The disk rolls without slipping and at a given instant has the angular motion shown. Determine the angular velocity and angular acceleration of the slotted link BC at this instant. The peg at A is fixed to the disk.

Prob. 16–98

Planar Kinetics of a Rigid Body: Force and Acceleration

17

In Chapter 16 the planar kinematics of rigid-body motion was presented in order of increasing complexity, that is, translation, rotation about a fixed axis, and general plane motion. The study of rigid-body kinetics in this chapter will be presented in somewhat the same order. The chapter begins by introducing a property of a body called the mass moment of inertia. Afterward, a derivation of the equations of general plane motion for a rigid body which has a plane of symmetry parallel to the plane of motion is given. These equations are then applied to specific problems of rigid-body translation, rotation about a fixed axis, and finally general plane motion. A kinetic study of these motions is referred to as the kinetics of planar motions or simply *planar kinetics*. The more complex study of three-dimensional rigid-body kinetics, which includes planar motion of unsymmetrical rigid bodies, is presented in Chapter 21.

Moment of Inertia 17.1

Since a body has a definite size and shape, an applied nonconcurrent force system may cause the body to both translate and rotate. The translational aspects of the motion were studied in Chapter 13 and are governed by the equation $\mathbf{F} = m\mathbf{a}$. It will be shown in Sec. 17.2 that the rotational aspects, caused by the moment $\mathbf{M}$, are governed by an equation of the form $\mathbf{M} = I\boldsymbol{\alpha}$. The symbol I in this equation is termed the moment of inertia. By comparison, the *moment of inertia* is a measure of the resistance of a body to *angular acceleration* ($\mathbf{M} = I\boldsymbol{\alpha}$) in the same way that *mass* is a measure of the body's resistance to *acceleration* ($\mathbf{F} = m\mathbf{a}$).

Fig. 17–1

We define the *moment of inertia* as the integral of the "second moment" about an axis of all the differential-size elements of mass dm which compose the body.* For example, consider the rigid body shown in Fig. 17–1. The body's moment of inertia about the z axis is

$$I = \int_m r^2 \, dm \qquad (17\text{--}1)$$

Here the "moment arm" r is the perpendicular distance from the axis to the arbitrary element dm. Since the formulation involves r, the value of I is different for each axis about which it is computed. In the study of planar kinetics, however, the axis which is generally chosen for analysis passes through the body's mass center G and is always perpendicular to the plane of motion. The moment of inertia computed about this axis will be denoted as I_G. Note that because r is squared in Eq. 17–1 the mass moment of inertia is always a positive quantity. Common units used for its measurement are $kg \cdot m^2$ or $slug \cdot ft^2$.

---

*Another property of the body, which measures the symmetry of the body's mass with respect to a coordinate system, is the mass product of inertia. This property applies to the three-dimensional motion of a body and will be discussed in Chapter 21.

PROCEDURE FOR ANALYSIS

In this treatment, we will only consider bodies having surfaces which are generated by revolving a curve about an axis. An example of such a body of which the volume of revolution is generated about the z axis is shown in Fig. 17–2. If the body consists of material having a variable mass density, $\rho = \rho\,(x, y, z)$, the elemental mass dm of the body may be expressed in terms of its density and volume as $dm = \rho\,dV$. Substituting in Eq. 17–1, the body's moment of inertia is then computed using *volume elements* for integration, i.e.,

$$I = \int_V r^2 \rho\,dV \qquad (17\text{–}2)$$

In the special case of ρ being a *constant*, this term may be factored out of the integral and the integration is then purely a function of geometry,

$$I = \rho \int_V r^2\,dV \qquad (17\text{–}3)$$

When the elemental volume chosen for integration has infinitesimal dimensions in all three directions, e.g., $dV = dx\,dy\,dz$, Fig. 17–2a, the moment of inertia of the body must be computed using "triple integration." The integration process can, however, be simplified to a *single integration* provided the chosen elemental volume has a differential size or thickness in only *one direction*. Shell or disk elements are often used for this purpose.

Shell Element. If a *shell element* having a height z, radius $r = y$, and thickness dy is chosen for integration, Fig. 17–2b, then the volume is $dV = (2\pi y)(z)\,dy$. This element may be used in Eq. 17–2 or 17–3 for computing the moment of inertia I_z of the body about the z axis, since the *entire element*, due to its "thinness," lies at the *same* perpendicular distance $r = y$ from the z axis (see Example 17–1).

Disk Element. If a disk element having a radius $r = y$ and a thickness dz is chosen for integration, Fig. 17–2c, then the volume is $dV = (\pi y^2)\,dz$. In this case, however, the element is *finite* in the radial direction, and consequently parts of it *do not* all lie at the *same radial distance* r from the z axis. As a result, Eq. 17–2 or 17–3 *cannot* be used to determine I_z. Instead, to perform the integration using this element, it is first necessary to determine the moment of inertia *of the element* about the z axis and then integrate this result (see Example 17–2).

(a)

(b)

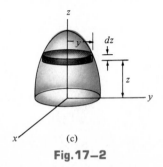

(c)

Fig. 17–2

Example 17-1

Determine the moment of inertia of the right circular cylinder shown in Fig. 17–3a about the z axis. The mass density ρ of the material is constant.

(a)   (b)

Fig. 17–3

Solution

Shell Element. This problem may be solved using the *shell element* in Fig. 17–3b and single integration. The volume of the element is $dV = (2\pi r)(h)\,dr$, so that the mass is $dm = \rho\,dV = \rho(2\pi hr\,dr)$. Since the *entire element* lies at the same distance r from the z axis, the moment of inertia *of the element* is

$$dI_z = r^2\,dm = \rho 2\pi hr^3\,dr$$

Integrating over the entire region of the cylinder yields

$$I_z = \int_m r^2\,dm = \rho 2\pi h \int_0^R r^3\,dr = \frac{\rho\pi}{2}R^4 h$$

The mass of the cylinder is

$$m = \int_m dm = \rho 2\pi h \int_0^R r\,dr = \rho\pi hR^2$$

so that

$$I_z = \frac{1}{2}mR^2 \qquad\qquad Ans.$$

Example 17–2

A solid is formed by revolving the shaded area shown in Fig. 17–4a about the y axis. If the mass density of the material is 5 slug/ft^3, determine the moment of inertia about the y axis.

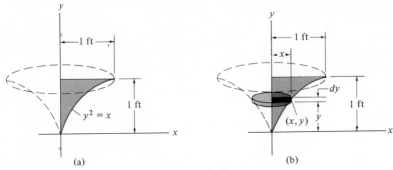

(a)   (b)

Fig. 17–4

Solution

Disk Element. The moment of inertia will be computed using a *disk element,* as shown in Fig. 17–4b. Here the element intersects the curve at the arbitrary point (x, y) and has a mass

$$dm = \rho \, dV = \rho(\pi x^2) \, dy$$

Although all portions of the element are *not* located at the same distance from the y axis, it is still possible to determine the moment of inertia dI_y *of the element* about the y axis. In the preceding example it was shown that the moment of inertia of a cylinder about its longitudinal axis is $I = \frac{1}{2}mR^2$, where m and R are the mass and radius of the cylinder. Since the height of the cylinder is not involved in this formula, the moment of inertia of the disk element in Fig. 17–4b is

$$dI_y = \tfrac{1}{2}(dm)x^2 = \tfrac{1}{2}[\rho(\pi x^2) \, dy]x^2$$

Substituting $x = y^2$, $\rho = 5$ slug/ft^3, and integrating with respect to y, from $y = 0$ to $y = 1$ ft, yields the moment of inertia for the entire solid.

$$I_y = \frac{\pi(5)}{2} \int_0^1 x^4 \, dy = \frac{\pi(5)}{2} \int_0^1 y^8 \, dy = 0.873 \text{ slug} \cdot \text{ft}^2 \quad \textit{Ans.}$$

Parallel-Axis Theorem. If the moment of inertia of the body about an axis passing through the body's mass center is known, then the body's moment of inertia may be determined about any other *parallel axis* by using the *parallel-axis theorem*. This theorem can be derived by considering the body shown in Fig. 17–5. The $\bar{z}$ axis passes through the body's mass center G, whereas the corresponding *parallel axis z* lies at a constant distance d away. Selecting the differential element of mass dm which is located at point (x, y) and using the Pythagorean theorem, $r^2 = (d + x)^2 + y^2$, we can express the moment of inertia of the body computed about the z axis as

$$I = \int_m r^2 \, dm = \int_m [(d + x)^2 + y^2] \, dm$$

$$= \int_m (x^2 + y^2) \, dm + 2d \int_m x \, dm + d^2 \int_m dm$$

Since $r_O^2 = x^2 + y^2$, the first integral represents I_G. The second integral equals *zero*, since the $\bar{z}$ axis passes through the body's mass center, i.e., $\int x \, dm = \bar{x} \int dm$ since $\bar{x} = 0$. Finally, the third integral represents the total mass m of the body. Hence, the moment of inertia about the z axis can be written as

$$I = I_G + md^2 \tag{17–4}$$

where

I_G = moment of inertia about the $\bar{z}$ axis passing through the mass center G
m = mass of the body
d = perpendicular distance between the parallel axes

Fig. 17–5

Radius of Gyration. Occasionally, the moment of inertia of a body about a specified axis is reported in handbooks using the *radius of gyration*. When this length, k, and the body's mass m are known, the body's moment of inertia is determined from the equation

$$I = mk^2 \quad \text{or} \quad k = \sqrt{\frac{I}{m}} \tag{17-5}$$

Note the *similarity* between the definition of k in this formula and r in the equation $dI = r^2\, dm$, which defines the moment of inertia of an elemental mass dm of the body about an axis.

Composite Bodies. If a body is constructed of a number of simple shapes such as disks, spheres, and rods, the moment of inertia of the body about any axis z can be determined by adding algebraically the moments of inertia of all the composite shapes computed about the z axis. Algebraic addition is necessary since a composite part must be considered as a negative quantity if it has already been counted as part of another part—for example a "hole" subtracted from a solid plate. The parallel-axis theorem is needed for the calculations if the center of gravity of each composite part does not lie on the z axis. For the calculation then $I = \Sigma(I_G + md^2)$. Here I_G for each of the composite parts is either known or can be computed by integration.*

*See the table given on the inside back cover of this book.

Example 17–3

If the plate shown in Fig. 17–6a has a density of 8000 kg/m³ and a thickness of 10 mm, compute its moment of inertia about an axis directed perpendicular to the page and passing through point O.

Fig. 17–6

Solution

The plate consists of two composite parts, the 250-mm-radius disk *minus* a 125-mm-radius disk, Fig. 17–6b. The moment of inertia about O can be determined by computing the moment of inertia of each of these parts about O and then *algebraically* adding the results. The computations are performed by using the parallel-axis theorem in conjunction with the data listed in the table on the inside back cover.

Disk. The moment of inertia of a thin disk about an axis perpendicular to the plane of the disk is $I_G = \frac{1}{2}mr^2$. The mass center of the disk is located at a distance of 0.25 m from point O. Thus,

$$m_d = \rho_d V_d = 8000 \text{ kg/m}^3 [\pi(0.25 \text{ m})^2(0.01 \text{ m})] = 15.71 \text{ kg}$$

$$(I_O)_d = \frac{1}{2}m_d r_d^2 + m_d d^2$$

$$= \frac{1}{2}(15.71 \text{ kg})(0.25 \text{ m})^2 + (15.71 \text{ kg})(0.25 \text{ m})^2$$

$$= 1.473 \text{ kg} \cdot \text{m}^2$$

Hole. For the 125-mm-radius disk (hole), we have

$$m_h = \rho_h V_h = 8000 \text{ kg/m}^3 [\pi(0.125 \text{ m})^2(0.01 \text{ m})] = 3.93 \text{ kg}$$

$$(I_O)_h = \frac{1}{2}m_h r_h^2 + m_h d^2$$

$$= \frac{1}{2}(3.93 \text{ kg})(0.125 \text{ m})^2 + (3.93 \text{ kg})(0.25 \text{ m})^2$$

$$= 0.276 \text{ kg} \cdot \text{m}^2$$

The moment of inertia of the plate about point O is therefore

$$I_O = (I_O)_d - (I_O)_h$$

$$= 1.473 - 0.276$$

$$= 1.20 \text{ kg} \cdot \text{m}^2 \qquad \qquad Ans.$$

Example 17–4

The pendulum consists of two thin rods, each having a weight of 10 lb and suspended from point O as shown in Fig. 17–7. Compute the pendulum's moment of inertia about an axis passing through (a) the pin at O, and (b) the mass center G of the pendulum.

Fig. 17–7

Solution

Part (a). Using the table on the inside back cover, the moment of inertia of rod OA about an axis perpendicular to the page and passing through the end point O of the rod is $I_O = \frac{1}{3}ml^2$. Hence,

$$(I_O)_{OA} = \frac{1}{3}ml^2 = \frac{1}{3}\left(\frac{10}{32.2}\right)(2)^2 = 0.414 \text{ slug} \cdot \text{ft}^2$$

The same value may be computed using $I_G = \frac{1}{12}ml^2$ and the parallel-axis theorem; i.e.,

$$(I_O)_{OA} = \frac{1}{12}ml^2 + md^2 = \frac{1}{12}\left(\frac{10}{32.2}\right)(2)^2 + \frac{10}{32.2}(1)^2$$

$$= 0.414 \text{ slug} \cdot \text{ft}^2$$

For rod BC we have

$$(I_O)_{BC} = \frac{1}{12}ml^2 + md^2 = \frac{1}{12}\left(\frac{10}{32.2}\right)(2)^2 + \left(\frac{10}{32.2}\right)(2)^2$$

$$= 1.346 \text{ slug} \cdot \text{ft}^2$$

The moment of inertia of the pendulum is therefore

$$I_O = 0.414 + 1.346 = 1.76 \text{ slug} \cdot \text{ft}^2 \qquad \textit{Ans.}$$

Part (b). The mass center G will be located relative to the pin at O. Assuming this distance to be $\bar{y}$, Fig. 17–7, and using the formula for determining the mass center, we have

$$\bar{y} = \frac{\Sigma \tilde{y}m}{\Sigma m} = \frac{1\left(\dfrac{10}{32.2}\right) + 2\left(\dfrac{10}{32.2}\right)}{\left(\dfrac{10}{32.2}\right) + \left(\dfrac{10}{32.2}\right)} = 1.5 \text{ ft}$$

The moment of inertia I_G may be computed in the same manner as I_O, which requires successive applications of the parallel-axis theorem in order to transfer the moments of inertia of rods OA and BC to G. A more direct solution, however, involves applying the parallel-axis theorem using the result for I_O, i.e.,

$$I_O = I_G + md^2; \qquad 1.76 = I_G + \left(\frac{20}{32.2}\right)(1.5)^2$$

$$I_G = 0.362 \text{ slug} \cdot \text{ft}^2 \qquad \textit{Ans.}$$

317

Problems

17–1. Determine the moment of inertia I_y for the slender rod. The rod's density ρ and cross-sectional area A are constant. Express the result in terms of the rod's total mass m.

Prob. 17–1

17–2. The right circular cone is formed by revolving the shaded area around the x axis. Determine the moment of inertia I_x and express the result in terms of the total mass m of the cone. The cone has a constant density ρ.

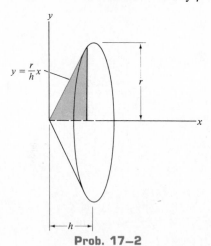

Prob. 17–2

17–3. The solid is formed by revolving the shaded area around the x axis. Determine the radius of gyration k_x. The mass density of the material is $\rho = 5 \text{ Mg/m}^3$.

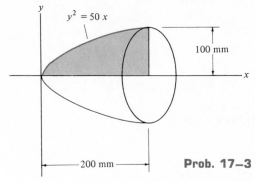

Prob. 17–3

*17–4. An ellipsoid is formed by rotating the shaded area about the x axis. Determine the moment of inertia of this body with respect to the x axis and express the result in terms of the mass m of the solid. The density of the material, ρ, is constant.

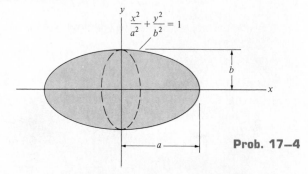

Prob. 17–4

17–5. The sphere is formed by revolving the shaded area around the x axis. Determine the moment of inertia I_x and express the result in terms of the total mass m of the sphere. The sphere has a constant density ρ.

Prob. 17–5

17-6. The semiellipsoid is formed by rotating the shaded area about the x axis. Determine the moment of inertia of this body with respect to the x axis, and express the result in terms of the mass m of the solid. The density of the material, ρ, is constant.

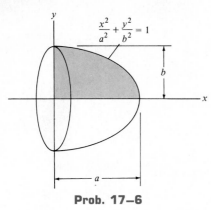

Prob. 17-6

17-7. Determine the moment of inertia of the homogeneous triangular prism with respect to the y axis. Express the result in terms of the mass m of the prism. *Hint:* For integration, use thin plate elements parallel to the x-y plane and having a thickness dz.

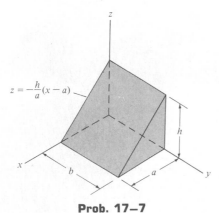

$$z = -\frac{h}{a}(x - a)$$

Prob. 17-7

*17-8. Determine the moment of inertia I_z of the torus. The mass of the torus is m and the density ρ is constant. *Suggestion:* Use a shell element.

Prob. 17-8

17-9. Determine the moment of inertia of the homogeneous pyramid of mass m with respect to the z axis. The density of the material is ρ. *Suggestion:* Use a rectangular plate element having a volume $dV = (2x)(2y)\, dz$.

Prob. 17-9

17–10. The concrete solid is formed by rotating the shaded area about the y axis. Determine the moment of inertia I_y. The specific weight of concrete is $\gamma = 150$ lb/ft^3.

$y = \frac{2}{9} x^2$

4 in.

←6 in.→

8 in.

Prob. 17–10

17–11. The wheel consists of a thin ring having a mass of 10 kg and four spokes made from slender rods each having a mass of 2 kg. Determine the wheel's moment of inertia about an axis perpendicular to the plane of the wheel and passing through point A.

500 mm

A

Prob. 17–11

***17–12.** The pendulum consists of a plate having a weight of 12 lb and a slender rod having a weight of 4 lb. Determine the radius of gyration of the pendulum about an axis directed perpendicular to the page and passing through point O.

1 ft

O

←1 ft→ ←— 3 ft —→ ←— 2 ft —→

Prob. 17–12

17–13. The pendulum consists of the 3-kg slender rod and the 5-kg thin plate. Determine the location $\bar{y}$ of the center of mass G of the pendulum, then calculate the moment of inertia of the pendulum about an axis perpendicular to the page and passing through G.

O

$\bar{y}$

2 m

G

A

0.5 m

←— 1 m —→

Prob. 17–13

17–14. Determine the moment of inertia about an axis perpendicular to the page and passing through the pin at O. The thin plate has a hole in its center. Its thickness is 50 mm and the material has a density of $\rho = 50$ kg/m^3.

O

150 mm

1.40 m   1.40 m

Prob. 17–14

17–15. Determine the location $\bar{y}$ of the center of mass G of the assembly, and then calculate the moment of inertia about an axis perpendicular to the page and passing through G. The block B has a mass of 3 kg and the mass of the semicylinder C is 5 kg. *Hint:* First show that the moment of inertia of the semicylinder about an axis perpendicular to the page and passing through the centroid is $I_z = 0.320mR^2$.

Prob. 17–15

*17–16. Determine the moment of inertia of the wheel about the z axis, which passes through the center point O. The material has a specific gravity $\gamma = 90$ lb/ft^3.

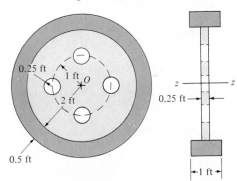

Prob. 17–16

17–17. Determine the moment of inertia about the z axis for the frustum of the cone which has a conical depression. The material has a density $\rho = 200$ kg/m^3.

Prob. 17–17

17–18. Locate the center of mass G of the baseball bat by determining $\bar{y}$. Then, compute the moment of inertia of the bat about an axis which passes through G and is perpendicular to the plane of the page. For the calculation, consider the bat to be composed of a truncated cone and cylinder. Neglect the size of the lip at A. The density of wood is $\rho_w = 750$ kg/m^3.

Prob. 17–18

17.2 Planar Kinetic Equations of Motion

In the following analysis we will limit our study of planar kinetics to rigid bodies which, along with their loadings, are considered to be *symmetrical with respect to a fixed reference plane.** The plane motion of such a rigid body may be analyzed in a fixed reference plane because the path of motion of each particle of the body is a plane curve parallel to the reference plane. Thus, for the analysis, the motion of the body may be represented by a *slab* having the same mass and exterior outline as the body. All the forces (and couples) acting on the body may then be projected onto the plane of this slab. An example of an arbitrary slab (body) of this type is shown in Fig. 17–8a. Here the *inertial frame of reference x, y, z* has its origin *coincident* with the arbitrary point *P* in the body. By definition, *these axes do not rotate and are either fixed or translate with constant velocity.*

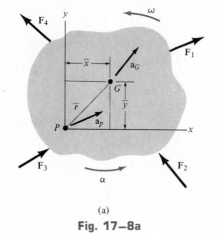

(a)

Fig. 17–8a

Equation of Translational Motion. The external forces shown on the body in Fig. 17–8a symbolically represent the effect of gravitational, electrical, magnetic, or contact forces between adjacent bodies. Since this force system has been considered previously in Sec. 13.3, for the analysis of a system of particles, the results may be used here, in which case the particles are contained within the boundary of the body (or slab). Hence, if the equation of motion is applied to each of the particles of the body, and the results added vectorially, it may be concluded that

$$\Sigma \mathbf{F} = m\mathbf{a}_G$$

This equation is referred to as the *equation of translational motion* for a rigid body. It states that *the sum of all the external forces acting on the body is equal to the body's mass times the acceleration of the mass center G.*

For motion of the body (or slab) in the *x-y* plane, the equation of translational motion may be written in the form of two independent scalar equations, namely,

$$\Sigma F_x = m(a_G)_x$$
$$\Sigma F_y = m(a_G)_y$$

Equation of Rotational Motion. We will now determine the effects caused by the moments of the external force system computed about an axis perpendicular to the plane of motion (the *z* axis) and passing through point *P*. As shown on the free-body diagram of the *i*th particle, Fig. 17–8b, $\mathbf{F}_i$ represents the *resultant external force* acting on the particle, and $\mathbf{f}_i$ is the *resultant of the internal forces* caused by interactions with adjacent particles. If the particle has a mass *dm* and at the instant considered its acceleration is $\mathbf{a}_i$, then the

---

*By doing this, the rotational equation of motion reduces to a rather simplified form. The more general case of body shape and loading is considered in Chapter 21.

kinetic diagram is constructed as shown in Fig. 17–8c. If moments of the forces acting on the particle are summed about point P, then we require

$$\mathbf{r} \times \mathbf{F}_i + \mathbf{r} \times \mathbf{f}_i = \mathbf{r} \times dm\mathbf{a}_i$$

or

$$d\mathbf{M}_P = \mathbf{r} \times dm\mathbf{a}_i$$

We will now write this equation in terms of the acceleration $\mathbf{a}_P$ of point P. If the body has an angular acceleration $\boldsymbol{\alpha}$ and angular velocity $\boldsymbol{\omega}$, Fig. 17–8a, then using Eq. 16–16 we have

$$d\mathbf{M}_P = dm\mathbf{r} \times (\mathbf{a}_P - \omega^2\mathbf{r} + \boldsymbol{\alpha} \times \mathbf{r})$$
$$= dm[\mathbf{r} \times \mathbf{a}_P - \omega^2(\mathbf{r} \times \mathbf{r}) + \mathbf{r} \times (\boldsymbol{\alpha} \times \mathbf{r})]$$

Particle free-body diagram

(b)

The second term on the right side is zero since $\mathbf{r} \times \mathbf{r} = \mathbf{0}$. Expressing the vectors with Cartesian components and carrying out the cross-product operations yields

$$dM_P\,\mathbf{k} = dm[(x\mathbf{i} + y\mathbf{j}) \times ((a_P)_x\mathbf{i} + (a_P)_y\mathbf{j})$$
$$+ (x\mathbf{i} + y\mathbf{j}) \times (\alpha\mathbf{k} \times (x\mathbf{i} + y\mathbf{j}))]$$
$$= dm(-y(a_P)_x + x(a_P)_y + \alpha x^2 + \alpha y^2)\mathbf{k}$$
$$dM_P = dm(-y(a_P)_x + x(a_P)_y + \alpha r^2)$$

$=$

By integrating with respect to the entire mass m of the body so as to include all the particles contained in the body, we obtain the resultant moment equation

$$\Sigma M_P = -\left(\int_m y\,dm\right)(a_P)_x + \left(\int_m x\,dm\right)(a_P)_y + \left(\int_m r^2\,dm\right)\alpha$$

Particle kinetic diagram

(c)

Fig. 17–8b, c

Here ΣM_P represents only the moment of the *external forces* acting on the body about point P. The resultant moment of the internal forces is zero, since for the entire body they occur in equal and opposite collinear pairs and thus the moment of each pair of forces about point P cancels. The integrals in the first and second terms on the right are used to locate the body's center of mass G with respect to P, since $\bar{y}m = \int y\,dm$ and $\bar{x}m = \int x\,dm$, Fig. 17–8a. Also, the last integral represents the body's moment of inertia computed about the z axis, i.e., $I_P = \int r^2\,dm$. Thus,

$$\downarrow+\Sigma M_P = -\bar{y}m(a_P)_x + \bar{x}m(a_P)_y + I_P\alpha \qquad (17\text{–}6)$$

It is possible to reduce this equation to a simpler form if point P coincides with the mass center G for the body.* If this is the case, then $\bar{x} = \bar{y} = 0$, and therefore

$$\Sigma M_G = I_G\alpha \qquad (17\text{–}7)$$

---

*It also reduces to this same simple form $\Sigma M_P = I_P\alpha$ if point P is a *fixed point* (see Eq. 17–16) or the acceleration of point P is directed along the line PG.

This rotational equation of motion states that the sum of the moments of all the external forces computed about the body's mass center G is equal to the product of the moment of inertia of the body about G and the magnitude of the body's angular acceleration.

Equation 17–6 can also be rewritten in terms of the x and y components of $\mathbf{a}_G$ and the body's moment of inertia I_G. With reference to Fig. 17–8a, using the parallel-axis theorem, $I_P = I_G + m(\bar{x}^2 + \bar{y}^2)$, we get

$$\zeta + \Sigma M_P = \bar{y}m[-(a_P)_x + \bar{y}\alpha] + \bar{x}m[(a_P)_y + \bar{x}\alpha] + I_G\alpha \qquad (17\text{–}8)$$

Expressing $\mathbf{a}_P$ in terms of $\mathbf{a}_G$ we have

$$\mathbf{a}_G = \mathbf{a}_P - \omega^2\bar{\mathbf{r}} + \boldsymbol{\alpha} \times \bar{\mathbf{r}}$$

$$(a_G)_x\mathbf{i} + (a_G)_y\mathbf{j} = (a_P)_x\mathbf{i} + (a_P)_y\mathbf{j} - \omega^2(\bar{x}\mathbf{i} + \bar{y}\mathbf{j}) + \alpha\mathbf{k} \times (\bar{x}\mathbf{i} + \bar{y}\mathbf{j})$$

Carrying out the cross product and equating the respective $\mathbf{i}$ and $\mathbf{j}$ components yields the two scalar equations.

$$(a_G)_x = (a_P)_x - \bar{x}\omega^2 - \bar{y}\alpha$$

$$(a_G)_y = (a_P)_y - \bar{y}\omega^2 + \bar{x}\alpha$$

From these equations, $[-(a_P)_x + \bar{y}\alpha] = [-(a_G)_x - \bar{x}\omega^2]$ and $[(a_P)_y + \bar{x}\alpha] = [(a_G)_y - \bar{x}\alpha]$. Substituting these results into Eq. 17–8 and simplifying gives

$$\zeta + \Sigma M_P = -\bar{y}m(a_G)_x + \bar{x}m(a_G)_y + I_G\alpha \qquad (17\text{–}9)$$

This important result indicates that when moments of the external forces acting on the free-body diagram are summed about point P, Fig. 17–8d, they are equivalent to the sum of the "kinetic moments" of the components of $m\mathbf{a}_G$ about P plus the "kinetic moment" of $I_G\boldsymbol{\alpha}$, Fig. 17–8e. In other words, when the "kinetic moments," $\Sigma(M_k)_P$, are computed, Fig. 17–8e, the vectors $m(\mathbf{a}_G)_x$ and $m(\mathbf{a}_G)_y$ are treated in the same manner as a force, that is, they can act at any point along their line of action. In a similar manner, $I_G\boldsymbol{\alpha}$ can be treated like a couple and can therefore act at any point. It is important to keep in mind, however, that $m\mathbf{a}_G$ and $I_G\boldsymbol{\alpha}$ are not the same as a force or a couple. Instead, they are caused by the effects of forces and moments acting on the body (see Sec. 13.2). With this in mind we can therefore write Eq. 17–9 in a more general form as

$$\Sigma M_P = \Sigma(M_k)_P \qquad (17\text{–}10)$$

General Application of the Equations of Motion. To summarize the above analysis, *three* scalar equations may be written to describe the general plane motion of a symmetrical rigid body.

$$\Sigma F_x = m(a_G)_x$$

$$\Sigma F_y = m(a_G)_y \qquad (17\text{–}11)$$

$$\Sigma M_G = I_G\alpha \quad \text{or} \quad \Sigma M_P = \Sigma(M_k)_P$$

Free–body diagram

(d)

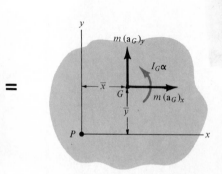

Kinetic diagram

(e)

Fig. 17–8d, e

$=$

When applying these equations, one should *always* draw a free-body diagram, Fig. 17–8*d*, in order to account for the terms involving ΣF_x, ΣF_y, ΣM_G or ΣM_P. In some problems it may also be helpful to draw the *kinetic diagram* for the body. This diagram graphically accounts for the terms $m(\mathbf{a}_G)_x$, $m(\mathbf{a}_G)_y$, and $I_G\boldsymbol{\alpha}$, and it is especially convenient when used to determine the components of ma_G and the moment terms in $\Sigma(M_k)_P$.*

Equations of Motion: Translation 17.3

When a rigid body undergoes a *translation*, Fig. 17–9*a*, all the particles of the body have the *same acceleration*, so that $\mathbf{a}_G = \mathbf{a}$. Furthermore, $\boldsymbol{\alpha} = \mathbf{0}$, in which case the rotational equation of motion applied at point G reduces to a simplified form, $\Sigma M_G = 0$. Application of this and the translational equations of motion will now be discussed for each of the two types of translation presented in Chapter 16.

Rectilinear Translation. When a body is subjected to *rectilinear translation*, all the particles of the body (slab) travel along parallel straight-line paths. The free-body and kinetic diagrams are shown in Fig. 17–9*b*. At the instant considered, the origin of the inertial reference is located at the center of mass. Since $I_G\boldsymbol{\alpha} = \mathbf{0}$, only ma_G is shown on the kinetic diagram. Hence, the equations of motion which apply in this case become

$$\Sigma F_x = m(a_G)_x$$
$$\Sigma F_y = m(a_G)_y \qquad\qquad (17\text{–}12)$$
$$\Sigma M_G = 0$$

*For this reason, the kinetic diagram will be used in the solution of an example problem whenever $\Sigma M_P = \Sigma(M_k)_P$ is applied.

(a)

(b)

Fig. 17–9a, b

The last equation requires that the sum of the moments of all the external forces computed about the body's center of mass be equal to zero. It is possible, of course, to sum moments about other points on or off the body, in which case the moment of ma_G must be taken into account. For example, if point A is chosen, which lies at a perpendicular distance d from the line of action of ma_G, the following moment equation applies:

$$\zeta+\Sigma M_A = \Sigma(M_k)_A; \qquad\qquad \Sigma M_A = (ma_G)d$$

Here the moment of the external forces about A (ΣM_A, free-body diagram) equals the moment of ma_G about A ($\Sigma(M_k)_A$, kinetic diagram).

Curvilinear Translation. When a rigid body is subjected to *curvilinear translation,* all the particles of the body travel along *parallel curved paths.* For analysis, it is often convenient to use an inertial coordinate system having an origin at G and axes which are oriented in the normal and tangential directions of the path of motion, Fig. 17–9c. The three scalar equations of motion are then

$$\Sigma F_n = m(a_G)_n$$
$$\Sigma F_t = m(a_G)_t \qquad\qquad (17\text{–}13)$$
$$\Sigma M_G = 0$$

where $(a_G)_t$ and $(a_G)_n$ represent, respectively, the magnitudes of the tangential and normal components of acceleration of point G.

If the moment equation $\Sigma M_G = 0$ is replaced by a moment summation about the arbitrary point B, Fig. 17–9c, it is necessary to account for the moments, $\Sigma(M_k)_B$, of the two components $m(\mathbf{a}_G)_n$ and $m(\mathbf{a}_G)_t$ about this point. From the kinetic diagram, h and e represent the perpendicular distances (or "moment arms") from B to the lines of action of the components. If positive moments are assumed to be clockwise, the required moment equation becomes

$$\zeta+\Sigma M_B = \Sigma(M_k)_B; \qquad \Sigma M_B = h[m(a_G)_n] - e[m(a_G)_t]$$

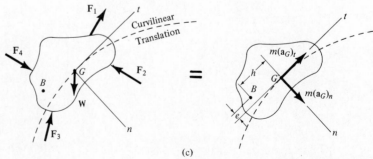

(c)

Fig. 17–9c

PROCEDURE FOR ANALYSIS

The following procedure provides a method for solving kinetic problems involving rigid-body *translation*.

Free-Body Diagram. Establish the inertial coordinate system and draw the free-body diagram in order to account for all the external forces ($\Sigma\mathbf{F}$) that act on the body. The direction and sense of the acceleration of the body's mass center $\mathbf{a}_G$ should also be established. (Recall that $\boldsymbol{\alpha} = \mathbf{0}$.) It may be sketched near, *but not on*, the free-body diagram. If it is decided that the rotational equation of motion $\Sigma M_P = \Sigma(M_k)_P$ is to be used, then it may also be beneficial to draw the kinetic diagram, since it graphically accounts for the components $m(\mathbf{a}_G)_x$, $m(\mathbf{a}_G)_y$ or $m(\mathbf{a}_G)_t$, $m(\mathbf{a}_G)_n$ and is thereby convenient for "visualizing" the terms needed in the moment sum $\Sigma(M_k)_P$.

Equations of Motion. Apply the three equations of motion, Eqs. 17–12 or Eqs. 17–13, with reference to an established x, y or n, t inertial coordinate system. To simplify the analysis, the moment equation $\Sigma M_G = 0$ can be replaced by the more general equation $\Sigma M_P = \Sigma(M_k)_P$, where point P is usually located at the intersection of the lines of action of as many unknown forces as possible.

If the body is in contact with a *rough surface* and slipping occurs, use the frictional equation $F = \mu_k N$ to relate the normal force $\mathbf{N}$ to its associated frictional force $\mathbf{F}$.

Kinematics. Use kinematics if a complete solution cannot be obtained strictly from the equations of motion. For *rectilinear translation* with *variable acceleration*, use

$$a_G = \frac{dv_G}{dt} \qquad a_G \, ds_G = v_G \, dv_G \qquad v_G = \frac{ds_G}{dt}$$

For *rectilinear translation* with *constant acceleration*, use

$$v_G = (v_G)_0 + a_G t$$
$$v_G^2 = (v_G)_0^2 + 2a_G[s_G - (s_G)_0]$$
$$s_G = (s_G)_0 + (v_G)_0 t + \tfrac{1}{2}a_G t^2$$

For *curvilinear translation*, use

$$(a_G)_n = \frac{v_G^2}{\rho} = \omega^2 \rho$$

$$(a_G)_t = \frac{dv_G}{dt} \qquad (a_G)_t \, ds_G = v_G \, dv_G \qquad (a_G)_t = \alpha\rho$$

The following examples numerically illustrate application of this procedure.

Example 17–5

The car shown in Fig. 17–10a has a mass of 2 Mg and a center of mass at G. Determine the car's acceleration if the "driving" wheels in the back are always slipping, whereas the front wheels freely rotate. Neglect the mass of the wheels. The coefficient of kinetic friction between the wheels and the road is $\mu_k = 0.25$.

(a)

Solution I

Free-Body Diagram. As shown in Fig. 17–10b, the rear-wheel frictional force $\mathbf{F}_B$ pushes the car forward, and since *slipping occurs* the magnitude of this force is related to the magnitude of its associated normal force $\mathbf{N}_B$ by $F_B = 0.25N_B$. The frictional forces acting on the *front wheels* are *zero*, since these wheels have negligible mass.* Here we will sum moments about the mass center. The car (point G) is assumed to accelerate to the left.

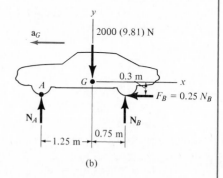

Equations of Motion

$$\xrightarrow{+} \Sigma F_x = m(a_G)_x; \qquad -0.25N_B = -2000a_G \qquad (1)$$

$$+\uparrow \Sigma F_y = m(a_G)_y; \qquad N_A + N_B - 2000(9.81) = 0 \qquad (2)$$

$$\downarrow+\Sigma M_G = 0; \quad -N_A(1.25) - 0.25N_B(0.3) + N_B(0.75) = 0 \qquad (3)$$

Solving,

$$a_G = 1.59 \text{ m/s}^2 \leftarrow \qquad \qquad Ans.$$

$$N_A = 6.88 \text{ kN} \uparrow$$

$$N_B = 12.7 \text{ kN} \uparrow$$

Solution II

Free-Body and Kinetic Diagrams. If the "moment" equation is applied about point A, then the unknown N_A will be eliminated from the equation. To "visualize" the moment of ma_G about A we will include the kinetic diagram as part of the analysis, Fig. 17–10c.

Equation of Motion. We require

$$\downarrow+\Sigma M_A = \Sigma(M_k)_A; \qquad N_B(2) - 2000(9.81)(1.25) = 2000a_G(0.3)$$

Solving this and Eq. (1) for a_G leads to a simpler solution than that obtained from Eqs. (1) to (3).

(c)

Fig. 17–10

---

*If the mass of the front wheels were to be included in the analysis, the frictional force acting at A would be *directed to the right* to create the necessary counterclockwise rotation of the wheels. The problem solution for this case would be more involved since a general-plane-motion analysis of the wheels would have to be considered (see Sec. 17.5).

Example 17–6

A 5-lb force is applied to the center of the roller at A, Fig. 17–11a. If the roller has negligible mass, determine the angle θ which the 6-lb uniform rod makes with the vertical and the speed of the roller after it moves 4 ft starting from rest.

(a)   (b)

Fig. 17–11

Solution

Free-Body Diagram. Since the 5-lb force acting on the roller is constant, the rod maintains a *constant angle* θ with the vertical and therefore the roller and rod have the *same* acceleration $\mathbf{a}_G$. The free-body diagram of the rod and roller is shown in Fig. 17–11b.

Equations of Motion

$$\xrightarrow{+} \Sigma F_x = m(a_G)_x; \qquad 5 = \frac{6}{32.2} a_G \qquad (1)$$

$$+\uparrow \Sigma F_y = m(a_G)_y; \qquad N_A - 6 = 0 \qquad (2)$$

$$\zeta+\Sigma M_G = 0; \qquad -5(1.5 \cos \theta) + N_A(1.5 \sin \theta) = 0 \qquad (3)$$

Solving yields

$$a_G = 26.8 \text{ ft/s}^2$$
$$N_A = 6 \text{ lb}$$
$$\theta = 39.8° \qquad\qquad\qquad Ans.$$

Kinematics. Since the acceleration is *constant,* the speed after 4 ft of displacement is

$$(\xrightarrow{+}) \qquad v_G^2 = (v_G)_0^2 + 2a_G[s_G - (s_G)_0]$$
$$v_G^2 = 0 + 2(26.8)[4 - 0]$$
$$v_G = 14.6 \text{ ft/s} \rightarrow \qquad\qquad Ans.$$

Example 17–7

A uniform 50-kg crate rests on a horizontal surface for which the coefficient of kinetic friction is $\mu_k = 0.2$. Determine the crate's acceleration if a
force of $P = 600$ N is applied to the crate as shown in Fig. 17–12a.

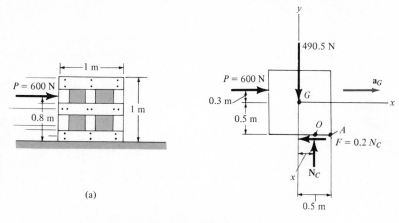

(a)

Fig. 17–12   (b)

Solution

Free-Body Diagram.   The force **P** can either cause the crate to slide or to
tip over. As shown in Fig. 17–12b, it is assumed that the crate slides, so
that $F = \mu_k N_C = 0.2N_C$. Also, the resultant normal force $\mathbf{N}_C$ acts at O, a
distance x (where $0 < x \le 0.5$ m) from the crate's center line.*

Equations of Motion

$$\overset{+}{\rightarrow}\Sigma F_x = m(a_G)_x; \qquad 600 - 0.2N_C = 50a_G \qquad (1)$$

$$+\uparrow \Sigma F_y = m(a_G)_y; \qquad N_C - 490.5 = 0 \qquad (2)$$

$$\underset{\downarrow}{+}\Sigma M_G = 0; \qquad -600(0.3) + N_C(x) - 0.2N_C(0.5) = 0 \qquad (3)$$

Solving, we obtain

$$N_C = 490.5 \text{ N}$$
$$x = 0.47 \text{ m}$$
$$a_G = 10.0 \text{ m/s}^2 \rightarrow \qquad\qquad Ans.$$

Since $x = 0.47$ m < 0.5 m, indeed the crate slides as originally assumed.
If the solution had given a value of $x > 0.5$ m, the problem would have to
be reworked with the assumption that tipping occurred. If this were the
case, $\mathbf{N}_C$ would act at the *corner point A* and $F \le 0.2N_C$, since in general
the crate would *not* be on the verge of sliding at the instant it begins to tip.

---

*The line of action of $\mathbf{N}_C$ does not necessarily pass through the mass center G ($x = 0$),
since $\mathbf{N}_C$ must counteract the effect of tipping caused by **P**. See Sec. 9.1 of *Engineering
Mechanics: Statics.*

Example 17-8

The 100-kg beam *BD* shown in Fig. 17–13a is supported by two rods having negligible mass. Determine the force created in each rod if at the instant $\theta = 30°$ the rods are both rotating with an angular velocity of $\omega = 6$ rad/s.

(a)

(b)

Fig. 17–13

Solution

Free-Body Diagram. The beam moves with *curvilinear translation* since points *B* and *D* and the center of mass *G* move along circular paths, each path having the same radius of 0.5 m. Using normal and tangential coordinates, the free-body diagram for the beam is shown in Fig. 17–13b. Because of the *translation, G* has the *same* motion as the pin at *B*, which is connected to both the rod and the beam. By studying the angular motion of rod *AB*, Fig. 17–13c, note that the tangential component of acceleration acts downward to the left due to the clockwise direction of $\boldsymbol{\alpha}$. Furthermore, the normal component of acceleration is *always* directed toward the center of curvature (toward point *A* for rod *AB*). Since the angular velocity of *AB* is 6 rad/s, then

(c)

$$a_n = \omega^2 r = (6)^2(0.5) = 18 \text{ m/s}^2$$

Equations of Motion

$$+\nwarrow \Sigma F_n = m(a_G)_n; \quad T_B + T_D - 981 \cos 30° = 100(18) \qquad (1)$$

$$+\swarrow \Sigma F_t = m(a_G)_t; \qquad 981 \sin 30° = 100(a_G)_t \qquad (2)$$

$$\big\downarrow + \Sigma M_G = 0; \quad -(T_B \cos 30°)0.4 + (T_D \cos 30°)0.4 = 0 \qquad (3)$$

Simultaneous solution of these three equations gives

$$T_B = T_D = 1.33 \text{ kN} \quad \nwarrow 30° \qquad \qquad Ans.$$

$$(a_G)_t = 4.90 \text{ m/s}^2$$

Problems

Except when stated otherwise, throughout this chapter assume that the coefficients of static and kinetic friction are equal, i.e., $\mu = \mu_s = \mu_k$.

17–19. The door has a weight of 200 lb and a center of gravity at G. Determine how far the door moves in 2 s, starting from rest, if a man pushes on it at C with a horizontal force of 30 lb. Also, find the vertical reactions at the rollers A and B.

Prob. 17–19

*17–20. The smooth 180-lb pipe has a length of 20 ft and negligible thickness. It is carried on a truck as shown. Determine the maximum acceleration which the truck can have without causing the normal reaction at A to be zero. Also determine the horizontal and vertical components of force which the truck exerts on the pipe at B.

Prob. 17–20

17–21. The 2-lb bottle rests on the checkout conveyor at a grocery store. If $\mu = 0.2$, determine the largest acceleration which the conveyor can have without causing the bottle to slip or tip. The center of mass is at G.

Prob. 17–21

17–22. The truck carries the 1200-lb safe which has a center of mass at G. Determine the largest acceleration of the truck so that the safe will not slip or tip on the truck bed. The coefficient of friction between the safe and the truck is $\mu = 0.5$.

Probs. 17–22 / 17–23

17–23. If the coefficient of friction between the wheels of the truck and the road is $\mu' = 0.6$, determine the deceleration of the safe if the driver slams on his brakes and all four wheels slide on the pavement. The truck has a weight of 3000 lb, the safe weighs 1200 lb, and the coefficient of friction between the safe and the truck is $\mu = 0.5$.

*17–24. A 75-kg man and 40-kg boy sit on the seesaw, which has negligible mass. At the instant the man lifts his feet from the ground, determine their accelerations if each sits upright, i.e., they undergo curvilinear translation. The centers of mass of the man and boy are at G_m and G_b, respectively. There is a pin at A.

Prob. 17–24

17–25. At the *start* of takeoff, the propeller on the 2-Mg plane exerts a horizontal thrust of 600 N on the plane. Determine the plane's acceleration and the vertical reactions at the nose wheel A and each of the other *two* wheels B. Neglect the lifting force of the wings since the plane is originally at rest. The mass center is at G.

Prob. 17–25

17–26 The jet aircraft has a total mass of 22 Mg and a center of mass at G. If a towing cable is attached to the upper portion of the nose wheel and exerts a force $T = 400$ N as shown, determine the acceleration of the plane and the normal reactions on the nose wheel and each of the *two* wing wheels located at B. Neglect the size of the wheels in the calculation and any lift caused by the wings.

Prob. 17–26

17–27. The sports car has been designed so that the 260-kg engine E and 165-kg transmission T have been placed over the front and rear wheels, respectively. Their mass centers are located at G_E and G_T. The remaining 512-kg mass of the body and frame is located at G_b. If power is supplied to the rear wheels only, determine the shortest time it takes for the car to reach a speed of 80 km/h, starting from rest. The front wheels are free rolling, and the coefficient of friction between the wheels and the road is $\mu = 0.4$. Neglect the mass of the wheels and driver.

Prob. 17–27

*17–28. The 1.6-Mg car shown has been "raked" by increasing the height $h = 0.2$ m of its center of mass. This was done by raising the springs on the rear axle. If the coefficient of friction between the rear wheels and the ground is $\mu = 0.3$, show that the car can accelerate slightly faster than its counterpart for which $h = 0$. Neglect the mass of the wheels and driver and assume the front wheels at B are free to roll.

Prob. 17–28

17–29. The bicycle and rider have a mass of 80 kg with center of mass located at G. If the coefficient of friction at the rear tire is $\mu_B = 0.8$, determine the normal reactions at the tires A and B, and the deceleration of the rider, when the rear wheel locks for braking. What is the normal reaction at the rear wheel when the bicycle is traveling at constant velocity and the brakes are not applied? Comparing the results, why is it a good idea to just apply the brakes on the rear wheel rather than just the front wheel when braking? Neglect the mass of the wheels.

Probs. 17–29 / 17–30

17–30. The bicycle and rider have a mass of 80 kg with center of mass located at G. Determine the minimum coefficient of friction between the road and the wheels which will cause the rider to "go over the handle bars" during braking of *both* wheels. This requires the rear wheel B to lift off the ground. Neglect the mass of the wheels.

17–31. The motorcycle has a mass of 125 kg and a center of mass at G_1, while the rider has a mass of 75 kg and a center of mass at G_2. If the coefficient of friction between the wheels and the pavement is $\mu = 0.8$, determine if it is possible for the rider to do a "wheely," i.e., lift the front wheel off the ground. What acceleration is necessary to do this? Neglect the mass of the wheels and assume that the front wheel is free to roll.

Probs. 17–31 / 17–32

*17–32. The motorcycle has a mass of 125 kg and a center of mass at G_1, while the rider has a mass of 75 kg and a center of mass at G_2. If no slipping occurs, determine the minimum driving force $\mathbf{F}_B$ which must be supplied to the rear wheel B in order to create an acceleration of $a = 3$ m/s^2. What are the normal reactions of the wheels on the ground? Neglect the mass of the wheels and assume that the front wheel is free to roll.

17–33. The beam has a length of 3 m and a mass of 500 kg. It is attached to the back of the truck using a 0.6-m-long chain AB. If the coefficient of friction at C is $\mu = 0.4$, determine the acceleration of the truck when the beam is supported at $\theta = 10°$ with the road as shown.

Prob. 17–33

17–34. The two 3-lb rods *EF* and *GH* are fixed (welded) to the link *AC* at *E*. Determine the internal axial force $\mathbf{E}_x$, shear force $\mathbf{E}_y$, and moment $\mathbf{M}_E$, which the bar *AC* exerts on *FE* at *E* if at the instant $\theta = 30°$ link *AB* has an angular velocity $\omega = 5$ rad/s and an angular acceleration $\alpha = 8$ rad/s² as shown.

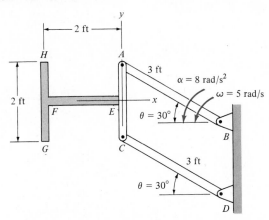

Prob. 17–34

17–35. Each uniform box on the stack of four boxes has a weight of 8 lb. The stack is being transported on the dolly which has a weight of 30 lb. Determine the maximum force **F** which the woman can exert on the handle in the direction shown so that no box on the stack will tip or slip. The coefficient of friction at all points of contact is $\mu = 0.5$. The dolly wheels are free to roll. Neglect their mass.

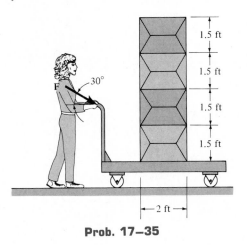

Prob. 17–35

***17–36.** The crate *C* has a weight of 150 lb and rests on the floor of a truck elevator for which $\mu = 0.4$. Determine the largest initial angular acceleration $\boldsymbol{\alpha}$, starting from rest, which the parallel links *AB* and *DE* can have without causing the crate to slip. No tipping occurs.

Prob. 17–36

17–37. A car having a weight of 4000 lb begins to skid and turn with the brakes applied to all four wheels. If the coefficient of friction between the wheels and the road is $\mu = 0.8$, determine the maximum critical height *h* of the center of gravity *G* such that the car does not overturn. Tipping will begin to occur after the car rotates 90° from its original direction of motion and, as shown in the figure, undergoes *translation* while skidding. *Hint:* Draw a free-body diagram of the car viewed from the front. When tipping occurs, the normal reactions of the wheels on the right side (or passenger side) are zero. Neglect the mass of the driver.

Prob. 17–37

335

17–38. The 50-kg uniform crate rests on the platform for which $\mu = 0.5$. If the supporting links have an angular velocity $\omega = 1$ rad/s, determine the greatest angular acceleration α they can have so that the crate does not slip or tip at the instant $\theta = 30°$ as shown.

Prob. 17–38

17–39. The car has a mass of 1500 kg and a center of mass at G. It is used to tow a crate having a mass of 600 kg. If the coefficient of friction is $\mu = 0.3$ between the crate and the ground, and $\mu' = 0.6$ between the rear tires of the car and the ground, determine the tension developed in the chain AB and the acceleration of the crate. Neglect the mass of the wheels and assume that the front wheels at C are free to roll.

Prob. 17–39

*17–40.** The uniform 100-lb block C rests on the 30-lb dolly AB. If the coefficient of friction between the dolly and the block is $\mu = 0.2$, determine the largest weight of the suspended block D so that block C does not slip or tip on the dolly when D is released. Neglect the mass of the pulley and the mass and size of the wheels at A and B.

Prob. 17–40

■**17–41.** The car, having a mass of 1.40 Mg and mass center at G_c, pulls a loaded trailer having a mass of 0.8 Mg and mass center at G_t. Determine the normal reactions at the car and trailer wheels if the driver applies the car's rear brakes C and causes the car to skid. Take $\mu_C = 0.4$ and assume the hitch at A is a pin or ball-and-socket joint. The wheels at B and D are free to roll. Neglect their mass and the mass of the driver.

Prob. 17–41

Equations of Motion: Rotation About a Fixed Axis **17.4**

Consider the rigid body (or slab) shown in Fig. 17–14a, which is constrained to rotate in the vertical plane about a fixed axis perpendicular to the page and passing through the pin at O. The angular velocity and angular acceleration are caused by the external force system acting on the body. Because the body's center of mass G moves in a *circular path*, the acceleration of this point is represented by its tangential and normal components. The *tangential component of acceleration* has a *magnitude* of $(a_G)_t = \alpha r_G$ and must act in a *direction* which is *consistent* with the angular acceleration $\boldsymbol{\alpha}$. The *magnitude* of the *normal component of acceleration* is $(a_G)_n = \omega^2 r_G$. This component is *always directed* from point G to O regardless of the direction of $\boldsymbol{\omega}$.

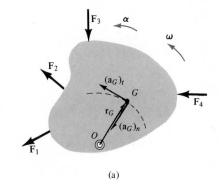

(a)

The free-body and kinetic diagrams for the body are shown in Fig. 17–14b. The weight of the body, $W = mg$, and the pin reaction $\mathbf{F}_O$ are included on the free-body diagram since they represent external forces acting on the body. The two components $m(\mathbf{a}_G)_t$ and $m(\mathbf{a}_G)_n$, shown on the kinetic diagram, are associated with the tangential and normal acceleration components of the body's mass center. These vectors act in the same *direction* as the acceleration components and have *magnitudes* of $m(a_G)_t$ and $m(a_G)_n$. The $I_G\boldsymbol{\alpha}$ vector acts in the same *direction* as $\boldsymbol{\alpha}$ and has a *magnitude* of $I_G\alpha$, where I_G is the body's moment of inertia calculated about an axis which is perpendicular to the page and passing through G. From the derivation given in Sec. 17.2, the equations of motion which apply to the body may be written in the form

$$\Sigma F_n = m(a_G)_n = m\omega^2 r_G$$
$$\Sigma F_t = m(a_G)_t = m\alpha r_G \qquad (17\text{–}14)$$
$$\Sigma M_G = I_G\alpha$$

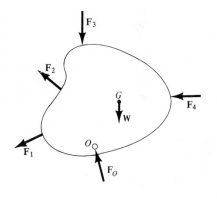

The moment equation may be replaced by a moment summation about any arbitrary point P on or off the body provided one accounts for the moments $\Sigma(M_k)_P$ produced by $I_G\boldsymbol{\alpha}$, $m(\mathbf{a}_G)_t$, and $m(\mathbf{a}_G)_n$ about the point. In many problems it is convenient to sum moments about the pin at O in order to eliminate the *unknown* force $\mathbf{F}_O$. From the kinetic diagram, Fig. 17–14b, this requires

$$\zeta+\Sigma M_O = \Sigma(M_k)_O; \qquad \Sigma M_O = r_G m(a_G)_t + I_G\alpha \qquad (17\text{–}15)$$

Note that the moment of $m(\mathbf{a}_G)_n$ is not included in the summation since the line of action of this vector passes through O. Substituting $(a_G)_t = r_G\alpha$, we may rewrite the above equation as $\zeta+\Sigma M_O = (I_G + mr_G^2)\alpha$. From the parallel-axis theorem, $I = I_G + md^2$, the term in parentheses represents the *moment of inertia of the body about the fixed axis of rotation passing through O*.

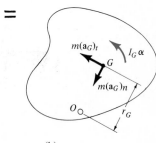

(b)

Fig. 17–14

337

Denoting this term by I_O, we can write the three equations of motion for the body as*

$$\Sigma F_n = m(a_G)_n = m\omega^2 r_G$$
$$\Sigma F_t = m(a_G)_t = m\alpha r_G \qquad (17\text{–}16)$$
$$\Sigma M_O = I_O\alpha$$

For applications, one should remember that "$I_O\alpha$" accounts for the "moment" of *both* $m(\mathbf{a}_G)_t$ *and* $I_G\boldsymbol{\alpha}$ about point O, Fig. 17–14b. In other words, $\Sigma M_O = \Sigma(M_k)_O = I_O\alpha$, as indicated by Eqs. 17–15 and 17–16.

PROCEDURE FOR ANALYSIS

The following procedure provides a method for solving kinetic problems which involve the rotation of a body about a fixed axis.

Free-Body Diagram. Draw the free-body diagram for the body and specify the direction and sense of the accelerations $(\mathbf{a}_G)_n$ and $(\mathbf{a}_G)_t$, and the angular acceleration $\boldsymbol{\alpha}$ of the body. Also compute the moment of inertia I_G or I_O. If it is decided that the rotational equation of motion $\Sigma M_P = \Sigma(M_k)_P$ is to be used, i.e., moments will *not* be summed about point G or O, then consider drawing the kinetic diagram in order to help "visualize" the "moment" developed by the components $m(\mathbf{a}_G)_n$, $m(\mathbf{a}_G)_t$, and $I_G\boldsymbol{\alpha}$ when writing the terms for the moment sum $\Sigma(M_k)_P$.

Equations of Motion. Apply the three equations of motion, Eqs. 17–14 or 17–16, with reference to an established x, y or n, t inertial coordinate system.

Kinematics. Use kinematics if a complete solution cannot be obtained strictly from the equations of motion. In this regard, if the *angular acceleration is variable*, use

$$\alpha = \frac{d\omega}{dt}, \qquad \alpha\, d\theta = \omega\, d\omega \qquad \omega = \frac{d\theta}{dt}$$

If the *angular acceleration is constant*, use

$$\omega = \omega_0 + \alpha_c t$$
$$\theta = \theta_0 + \omega_0 t + \tfrac{1}{2}\alpha_c t^2$$
$$\omega^2 = \omega_0^2 + 2\alpha_c(\theta - \theta_0)$$

The following examples numerically illustrate application of this procedure.

*The result $\Sigma M_O = I_O\alpha$ can also be obtained *directly* from Eq. 17–6 by selecting point P to coincide with O, realizing that $(a_P)_x = (a_P)_y = 0$.

Example 17–9

The 30-kg disk shown in Fig. 17–15*a* is pin-supported at its center. If it starts from rest, determine the number of revolutions it must make to attain an angular velocity of 20 rad/s. Also, what are the reactions at the pin? The disk is acted upon by a constant force $F = 10$ N, which is applied to a cord wrapped around its periphery, and a constant couple $M = 5$ N $\cdot$ m. Neglect the mass of the cord in the calculation.

Fig. 17–15

(a)   (b)

Solution

Free-Body Diagram. (Fig. 17–15*b*). Note that the mass center is not subjected to an acceleration; however, the disk has an angular acceleration.

The moment of inertia of the disk is

$$I_O = \tfrac{1}{2}mr^2 = \frac{1}{2}(30)(0.2)^2 = 0.6 \text{ kg} \cdot \text{m}^2$$

Equations of Motion

$$\xrightarrow{+}\Sigma F_x = m(a_G)_x; \qquad\qquad O_x = 0 \qquad\qquad\qquad \textit{Ans.}$$

$$+\uparrow\Sigma F_y = m(a_G)_y; \qquad O_y - 294.3 - 10 = 0$$

$$O_y = 304.3 \text{ N} \qquad\qquad\qquad \textit{Ans.}$$

$$\zeta+\Sigma M_O = I_O\alpha; \qquad -10(0.2) - 5 = -0.6\alpha \qquad \alpha = 11.7 \text{ rad/s}^2$$

Kinematics. Since α is constant, the number of radians the disk must turn to obtain an angular velocity of 20 rad/s is

$$\gamma+ \qquad\qquad \omega^2 = \omega_0^2 + 2\alpha_c(\theta - \theta_0)$$

$$(20)^2 = 0 + 2(11.7)(\theta - 0)$$

$$\theta = 17.1 \text{ rad}$$

Hence,

$$\theta = 17.1 \text{ rad}\left(\frac{1 \text{ rev}}{2\pi \text{ rad}}\right) = 2.73 \text{ rev} \downarrow \qquad\qquad \textit{Ans.}$$

Example 17–10

The drum shown in Fig. 17–16a has a mass of 60 kg and a radius of gyration $k_O = 0.25$ m. A cord of negligible mass is wrapped around the periphery of the drum and attached to a crate having a mass of 20 kg. If the crate is released, determine the drum's angular acceleration.

(a)

Solution I

Free-Body Diagram. Here we will consider the drum and crate separately, Fig. 17–16b. Note that the downward acceleration of the crate, **a**, creates a counterclockwise angular acceleration **α** of the drum.

The moment of inertia of the drum is

$$I_O = mk_O^2 = (60)(0.25)^2 = 3.75 \text{ kg} \cdot \text{m}^2$$

Equations of Motion. By observation, applying the translational equations of motion $\Sigma F_x = m(a_G)_x$ and $\Sigma F_y = m(a_G)_y$ to the drum is of no consequence to the solution, since these equations involve the unknowns O_x and O_y. Thus, for the drum and crate, respectively

$$\zeta + \Sigma M_O = I_O\alpha; \qquad T(0.4) = 3.75\alpha \tag{1}$$
$$+ \downarrow \Sigma F_y = m(a_G)_y; \qquad 20(9.81) - T = 20a \tag{2}$$

Kinematics. Since the point of contact A between the cord and drum has the same tangential component of acceleration, Fig. 17–16a, this requires that

$$\zeta + a = \alpha r; \qquad a = \alpha(0.4) \tag{3}$$

Solving the above equations

$$T = 105.9 \text{ N}$$
$$a = 4.52 \text{ m/s}^2$$
$$\alpha = 11.3 \text{ rad/s}^2 \nwarrow \qquad \textit{Ans.}$$

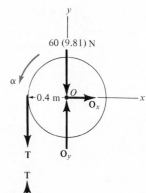

(b)

Solution II

Free-Body and Kinetic Diagrams. The cable tension T can be eliminated from the analysis by considering the drum and crate as a single system, Fig. 17–16c.

Equations of Motion. Using Eq. (3) and applying the moment equation about point O to eliminate the unknowns O_x and O_y, we have

$$\zeta + \Sigma M_O = \Sigma(M_k)_O; \qquad 20(9.81)(0.4) = 3.75\alpha + [20(0.4\alpha)]0.4$$
$$\alpha = 11.3 \text{ rad/s}^2 \nwarrow \qquad \textit{Ans.}$$

(c)

Fig. 17–16

Example 17–11

The unbalanced 50-lb flywheel shown in Fig. 17–17a has a radius of gyration of $k_G = 0.6$ ft about an axis passing through its mass center G. If it has a clockwise angular velocity of 8 rad/s at the instant shown, determine the horizontal and vertical components of reaction at the pin O.

(a)

Solution I

Free-Body Diagram. Since G moves in a *circular path*, $(a_G)_x = \alpha r_G$ and $(a_G)_y = \omega^2 r_G$. By inspection, the weight creates a clockwise angular acceleration $\boldsymbol{\alpha}$, so $(\mathbf{a}_G)_y$ acts downward, in accordance with the direction of $\boldsymbol{\alpha}$, Fig. 17–17b. Furthermore, $(\mathbf{a}_G)_x$ acts to the left, toward the center of curvature O.

The moment of inertia of the flywheel about its mass center is determined from the radius of gyration and the mass; i.e., $I_G = mk_G^2 = (50/32.2)(0.6)^2 = 0.559$ slug · ft².

Equations of Motion

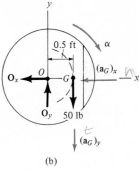

$$\xrightarrow{+} \Sigma F_x = m\omega^2 r_G; \qquad -O_x = -\left(\frac{50}{32.2}\right)(8)^2(0.5) \qquad (1)$$

$$+\uparrow \Sigma F_y = m\alpha r_G; \qquad O_y - 50 = -\left(\frac{50}{32.2}\right)(\alpha)(0.5) \qquad (2)$$

$$\zeta + \Sigma M_G = I_G \alpha; \qquad -O_y(0.5) = -0.559\alpha \qquad (3)$$

Solving,

$$\alpha = 26.4 \text{ rad/s}^2 \zeta, \quad O_x = 49.7 \text{ lb} \leftarrow, \quad O_y = 29.5 \text{ lb} \uparrow \qquad Ans.$$

(b)

Fig. 17–17

Solution II

If $\Sigma M_O = I_O \alpha$ is applied, then by the parallel-axis theorem the moment of inertia of the flywheel about O is

$$I_O = I_G + mr_G^2 = 0.559 + \left(\frac{50}{32.2}\right)(0.5)^2 = 0.947 \text{ slug} \cdot \text{ft}^2$$

Hence, from the free-body diagram, Fig. 17–17b, we require

$$\zeta + \Sigma M_O = I_O \alpha; \qquad -50(0.5) = -0.947\alpha$$

Solving for α and substituting into Eq. (2) or (3) yields the answer for O_y obtained previously.

Example 17–12

The slender rod shown in Fig. 17–18a has a mass m and length l and is released from rest when $\theta = 0°$. Determine the horizontal and vertical components of force which the pin at O exerts on the rod at the instant $\theta = 90°$.

(a)

Solution

Free-Body Diagram. The free-body diagram for the rod is shown when the rod is in the general position θ, Fig. 17–18b. For convenience, the force components at A are shown acting in the n and t directions. Note that α acts clockwise and $(\mathbf{a}_G)_t$ acts downward to the left in accordance with α.

The moment of inertia of the rod about point A is

$$I_A = \tfrac{1}{3}ml^2$$

Equations of Motion. Moments will be summed about A in order to eliminate the reactive forces there.*

$$+\nwarrow \Sigma F_n = m\omega^2 r_G; \qquad A_n - mg \sin \theta = m\omega^2(l/2) \qquad (1)$$

$$+\swarrow \Sigma F_t = m\alpha r_G; \qquad A_t + mg \cos \theta = m\alpha(l/2) \qquad (2)$$

$$\curvearrowright + \Sigma M_A = I_A \alpha; \qquad mg \cos \theta(l/2) = (\tfrac{1}{3}ml^2)\alpha \qquad (3)$$

(b)

Fig. 17–18

Kinematics. For a given angle θ there are four unknowns in the above three equations: A_n, A_t, ω, and α. As shown by Eq. (3) α is *not constant;* rather, it depends upon the position θ of the rod. Using kinematics, α and ω can be related to θ by the equation

$$\curvearrowright + \qquad\qquad \omega \, d\omega = \alpha \, d\theta \qquad\qquad (4)$$

In order to solve for ω at $\theta = 90°$, eliminate α from Eqs. (3) and (4), which yields

$$\omega \, d\omega = (1.5 \, g/l) \cos \theta \, d\theta$$

Since $\omega_0 = 0$ at $\theta_0 = 0°$, we have

$$\int_0^\omega \omega \, d\omega = (1.5 \, g/l) \int_{0°}^{90°} \cos \theta \, d\theta$$

$$\omega^2 = 3 \, g/l$$

Substituting this value into Eq. (1) with $\theta = 90°$ and solving Eqs. (1) to (3) yields

$$\alpha = 0, \quad A_t = 0, \quad A_n = 2.5 \, mg \qquad\qquad Ans.$$

*If $\Sigma M_A = \Sigma(M_k)_A$ is used, one must account for the moments of $I_G\alpha$ and $m(\mathbf{a}_G)_t$ about A. Refer to Eq. 17–15.

Problems

17–42. A motor supplies a constant torque $M = 2\,\text{N} \cdot \text{m}$ to a 50-mm-diameter shaft O connected to the center of the 30-kg flywheel. The resultant bearing friction $\mathbf{F}$, which the pin exerts on the shaft, acts tangent to the supporting shaft and has a magnitude of 50 N. Determine how long the torque must be applied to the shaft to increase the flywheel's rotational speed from 4 to 15 rad/s. The flywheel has a radius of gyration $k_O = 0.15$ m about its center.

Probs. 17–42 / 17–43

17–43. If the motor in Prob. 17–42 is disengaged from the shaft once the flywheel is rotating at 15 rad/s, so $M = 0$, determine how long it will take before the resultant bearing frictional force $F = 50$ N stops the flywheel from rotating.

*17–44. The 10-lb rod is pin-connected to its support and has an angular velocity $\omega = 4$ rad/s when it is in the horizontal position shown. Determine its angular acceleration and the horizontal and vertical components of reaction which the pin exerts on the rod at this instant.

Prob. 17–44

17–45. The 10-kg wheel has a radius of gyration $k_A = 200$ mm. If the wheel is subjected to a moment $M = (5t)\,\text{N} \cdot \text{m}$, where t is in seconds, determine its angular velocity when $t = 3$ s starting from rest. Also, compute the reactions which the fixed pin A exerts on the wheel during the motion.

Prob. 17–45

17–46. Determine the force $\mathbf{T}_A$ which must be applied to the cable at A in order to give the 10-kg block B an upward acceleration of 200 mm/s². Assume that the cable does not slip over the surface of the 20-kg disk. Compute the tension in the vertical segment of the cord that supports the block and explain why this tension is different from that at A. The disk is pinned at its center C and is free to rotate. Neglect the mass of the cable.

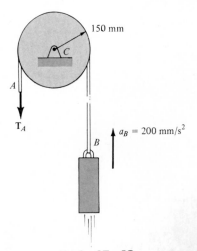

Prob. 17–46

343

17–47. The single blade PB of the fan has a mass of 2 kg and a moment of inertia $I_G = 0.18$ kg · m² about an axis passing through its center of mass G. If the blade is subjected to an angular acceleration $\alpha = 5$ rad/s², and at a given instant it has an angular velocity $\omega = 6$ rad/s when it is in the vertical position shown, determine the internal axial force A, shear force V, and bending moment M, which the hub exerts on the blade at point P.

$\omega = 6$ rad/s
$\alpha = 5$ rad/s²

Prob. 17–47

17–49. The cord is wrapped around the inner core of the spool. If a 5-lb weight B is suspended from the cord and released from rest, determine the spool's angular velocity when the weight has descended 8 ft. Neglect the mass of the cord. The spool has a weight of 180 lb and the radius of gyration about the axle A is $k_A = 1.25$ ft.

Prob. 17–49

*17–48. The 20-kg roll of paper has a radius of gyration $k_A = 90$ mm about an axis passing through point A. It is pin-supported at both ends by two brackets AB. If the roll rests against a wall for which $\mu_C = 0.2$ and a vertical force of 30 N is applied to the end of the paper, determine the angular acceleration of the roll as the paper unrolls.

Prob. 17–48

30 N

17–50. Two blocks A and B, having a weight of 10 lb and 5 lb, respectively, are attached to the ends of a cord which passes over a 3-lb pulley (disk). If the blocks are released from rest, determine their speed in $t = 0.5$ s. The cord does not slip on the pulley. Neglect the mass of the cord. *Suggestion:* Analyze the "system" consisting of both the blocks and the pulley.

Prob. 17–50

17–51. Cable is unwound from a spool supported on small rollers at A and B by exerting a force $T = 300$ N on the cable in the direction shown. Compute the time needed to unravel 5 m of cable from the spool if the spool and cable have a total mass of 600 kg and a centroidal radius of gyration $k_O = 1.2$ m. For the calculation, neglect the mass of the cable being unwound and the mass of the rollers at A and B.

Prob. 17–51

*17–52. Disk A has a weight of 5 lb and disk B has a weight of 10 lb. If no slipping occurs between them, determine the couple M which must be applied to disk A to give it an angular acceleration of 4 rad/s^2.

Prob. 17–52

17–53. The pendulum consists of a 15-lb disk and a 10-lb slender rod. Compute the horizontal and vertical components of reaction that the pin O exerts on the rod just as it passes the horizontal position, at which time its angular velocity is $\omega = 8$ rad/s.

Prob. 17–53

17–54. The door will close automatically using torsional springs mounted on the hinges. Each spring has a stiffness $k = 50$ N · m/rad so that the torque M on each hinge is $M = (50\,\theta)$ N · m, where θ is measured in radians. If the door is released from rest when it is open at $\theta = 90°$, determine its angular velocity at the instant $\theta = 0°$. For the calculation, treat the door as a thin plate having a mass of 70 kg.

Prob. 17–54

17–55. A force $F = 2$ lb is applied perpendicular to the axis of the 5-lb rod and moves from O to A at a constant rate of 4 ft/s. If the rod is at rest when $\theta = 0°$ and F is at O when $t = 0$, determine the rod's angular velocity at the instant the force is at A. How far has the rod rotated when this occurs? The rod rotates in the *horizontal plane*.

Prob. 17–55

345

*17–56. The furnace cover has a mass of 20 kg and a radius of gyration $k_G = 0.25$ m about its mass center G. If an operator applies a force $F = 120$ N to the handle in order to open the cover, determine the cover's initial angular acceleration and the horizontal and vertical components of reaction which the pin at A exerts on the cover at the instant the cover begins to open. Neglect the mass of the handle BAC in the calculation.

Prob. 17–56

17–57. If the support at B is suddenly removed, determine the initial horizontal and vertical components of reaction which the pin A exerts on the rod ACB. Segments AC and CB each have a weight of 10 lb.

Prob. 17–57

17–58. The wheel has a mass of 25 kg and a radius of gyration $k_B = 0.15$ m. It is originally spinning at $\omega_1 = 40$ rad/s. If it is placed on the ground, for which $\mu_C = 0.5$, determine the time required for the motion to stop. What are the horizontal and vertical components of reaction which the pin at A exerts on AB during this time? Neglect the mass of AB.

Prob. 17–58

17–59. The slender rod of mass m is released from rest when $\theta = 45°$. At the same instant a ball B having the same mass m is released. Will the ball B or the end A of the rod have the greatest speed when they pass the horizontal ($\theta = 0°$)? What is the difference in their speeds?

Prob. 17–59

*17–60. Determine the angular acceleration of the 25-kg diving board and the horizontal and vertical components of reaction at the pin A the instant the man jumps off. Assume that the board is uniform and rigid, and that at the instant he jumps off the spring is compressed a maximum amount of 200 mm, $\omega = 0$, and the board is horizontal.

Prob. 17–60

17–61. The kinetic diagram representing the general rotational motion of a rigid body about a fixed axis at O is shown in the figure. Show that $I_G\alpha$ may be eliminated by moving the vectors $m(\mathbf{a}_G)_t$ and $m(\mathbf{a}_G)_n$ to point P, located a distance $r_{GP} = k_G^2/r_{OG}$ from the center of mass G of the body. Here k_G represents the radius of gyration of the body about G. The point P is called the *center of percussion* of the body.

Prob. 17–61

17–62. If $\bar{y} = 393$ mm and $I_G = 0.0885$ kg $\cdot$ m^2, determine the distance h from the center of the grip O to the center of percussion P of the baseball bat. The bat has a mass of 1.65 kg. When the bat strikes a ball at the center of percussion, no stinging effect is felt in the hands of the batter at O. Explain why this is so. *Hint:* See Prob. 17–61 and assume that the point of rotation is at O.

Prob. 17–62

17–63. The relay switch consists of an electromagnet E and a 20-g armature AB (slender bar) which is pinned at A and lies in the vertical plane. When the current is turned off, the armature is held open against the smooth stop at B by the spring CD, which exerts an upward vertical force $F_s = 0.85$ N on the armature at C. When the current is turned on, the electromagnet attracts the armature at E with a vertical force $F = 0.8$ N. Determine the initial angular acceleration of the armature when the contact BF begins to close.

Prob. 17–63

*17–64.** The armature (slender rod) AB has a mass of 0.2 kg and can pivot about the pin at A. Movement is controlled by the electromagnet E, which exerts a horizontal attractive force on the armature at B of $F_B = (2(10^{-4})/l^2)$ N, where l in meters is the gap between the armature and the magnet at any instant. If the armature lies in the horizontal plane, and is originally at rest, determine the speed of the contact at B the instant $l = 0.01$ m. Originally $l = 0.02$ m.

Prob. 17–64

17–65. Block A has a mass m and rests on a surface having a coefficient of friction μ. The cord attached to A passes over a pulley at C and is attached to a block B having a mass $2m$. If B is released, determine the acceleration of A. Assume that the cord does not slip over the pulley. The pulley can be approximated as a thin disk of radius r and mass $\frac{1}{4}m$. Neglect the mass of the cord.

Prob. 17–65

17–66. A cord having negligible mass is wrapped over the 15-lb disk at A and passes over the 5-lb disk at B. If a 3-lb block C is attached to its end and released from rest, determine the speed of the block after it descends 3 ft. Also, what is the tension in the horizontal and vertical segments of the cord? Assume no slipping of the cord over the disk at B. Neglect friction at the pins D and E.

Prob. 17–66

17–67. The ladder is released from rest when $\theta \approx 0°$. Determine as a function of θ the normal and frictional forces which are exerted on the ladder by the ground as it falls downward. Is it possible for the ladder to slip as it falls? Why? For the calculation, treat the ladder as a slender rod of length l and mass m.

Prob. 17–67

*17–68. The drum has a weight of 50 lb and a radius of gyration $k_A = 0.4$ ft. A 35-ft-long chain having a weight of 2 lb/ft is wrapped around the outer surface of the drum so that $s = 3$ ft is suspended as shown. If the drum is originally at rest, determine its angular velocity after the end B has descended $s = 13$ ft. Neglect the thickness of the chain.

Prob. 17–68

Equations of Motion: General Plane Motion 17.5

The rigid body (or slab) shown in Fig. 17–19a is subjected to general plane motion caused by the externally applied force system. The free-body and kinetic diagrams for the body are shown in Fig. 17–19b. The vector $m\mathbf{a}_G$ (shown dashed) has the same *direction* as the acceleration of the body's mass center, and $I_G\boldsymbol{\alpha}$ acts in the same *direction* as the angular acceleration. If an x and y inertial coordinate system is chosen as shown, the three equations of motion may be written as

$$\Sigma F_x = m(a_G)_x$$
$$\Sigma F_y = m(a_G)_y \qquad (17\text{–}17)$$
$$\Sigma M_G = I_G\alpha$$

In some problems it may be convenient to sum moments about some point P other than G. This is usually done in order to eliminate unknown forces from the moment summation. When used in this more general sense, the three equations of motion become

$$\Sigma F_x = m(a_G)_x$$
$$\Sigma F_y = m(a_G)_y \qquad (17\text{–}18)$$
$$\Sigma M_P = \Sigma(M_k)_P$$

where $\Sigma(M_k)_P$ represents the moment sum of $I_G\boldsymbol{\alpha}$ and $m\mathbf{a}_G$ (or its components) about P as determined by the data on the kinetic diagram.

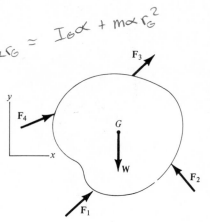

(handwritten annotations:)
$= I_G\alpha + (ma_G)(r_G)$
$a_G = \alpha r_G = I_G\alpha + m\alpha r_G^2$

PROCEDURE FOR ANALYSIS

The following procedure provides a method for solving kinetic problems involving general plane motion of a rigid body.

Free-Body Diagram. Draw the free-body diagram for the body and specify the direction and sense of the acceleration of the mass center, $\mathbf{a}_G$, and the angular acceleration $\boldsymbol{\alpha}$ of the body. Also, compute the moment of inertia I_G. If it is decided that the rotational equation of motion $\Sigma M_P = \Sigma(M_k)_P$ is to be used, then consider drawing the kinetic diagram in order to help "visualize" the "moments" developed by the components $m(\mathbf{a}_G)_x$, $m(\mathbf{a}_G)_y$, and $I_G\boldsymbol{\alpha}$ when writing the terms in the moment sum $\Sigma(M_k)_P$.

Equations of Motion. Apply the three equations of motion, Eqs. 17–17 or 17–18, with reference to an established x, y inertial coordinate system.

Kinematics. Use kinematics if a complete solution cannot be obtained strictly from the equations of motion. In particular, if the body's motion is *constrained* due to its supports, additional equations may be obtained by using $\mathbf{v}_B = \mathbf{v}_A + \mathbf{v}_{B/A}$ and $\mathbf{a}_B = \mathbf{a}_A + \mathbf{a}_{B/A}$, which relate the motions of any two points A and B on the body (see Example 17–16).

(b)

Fig. 17–19

(a)

(b)

Fig. 17–20

Frictional Rolling Problems. There is a class of planar kinetics problems which deserves special mentioning. These problems involve wheels, cylinders, or bodies of similar shape, which roll on a *rough* plane surface. Because of the applied loadings, it may not be known if the body *rolls without slipping,* or if it *slides as it rolls.* For example, consider the homogeneous disk shown in Fig. 17–20a, which has a mass m and is subjected to a known horizontal force **P**. Following the procedure outlined above, the free-body diagram is shown in Fig. 17–20b. Applying the three equations of motion yields

$$\xrightarrow{+}\Sigma F_x = m(a_G)_x; \qquad\qquad P - F = ma_G \qquad\qquad (17\text{–}19)$$

$$+\uparrow \Sigma F_y = m(a_G)_y; \qquad\qquad N - mg = 0 \qquad\qquad (17\text{–}20)$$

$$\zeta + \Sigma M_G = I_G\alpha; \qquad\qquad Fr = I_G\alpha \qquad\qquad (17\text{–}21)$$

A fourth equation is needed since these *three equations* contain *four unknowns; F, N, α,* and a_G.

No Slipping. If the frictional force **F** is great enough to allow the disk to roll *without slipping,* then a_G may be related to α by the *kinematic equation**

$$a_G = \alpha r \qquad\qquad (17\text{–}22)$$

The four unknowns are determined by *solving simultaneously* Eqs. 17–19 to 17–22. When the solution is obtained, the assumption of no slipping must be *checked.* In this regard, recall that no slipping occurs provided $F \le \mu_s N$, where μ_s is the static coefficient of friction. If the inequality is satisfied, the problem is solved. However, if $F > \mu_s N$, the problem must be *reworked,* since then the disk slips as it rolls.

Slipping. In the case of slipping, α and a_G are *independent of one another* so that Eq. 17–22 does not apply. Instead, the magnitude of the frictional force is related to the magnitude of the normal force using the coefficient of kinetic friction μ_k, i.e.,

$$F = \mu_k N \qquad\qquad (17\text{–}23)$$

Here **F** acts to the left on the disk to prevent the slipping motion to the right. In this case Eqs. 17–19 to 17–21 and 17–23 are used for the solution. Examples 17–14 and 17–15 illustrate these concepts numerically.

---

*Note that the contact point at the ground is the instantaneous center of zero velocity; it does *not* have zero acceleration. Equation 17–22 is derived in Example 16–3.

Example 17–13

The spool in Fig. 17–21a has a mass of 8 kg and a radius of gyration of $k_G = 0.35$ m. If cords of negligible mass are wrapped around its inner hub and outer rim as shown, determine the spool's angular velocity 3 s after it is released from rest.

(a)

Solution

Free-Body Diagram. Since no forces act on the spool in the horizontal direction, $\mathbf{a}_G$ acts upward, and consequently, $\boldsymbol{\alpha}$ acts clockwise, Fig. 17–21b.

The moment of inertia of the spool about its mass center is

$$I_G = mk_G^2 = 8(0.35)^2 = 0.980 \text{ kg} \cdot \text{m}^2$$

Equations of Motion

$$+\uparrow \Sigma F_y = m(a_G)_y; \qquad T + 100 - 78.48 = 8a_G \qquad (1)$$

$$\zeta + \Sigma M_G = I_G\alpha; \qquad 100(0.2) - T(0.5) = 0.980\alpha \qquad (2)$$

(b)

Fig. 17–21

Kinematics. A complete solution is obtained if kinematics is used to relate a_G to α. In this case the spool "rolls without slipping" on the cord at A. Hence, we can use the results of Example 16–3, so that

$$\zeta + a_G = r\alpha; \qquad a_G = 0.5\alpha \qquad (3)$$

Solving Eqs. (1) to (3), we have

$$\alpha = 10.3 \text{ rad/s}^2$$
$$a_G = 5.16 \text{ m/s}^2$$
$$T = 19.8 \text{ N}$$

Since α is constant, the angular velocity in 3 s is

$\zeta +$
$$\omega = \omega_0 + \alpha_c t$$
$$= 0 + 10.3(3)$$
$$= 30.9 \text{ rad/s} \downarrow \qquad \qquad Ans.$$

Example 17–14

The 50-lb wheel shown in Fig. 17–22a has a radius of gyration $k_G = 0.70$ ft. If a 35-lb · ft couple moment is applied to the wheel, determine the acceleration of its mass center G. The static and kinetic coefficients of friction between the wheel and the plane at A are $\mu_s = 0.3$ and $\mu_k = 0.25$, respectively.

$M = 35$ lb · ft

1.25 ft

(a)

Solution

Free-Body Diagram. By inspection, it is seen that the frictional force $\mathbf{F}_A$, acting to the right, gives the wheel its forward acceleration, Fig. 17–22b.
The moment of inertia is

$$I_G = mk_G^2 = \frac{50}{32.2}(0.70)^2 = 0.761 \text{ slug} \cdot \text{ft}^2$$

Equations of Motion

$$\xrightarrow{+}\Sigma F_x = m(a_G)_x; \qquad F_A = \frac{50}{32.2}a_G \tag{1}$$

$$+\uparrow \Sigma F_y = m(a_G)_y; \qquad N_A - 50 = 0 \tag{2}$$

$$\zeta+\Sigma M_G = I_G\alpha; \qquad 35 - 1.25(F_A) = 0.761\alpha \tag{3}$$

50 lb

35 lb · ft

1.25 ft

a_G

F_A

N_A

(b)

Fig. 17–22

Kinematics (No Slipping). If this assumption is made then

$$a_G = 1.25\alpha \tag{4}$$

Solving Eqs. (1) to (4),

$$N_A = 50.0 \text{ lb} \qquad F_A = 21.3 \text{ lb}$$
$$\alpha = 11.0 \text{ rad/s}^2 \qquad a_G = 13.7 \text{ ft/s}^2$$

The original assumption of no slipping requires $F_A \le \mu_s N_A$. However, since 21.3 lb > 0.3(50 lb) = 15 lb, the wheel slips as it rolls.

(Slipping). This requires $F_A = \mu_k N_A$, or

$$F_A = 0.25N_A \tag{5}$$

Solving Eqs. (1) to (3) and (5) yields

$$N_A = 50.0 \text{ lb} \qquad F_A = 12.5 \text{ lb}$$
$$\alpha = 25.5 \text{ rad/s}^2$$
$$a_G = 8.0 \text{ ft/s}^2 \rightarrow \qquad\qquad Ans.$$

Example 17–15

The uniform slender beam shown in Fig. 17–23a has a mass of 100 kg and a moment of inertia $I_G = 75$ kg $\cdot$ m². If the coefficients of static and kinetic friction between the end of the beam and the surface are both equal to $\mu_A = 0.25$, determine the beam's angular acceleration at the instant the 400-N horizontal force is applied. The beam is originally at rest.

Solution

Free-Body Diagram. Figure 17–23b. The path of motion of the mass center G will be along an unknown curved path having a radius of curvature which is initially parallel to the x axis. There is no normal or y component of acceleration since the beam is originally at rest, i.e., $\mathbf{v}_G = \mathbf{0}$, so that $(a_G)_y = v_G^2/\rho = 0$.

Equations of Motion

$$\xrightarrow{+} \Sigma F_x = m(a_G)_x; \qquad 400 - F_A = 100a_G \qquad (1)$$

$$+\uparrow \Sigma F_y = m(a_G)_y; \qquad N_A - 981 = 0 \qquad (2)$$

$$\curvearrowright + \Sigma M_G = I_G \alpha; \qquad F_A(1.5) - 400(1) = 75\alpha \qquad (3)$$

A fourth equation is needed for a complete solution.

Kinematics (No Slipping). In this case point A acts as a "pivot" so that

$$\curvearrowright + a_G = \alpha r_{AG}; \qquad a_G = 1.5\alpha \qquad (4)$$

Solving Eqs. (1) to (4) yields

$$N_A = 981 \text{ N} \qquad F_A = 300 \text{ N}$$
$$a_G = 1.0 \text{ m/s}^2 \qquad \alpha = 0.667 \text{ rad/s}^2$$

Testing the original assumption of no slipping requires $F_A < \mu_A N_A$. However, $300 > 0.25(981) = 245.3$ N. (Slips at A.)

(Slipping). For this case Eq. (4) does *not* apply. Instead, the frictional equation $F_A = \mu_A N_A$ is used, i.e.,

$$F_A = 0.25N_A \qquad (5)$$

Solving Eqs. (1) to (3) and (5) simultaneously yields

$$N_A = 981 \text{ N} \qquad F_A = 245.3 \text{ N} \qquad a_G = 1.55 \text{ m/s}^2$$
$$\alpha = -0.427 \text{ rad/s}^2 = 0.427 \text{ rad/s}^2 \curvearrowleft \qquad \textit{Ans.}$$

(a)

(b)

Fig. 17–23

Example 17–16

The 30-kg wheel shown in Fig. 17–24*a* has a mass center at *G* and a radius of gyration $k_G = 0.15$ m. If the wheel is originally at rest and released from the position shown, determine its angular acceleration. *No slipping occurs.*

(a)

Solution

Free-Body and Kinetic Diagrams. The two unknowns $\mathbf{F}_A$ and $\mathbf{N}_A$ shown on the free-body diagram, Fig. 17–24*b*, can be eliminated from the analysis by summing moments about point *A*. The kinetic diagram accompanies the solution in order to illustrate application of $\Sigma(M_k)_A$. Since the *path of motion* of *G* is *unknown*, the two components $m(\mathbf{a}_G)_x$ and $m(\mathbf{a}_G)_y$ must be shown on the kinetic diagram, Fig. 17–24*b*.

The moment of inertia is

$$I_G = mk_G^2 = 30(0.15)^2 = 0.675 \text{ kg} \cdot \text{m}^2$$

Equation of Motion. Applying the rotational equation of motion about point *A*,

$$\zeta + \Sigma M_A = \Sigma(M_k)_A;$$
$$30(9.81)(0.1) = 0.675\alpha + 30(a_G)_x(0.25) + 30(a_G)_y(0.1) \quad (1)$$

There are three unknowns in this equation: $(a_G)_x$, $(a_G)_y$, and α.

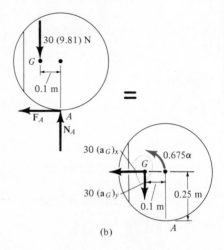

(b)

Kinematics. Using kinematics, $(a_G)_x$, $(a_G)_y$, and α can be related. Since no slipping occurs, $a_O = \alpha r = \alpha(0.25)$, directed to the left, Fig. 17–24*c*. Also, $\boldsymbol{\omega} = \mathbf{0}$ since the wheel is originally at rest. Using Eq. 16–15, with point *O* as the base point, Fig. 17–24*d*, we have

$$\mathbf{a}_G = \mathbf{a}_O + (\mathbf{a}_{G/O})_t + (\mathbf{a}_{G/O})_n$$
$$\underset{\leftarrow}{(a_G)_x} + \underset{\downarrow}{(a_G)_y} = \underset{\leftarrow}{\alpha(0.25)} + \underset{\downarrow}{\alpha(0.1)} + 0$$

$\omega = 0$
$a_O = \alpha(0.25)$

Absolute motion

(c)

Equating the respective horizontal and vertical components, we have

$$(a_G)_x = \alpha(0.25) \quad (2)$$
$$(a_G)_y = \alpha(0.1) \quad (3)$$

Solving Eqs. (1) to (3) yields

$$\alpha = 10.33 \text{ rad/s}^2 \, \zeta \qquad \textit{Ans.}$$
$$(a_G)_x = 2.58 \text{ m/s}^2$$
$$(a_G)_y = 1.03 \text{ m/s}^2$$

As an exercise, show that $F_A = 77.4$ N and $N_A = 263.3$ N.

$(a_{G/O})_t = \alpha(0.1)$

Relative motion

(d)

Fig. 17–24

Problems

17–69. If the disk in Fig. 17–20a *rolls without slipping,* show that when moments are summed about the instantaneous center of zero velocity, *IC,* it is possible to use the moment equation $\Sigma M_{IC} = I_{IC}\alpha$, where I_{IC} represents the moment of inertia of the disk calculated about the instantaneous axis of zero velocity.

17–70. Two men exert constant vertical forces of 40 lb and 30 lb at the ends *A* and *B* of a uniform plank which has a weight of 50 lb. If the plank is originally at rest, determine the acceleration of its center and its angular acceleration. Assume the plank to be a thin plate.

Prob. 17–70

17–71. The 500-lb beam is supported at *A* and *B* when it is subjected to a force of 1000 lb as shown. If the pin support at *A* suddenly fails, determine the beam's initial angular acceleration and the force of the roller support on the beam. For the calculation, assume that the beam is a slender rod so that its thickness can be neglected.

Prob. 17–71

*17–72. A woman sits in a rigid position in the middle of the swing. The combined weight of the woman and swing is 180 lb and the radius of gyration about the center of mass *G* is $k_G = 2.5$ ft. If a man pushes on the swing with a horizontal force $F = 20$ lb as shown, determine the initial angular acceleration and the tension in each of the two supporting chains *AB*. During the motion, assume that the chain segment *CAD* remains rigid. The swing is originally at rest.

Prob. 17–72

17–73. A uniform rod having a weight of 10 lb is pin-supported at A from a roller which rides on a horizontal track. If the rod is originally at rest, and a horizontal force $F = 15$ lb is applied to the roller, determine the initial acceleration of the roller. Neglect the mass of the roller and its size d in the computations.

Probs. 17–73 / 17–74

17–74. Solve Prob. 17–73 assuming that the roller at A is replaced by a slider block having a negligible mass. The coefficient of friction between the block and the track is $\mu = 0.2$. Neglect the dimension d and the size of the block in the computations.

17–75. The spool has a mass of 500 kg and a radius of gyration $k_G = 1.30$ m. It rests on the surface of a conveyor belt for which $\mu = 0.4$. If the conveyor accelerates at $a = 1$ m/s^2, determine the initial tension in the wire and the angular acceleration of the spool. The spool is originally at rest.

Prob. 17–75

*17–76. The spool and wire wrapped around its core have a mass of 20 kg and a centroidal radius of gyration $k_G = 250$ mm. If the coefficient of friction at the ground is $\mu_B = 0.1$, determine the angular acceleration of the spool when the 30-N · m couple is applied.

Prob. 17–76

17–77. The lawn roller has a mass of 80 kg and a radius of gyration $k_G = 0.175$ m. If it is pushed forward with a force of 200 N when the handle is at 45°, determine its angular acceleration. The coefficient of friction between the ground and the roller is $\mu_A = 0.12$.

Prob. 17–77

17–78. A boy sits snugly inside a large tire such that both the boy and tire have a total weight of 185 lb, a center of mass at G, and a radius of gyration $k_G = 1.65$ ft about G. If the tire rolls freely down the incline, determine the normal and frictional forces it exerts on the ground when it is in the position shown and has an angular velocity of 6 rad/s. Assume that the tire does not slip as it rolls.

Prob. 17–78

17–79. A bowling ball has a weight of 16 lb and a radius of 0.375 ft. It is cast horizontally onto an alley such that initially $\omega = 0$ and its mass center has a velocity $v = 8$ ft/s. If the coefficient of friction between the floor and the ball is $\mu = 0.12$, determine the distance the ball travels before it rolls without slipping. For the calculation, neglect the finger holes in the ball and assume the ball has a uniform density.

Prob. 17–79

*17–80. By pressing down with the finger at B, a thin ring having a mass m is given an initial velocity v_1 and a backspin ω_1 when the finger is released. If the coefficient of friction between the table and the ring is μ, determine the distance the ring travels forward before backspinning stops.

Prob. 17–80

17–81. The 10-lb disks are connected together using a rod having a negligible mass. If they roll on their contacting surfaces without slipping, determine the initial angular acceleration of each disk if the system is released from rest from the position shown.

Prob. 17–81

17–82. The 150-lb slender bar is supported by two cords AB and AC. If cord AC suddenly breaks, determine the initial angular acceleration of the bar and the tension in cord AB.

Prob. 17–82

17–83. The 15-lb disk rests on a 5-lb plate. A cord is wrapped around the periphery of the disk and attached to the wall at B. If a torque $M = 40$ lb · ft is applied to the disk, determine the angular acceleration of the disk and the time needed for the end C of the plate to travel 3 ft and strike the wall. The coefficient of friction between the plate and disk is $\mu = 0.2$, and between the plate and floor $\mu' = 0.1$. Neglect the mass of the cord.

Prob. 17–83

***17–84.** The 10-lb disk and 4-lb block are released from rest. Determine the velocity of the block when $t = 3$ s. The coefficient of friction at A is $\mu_A = 0.2$. Neglect the mass of the cord and pulleys.

Probs. 17–84 / 17–85

17–85. The 10-lb disk and 10-lb block are released from rest. Determine the velocity of the block when $t = 3$ s. The coefficient of friction at A is $\mu_A = 0.2$. Neglect the mass of the cord and pulleys.

17–86. A long strip of paper is wrapped into two rolls, each having a mass of 10 kg. The rolls are placed on a *smooth surface*. The core of roll A is subjected to a horizontal force of 60 N. Determine the initial tension in the paper between the rolls and the angular acceleration of each roll. For the calculation, assume the rolls to be approximated by cylinders. Neglect the mass of the paper between the rolls.

Prob. 17–86

17–87. The tractor has a mass of 900 kg, excluding the rear wheels. The mass center is located at G. Each of the rear wheels has a mass of 50 kg and a radius of gyration $k_A = 0.7$ m. If the tractor coasts freely down the 10° slope, determine its velocity when $t = 5$ s starting from rest. For the calculation, neglect the mass of the front wheels B and the effects of friction at these wheels. The rear wheels roll without slipping at C. Also compute the normal reactions acting on each of the front and rear wheels.

Prob. 17–87

*17–88. A biomechanical model of the head and neck of a restrained passenger in an automobile is shown in the figure. Measurements indicate that the head plus half the neck, assumed pinned at A, have a mass of 0.28 slug and a moment of inertia about an axis through the center of gravity $I_G = 0.024$ slug $\cdot$ ft^2. If an automobile and the passenger's torso are subjected to a horizontal deceleration $a_A = 15g = 483$ ft/s^2, when the head is in the position shown, determine the horizontal and vertical components of force which the neck exerts on the head at A at this instant. The restraining moment M_A is required to hold the head in the *nonrotating* equilibrium position shown, prior to the deceleration. This moment remains constant during the deceleration.

Prob. 17–88

17–89. The wheel has a mass of 50 kg and a radius of gyration $k_G = 0.4$ m. If it rolls without slipping down the inclined plank, determine the forces which the supports at A and B exert on the plank at the instant the wheel is located at the midpoint of the plank. The plank is uniform and has a mass of 20 kg. Assume that horizontal and vertical reactions occur at A, and a normal reaction occurs at B.

Prob. 17–89

17–90. The 25-lb slender rod has a length of 6 ft. Using a collar of negligible mass, its end A is confined to move along the smooth circular bar of radius $3\sqrt{2}$ ft. End B rests on the floor, for which $\mu_B = 0.4$. If the bar is released from rest when $\theta = 30°$, determine the angular acceleration of the bar at this instant.

Prob. 17–90

17–91. Each of the two links (slender rods) has a mass of 2 kg/m. If the piston has a mass of 3 kg, determine the torque M that must be applied to link AB so that the piston has an acceleration of 600 m/s^2 at the instant shown. Originally the system is at rest. Motion occurs in the horizontal plane.

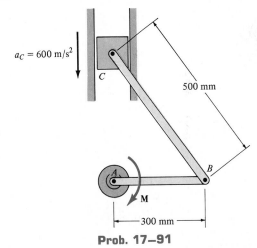

Prob. 17–91

*17–92. The ladder has a weight W and rests against the smooth wall and ground. Determine its angular acceleration as a function of θ when it is released and allowed to slide downward. For the calculation, treat the ladder as a slender rod.

■17–93. The cylinder has a mass m and is released from rest at $\theta = 0°$. If it rolls down the surface of the fixed semi-circular track, determine the angle θ at which the cylinder just begins to slip on the surface. The coefficient of friction between the cylinder and track is $\mu = 0.3$.

Prob. 17–92

Prob. 17–93

Planar Kinetics of a Rigid Body: Work and Energy

It was shown in Chapter 14 that problems which involve force, velocity, and displacement can conveniently be solved by using the principle of work and energy, or if the force system is "conservative," the conservation of energy theorem. In this chapter we will apply work and energy methods to solve problems involving the planar motion of a rigid body. The more general discussion as applied to the three-dimensional motion of a rigid body is presented in Chapter 21.

Before discussing the principle of work and energy for a body, however, the methods for obtaining the body's kinetic energy when it is subjected to translation, rotation about a fixed axis, or general plane motion will be developed.

Kinetic Energy 18.1

Consider the rigid body shown in Fig. 18–1, which is represented here by a *slab* moving in the inertial *x-y* reference plane. An arbitrary *i*th particle of the body, having a mass dm, is located at r from the arbitrary point P. If at the *instant* shown the particle has a velocity $\mathbf{v}_i$, then the particle's kinetic energy is $T_i = \frac{1}{2} dm \, v_i^2$. The kinetic energy of the entire body is determined by writing similar expressions for each particle of the body and integrating the results, i.e.,

$$T = \frac{1}{2} \int_m dm \, v_i^2$$

Fig. 18–1

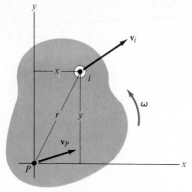

Fig. 18–1

This equation may be written in another manner by expressing $\mathbf{v}_i$ in terms of the velocity of point P, i.e., $\mathbf{v}_P$. If the body has an angular velocity $\boldsymbol{\omega}$, then from Fig. 18–1 we have

$$\mathbf{v}_i = \mathbf{v}_P + \mathbf{v}_{i/P}$$
$$= (v_P)_x\mathbf{i} + (v_P)_y\mathbf{j} + \omega\mathbf{k} \times (x\mathbf{i} + y\mathbf{j})$$
$$= ((v_P)_x - \omega y)\mathbf{i} + ((v_P)_y + \omega x)\mathbf{j}$$

The square of the magnitude of $\mathbf{v}_i$ is thus

$$\mathbf{v}_i \cdot \mathbf{v}_i = v_i^2 = [(v_P)_x - \omega y]^2 + [(v_P)_y + \omega x]^2$$
$$= (v_P^2)_x - 2(v_P)_x\omega y + \omega^2 y^2 + (v_P^2)_y + 2(v_P)_y\omega x + \omega^2 x^2$$
$$= v_P^2 - 2(v_P)_x\omega y + 2(v_P)_y\omega x + \omega^2 r^2$$

Substituting into the equation of kinetic energy yields

$$T = \tfrac{1}{2}\left(\int_m dm\right)v_P^2 - (v_P)_x\omega\left(\int_m y\,dm\right) + (v_P)_y\omega\left(\int_m x\,dm\right) + \tfrac{1}{2}\omega^2\left(\int_m r^2\,dm\right)$$

The first integral on the right represents the entire mass m of the body. Since $\bar{x}m = \int x\,dm$ and $\bar{y}m = \int y\,dm$, the second and third integrals locate the body's center of mass G with respect to P. The last integral represents the body's moment of inertia I_P, computed about the z axis passing through point P. Thus,

$$T = \tfrac{1}{2}mv_P^2 - (v_P)_x\,\omega\bar{y}\,m + (v_P)_y\,\omega\bar{x}\,m + \tfrac{1}{2}I_P\,\omega^2 \qquad (18\text{–}1)$$

This equation reduces to a simpler form if point P coincides with the mass center G for the body, in which case $\bar{x} = \bar{y} = 0$, and therefore

$$T = \tfrac{1}{2}mv_G^2 + \tfrac{1}{2}I_G\omega^2 \qquad (18\text{–}2)$$

Here I_G is the mass moment of inertia of the body about an axis which is perpendicular to the plane of motion and passes through the mass center. Both terms on the right side are always *positive,* since the velocities are squared. Furthermore, it may be verified that these terms have units of length times force, common units being m · N or ft · lb. Recall, however, that in the SI system the unit of energy is the joule (J), where 1 J = 1 m · N.

Translation. When a rigid body of mass m is subjected to either rectilinear or curvilinear *translation,* the kinetic energy due to rotation is zero, since $\boldsymbol{\omega} = \mathbf{0}$. From Eq. 18–2, the kinetic energy of the body is therefore

$$T = \tfrac{1}{2}mv_G^2$$

where v_G is the magnitude of the body's translational velocity at the instant considered, Fig. 18–2.

Translation

Fig. 18–2

Rotation About a Fixed Axis.   When a rigid body is *rotating about a fixed axis* passing through point O, Fig. 18–3, the body has both *translational* and *rotational* kinetic energy as defined by Eq. 18–2, i.e.,

$$T = \tfrac{1}{2}mv_G^2 + \tfrac{1}{2}I_G\omega^2$$

The body's kinetic energy may be formulated in another manner by noting that $v_G = r_{G/O}\omega$, in which case $T = \tfrac{1}{2}(I_G + mr_{G/O}^2)\omega^2$. By the parallel-axis theorem, the terms inside the parentheses represent the moment of inertia I_O of the body about an axis perpendicular to the plane of motion and passing through point O. Hence,*

$$T = \tfrac{1}{2}I_O\omega^2 \tag{18–3}$$

From the derivation, this equation may be substituted for Eq. 18–2, since it accounts for *both* the translational kinetic energy of the body's mass center and the rotational kinetic energy of the body computed about the mass center.

General Plane Motion.   When a rigid body is subjected to general plane motion, Fig. 18–4, it has an angular velocity $\boldsymbol{\omega}$ and its mass center has a velocity $\mathbf{v}_G$. Hence, the kinetic energy is defined by Eq. 18–2, i.e.,

$$T = \tfrac{1}{2}mv_G^2 + \tfrac{1}{2}I_G\omega^2$$

Here it is seen that the total kinetic energy of the body consists of the *scalar* sum of the body's *translational* kinetic energy, $\tfrac{1}{2}mv_G^2$, and *rotational* kinetic energy about its mass center, $\tfrac{1}{2}I_G\omega^2$.

Because energy is a scalar quantity, the total kinetic energy for a system of *connected* rigid bodies is the sum of the kinetic energies of all the moving parts. Depending upon the type of motion, the kinetic energy of *each body* is found by applying Eq. 18–2 or the alternative forms mentioned above.†

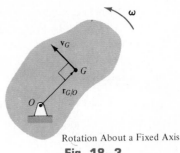

Rotation About a Fixed Axis
Fig. 18–3

General plane motion
Fig. 18–4

---

*The similarity between this derivation and that of $\Sigma M_O = I_O\alpha$, Eq. 17–16, should be noted. Also note that the same result can be obtained from Eq. 18–1 by selecting point P at O, realizing that $\mathbf{v}_O = \mathbf{0}$.

†A brief review of Secs. 16.5 to 16.7 may prove helpful in solving problems, since computations for kinetic energy require a kinematic analysis of velocity.

Example 18-1

The system of three elements shown in Fig. 18-5a consists of a 6-kg block B, a 10-kg disk D, and a 12-kg cylinder C. A continuous cord of negligible mass is wrapped around the cylinder, passes over the disk, and is then attached to the block. If the block is moving downward with a speed of 0.8 m/s and the cylinder rolls without slipping, determine the total kinetic energy of the system at this instant.

(a)

(b)

Fig. 18-5

Solution

By inspection, the block is translating, the disk rotates about a fixed axis, and the cylinder has general plane motion. Hence, in order to compute the kinetic energy of the disk and cylinder, it is first necessary to determine ω_D, ω_C, and v_G, Fig. 18-5a. From the *kinematics* of the disk,

$$v_B = r_D\omega_D; \qquad 0.8 \text{ m/s} = (0.1 \text{ m})\omega_D; \qquad \omega_D = 8 \text{ rad/s}$$

Since the cylinder rolls without slipping, the instantaneous center of zero velocity is at the point of contact with the ground, Fig. 18-5b, hence,

$$v_E = r_{E/IC}\omega_C; \qquad 0.8 \text{ m/s} = (0.2 \text{ m})\omega_C; \qquad \omega_C = 4 \text{ rad/s}$$
$$v_G = r_{G/IC}\omega_C; \qquad v_G = (0.1 \text{ m})(4 \text{ rad/s}) = 0.4 \text{ m/s}$$

The kinetic energy of the block is

$$T_B = \tfrac{1}{2}m_B v_B^2; \quad T_B = \frac{1}{2}(6 \text{ kg})(0.8 \text{ m/s})^2 = 1.92 \text{ J}$$

The kinetic energy of the disk is

$$T_D = \tfrac{1}{2}I_D\omega_D^2; \quad T_D = \tfrac{1}{2}(\tfrac{1}{2}m_D r_D^2)\omega_D^2$$
$$= \frac{1}{2}\left[\frac{1}{2}(10 \text{ kg})(0.1 \text{ m})^2\right](8 \text{ rad/s})^2 = 1.60 \text{ J}$$

Finally, the kinetic energy of the cylinder is

$$T_C = \tfrac{1}{2}mv_G^2 + \tfrac{1}{2}I_G\omega_C^2; \quad T_C = \tfrac{1}{2}mv_G^2 + \tfrac{1}{2}(\tfrac{1}{2}m_C r_C^2)\omega_C^2$$
$$= \frac{1}{2}(12 \text{ kg})(0.4 \text{ m/s})^2$$
$$+ \frac{1}{2}\left[\frac{1}{2}(12 \text{ kg})(0.1 \text{ m})^2\right](4 \text{ rad/s})^2 = 1.44 \text{ J}$$

The total kinetic energy of the system is therefore

$$T = T_B + T_D + T_C$$
$$= 1.92 \text{ J} + 1.60 \text{ J} + 1.44 \text{ J} = 4.96 \text{ J} \qquad \textit{Ans.}$$

The Work of a Force 18.2

Several types of forces are often encountered in planar kinetics problems involving a rigid body. The work of each of these forces has been presented in Sec. 14.1 and is listed below as a summary.

Work of a Variable Force. If an external force **F** acts on a rigid body the work done by the force when it moves along the path s, Fig. 18–6, is defined as

$$U = \int_s F \cos \theta \, ds \qquad (18\text{–}4)$$

Here θ is the angle between the "tails" of the force vector and the differential displacement. In general, the integration must account for the variation of the force's direction and magnitude.

Fig. 18–6

Work of a Constant Force. If an external force $\mathbf{F}_c$ acts on a rigid body, Fig. 18–7, and maintains a constant magnitude F_c and constant direction θ, while the body undergoes a translation s, Eq. 18–4 can be integrated so that the work becomes

$$U_{F_c} = (F_c \cos \theta)s \qquad (18\text{–}5)$$

Here $F_c \cos \theta$ represents the magnitude of the component of force in the direction of displacement.

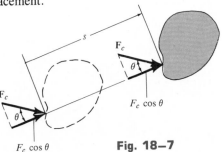

Fig. 18–7

Work of a Weight. The weight of a body does work only when the body's center of mass G undergoes a *vertical displacement* Δy. If this displacement is *upward*, Fig. 18–8, the work is negative, since the weight and displacement are in the opposite directions.

$$U_W = -W \, \Delta y \qquad (18\text{–}6)$$

Likewise, if the displacement is downward $(-\Delta y)$ the work becomes positive. Here the elevation change is considered to be small so that **W**, which is caused by gravitation, is constant.*

*If the difference in elevation is large, use the results of Prob. 14–25.

Fig. 18–8

Fig. 18-9

Work of a Spring Force. If a linear elastic spring is attached to a body, the spring force $F_s = ks$ *acting on the body* does work when the spring either stretches or compresses from s_1 to a *further* position s_2. In both cases the work will be *negative* since the *displacement of the body* is always in the opposite direction to the force, Fig. 18-9. The work done is

$$U_s = -(\tfrac{1}{2}ks_2^2 - \tfrac{1}{2}ks_1^2) \qquad (18\text{-}7)$$

where $|s_2| > |s_1|$.

Forces That Do No Work. There are some external forces that do no work when the body is displaced. These forces can act either at *fixed points* on the body or they can have a direction *perpendicular to their displacement.* Examples include the reactions at a pin support about which a body rotates, the normal reaction acting on a body that moves along a surface, and the weight of a body when the center of gravity of the body moves in a *horizontal plane.* A frictional force $\mathbf{F}_r$ acting on a body as it *rolls without slipping* over a rough surface also does no work, Fig. 18-10.* This is because, during any *instant of time dt*, $\mathbf{F}_r$ acts at a point on the body which has *zero velocity* (instantaneous center, IC). In other words, for any differential rolling movement ds_{IC} of the body's IC, the work is $dU = F_r\,ds_{IC} = F_r(v_{IC}dt) = 0$, since $v_{IC} = 0$.

Fig. 18-10

18.3 The Work of a Couple

Recall that a *couple* consists of a pair of noncollinear forces which have equal magnitudes and opposite directions. When a body subjected to a couple undergoes general plane motion, the two forces do work *only* when the body

---

*The work done by the frictional force *when the body slips* has been discussed in Sec. 14.3.

undergoes a *rotation*. To show this, consider the body in Fig. 18–11a, which is subjected to a couple having a magnitude of $M = Fr$. Any general differential displacement of the body can be considered as a separate translation and rotation. When the body *translates* such that the *component of displacement* along the line of action of the forces is ds_t, Fig. 18–11b, clearly the "positive" work of one force *cancels* the "negative" work of the other. Consider now a differential rotation $d\theta$ of the body about an axis which is perpendicular to the plane of the couple and intersects the plane at the midpoint O, Fig. 18–11c. (For the derivation any other point in the plane may also be considered.) As shown, each force undergoes a displacement $ds_\theta = (r/2)\,d\theta$ in the direction of the force; hence, the total work done is

$$dU_M = F\left(\frac{r}{2}\,d\theta\right) + F\left(\frac{r}{2}\,d\theta\right) = (Fr)\,d\theta$$

$$= M\,d\theta$$

Here the line of action of $d\theta$ is parallel to the line of action of M. This is *always the case for general plane motion*, since M and $d\theta$ are perpendicular to the plane of motion. Furthermore, the resultant work is *positive* when M and $d\theta$ are in the *same direction* and *negative* if these vectors are in *opposite directions*.

When the body rotates in the plane through a finite angle θ measured in radians, from θ_1 to θ_2, the work of a couple is

$$U_M = \int_{\theta_1}^{\theta_2} M\,d\theta \qquad (18\text{–}8)$$

If **M** has a *constant magnitude*, then

$$U_M = M(\theta_2 - \theta_1) \qquad (18\text{–}9)$$

Here the work is *positive* provided M and $(\theta_2 - \theta_1)$ are in the same direction.

(a)

Translation
(b)

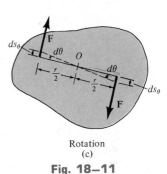

Rotation
(c)

Fig. 18–11

Principle of Work and Energy 18.4

In Sec. 14.2 the principle of work and energy was developed for a particle. By applying this principle to each of the particles of a rigid body and adding the results algebraically, since energy is a scalar, the principle of work and energy for a rigid body may be developed. In this regard, the body's initial and final kinetic energies have been defined by the formulations in Sec. 18.1, and the work done by the external forces and couples is defined by formulations in Secs. 18.2 and 18.3. Note that the work of the body's *internal forces* does not have to be considered. These forces occur in equal but opposite collinear pairs, so as a result, when the body moves, the work of one force cancels that of its counterpart. Furthermore, since the body is rigid, *no relative movement* between these forces occurs, so that no internal work is done. Thus the principle of work and energy for a rigid body may be written as

$$T_1 + \Sigma U_{1-2} = T_2 \tag{18-10}$$

This equation states that the body's initial translational *and* rotational kinetic energy plus the work done by all the external forces and couples acting on the body as the body moves from its initial to its final position is equal to the body's final translational *and* rotational kinetic energy.

When several rigid bodies are pin-connected, connected by inextensible cables, or in mesh with one another, this equation may be applied to the entire system of connected bodies. In all these cases the internal forces, which hold the various members together, do no work and hence are eliminated from the analysis.

PROCEDURE FOR ANALYSIS

The principle of work and energy is used to solve kinetics problems that involve *velocity, force,* and *displacement,* since these terms are involved in the formulation. For application, it is suggested that the following procedure be used.

Kinetic Energy (Kinematic Diagrams). Determine the kinetic-energy terms T_1 and T_2 by applying the equation $T = \frac{1}{2}mv_G^2 + \frac{1}{2}I_G\omega^2$ or an appropriate form of this equation developed in Sec. 18.1. In this regard, *kinematic diagrams* for velocity may be useful for determining v_G and ω, or for establishing a *relationship* between v_G and ω.

Work (Free-Body Diagram). Draw a free-body diagram of the body when it is located at an intermediate point along the path, in order to account for all the forces and couples which do work on the body. The work of each force and couple can be computed using the appropriate formulations outlined in Secs. 18.2 and 18.3. Since *algebraic addition* of the work terms is required, it is important that the proper sign of each term be specified. Specifically, work is *positive* when the force (couple) is in the *same direction* as its displacement (rotation); otherwise, it is negative.

Principle of Work and Energy. Apply the principle of work and energy, $T_1 + \Sigma U_{1-2} = T_2$. Since this is a scalar equation, it can be used to solve for only one unknown when it is applied to a single rigid body. This is in contrast to the three scalar equations of motion, Eqs. 17-11, which may be written for the same body.

The following example problems numerically illustrate application of this procedure.

Example 18–2

The 30-kg disk shown in Fig. 18–12a is pin-supported at its center. Determine the number of revolutions it must make to attain an angular velocity of 20 rad/s starting from rest. It is acted upon by a constant force $F = 10$ N, which is applied to a cord wrapped around its periphery, and a constant couple $M = 5$ N $\cdot$ m. Neglect the mass of the cord in the calculation.

(a) $F = 10$ N   **Fig. 18–12**   (b) $F = 10$ N

Solution

Kinetic Energy. Since the disk rotates about a fixed axis, the kinetic energy can be computed using $T = \frac{1}{2}I_O\omega^2$, where the moment of inertia is $I_O = \frac{1}{2}mr^2$. Initially, the disk is at rest, so that

$$T_1 = 0$$

$$T_2 = \tfrac{1}{2}I_O\omega_2^2 = \frac{1}{2}\left[\frac{1}{2}(30)(0.2)^2\right](20)^2 = 120 \text{ J}$$

Work (Free-Body Diagram). As shown in Fig. 18–12b, the pin reactions O_x and O_y and the weight (294.3 N) do no work, since they are not displaced. The *couple,* having a constant magnitude, does positive work $U_M = M\theta$ as the disk *rotates* through a clockwise angle of θ rad, and the *constant force* **F** does positive work $U_{F_c} = Fs$ as the cord *moves* downward $s = \theta r = \theta(0.2)$ m.

Principle of Work and Energy

$$\{T_1\} + \{\Sigma U_{1-2}\} = \{T_2\}$$
$$\{T_1\} + \{M\theta + Fs\} = \{T_2\}$$
$$\{0\} + \{5\theta + (10)\theta(0.2)\} = \{120\}$$

$$\theta = 17.1 \text{ rad} = 17.1 \text{ rad}\left(\frac{1 \text{ rev}}{2\pi \text{ rad}}\right) = 2.73 \text{ rev} \qquad \textit{Ans.}$$

This problem has also been solved in Example 17–9. Compare the two methods of solution and note that since force, velocity, and displacement θ are involved, a work-energy approach yields a more direct solution.

Example 18–3

The uniform 5-kg bar shown in Fig. 18–13a is acted upon by the 30-N force which always acts perpendicular to the bar as shown. If the bar has an initial clockwise angular velocity $\omega_1 = 10$ rad/s when $\theta = 0°$, determine its angular velocity at the instant $\theta = 90°$.

30 N

θ

0.6 m

A

(a)

Fig. 18–13

Solution

Kinetic Energy (Kinematic Diagrams). Two kinematic diagrams of the bar when $\theta = 0°$ (position 1) and $\theta = 90°$ (position 2) are shown in Fig. 18–13b. The initial kinetic energy may be computed with reference to either the fixed point of rotation A or the center of mass G. If A is considered, then

$$T_1 = \tfrac{1}{2}I_A\omega_1^2 = \frac{1}{2}\left[\frac{1}{3}(5)(0.6)^2\right](10)^2 = 30 \text{ J}$$

If point G is considered, then

$$T_1 = \tfrac{1}{2}m(v_G)_1^2 + \tfrac{1}{2}I_G\omega^2 = \frac{1}{2}(5)(3)^2 + \frac{1}{2}\left[\frac{1}{12}(5)(0.6)^2\right](10)^2 = 30 \text{ J}$$

In the final position

$$T_2 = \frac{1}{2}\left[\frac{1}{3}(5)(0.6)^2\right]\omega_2^2 = 0.3\omega_2^2$$

G $(v_G)_1 = 3$ m/s

0.3 m $\omega_1 = 10$ rad/s

A ω_2 G

$(v_G)_2$

(b)

Work (Free-Body Diagram) Fig. 18–13c. The reactions A_x and A_y do no work, since these forces do not move. The 49.05-N weight, centered at G, moves downward through a vertical distance $\Delta y = 0.3$ m. The 30-N force moves tangent to its path of length $\tfrac{1}{2}\pi(0.6)$ m.

30 N

θ

G

A_x

49.05 N

A_y

(c)

Principle of Work and Energy

$$\{T_1\} + \{\Sigma U_{1-2}\} = \{T_2\}$$
$$\{30\} + \{49.05(0.3) + 30[\tfrac{1}{2}\pi(0.6)]\} = \{0.3\omega_2^2\}$$
$$\omega_2 = 15.6 \text{ rad/s} \downarrow \qquad\qquad Ans.$$

Example 18–4

The wheel shown in Fig. 18–14a weighs 40 lb and has a radius of gyration $k_G = 0.6$ ft about its mass center G. If it is subjected to a clockwise couple of 15 lb · ft and rolls from rest without slipping, determine its angular velocity after its center G moves 0.5 ft. The spring has a stiffness $k = 10$ lb/ft and is initially unstretched when the couple is applied.

Fig. 18–14

(a)

(b)

Solution

Kinetic Energy (Kinematic Diagram). Since the wheel is initially at rest,

$$T_1 = 0$$

The kinematic diagram of the wheel when it is in the final position is shown in Fig. 18–14b. The velocity of the mass center $(\mathbf{v}_G)_2$ can be related to the angular velocity ω_2 by using the instantaneous center of zero velocity (IC), i.e., $(v_G)_2 = 0.8\omega_2$. Hence, the final kinetic energy is

$$T_2 = \tfrac{1}{2}m(v_G)_2^2 + \tfrac{1}{2}I_G(\omega_2)^2$$

$$= \frac{1}{2}\left(\frac{40}{32.2}\right)(0.8\omega_2)^2 + \frac{1}{2}\left[\frac{40}{32.2}(0.6)^2\right](\omega_2)^2$$

$$= 0.621(\omega_2)^2$$

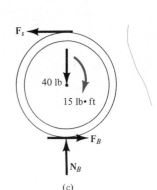

(c)

Work (Free-Body Diagram). As shown in Fig. 18–14c, only the spring force $\mathbf{F}_s$ and the couple do work. The normal force does not move along its line of action and the frictional force does *no work*, since the wheel does not slip as it rolls.

The work of $\mathbf{F}_s$ may be computed using $U_s = -\tfrac{1}{2}ks^2$. (Why is the work negative?) Since the wheel does not slip when the center G moves 0.5 ft, the wheel rotates $\theta = s_G/r_{G/IC} = 0.5/0.8 = 0.625$ rad, Fig. 18–14b. Hence, the spring stretches $s_A = \theta\, r_{A/IC} = 0.625(1.6) = 1$ ft.

Principle of Work and Energy

$$\{T_1\} + \{\Sigma U_{1-2}\} = \{T_2\}$$

$$\{T_1\} + \{M\theta - \tfrac{1}{2}ks^2\} = \{T_2\}$$

$$\{0\} + \{15(0.625) - \frac{1}{2}(10)(1)^2\} = \{0.621(\omega_2)^2\}$$

$$\omega_2 = 2.65 \text{ rad/s} \downarrow \qquad Ans.$$

Example 18–5

The 10-kg rod shown in Fig. 18–15a is constrained so that its ends move along the grooved slots. The rod is initially at rest when $\theta = 0°$. If the slider block at B is acted upon by a horizontal force $P = 50$ N, determine the angular velocity of the rod at the instant $\theta = 45°$. Neglect the mass of blocks A and B. (Why can the principle of work and energy be used to solve this problem?)

(a)

Solution

Kinetic Energy (Kinematic Diagrams). Two kinematic diagrams of the rod, when it is in the initial position 1 and final position 2, are shown in Fig. 18–15b. When the rod is in position 1, $T_1 = 0$ since $(\mathbf{v}_G)_1 = \boldsymbol{\omega}_1 = \mathbf{0}$. In position 2 the angular velocity is $\boldsymbol{\omega}_2$ and the velocity of the mass center is $(\mathbf{v}_G)_2$. Hence, the kinetic energy is

$$T_2 = \tfrac{1}{2}m(v_G)_2^2 + \tfrac{1}{2}I_G(\omega_2)^2$$
$$= \frac{1}{2}(10)(v_G)_2^2 + \frac{1}{2}\left[\frac{1}{12}(10)(0.8)^2\right](\omega_2)^2$$
$$= 5(v_G)_2^2 + 0.267(\omega_2)^2 \qquad (1)$$

The two unknowns $(v_G)_2$ and ω_2 may be related via the instantaneous center of zero velocity for the rod, Fig. 18–15b. It is seen that as A moves downward with a velocity $(\mathbf{v}_A)_2$, B moves horizontally to the left with a velocity $(\mathbf{v}_B)_2$. Knowing these directions, the IC may be determined as shown in the figure. Hence,

$$(v_G)_2 = r_{G/IC}\omega_2 = (0.4\tan 45°)\omega_2$$
$$= 0.4\omega_2$$

Substituting into Eq. (1), we have

$$T_2 = 5(0.4\omega_2)^2 + 0.267(\omega_2)^2 = 1.067(\omega_2)^2$$

(b)

Work (Free-Body Diagram). Fig. 18–15c. The normal forces $\mathbf{N}_A$ and $\mathbf{N}_B$ do no work as the rod is displaced. Why? The 98.1-N weight is displaced a vertical distance of $\Delta y = (0.4 - 0.4\cos 45°)$ m; whereas the 50-N force moves a horizontal distance of $s = (0.8 \sin 45°)$ m.

Principle of Work and Energy

$$\{T_1\} + \{\Sigma U_{1-2}\} = \{T_2\}$$
$$\{T_1\} + \{W(\Delta y) + Ps\} = \{T_2\}$$
$$\{0\} + \{98.1(0.4 - 0.4\cos 45°) + 50(0.8\sin 45°)\} = \{1.067(\omega_2)^2\}$$

Solving for ω_2 gives

$$\omega_2 = 6.11 \text{ rad/s} \downarrow \qquad\qquad Ans.$$

(c)

Fig. 18–15

Problems

Except when stated otherwise, throughout this chapter, assume that the coefficients of static and kinetic friction are equal, i.e., $\mu = \mu_s = \mu_k$.

18–1. At a given instant the body of mass m has an angular velocity $\boldsymbol{\omega}$ and its mass center has a velocity $\mathbf{v}_G$. Show that its kinetic energy can be represented as $T = \frac{1}{2}I_{IC}\omega^2$, where I_{IC} is the moment of inertia of the body computed about the instantaneous axis of zero velocity, located at a distance $r_{G/IC}$ from the mass center as shown.

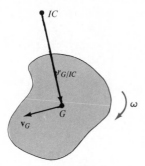

Prob. 18–1

18–2. A yo-yo has a weight of 0.3 lb and a radius of gyration $k_O = 0.06$ ft. If it is released from rest, determine how far it must descend in order to attain an angular velocity $\omega = 70$ rad/s. Neglect the mass of the string and assume that the string is wound around the central peg such that the mean radius at which it unravels is $r = 0.02$ ft.

Prob. 18–2

18–3. The tub of the mixer has a weight of 70 lb and a radius of gyration $k_G = 1.3$ ft about its center of gravity. If a constant torque of 60 lb · ft is applied to the dumping wheel, determine the angular velocity of the tub when it has turned to $\theta = 90°$. Originally the tub is at rest when $\theta = 0°$.

Prob. 18–3

*18–4. The revolving door consists of four doors which are attached to an axle AB. Each door can be assumed to be a 50-lb thin plate. Friction at the axle contributes a moment of 2 lb · ft which resists the rotation of the doors. If a man passes through one door by always pushing with a force $P = 15$ lb perpendicular to the plane of the door as shown, determine the door's angular velocity after it has rotated 90°. The doors are originally at rest.

Prob. 18–4

373

18–5. The pulley has a mass of 20 kg and a radius of gyration $k_O = 350$ mm. A motor supplies a torque $M = (150 + 20\theta)$ N · m, where θ is in radians, about the drive shaft at O. Determine the speed of the 15-kg crate after it rises $s = 3$ m starting from rest. Neglect the mass of the attached cable.

400 mm

M

O

Prob. 18–5

18–6. The cylinder has a mass of 80 kg. A horizontal force $P = 750$ N is applied to a cable wrapped around its outer surface. If the cylinder is originally at rest, determine its angular velocity after the center of the cylinder O has moved $s = 2$ m. The cylinder rolls without slipping. Neglect the mass of the cable.

$P = 750$ N

O

0.75 m

A

Prob. 18–6

18–7. The 50-kg uniform door is closed by applying a constant handle force of 20 N, which is always directed perpendicular to the plane of the door. If the door is originally at rest and open at $\theta = 90°$, determine the door's angular velocity just before it closes ($\theta = 0°$). Neglect friction at the hinges. For the calculation, assume the door to be approximated by a thin plate.

z

θ

1 m

0.1 m

2 m

$P = 20$ N

y

x

Prob. 18–7

*18–8. The soap-box car has a weight of 110 lb, including the passenger but *excluding* its four wheels. Each wheel has a weight of 5 lb and a radius of gyration $k = 0.3$ ft, computed about an axis passing through the wheel's axle. Determine the car's speed after it has traveled 100 ft starting from rest. The wheels roll without slipping. Neglect air resistance.

0.5 ft

0.5 ft

B

A

100 ft

30°

Prob. 18–8

18–9. The 80-kg wheel has a radius of gyration $k_O = 125$ mm and is rotating at 20 rad/s when the vertical force $P = 200$ N is applied to the brake handle. If the coefficient of friction between the brake at B and the wheel is $\mu = 0.35$, determine the number of revolutions the wheel makes before it stops. What are the horizontal and vertical components of reaction at A while the wheel is stopping?

Prob. 18–9

18–10. The gear has a weight of 15 lb and a radius of gyration $k_G = 0.375$ ft. If the spring is unstretched when the torque $M = 6$ lb · ft is applied, determine the gear's angular velocity after its mass center G has moved to the left $s = 2$ ft.

Prob. 18–10

18–11. The 100-lb block is transported a short distance by using two cylindrical rollers, each having a weight of 35 lb. If a horizontal force $P = 25$ lb is applied to the block, determine the block's speed after it has been displaced $s = 2$ ft. Originally the block is at rest. No slipping occurs.

Prob. 18–11

*18–12. The linkage consists of two 8-lb rods AB and CD and a 10-lb rod AD. When $\theta = 0°$, rod AB is rotating with an angular velocity $\omega_{AB} = 2$ rad/s. If at this instant rod CD is subjected to a couple moment $M = 15$ lb · ft and rod AD is subjected to a horizontal force $P = 20$ lb as shown, determine ω_{AB} at the instant $\theta = 90°$.

Prob. 18–12

18–13. The wheel has a mass of 20 kg and a radius of gyration $k_A = 0.13$ m. The attached spring is unstretched and is restricted from stretching more than 300 mm due to the chain. If the wheel is originally at rest and then subjected to a torque $M = 40$ N · m, determine its angular velocity just before the chain becomes taut. Assume no slipping at B. Neglect the mass of the chain.

Prob. 18–13

18–14. The two 2-kg gears A and B are attached to the ends of a 3-kg slender bar. The gears roll within the fixed ring gear C, which lies in the horizontal plane. If a 10-N · m torque is applied to the center of the bar as shown, determine the number of revolutions the bar must rotate starting from rest in order for it to have an angular velocity of $\omega_{AB} = 20$ rad/s. For the calculation, assume the gears can be approximated by thin disks. What is the result if the gears lie in the vertical plane?

Prob. 18–14

18–15. The gear has a mass of 2 kg and a radius of gyration $k_A = 0.15$ m. The connecting link (slender rod) and slider block have a mass of 4 kg and 1 kg, respectively. If the gear has an angular velocity $\omega = 8$ rad/s at the instant $\theta = 45°$, determine the gear's angular velocity when $\theta = 0°$. Neglect friction.

Prob. 18–15

*18–16. A man having a weight of 180 lb sits in a chair of the Ferris wheel, which has a weight of 15,000 lb and a radius of gyration $k_O = 37$ ft excluding the man. If a torque $M = 80(10^3)$ lb · ft is applied about O, determine the angular velocity of the wheel after it has rotated 180°. Neglect the weight of the chairs and note that the man remains in an upright position as the wheel rotates. The wheel starts from rest in the position shown.

Prob. 18–16

18–17. A man having a weight of 150 lb crouches down on the end of a diving board as shown. In this position the radius of gyration about his center of gravity is $k_G = 1.2$ ft. While holding this position at $\theta = 0°$, he rotates about his toes at A until he loses contact with the board when $\theta = 90°$. If he remains rigid, determine approximately how many revolutions he then makes before striking the water after falling 30 ft.

Prob. 18–17

18–18. The 10-kg rod AB is pin-connected at A and subjected to a couple $M = 15$ N · m. If the rod is released from rest when the spring is unstretched, at $\theta = 30°$, determine the rod's angular velocity at the instant $\theta = 60°$. As the rod rotates, the spring always remains horizontal, because of the roller support at C.

k = 40 N/m

0.75 m

$M = 15$ N · m

Prob. 18–18

18–19. A vertical force $P = 1.5$ kN is used to lift the 20-kg wheel and the 50-kg crate B. Assuming that the rope does not slip over the wheel, determine the speed of B after the center O of the wheel rises 2 m starting from rest. The radius of gyration of the wheel is $k_O = 125$ mm. Neglect the mass of the rope.

$P = 1.5$ kN

$r = 150$ mm

Prob. 18–19

*18–20. The 50-lb crate C is lifted using the hoist shown. A motor delivers 50 lb · ft of torque to the pinion gear D, which is in mesh with the larger gear A and the attached drum over which the cable winds. If A has a weight of 120 lb and a radius of gyration $k_A = 0.6$ ft, determine the velocity of the crate when it has risen 5 ft starting from rest. Neglect the mass of the pinion gear and the cable and pulley at B.

$M = 50$ lb · ft

1 ft

1 ft

1.25 ft

0.25 ft

B

C

Probs. 18–20 / 18–21

18–21. Solve Prob. 18–20 assuming that pulley B is a disk having a weight of 20 lb.

377

18.5 Conservation of Energy

When a force system acting on a rigid body consists only of *conservative forces*, the conservation of energy theorem may be used to solve a problem which otherwise would be solved using the principle of work and energy. This theorem is often easier to apply since the work of a conservative force is *independent of the path* and only depends upon the initial and final positions of the body. It was shown in Sec. 14.5 that the work of a conservative force may be expressed as the difference in the body's potential energy measured from an arbitrarily selected reference or datum.

Gravitational Potential Energy. Since the total weight of a body can be considered concentrated at its center of gravity, the *gravitational potential energy* of the body is determined by knowing the height of the body's center of gravity above or below a horizontal datum. Measuring y_G as *positive upward*, the gravitational potential energy of the body is thus

$$V_g = W\, y_G \tag{18–11}$$

Here the potential energy is *positive* when y_G is positive, since the weight has the ability to do *positive work* when the body is moved back to the datum, Fig. 18–16. Likewise, if the body is located *below* the datum ($-y_G$), the gravitational potential energy is *negative*, since the weight does *negative work* when the body is moved back to the datum.

Gravitational potential energy

Fig. 18–16

Elastic Potential Energy. The force developed by an elastic spring is also a conservative force. The *elastic potential energy* which a spring imparts to an attached body when the spring is elongated or compressed from an initial unstretched position ($s = 0$) to a final position s, Fig. 18–17, is

Elastic potential energy

Fig. 18–17

$$V_e = +\tfrac{1}{2}ks^2 \qquad\qquad (18\text{–}12)$$

In the deformed position, the spring force acting *on the body* always has the capacity for doing positive work when the spring is returned back to its original undeformed position (see Sec. 14.5).

In general, if a body is subjected to both gravitational and elastic forces, the total *potential energy* is expressed as a potential function V represented as the algebraic sum

$$V = V_g + V_e \qquad\qquad (18\text{–}13)$$

Here measurement of V depends upon the location of the body with respect to a selected datum in accordance with Eqs. 18–11 and 18–12.

Conservation of Energy. Realizing that the work of conservative forces can be written as a difference in their potential energies, i.e., $(\Sigma U_{1-2})_{\text{cons}} = V_1 - V_2$, Eq. 14–21, we can rewrite the principle of work and energy for a rigid body as

$$T_1 + V_1 + (\Sigma U_{1-2})_{\text{noncons}} = T_2 + V_2 \qquad\qquad (18\text{–}14)$$

Here $(\Sigma U_{1-2})_{\text{noncons}}$ represents the work of the nonconservative forces, such as friction, acting on the body. If this term is zero, then

$$T_1 + V_1 = T_2 + V_2 \qquad\qquad (18\text{–}15)$$

This is the equation of the conservation of mechanical energy for the body. It states that the *sum* of the potential and kinetic energies of the body remains *constant* when the body moves from one position to another. It also applies to a system of smooth, pin-connected rigid bodies, bodies connected by inextensible cords, and bodies in mesh with other bodies. In all these cases the forces acting at the points of contact are *eliminated* from the analysis, since they occur in equal and opposite collinear pairs and each pair of forces moves through an equal distance when the system undergoes a displacement.

PROCEDURE FOR ANALYSIS

The conservation of energy equation is used to solve problems involving *velocity, displacement,* and *conservative force systems.* For application it is suggested that the following procedure be used.

Potential Energy. Draw two diagrams showing the body located at its initial and final positions along the path. If the center of mass of the body is subjected to a *vertical displacement,* determine where to establish the fixed horizontal datum from which to measure the body's gravitational potential energy V_g. Although the location of the datum is arbitrary, it is advantageous to place it through G when the body is either at the initial or final point of its path, since at the datum $V_g = 0$. Data pertaining to the elevation y of the body's mass center from the datum and the extension or compression of any connecting springs can be determined from the geometry associated with the two diagrams. Recall that the potential energy $V = V_g + V_e$, where $V_g = \pm Wy$ and $V_e = \frac{1}{2}ks^2$.

Kinetic Energy. The kinetic-energy terms T_1 and T_2 are determined from $T = \frac{1}{2}mv_G^2 + \frac{1}{2}I_G\omega^2$ or an appropriate form of this equation as developed in Sec. 18.1. In this regard, kinematic diagrams for velocity may be useful for determining v_G and ω, or for establishing a *relationship* between these quantities.

Conservation of Energy. Apply the conservation of energy equation $T_1 + V_1 = T_2 + V_2$.

It is important to remember that *only problems involving conservative force systems may be solved by using this equation.* As stated in Sec. 14.5, friction or other drag-resistant forces, which depend upon velocity or acceleration, are nonconservative. The work of such forces is transformed into thermal energy used to heat up the surfaces of contact, and consequently this energy is dissipated into the surroundings and may not be recovered. Therefore, problems involving frictional forces can either be solved by using the principle of work and energy written in the form of Eq. 18–14, if it applies, or the equations of motion.

The following example problems numerically illustrate application of the above procedure.

Example 18-6

The 10-kg rod AB shown in Fig. 18–18a is confined so that its ends move in the horizontal and vertical slots. The spring has a stiffness of $k = 800$ N/m and is unstretched when $\theta = 0°$. Determine the angular velocity of AB when $\theta = 0°$, if AB is released from rest when $\theta = 30°$. Neglect the mass of the slider blocks.

(a)

Solution

Potential Energy. The two diagrams of the rod, when it is located at its initial and final positions, are shown in Fig. 18–18b. The datum, used to measure the gravitational potential energy, is placed in line with the rod when $\theta = 0°$.

When the rod is in position 1, the center of mass G is located *below the datum* so that the gravitational potential energy is *negative*. Furthermore, (positive) elastic potential energy is stored in the spring, since it is stretched a distance of $s_1 = (0.4 \sin 30°)$ m. Thus,

$$V_1 = -Wy_1 + \tfrac{1}{2}ks_1^2$$

$$= -98.1(0.2 \sin 30°) + \frac{1}{2}(800)(0.4 \sin 30°)^2 = 6.19 \text{ J}$$

When the rod is in position 2, the potential energy of the rod is zero, since the spring is unstretched, $s_2 = 0$, and the center of mass G is located at the datum. Thus,

$$V_2 = 0$$

Kinetic Energy. The rod is released from rest from position 1, thus $(\mathbf{v}_G)_1 = \mathbf{0}$ and $\boldsymbol{\omega}_1 = \mathbf{0}$, and

$$T_1 = 0$$

In position 2 the angular velocity is $\boldsymbol{\omega}_2$ and the rod's mass center has a velocity of $(\mathbf{v}_G)_2$. Using *kinematics*, $(\mathbf{v}_G)_2$ can be related to $\boldsymbol{\omega}_2$ as shown in Fig. 18–18c. At the instant considered, the instantaneous center of zero velocity (IC) for the rod is at point A; hence, $(v_G)_2 = (r_{G/IC})\omega_2 = (0.2)\omega_2$. Thus,

$$T_2 = \tfrac{1}{2}m(v_G)_2^2 + \tfrac{1}{2}I_G(\omega_2)^2$$

$$= \frac{1}{2}(10)(0.2\omega_2)^2 + \frac{1}{2}\left[\frac{1}{12}(10)(0.4)^2\right](\omega_2)^2 = 0.267\omega_2^2$$

Conservation of Energy

$$\{T_1\} + \{V_1\} = \{T_2\} + \{V_2\}$$

$$\{0\} + \{6.19\} = \{0.267\omega_2^2\} + \{0\}$$

$$\omega_2 = 4.82 \text{ rad/s} \ \text{\Large ↰}$$

Ans.

Fig. 18–18

Example 18–7

The disk shown in Fig. 18–19a has a weight of 30 lb and is attached to a spring which has a stiffness $k = 2$ lb/ft and an unstretched length of 1 ft. If the disk is released from rest in the position shown and rolls without slipping, determine its angular velocity at the instant it is displaced 3 ft.

Solution

Potential Energy. Two diagrams of the disk, when it is located in its initial and final positions, are shown in Fig. 18–19b. A gravitational datum is not needed here since the weight is not displaced vertically. From the problem geometry the spring is stretched $s_1 = (\sqrt{3^2 + 4^2} - 1) = 4$ ft and $s_2 = (4 - 1) = 3$ ft in the initial and final positions, respectively. Hence,

$$V_1 = \tfrac{1}{2}ks_1^2 = \frac{1}{2}(2)(4)^2 = 16 \text{ J}$$

$$V_2 = \tfrac{1}{2}ks_2^2 = \frac{1}{2}(2)(3)^2 = 9 \text{ J}$$

Kinetic Energy. The disk is released from rest so that $(\mathbf{v}_G)_1 = \mathbf{0}$, $\boldsymbol{\omega}_1 = \mathbf{0}$, and

$$T_1 = 0$$

Since the disk rolls without slipping, $(\mathbf{v}_G)_2$ can be related to $\boldsymbol{\omega}_2$ from the instantaneous center of zero velocity, Fig. 18–19c. Hence, $(v_G)_2 = 0.75\omega_2$. Thus,

$$T_2 = \tfrac{1}{2}m(v_G)_2^2 + \tfrac{1}{2}I_G(\omega_2)^2$$

$$= \frac{1}{2}\left(\frac{30}{32.2}\right)(0.75\omega_2)^2 + \frac{1}{2}\left[\frac{1}{2}\left(\frac{30}{32.2}\right)(0.75)^2\right](\omega_2)^2$$

$$= 0.393(\omega_2)^2$$

Conservation of Energy

$$\{T_1\} + \{V_1\} = \{T_2\} + \{V_2\}$$
$$\{0\} + \{16\} = \{0.393(\omega_2)^2\} + \{9\}$$
$$\omega_2 = 4.22 \text{ rad/s} \;\curvearrowleft \qquad\qquad Ans.$$

(a)

(b)

(c)

Fig. 18–19

Example 18–8

The 10-kg homogeneous disk shown in Fig. 18–20a is attached to a uniform 5-kg rod AB. If the assembly is released from rest when $\theta = 60°$, determine the angular velocity of the rod when $\theta = 0°$. Assume that the disk rolls without slipping. Neglect friction along the guide and the mass of the collar at B.

0.6 m

0.1 m

(a)

Fig. 18–20

Solution

Potential Energy. Two diagrams for the rod and disk, when they are located at their initial and final positions, are shown in Fig. 18–20b. For convenience the datum, which is horizontally fixed, passes through point A.

When the system is in position 1, the rod's weight has positive potential energy. Thus,

$$V_1 = W_R y_1 = 49.05(0.3 \sin 60°) = 12.74 \text{ J}$$

When the system is in position 2, both the weight of the rod and the weight of the disk have zero potential energy. Why? Thus,

$$V_2 = 0$$

(b)

Kinetic Energy. Since the entire system is at rest in the initial position,

$$T_1 = 0$$

In the final position the rod has an angular velocity $(\omega_R)_2$ and its mass center has a velocity $(\mathbf{v}_G)_2$, Fig. 18–20c. Since the rod is *fully extended* in this position, the disk is momentarily at rest, so $(\omega_D)_2 = \mathbf{0}$ and $(\mathbf{v}_A)_2 = \mathbf{0}$. For the rod $(\mathbf{v}_G)_2$ can be related to $(\omega_R)_2$ from the instantaneous center of zero velocity, which is located at point A, Fig. 18–20c. Hence, $(v_G)_2 = r_{G/IC}(\omega_R)_2$ or $(v_G)_2 = 0.3(\omega_R)_2$. Thus,

$$T_2 = \tfrac{1}{2}m_R(v_G)_2^2 + \tfrac{1}{2}I_G(\omega_R)_2^2 + \tfrac{1}{2}m_D(v_A)_2^2 + \tfrac{1}{2}I_A(\omega_D)_2^2$$

$$= \frac{1}{2}(5)(0.3(\omega_R)_2)^2 + \frac{1}{2}\left[\frac{1}{12}(5)(0.6)^2\right](\omega_R)_2^2 + 0 + 0$$

$$= 0.3(\omega_R)_2^2$$

(c)

Conservation of Energy

$$\{T_1\} + \{V_1\} = \{T_2\} + \{V_2\}$$

$$\{0\} + \{12.74\} = \{0.3(\omega_R)_2^2\} + \{0\}$$

$$(\omega_R)_2 = 6.52 \text{ rad/s} \downarrow \qquad\qquad Ans.$$

Problems

18–22. Solve Prob. 18–2 using the conservation of energy equation.

18–23. Solve Prob. 18–15 using the conservation of energy equation.

*18–24. Solve Prob. 18–8 using the conservation of energy equation.

18–25. Solve Prob. 18–17 using the conservation of energy equation.

18–26. The link is rotating about point O in the vertical plane with an angular velocity $\omega = 5$ rad/s when $\theta = 0°$. If it has a mass of 20 kg, a mass center at G, and a radius of gyration $k_G = 300$ mm, determine its angular velocity at the instant $\theta = 90°$.

Prob. 18–26

18–27. The spool has a mass of 200 kg and a radius of gyration $k_O = 525$ mm. If a man having a mass of 70 kg grabs the end of the cable and steps off the ledge from an at-rest position, determine his velocity when he strikes the ground. Neglect the mass of the cable.

Prob. 18–27

*18–28. An automobile tire has a mass of 7 kg and radius of gyration $k_G = 0.3$ m. If it is released from rest at A on the incline, determine its angular velocity when it reaches the horizontal plane. The tire rolls without slipping.

Prob. 18–28

18–29. A girl having a weight of 110 lb is sitting on a swing such that when she is at A, $\theta = 30°$, her velocity is equal to zero and her center of gravity is at G. If she keeps this same fixed (rigid) position as she swings forward, determine the angular velocity of the swing when she reaches the lowest point at B ($\theta = 0°$). Her radius of gyration about an axis passing through G is 1.8 ft.

Prob. 18–29

18–30. Pulley A and the attached drum B have a weight of 20 lb and a centroidal radius of gyration $k_B = 0.6$ ft. Determine the speed of the 20-lb crate C at the instant $s = 10$ ft. Initially, the crate is released from rest when $s = 5$ ft. The pulley at P "rolls" downward on the cord without slipping. For the calculation, neglect the mass of this pulley and the cord as it unravels from A and B.

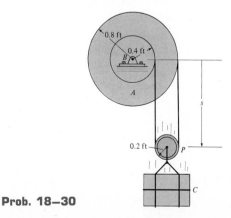

Prob. 18–30

18–31. The uniform 150-lb stone (rectangular block) is being turned over on its side by pulling the vertical cable *slowly* upward until the stone begins to tip. If it then falls freely ($T = 0$) from an essentially balanced at-rest position, determine the speed at which the corner A strikes the pad at B. The stone does not slip at its corner C as it falls.

Prob. 18–31

*****18–32.** The jeep and passenger have a total weight of 1600 lb, not including the four wheels. Each wheel has a weight of 20 lb and a radius of gyration $k_G = 0.4$ ft about an axis passing through its center. If the jeep starts from rest and coasts down the hill in neutral gear, determine the distance it must travel to attain a speed of 15 ft/s. The tires do not slip on the ground. Neglect the effects of friction and air resistance.

Prob. 18–32

18–33. The system consists of a 20-lb disk A, 4-lb slender rod BC, and a 1-lb smooth collar C. If the disk rolls without slipping, determine the velocity of the collar at the instant the rod becomes horizontal, i.e., $\theta = 0°$. The system is released from rest when $\theta = 45°$.

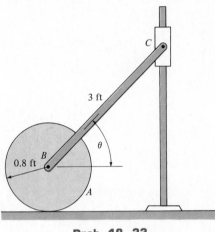

Prob. 18–33

18–34. The beam has a weight of 1500 lb and is being raised to a vertical position by pulling on its bottom end A. If the man lets go of the cord when $\theta = 60°$ and the beam is at rest, determine the speed of A at the instant cord BC becomes vertical. Neglect friction and the mass of the cords, and treat the beam as a slender rod.

Prob. 18–34

18–35. The window AB, shown in side view, has a mass of 10 kg and a radius of gyration $k_G = 0.2$ m. The attached elastic cable BC has a stiffness $k = 50$ N/m and an unstretched length of 0.4 m. If the window is released from rest when $\theta \approx 0°$, determine its angular velocity just before it strikes the wall at D, $\theta = 90°$.

Prob. 18–35

*18–36. The uniform window shade AB has a total weight of 0.4 lb. When it is released, it winds around the spring-loaded core O. Motion is caused by a spring within the core, which is coiled so that it exerts a torque $M = (0.3(10^{-3})\theta)$ lb · ft, where θ is in radians, on the core. If the shade is released from rest, determine the angular velocity of the core at the instant the shade is completely rolled up, i.e., after 12 revolutions. When this occurs, the spring becomes uncoiled and the radius of gyration of the shade about the axle at O is $k_O = 0.9$ in. *Note:* The elastic potential energy of a torsional spring is $V_e = \frac{1}{2}k\theta^2$, where $M = k\theta$. Here $k = 0.3(10^{-3})$ lb · ft/rad.

Prob. 18–36

18–37. The window AB, shown in side view, has a weight of 8 lb, a center of gravity at G, and a radius of gyration $k_A = 0.4$ ft about its hinged axis at A. If the spring is unstretched when $\theta \approx 0°$ and the window is released from rest from this position, determine the speed with which the end B strikes the wall D, $\theta = 90°$. Neglect friction at the hinge A and at pulley C.

Prob. 18–37

18–38. The pendulum consists of a 0.5-lb rod AB, a 2-lb rod CD, and a 6-lb disk. The spring is stretched 0.3 ft when rod CD is horizontal as shown. If the pendulum is released from rest, determine its angular velocity at the instant CD becomes vertical. The roller at F allows a segment of the spring to remain vertical as another portion wraps around the rim of the disk.

Prob. 18–38

18–39. The spring is stretched 0.5 ft when the 20-lb bar is in the horizontal position. If in this position it has an angular velocity $\omega = 4$ rad/s, determine the bar's angular velocity at the instant it swings down to the position $\theta = 90°$.

Prob. 18–39

*18–40. The spool at A has a mass of 12 kg and a radius of gyration $k_C = 225$ mm. Pulley B has a mass of 7 kg and may be considered as a disk. If the system is released from rest and the cord does not slip, determine the angular velocity of the pulley after the spool descends 0.5 m. Neglect the mass of the cord.

Prob. 18–40

18–41. The slender rod has a mass of 10 kg and is released from rest when $\theta = 0°$. Determine its angular velocity at the instant $\theta = 30°$. The attached block at D has a mass of 4 kg. The mass of the cable and pulley at C can be neglected.

Prob. 18–41

18–42. The garage door AB has a weight of 80 lb and can be approximated as a thin plate. During operation, a small roller at B is guided along the fixed horizontal track, while the two arms AC located on each side of the door are used to guide the bottom of the door along a circular arc having a center at D and a radius of 4 ft. If the end A of the door is given a horizontal velocity of 3 ft/s to the left and the door has a clockwise angular velocity $\omega_{AB} = 2$ rad/s when $\theta = 0°$ and ADC is vertical, determine the velocity of A at the instant the door becomes fully open, i.e., $\theta = 90°$. Each of the two springs has a stiffness of 50 lb/ft and an unstretched length of 1 ft. Neglect the mass of the arms ADC.

Prob. 18–42

Planar Kinetics of a Rigid Body: Impulse and Momentum

In Chapter 15 it was shown that problems which involve force, velocity, and time can conveniently be solved by using the principle of linear or angular impulse and momentum. In this chapter we will extend these concepts somewhat in order to determine the impulse and momentum relationships that apply to a rigid body. The three planar motions which will be considered are translation, rotation about a fixed axis, and general plane motion. The more general discussion of momentum principles applied to the three-dimensional motion of a rigid body is presented in Chapter 21.

Linear and Angular Momentum   19.1

In this section we will develop formulations for the linear and angular momentum of a rigid body which is symmetrical with respect to an inertial x-y reference plane. As a result, we will represent the body as a slab having the same mass as the body and moving in the reference plane.

Linear Momentum. The linear momentum of a rigid body is determined by summing vectorially the linear momenta of all the particles of the body, i.e., $\mathbf{L} = \Sigma m_i \mathbf{v}_i$. We can simplify this expression by noting that $\Sigma m_i \mathbf{v}_i = m \mathbf{v}_G$ (see Sec. 15.2) so that

$$\mathbf{L} = m\mathbf{v}_G \qquad (19\text{–}1)$$

This equation states that the body's linear momentum is a vector quantity having a *magnitude* mv_G, which is commonly measured in units of kg · m/s or slug · ft/s, and a *direction* defined by $\mathbf{v}_G$, the velocity of the mass center.

Particle momentum
diagram

(a)

Body momentum
diagram

(b)

Fig. 19–1

Angular Momentum. Consider the body in Fig. 19–1*a*, which is subjected to general plane motion. At the instant shown, the arbitrary point P has a velocity $\mathbf{v}_P$ and the body has an angular velocity $\boldsymbol{\omega}$. If the velocity of the ith particle of the body is to be determined, Fig. 19–1*a*, then

$$\mathbf{v}_i = \mathbf{v}_P + \mathbf{v}_{i/P}$$

The angular momentum of particle i about point P is equal to the "moment" of the particle's linear momentum about P, Fig. 19–1*a*. Thus,

$$d\mathbf{H}_P = \mathbf{r} \times dm\,\mathbf{v}_i$$

Expressing $\mathbf{v}_i$ in terms of $\mathbf{v}_P$ and using Cartesian vectors, we have

$$dH_P\,\mathbf{k} = dm(x\mathbf{i} + y\mathbf{j}) \times [(v_P)_x\mathbf{i} + (v_P)_y\mathbf{j} + \omega\mathbf{k} \times (x\mathbf{i} + y\mathbf{j})]$$
$$dH_P = -dm\,y(v_P)_x + dm\,x(v_P)_y + dm\,\omega r^2$$

Integrating over the entire mass m of the body, we obtain

$$\zeta + H_P = -\left(\int_m y\,dm\right)(v_P)_x + \left(\int_m x\,dm\right)(v_P)_y + \left(\int_m r^2\,dm\right)\omega$$

Here H_P represents the angular momentum of the body about an axis (the z axis) perpendicular to the plane of motion and passing through point P. Since $\bar{x}m = \int x\,dm$ and $\bar{y}m = \int y\,dm$, the integrals for the first and second terms on the right are used to locate the body's center of mass G with respect to P. Also, the last integral represents the body's moment of inertia computed about the z axis, i.e., $I_P = \int r^2\,dm$. Thus,

$$\zeta + H_P = -\bar{y}m(v_P)_x + \bar{x}m(v_P)_y + I_P\omega \qquad (19–2)$$

This equation reduces to a simpler form if point P coincides with the mass center G for the body,* in which case $\bar{x} = \bar{y} = 0$ and therefore

$$H_G = I_G\omega$$

where I_G is the mass moment of inertia of the body about an axis which is perpendicular to the plane of motion and passes through the mass center, and $\boldsymbol{\omega}$ is the instantaneous angular velocity of the body. It is important to realize that $\mathbf{H}_G$ is a vector quantity having a *magnitude* $I_G\omega$, which is commonly measured in units of kg · m²/s or slug · ft²/s, and a *direction* defined by $\boldsymbol{\omega}$, which is always perpendicular to the slab. Since $\boldsymbol{\omega}$ is a free vector, $\mathbf{H}_G$ *can act at any point on the body* (or slab) provided it preserves its same magnitude and direction. Furthermore, since angular momentum is equal to the "moment" of the linear momentum, the *line of action of* $\mathbf{L}$ *must pass through the body's mass center* G in order to preserve the correct magnitude of $\mathbf{H}_G$ when "moments" are computed about G, Fig. 19–1*b*.† In other words, $\mathbf{L}$ creates *zero* angular momentum about G, so that simply $\mathbf{H}_G = I_G\boldsymbol{\omega}.$

*Note that it *also* reduces to the same simple form, $H_P = I_P\omega$, if point P is a *fixed point* (see Eq. 19–6) or the velocity of point P is directed along the line PG.
†A formalized proof of this statement is given as Prob. 19–3.

Translation. When a rigid body of mass m is subjected to either rectilinear or curvilinear *translation,* Fig. 19–2a, its mass center has a velocity of $\mathbf{v}_G = \mathbf{v}$ and $\boldsymbol{\omega} = \mathbf{0}$. Hence, the linear momentum and the angular momentum, computed about G, become

$$\mathbf{L} = m\mathbf{v}_G$$
$$\mathbf{H}_G = \mathbf{0} \qquad\qquad (19\text{--}3)$$

If the angular momentum is computed about any other point A on or off the body, Fig. 19–2a, the "moment" of the linear momentum $\mathbf{L}$ must be computed about the point. Since d is the "moment arm" as shown in the figure, then $\zeta + H_A = (d)(mv_G)$.

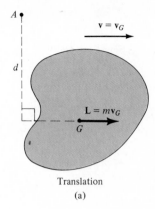
Translation
(a)

Rotation about a Fixed Axis. When a rigid body is *rotating about a fixed axis* passing through point O, Fig. 19–2b, the linear momentum and the angular momentum computed about G are

$$L = mv_G$$
$$H_G = I_G\omega \qquad\qquad (19\text{--}4)$$

It is sometimes convenient to compute the angular momentum of the body about point O. In this case it is necessary to account for the "moments" of *both* $\mathbf{L}$ and $\mathbf{H}_G$ about O. Noting that $\mathbf{L}$ (or $\mathbf{v}_G$) is always *perpendicular* to $\mathbf{r}_{G/O}$, we have

$$\zeta + H_O = I_G\omega + r_{G/O}(mv_G) \qquad\qquad (19\text{--}5)$$

This equation may be *simplified* by first substituting $v_G = r_{G/O}\omega$, in which case $H_O = (I_G + mr_{G/O}^2)\omega$, and, by the parallel-axis theorem, noting that the terms inside the parentheses represent the moment of inertia I_O of the body about an axis perpendicular to the plane of motion and passing through point O. Hence,*

$$H_O = I_O\omega \qquad\qquad (19\text{--}6)$$

For the computation, then, either Eq. 19–5 or 19–6 can be used.

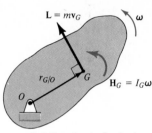
Rotation about a fixed axis
(b)

General Plane Motion. When a rigid body is subjected to general plane motion, Fig. 19–2c, the linear momentum and the angular momentum computed about G become

$$L = mv_G$$
$$H_G = I_G\omega \qquad\qquad (19\text{--}7)$$

General plane motion
(c)
Fig. 19–2

*The similarity between this derivation and that of Eq. 17–16 ($\Sigma M_O = I_O\alpha$) and Eq. 18–3 ($T = \frac{1}{2}I_O\omega^2$) should be noted. Also note that the same result can be obtained from Eq. 19–2 by selecting point P at O, realizing that $(v_O)_x = (v_O)_y = 0$.

If the angular momentum is computed about a point A located either on or off the body, Fig. 19–2c, it is necessary to compute the moments of *both* $\mathbf{L}$ and $\mathbf{H}_G$ about this point. In this case,

$$\zeta + H_A = I_G\omega + (d)(mv_G)$$

Here d is the moment arm, as shown in the figure.

19.2 Principle of Impulse and Momentum

Like the case for particle motion, the principle of impulse and momentum for a rigid body is developed by *combining* the equation of motion with kinematics. The resulting equation will allow a *direct solution to problems involving force, velocity, and time*.

Principle of Linear Impulse and Momentum. The equation of translational motion for a rigid body can be written as $\Sigma\mathbf{F} = m\mathbf{a}_G = m\,(d\mathbf{v}_G/dt)$. Since the mass of the body is constant,

$$\Sigma\mathbf{F} = \frac{d}{dt}(m\mathbf{v}_G)$$

Multiplying both sides by dt and integrating from $t = t_1$, $\mathbf{v}_G = (\mathbf{v}_G)_1$ to $t = t_2$, $\mathbf{v}_G = (\mathbf{v}_G)_2$ yields

$$\Sigma\int_{t_1}^{t_2} \mathbf{F}\,dt = m(\mathbf{v}_G)_2 - m(\mathbf{v}_G)_1 \tag{19–8}$$

This equation is referred to as the *principle of linear impulse and momentum*. It states that the sum of all the impulses created by the *external force system* which acts on the body during the time interval t_1 to t_2 is equal to the change in the linear momentum of the body during the time interval.

Principle of Angular Impulse and Momentum. If the body is subjected to *general plane motion* we can write $\Sigma M_G = I_G\alpha = I_G(d\omega/dt)$. Since the moment of inertia is constant,

$$\Sigma M_G = \frac{d}{dt}(I_G\omega)$$

Multiplying both sides by dt and integrating from $t = t_1$, $\omega = \omega_1$ to $t = t_2$, $\omega = \omega_2$ gives

$$\Sigma\int_{t_1}^{t_2} M_G\,dt = I_G\omega_2 - I_G\omega_1 \tag{19–9}$$

In a similar manner, for *rotation about a fixed axis* passing through point O, Eq. 17–16 ($\Sigma M_O = I_O\alpha$) when integrated becomes

$$\Sigma\int_{t_1}^{t_2} M_O\,dt = I_O\omega_2 - I_O\omega_1 \tag{19–10}$$

Equations 19–9 and 19–10 are referred to as the *principle of angular impulse and momentum*. Both equations state that the sum of the angular impulses acting on the body during the time interval t_1 to t_2 is equal to the change in the body's angular momentum during this time interval. In particular, the angular impulse considered is determined by integrating the moments about point G or O of all the external forces and couples applied to the body.

To summarize the preceding concepts, if motion is occurring in the x-y plane, using impulse and momentum principles the following *three scalar equations* may be written which describe the *planar motion* of the body:

$$m(v_{Gx})_1 + \Sigma \int_{t_1}^{t_2} F_x \, dt = m(v_{Gx})_2$$

$$m(v_{Gy})_1 + \Sigma \int_{t_1}^{t_2} F_y \, dt = m(v_{Gy})_2 \qquad (19\text{–}11)$$

$$I_G \omega_1 + \Sigma \int_{t_1}^{t_2} M_G \, dt = I_G \omega_2$$

The first two of these equations represent the principle of linear impulse and momentum in the x-y plane, Eq. 19–8, and the third equation represents the principle of angular impulse and momentum about the z axis, Eq. 19–9.

The terms in Eqs. 19–11 can be graphically accounted for by drawing a set of impulse and momentum diagrams for the body, Fig. 19–3. Note that the linear momenta $m\mathbf{v}_G$ are applied at the body's mass center, Fig. 19–3a and c; whereas the angular momenta $I_G\boldsymbol{\omega}$ are free vectors, and therefore, like a couple, they may be applied at any point on the body. When the impulse diagram is constructed, Fig. 19–3b, vectors $\mathbf{F}$ and $\mathbf{M}$ which vary with time are indicated by the integrals. However, if $\mathbf{F}$ and $\mathbf{M}$ are *constant* from t_1 to t_2, integration of the impulses yields $\mathbf{F}(t_2 - t_1)$ and $\mathbf{M}(t_2 - t_1)$, respectively. Such is the case for the body's weight $\mathbf{W}$, Fig. 19–3b.

Equations 19–11 may also be applied to an entire system of connected bodies rather than to each body separately. Doing this eliminates the need to include reactive impulses which occur at the connections since they are *internal* to the system. The resultant equations may be written in symbolic form as

$$\left(\Sigma \frac{\text{syst. linear}}{\text{momentum}} \right)_{x1} + \left(\Sigma \frac{\text{syst. linear}}{\text{impulse}} \right)_{x(1-2)} = \left(\Sigma \frac{\text{syst. linear}}{\text{momentum}} \right)_{x2}$$

$$\left(\Sigma \frac{\text{syst. linear}}{\text{momentum}} \right)_{y1} + \left(\Sigma \frac{\text{syst. linear}}{\text{impulse}} \right)_{y(1-2)} = \left(\Sigma \frac{\text{syst. linear}}{\text{momentum}} \right)_{y2}$$

$$\left(\Sigma \frac{\text{syst. angular}}{\text{momentum}} \right)_{O1} + \left(\Sigma \frac{\text{syst. angular}}{\text{impulse}} \right)_{O(1-2)} = \left(\Sigma \frac{\text{syst. angular}}{\text{momentum}} \right)_{O2}$$

$$(19\text{–}12)$$

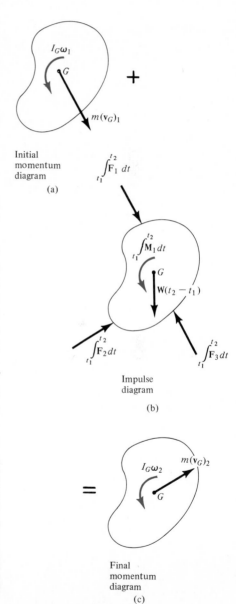

Initial momentum diagram

(a)

Impulse diagram

(b)

Final momentum diagram

(c)

Fig. 19–3

As indicated, the system's angular momentum and angular impulse must be computed with respect to the *same fixed reference point O* for all the bodies of the system.

PROCEDURE FOR ANALYSIS

Impulse and momentum principles are used to solve kinetics problems that involve *velocity, force,* and *time* since these terms are involved in the formulation. For application it is suggested that the following procedure be used.

Free-Body Diagram. Establish the x, y, z inertial frame of reference and draw the free-body diagram in order to account for all the forces that produce impulses on the body. The direction and sense of the initial and final velocity of the body's mass center, $\mathbf{v}_G$, and the body's angular velocity $\boldsymbol{\omega}$ should also be established. These vectors may be sketched near, but not on, the free-body diagram. Also compute the moment of inertia I_G or I_O.

As an alternative procedure draw the impulse and momentum diagrams for the body or system of bodies. Each of these diagrams represents an outlined shape of the body which graphically accounts for the data required for each of the three terms in Eqs. 19–11 or 19–12, Fig. 19–3. Note that these diagrams are particularly helpful in order to visualize the "moment" terms used in the principle of angular impulse and momentum, when it has been decided that application of this equation is to be about a point other than the body's mass center G or a fixed point O.

Principle of Impulse and Momentum. Apply the three scalar equations 19–11 (or 19–12) with respect to an established x, y inertial coordinate system. In cases where the body is rotating about a fixed axis, Eq. 19–10 may be substituted for the third of Eqs. 19–11.

Kinematics. If more than three equations are needed for a complete solution, it may be possible to relate the velocity of the body's mass center to the body's angular velocity using *kinematics*. If the motion appears to be complicated, kinematic (velocity) diagrams may be helpful in obtaining the necessary relation.

In general, a method to be used for the solution of a particular type of problem should be decided upon *before* attempting to solve the problem. As stated above, the *principle of impulse and momentum* is most suitable for solving problems which involve *velocity, force,* and *time.* For some problems, however, a combination of the equation of motion and its two integrated forms, the principle of work and energy and the principle of impulse and momentum, will yield the most direct solution to the problem.

Example 19–1

The disk shown in Fig. 19–4a weighs 20 lb and is pin-supported at its center. If it is acted upon by a constant couple of 4 lb · ft and a force of 10 lb which is applied to a cord wrapped around its periphery, determine the angular velocity of the disk two seconds after starting from rest. What are the force components of reaction at pin A?

Fig. 19–4

(a)

(b)

Solution

Free-Body Diagram (Fig. 19–4b). Note that all the forces acting on the disk are constant since the force and couple causing the motion are constant.

The moment of inertia of the disk about its fixed axis of rotation is

$$I_A = \tfrac{1}{2}mr^2 = \frac{1}{2}\left(\frac{20}{32.2}\right)(0.75)^2 = 0.175 \text{ slug} \cdot \text{ft}^2$$

Principle of Impulse and Momentum

$(\xrightarrow{+})$

$$m(v_{Ax})_1 + \Sigma \int_{t_1}^{t_2} F_x \, dt = m(v_{Ax})_2$$

$$0 + A_x(2) = 0$$

$(+\uparrow)$

$$m(v_{Ay})_1 + \Sigma \int_{t_1}^{t_2} F_y \, dt = m(v_{Ay})_2$$

$$0 + A_y(2) - 20(2) - 10(2) = 0$$

$(\curvearrowright +)$

$$I_A\omega_1 + \Sigma \int_{t_1}^{t_2} M_A \, dt = I_A\omega_2$$

$$0 + 4(2) + [10(2)](0.75) = 0.175\omega_2$$

Solving these equations yields

$$A_x = 0 \qquad\qquad\qquad \textit{Ans.}$$
$$A_y = 30 \text{ lb} \uparrow \qquad\qquad \textit{Ans.}$$
$$\omega_2 = 132 \text{ rad/s} \downarrow \qquad\qquad \textit{Ans.}$$

Example 19–2

The block shown in Fig. 19–5a has a mass of 6 kg. It is attached to a cord which is wrapped around the periphery of a 20-kg disk that has a moment of inertia $I_A = 0.40$ kg $\cdot$ m². If the block is initially moving downward with a speed of 2 m/s, determine its speed in 3 s. Neglect the mass of the cord in the calculation.

(a)   **Fig. 19–5a, b**   (b)

Solution I

Free-Body Diagram. The free-body diagrams of the block and disk are shown in Fig. 19–5b. All the forces are constant since the weight of the block causes the motion.

Principle of Impulse and Momentum. We can eliminate $\mathbf{A}_x$ and $\mathbf{A}_y$ from the analysis by applying the principle of angular momentum about point A. Hence for the block and disk we have

$(+\uparrow)$
$$m_B(v_B)_1 + \Sigma \int F_y \, dt = m_B(v_B)_2$$

$$-6(2) + T(3) - 58.86(3) = -6(v_B)_2 \qquad (1)$$

$(\curvearrowleft+)$
$$I_A\omega_1 + \Sigma \int M_A \, dt = I_A\omega_2$$

$$0.40(\omega_1) + T(3)(0.2) = 0.4\omega_2 \qquad (2)$$

Kinematics. Since $\omega = v_B/r$, then $\omega_1 = 2/0.2 = 10$ rad/s and $\omega_2 = (v_B)_2/0.2 = 5(v_B)_2$. Substituting and solving for $(v_B)_2$ yields

$$(v_B)_2 = 13.0 \text{ m/s} \downarrow \qquad\qquad Ans.$$

Solution II

Impulse and Momentum Diagrams. We can eliminate solving for the cable tension T by considering the *system* consisting of the block, the cable, and the disk. The impulse and momentum diagrams have been drawn to clarify application of the principle of angular momentum about point A, Fig. 19–5c.

Principle of Angular Impulse and Momentum. Realizing that $\omega_1 = 10$ rad/s and $\omega_2 = 5(v_B)_2$, we have

$$\left(\sum \frac{\text{syst. angular}}{\text{momentum}}\right)_{A1} + \left(\sum \frac{\text{syst. angular}}{\text{impulse}}\right)_{A(1-2)} = \left(\sum \frac{\text{syst. angular}}{\text{momentum}}\right)_{A2}$$

$$(\curvearrowright +) \quad 6(2)(0.2) + 0.4(10) + 58.86(3)(0.2) = 6(v_B)_2(0.2) + 0.4(5(v_B)_2)$$

$$(v_B)_2 = 13.0 \text{ m/s} \downarrow \qquad\qquad Ans.$$

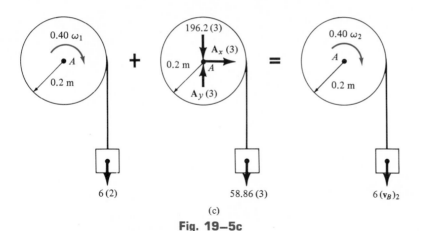

(c)

Fig. 19–5c

Example 19–3

The 100-kg spool shown in Fig. 19–6a has a radius of gyration $k_G = 0.35$ m. A cable is wrapped around the central hub of the spool and a variable horizontal force having a magnitude of $P = (t + 10)$ N is applied, where t is measured in seconds. If the spool is initially at rest, determine its angular velocity in 5 s. Assume that the spool rolls without slipping at A.

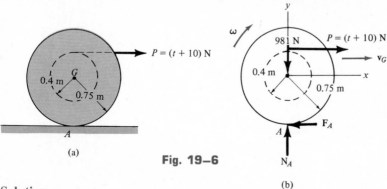

(a)

Fig. 19–6

(b)

Solution

Free-Body Diagram. By inspection of the free-body diagram, the *variable* force **P** will cause the friction force $\mathbf{F}_A$ to be variable, and thus the impulses created by both **P** and $\mathbf{F}_A$ must be determined by integration.

The moment of inertia of the spool about its mass center is

$$I_G = mk_G^2 = 100(0.35)^2 = 12.25 \text{ kg} \cdot \text{m}^2$$

Principle of Impulse and Momentum

$(\overset{+}{\rightarrow})$
$$m(v_G)_1 + \Sigma \int F_x \, dt = m(v_G)_2$$

$$0 + \int_0^5 (t + 10) \, dt - \int F_A \, dt = 100(v_G)_2$$

$(\curvearrowleft +)$
$$I_G\omega_1 + \Sigma \int M_G \, dt = I_G\omega_2$$

$$0 + \left[\int_0^5 (t + 10) \, dt \right](0.4) + \left(\int F_A \, dt \right)(0.75) = 100(0.35)^2\omega_2$$

Kinematics. Since the spool does not slip, the instantaneous center of zero velocity is at point A. Hence, the velocity of G can be expressed in terms of the spool's angular velocity as $(v_G)_2 = 0.75\omega_2$, Fig. 19–6b. Substituting this, evaluating the integral for the impulse caused by **P** (i.e., 62.5 N · s), and eliminating the unknown impulse $\int F_A \, dt$, we obtain

$$\omega_2 = 1.05 \text{ rad/s} \downarrow \qquad \textit{Ans.}$$

Example 19–4

The Charpy impact test is used in materials testing to determine the energy absorption characteristics of a material during impact. The test is performed using the pendulum shown in Fig. 19–7a that has a mass m, mass center at G, and a radius of gyration k_G about G. Determine the distance r_P from the pin at A to the point P where the impact with the specimen S should occur so that the horizontal force at the pin is essentially zero during the impact. For the computation, assume the specimen absorbs all the pendulum's kinetic energy during the time it falls and thereby stops the pendulum from swinging at $\theta = 0°$.

Solution

Free-Body Diagram. As shown on the free-body diagram, the conditions of the problem require the horizontal impulse at A to be zero. Just before impact, we assume the pendulum has an angular velocity ω_1 and the mass center of the pendulum is moving at $(v_G)_1 = \bar{r}\omega_1$.

Principle of Impulse and Momentum. We will apply the principle of angular impulse and momentum about point A. Thus,

$(\circlearrowleft +)$ $\qquad I_A\omega_1 - \left(\int F\,dt\right)r_P = 0$

$(\xrightarrow{+})$ $\qquad -m(\bar{r}\omega_1) + \int F\,dt = 0$

Eliminating the impulse $\int F\,dt$ and substituting $I_A = mk_G^2 + m\bar{r}^2$ yields

$$[mk_G^2 + m\bar{r}^2]\omega_1 - m(\bar{r}\omega_1)r_P = 0$$

Factoring out $m\omega_1$ and solving for r_P yields

$$r_P = \bar{r} + \frac{k_G^2}{\bar{r}} \qquad\qquad Ans.$$

The point P, so defined, is called the *center of percussion*. By placing the striking point at P, the force developed at the pin will be minimized. Many sports rackets, clubs, etc. are designed so that the object being struck occurs at the center of percussion. As a consequence, no "sting" or little sensation occurs in the hand of the player. (Also see Probs. 17–61 and 19–4.)

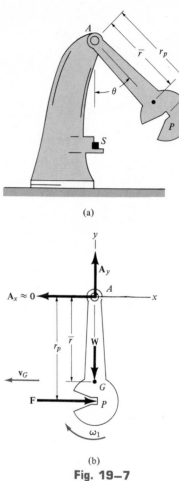

(a)

(b)

Fig. 19–7

Problems

Except when stated otherwise, throughout this chapter, assume that the coefficients of static and kinetic friction are equal, i.e., $\mu = \mu_s = \mu_k$.

19-1. Show that if a slab is rotating about a fixed axis perpendicular to the slab and passing through its mass center G, the angular momentum is the same when computed about any other point P on the slab.

Prob. 19-1

19-2. At a given instant, the body has a linear momentum $\mathbf{L} = m\mathbf{v}_G$ and an angular momentum $\mathbf{H}_G = I_G\boldsymbol{\omega}$ computed about its mass center. Show that the angular momentum of the body computed about the instantaneous center of zero velocity IC can be expressed as $\mathbf{H}_{IC} = I_{IC}\boldsymbol{\omega}$, where I_{IC} represents the body's moment of inertia computed about the instantaneous axis of zero velocity. As shown, the IC is located at a distance of $r_{G/IC}$ away from the mass center G.

Prob. 19-2

19-3. Show that the line of action of the body's linear momentum $\mathbf{L}$ passes through the mass center of the body as indicated in Fig. 19-1b. *Hint:* Use Eq. 19-2 and parallel the development used to derive Eq. 17-9 from Eq. 17-6.

*19-4. The rigid body (slab) has a mass m and is rotating with an angular velocity $\boldsymbol{\omega}$ about an axis passing through the fixed point O. Show that the momenta of all the particles composing the body can be represented by a single vector having a magnitude mv_G and acting through point P, called the *center of percussion*, which lies at a distance $r_{P/G} = k_G^2/r_{G/O}$ from the mass center G. Here k_G is the radius of gyration of the body, computed about an axis perpendicular to the plane of motion and passing through G.

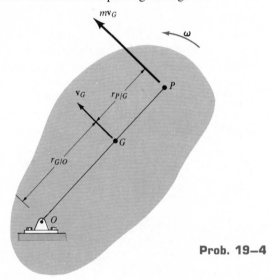

Prob. 19-4

19-5. The pulley has a weight of 8 lb and may be treated as a thin disk. A belt wrapped over its surface is subjected to forces $T_A = 4$ lb and $T_B = 5$ lb. Determine the angular velocity of the pulley when $t = 4$ s if it starts from rest when $t = 0$. Neglect the mass of the belt.

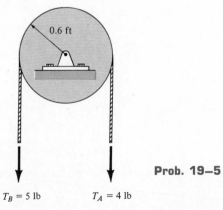

Prob. 19-5

$T_B = 5$ lb $T_A = 4$ lb

19–6. A sphere and cylinder are released from rest on the ramp at $t = 0$. If each has a mass m and a radius r, determine their angular velocities at time t. Assume no slipping occurs.

Prob. 19–6

19–7. The 4-kg slender rod rests on a smooth floor. If it is kicked so as to receive a horizontal impulse $I = 8 \, \text{N} \cdot \text{s}$ at point A as shown, determine its angular velocity and the velocity of the mass center.

Prob. 19–7

***19–8.** The spacecraft has a mass of 1200 kg and a moment of inertia $I_G = 900 \, \text{kg} \cdot \text{m}^2$ about an axis passing through G and directed perpendicular to the page. If it is traveling forward with a speed $v_G = 800$ m/s and executes a turn by means of two jets, which provide a constant thrust of 400 N for 0.3 s, determine the spacecraft's angular velocity just after the jets are turned off.

Prob. 19–8

19–9. The double pulley consists of two wheels which are attached to one another and turn at the same rate. The pulley has a mass of 30 kg and a radius of gyration $k_O = 250$ mm. If two men A and B grab the suspended ropes and step off the ledges at the same time, determine their speeds in 4 s starting from rest. The men A and B have a mass of 60 kg and 70 kg, respectively. Assume they do not move relative to the rope during the motion. Neglect the mass of the rope.

Prob. 19–9

19–10. Spool B is at rest and spool A is rotating at 6 rad/s when the slack in the cord connecting them is taken up. Determine the angular velocity of each spool immediately after the cord is jerked tight by the spinning of spool A. A and B have weights and radii of gyration $W_A = 30$ lb, $k_A = 0.8$ ft and $W_B = 15$ lb, $k_B = 0.6$ ft, respectively.

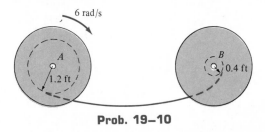

Prob. 19–10

401

19–11. The hoop (thin ring) has a mass of 5 kg and is released down the inclined plane such that it has a backspin $\omega = 8$ rad/s and its center has a velocity $v_G = 3$ m/s as shown. If the coefficient of friction between the hoop and the plane is $\mu_k = 0.6$, determine how long the hoop rolls before it stops slipping.

Prob. 19–11

*19–12. The space shuttle is located in "deep space," where the effects of gravity can be neglected. It has a mass of 120 Mg, a center of mass at G, and a radius of gyration $k_G = 14$ m about the x axis. It is originally traveling straight at $v = 3$ km/s when the pilot turns on the engine at A, creating a thrust $T = 600(1 - e^{-0.3t})$ kN, where t is in seconds. Determine the shuttle's angular velocity 2 s later.

Prob. 19–12

19–13. The drum of mass m, radius r, and radius of gyration k_O rolls along an inclined plane for which the coefficient of friction is μ. If the drum is released from rest, determine the maximum angle θ for the incline so that it rolls without slipping at A.

Prob. 19–13

19–14. The double pulley consists of two wheels which are attached to one another and turn at the same rate. The pulley has a mass of 15 kg and a radius of gyration of $k_O = 110$ mm. If the block at A has a mass of 40 kg, determine the speed of the block in 3 s after a constant force of 2 kN is applied to the rope wrapped around the inner hub of the pulley. The block is originally at rest. Neglect the mass of the rope.

Prob. 19–14

19–15. The spool has a weight of 30 lb and a radius of gyration $k_O = 0.45$ ft. A cord is wrapped around its inner hub and its end subjected to a horizontal force $P = 5$ lb. Determine the spool's angular velocity in 4 s starting from rest. Assume the spool rolls without slipping.

0.9 ft

O

$P = 5$ lb

0.3 ft

A

Probs. 19–15 / 19–16

***19–16.** Work Prob. 19–15 if the coefficient of friction is $\mu_A = 0.3$.

19–17. A cord is wrapped around the rim of each 10-lb disk. If disk B is released from rest, determine the angular velocity of disk A in 2 s. Neglect the mass of the cord.

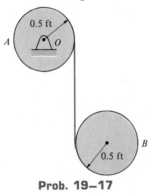

0.5 ft

A

O

B

0.5 ft

Prob. 19–17

19–18. If the ball has a weight W and radius r and is thrown onto a *rough surface* with a velocity $\mathbf{v}_O$ parallel to the surface, determine the amount of backspin, ω, it must be given so that it stops spinning at the same instant that its forward velocity is zero. It is not necessary to know the coefficient of friction at A for the calculation.

ω

$\mathbf{v}_O$

r

A

Prob. 19–18

19–19. The flywheel A has a mass of 60 kg and a radius of gyration $k_C = 160$ mm. Gear B has a mass of 25 kg and a radius of gyration $k_D = 125$ mm. If a motor supplies a clockwise torque having a magnitude $M = (1.2t)$ N · m, where t is in seconds, to the flywheel, determine the angular velocity of the gear in 3 s. Initially the flywheel is rotating at $\omega_1 = 2$ rad/s.

$\omega_1 = 2$ rad/s

$M = 1.2t$

C

200 mm

150 mm

D

A

B

Prob. 19–19

***19–20.** The inner hub of the wheel rests on the horizontal track. If it does not slip at A, determine the speed of the 10-lb block when $t = 2$ s after it is released from rest. The wheel has a weight of 30 lb and a radius of gyration $k_G = 1.3$ ft. Neglect the mass of the pulley and cord.

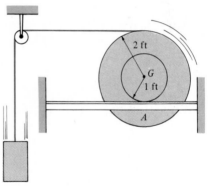

2 ft

G

1 ft

A

Prob. 19–20

19–21. A torque $M = (0.05e^{0.5t})$ N · m, where t is in seconds, is applied to gear A. If gear A is originally rotating at 5 rad/s, determine its angular velocity when $t = 2$ s. The two smaller *equivalent* gears B and C are pinned at their centers. The mass and centroidal radii of gyration of the gears about axes perpendicular to the page and passing through their centers are given in the figure.

$(\omega_A)_1 = 5$ rad/s

$m' = 0.3$ kg
$k' = 28$ mm

M

120 mm

40 mm

$m = 0.8$ kg
$k = 80$ mm

Prob. 19–21

19–22. The slender rod has a mass m and is suspended at its end A by a cord. If the rod receives a horizontal blow giving it an impulse $\mathbf{I}$ at its bottom B, determine the location y of the point P about which the rod appears to rotate during the impact.

A

P

l

y

$\mathbf{I}$ B

Prob. 19–22

19–23. The two rods each have a mass m, a length l, and lie on the smooth horizontal plane. If an impulse $\mathbf{I}$ is applied at an angle of 45° to one of the rods at mid-length as shown, determine the angular velocity of each rod just after the impulse. The rods are pin-connected at B.

l $l/2$ $l/2$

A B 45° C

$\mathbf{I}$

Prob. 19–23

Conservation of Momentum 19.3

Conservation of Linear Momentum. If the sum of all the *linear impulses* acting on a system of connected rigid bodies is *zero,* the linear momentum of the system is constant or conserved. Consequently, the first two of Eqs. 19–12 reduce to the form

$$\left(\sum \frac{\text{syst. linear}}{\text{momentum}}\right)_1 = \left(\sum \frac{\text{syst. linear}}{\text{momentum}}\right)_2 \qquad (19\text{--}13)$$

This equation is referred to as the *conservation of linear momentum.*

Without inducing appreciable errors in the computations, it may be possible to apply Eq. 19–13 in a specified direction for which the linear impulses are small or *nonimpulsive.* Specifically, nonimpulsive forces occur when small forces act over very short periods of time. For example, the impulse created by the force of a tennis racket hitting a ball during a very short time interval Δt is large, whereas the impulse of the weight of the ball during this time is small by comparison and may therefore be neglected in the motion analysis of the ball during Δt.

Conservation of Angular Momentum. The angular momentum of a system of connected rigid bodies is conserved about the system's center of mass G, or a fixed point O, when the sum of all the angular impulses created by the external forces acting on the system is zero or appreciably small (nonimpulsive) when computed about these points. The third of Eqs. 19–12 then becomes

$$\left(\sum \frac{\text{syst. angular}}{\text{momentum}}\right)_{O1} = \left(\sum \frac{\text{syst. angular}}{\text{momentum}}\right)_{O2} \qquad (19\text{--}14)$$

This equation is referred to as the *conservation of angular momentum.* In the case of a single rigid body, Eq. 19–14 applied to point G becomes $(I_G\omega)_1 = (I_G\omega)_2$. To illustrate an application of this equation, consider a swimmer who executes a somersault after jumping off a diving board. By tucking his arms and legs in close to his chest, he *decreases* his body's moment of inertia and thus *increases* his angular velocity ($I_G\omega$ must be constant). If he straightens out just before entering the water, his body's moment of inertia is *increased* and his angular velocity *decreases*. Since the weight of his body creates a linear impulse during the time of motion, this example also illustrates that the angular momentum of a body is conserved and yet the linear momentum is *not*. Such cases occur whenever the external forces creating the linear impulse pass through either the center of mass of the body or a fixed axis of rotation.

PROCEDURE FOR ANALYSIS

Provided the initial velocity of the body is known, the conservation of linear or angular momentum is used to determine the final velocity of a body *just after* the time period considered. Furthermore, by applying these equations to a *system* of bodies, the internal impulses acting within the system, which may be unknown, are eliminated from the analysis, since they occur in equal but opposite collinear pairs. For application it is suggested that the following procedure be used.

Free-Body Diagram. Establish the x, y, z inertial frame of reference and draw the free-body diagram for the body or system of bodies during the time of impact. From this diagram classify each of the applied forces as being either "impulsive" or "nonimpulsive." In general, for short times, "nonimpulsive forces" consist of the weight of a body, the force of a slightly deformed spring, or any force that is *known to be small* when compared to the impulsive force. It will also be possible to tell if the conservation of linear or angular momentum can be applied. Specifically, the *conservation of linear momentum* applies in a given direction when *no* external impulsive forces act on the body or system in that direction; whereas the *conservation of angular momentum* applies about a fixed point O or at the mass center G of a body or system of bodies when all the external impulsive forces acting on the body or system create zero moment (or zero angular impulse) about O or G.

As an alternative procedure, consider drawing the impulse and momentum diagrams for the body or system of bodies. These diagrams are particularly helpful in order to visualize the "moment" terms used in the conservation of angular momentum equation, when it has been decided that angular momenta are to be computed about a point other than the body's mass center G.

Conservation of Momentum. Apply the conservation of linear or angular momentum in the appropriate directions.

Kinematics. Use *kinematics* if further equations are necessary for the solution of a problem. If the motion appears to be complicated, kinematic (velocity) diagrams may be helpful in obtaining the necessary kinematic relations.

If it is necessary to determine an *internal impulsive force* acting on only one body of a system, the body must be *isolated* (free-body diagram) and the principle of linear or angular impulse and momentum must be applied *to the body.* After the impulse $\int F\, dt$ is calculated, then, provided the time Δt for which the impulse acts is known, the *average impulsive force* F_{avg} can be determined from $F_{\text{avg}} = \int F\, dt / \Delta t$.

The following examples numerically illustrate application of this procedure.

Example 19-5

The 10-kg wheel shown in Fig. 19-8a has a moment of inertia $I_G = 0.156$ kg $\cdot$ m². Assuming that the wheel does not slip or rebound, determine the minimum velocity $\mathbf{v}_G$ it must have to just roll over the obstruction at A.

(a)

Solution

Free-Body Diagram. Since no slipping or rebounding occurs, the wheel essentially *pivots* about point A during contact. Note that only two forces (impulses) act on the wheel, Fig. 19-8b. By comparison, the force at A is much greater than that caused by the weight ($W = 10(9.81) = 98.1$ N), and since the time of impact is very short, the weight can be considered nonimpulsive. The impulsive force $\mathbf{F}$ at A has both an unknown magnitude and an unknown direction θ. To eliminate this force from the analysis, note that during impact *only* the nonimpulsive weight force creates an *angular impulse about point A*. Furthermore, since $(98.1\Delta t)d \approx 0$, angular momentum about A is essentially *conserved*. Since this is *not* a fixed point, the velocities of the mass center and the angular velocities of the body are shown in Fig. 19-8c.*

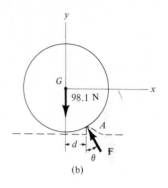

(b)

Conservation of Angular Momentum. With reference to Fig. 19-8c,

$$(\curvearrowright +) \qquad (\mathbf{H}_A)_1 = (\mathbf{H}_A)_2$$

$$r'm(v_G)_1 + I_G\omega_1 = rm(v_G)_2 + I_G\omega_2$$

$$(0.2 - 0.03)(10)(v_G)_1 + (0.156)(\omega_1) = (0.2)(10)(v_G)_2 + (0.156)(\omega_2)$$

Kinematics. Since no slipping occurs, $\omega = v_G/r = v_G/0.2 = 5v_G$. Substituting this into the above equation and simplifying yields

$$(v_G)_2 = 0.892(v_G)_1 \qquad (1)$$

(c)

Fig. 19-8

Conservation of Energy. In order to roll over the obstruction, the wheel must pass the dashed position 3 shown in Fig. 19-8d. Hence, if $(v_G)_2$ (or $(v_G)_1$) is to be a minimum, it is necessary that the kinetic energy of the wheel at position 2 be equal to the potential energy at position 3. Constructing the datum through the center of mass, as shown in the figure, and applying the conservation of energy equation, we have

$$\{T_2\} + \{V_2\} = \{T_3\} + \{V_3\}$$

$$\left\{ \frac{1}{2}(10)(v_G)_2^2 + \frac{1}{2}(0.156)(\omega_2)^2 \right\} + \{0\} = \{0\} + \{(98.1)(0.03)\}$$

Substituting $\omega_2 = 5(v_G)_2$ and Eq. (1) into this equation, and simplifying, yields

$$3.98(v_G)_1^2 + 1.55(v_G)_1^2 = 2.94$$

Thus,

$$(v_G)_1 = 0.729 \text{ m/s} \rightarrow$$

(d)

*If preferred, the impulse and momentum diagrams can be drawn.

Example 19–6

The 5-kg slender rod shown in Fig. 19–9a is pinned at O and is initially at rest. If a 4-g bullet is fired into the rod with a velocity of 400 m/s, as shown in the figure, determine the angular velocity of the rod just after the bullet becomes embedded in it.

(a)

Solution

Impulse and Momentum Diagrams. The impulse which the bullet exerts on the rod can be eliminated from the analysis and the angular velocity of the rod just after impact can be determined by considering the bullet and rod as a single system. To clarify the principles involved, the impulse and momentum diagrams are shown in Fig. 19–9b. The momentum diagrams are drawn *just before and just after impact*. During impact, the bullet and rod exchange equal but opposite *internal impulses* at A. As shown on the impulse diagram, the impulses that are external to the system are due to the reactions at O and the weights of the bullet and rod. Since the time of impact, Δt, is very short, the rod moves only a slight amount and so the "moments" of these impulses about point O are essentially zero, and therefore angular momentum is conserved about this point.

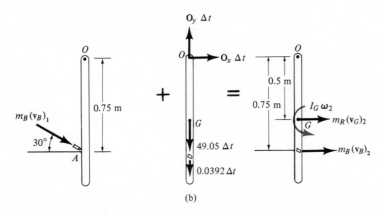

Fig. 19–9

(b)

Conservation of Angular Momentum. From Fig. 19–9b, we have

$$(\zeta+) \qquad \Sigma(H_O)_1 = \Sigma(H_O)_2$$

$$m_B(v_B)_1 \cos 30°(0.75 \text{ m}) = m_B(v_B)_2(0.75 \text{ m}) + m_R(v_G)_2(0.5 \text{ m}) + I_G\omega_2$$

$$(0.004)(400 \cos 30°)(0.75) = (0.004)(v_B)_2(0.75) + (5)(v_G)_2(0.5) + \left[\frac{1}{12}(5)(1)^2\right]\omega_2$$

or

$$1.039 = 0.003(v_B)_2 + 2.50(v_G)_2 + 0.417\omega_2 \qquad (1)$$

Kinematics. Since the rod is pinned at O, from Fig. 19–9c we have

$$(v_G)_2 = 0.5\omega_2 \qquad (v_B)_2 = 0.75\omega_2$$

Substituting into Eq. (1) and solving yields

$$\omega_2 = 0.622 \text{ rad/s} \uparrow \qquad\qquad \textit{Ans.}$$

(c)

Eccentric Impact **19.4**

Impact occurs when two bodies collide with each other during a very *short* interval of time. During the collision very large equal but opposite impulsive forces are created between the two bodies at their point of contact. The concepts involving central and oblique impact of particles have been presented in Sec. 15.4. We will expand this treatment and discuss the eccentric impact of two bodies. *Eccentric impact* occurs when the line connecting the mass centers of the two bodies does not coincide with the line of impact. This type of impact often occurs when one or both of the bodies is constrained to rotate about a fixed axis. Consider, for example, the collision between the two bodies A and B, shown in Fig. 19–10a, which collide at point C. Body B rotates about an axis passing through point O, whereas body A both rotates and translates. It is assumed that just before collision body B is rotating counterclockwise with an angular velocity $(\boldsymbol{\omega}_B)_1$, and the velocity of the contact point C located on body A is $(\mathbf{u}_A)_1$. Kinematic diagrams for both bodies just before collision are shown in Fig. 19–10b. Provided the bodies are smooth at point C, the *impulsive forces* they exert on each other *are directed along the line of impact*. Hence, the component of velocity of point C on body B, which is directed along the line of impact, is $(v_B)_1 = (\omega_B)_1 r$, Fig. 19–10b. Likewise, on body A the component of velocity $(\mathbf{u}_A)_1$ along the line of impact is $(\mathbf{v}_A)_1$. In order for a collision to occur, $(v_A)_1 > (v_B)_1$.

Line of impact

Plane of impact

(a)

$(\omega_B)_1$

r

C

$(v_B)_1 = (\omega_B)_1 r$

B

$(\mathbf{u}_A)_1$

$(\mathbf{v}_A)_1$

C

A

Velocity
before collision

(b)

Fig. 19–10a, b

Velocity at maximum
deformation

(d)

Deformation
impulse

(c)

Fig. 19–10c–f

Restitution impulse

(e)

Velocity
after collision

(f)

During the impact an equal but opposite impulsive force **P** is exerted between the bodies which *deforms* their shapes at the point of contact. The resulting impulse is shown on the impulse-vector diagrams for both bodies, Fig. 19–10c. Note that the impulsive force created at point C, on the rotating body B, creates impulsive pin reactions at the supporting pin O. On these diagrams it is assumed the impact creates forces which are much larger than the weights of the bodies. Hence, the impulses created by the weights are not shown since they are negligible compared to $\int \mathbf{P}\, dt$ and the reactive impulses at O. When the deformation of point C is a maximum, C on both the bodies moves with a common velocity **v** along the line of impact, Fig. 19–10d. A period of *restitution* then occurs in which the bodies tend to regain their original shapes. The restitution phase creates an equal but opposite impulsive force **R** acting between the bodies as shown on the impulse-vector diagram, Fig. 19–10e. After restitution the bodies move apart such that point C on body B has a velocity $(\mathbf{v}_B)_2$ and point C on body A has a velocity $(\mathbf{u}_A)_2$, Fig. 19–10f, where $(v_B)_2 > (v_A)_2$.

In general, a problem involving the impact of two bodies requires determining the *two unknowns* $(v_A)_2$ and $(v_B)_2$, assuming $(v_A)_1$ and $(v_B)_1$ are known (or can be determined using kinematics, energy methods, Newton's second law of motion, etc.) To solve this problem two equations must be written. The *first equation* generally involves application of *the conservation of angular momentum to the two bodies*. In the case of bodies A and B, we can state that angular momentum is conserved about point O since the impulses at C are internal to the system and the impulses at O create zero moment (or zero angular impulse) about point O. The *second equation* is obtained using the definition of the *coefficient of restitution, e,* which is a ratio of the restitution impulse to the deformation impulse. To establish a useful form of this equation we must first apply the principle of angular impulse and momentum about point O to bodies B and A separately. Combining the results, we then obtain the necessary equation. Proceeding in this manner, the principle of impulse and momentum applied to body B just before the collision to the time of maximum deformation, Fig. 19–10b, c, and d, becomes

$$(\downarrow +) \qquad\qquad I_O(\omega_B)_1 + r \int P \, dt = I_O\omega \qquad\qquad (19\text{--}15)$$

Here I_O is the moment of inertia of body B about point O. Similarly, applying the principle of angular impulse and momentum from the instant of maximum deformation to the time just after the impact, Fig. 19–10d, e and f, yields

$$(\downarrow +) \qquad\qquad I_O\omega + r \int R \, dt = I_O(\omega_B)_2 \qquad\qquad (19\text{--}16)$$

Solving Eqs. 19–15 and 19–16 for $\int P \, dt$ and $\int R \, dt$, respectively, and formulating e, we have

$$e = \frac{\displaystyle\int R \, dt}{\displaystyle\int P \, dt} = \frac{r(\omega_B)_2 - r\omega}{r\omega - r(\omega_B)_1} = \frac{(v_B)_2 - v}{v - (v_B)_1}$$

In the same manner, we may write an equation which relates the magnitudes of velocities $(v_A)_1$ and $(v_A)_2$ of body A. The result is

$$e = \frac{(v_A)_2 - v}{v - (v_A)_1}$$

Combining the above equations by eliminating the common velocity v yields the desired result, i.e.,

$$e = \frac{(v_B)_2 - (v_A)_2}{(v_A)_1 - (v_B)_1} \qquad\qquad (19\text{--}17)$$

This equation is identical to Eq. 15–11, which was derived for the central impact occurring between two particles. Equation 19–17, however, states that the coefficient of restitution, measured at the points of contact (C) between two colliding bodies, is equal to the ratio of the relative velocity of *separation* of the points *just after impact* to the relative velocity at which the points *approach* one another *just* before impact. In deriving Eq. 19–17, we assumed that the points of contact for both bodies move to the right *both* before and after impact. If motion of any one of the contacting points occurs to the left, the velocity of this point is considered a negative quantity in Eq. 19–17.

As stated previously, when Eq. 19–17 is used in conjunction with the conservation of angular momentum for the bodies, it provides a useful means of obtaining the velocities of two colliding bodies just after collision.

Example 19–7

The 10–lb slender rod is suspended from the pin at A, Fig. 19–11a. If a 2-lb ball B is thrown at the rod and strikes its center with a horizontal velocity of 30 ft/s, determine the angular velocity of the rod just after impact. The coefficient of restitution is $e = 0.4$.

Fig. 19–11

(a)

Solution

Conservation of Angular Momentum.   Consider the ball and rod as a system, Fig. 19–11b. Angular momentum is conserved about point A since the impulsive force between the rod and ball is *internal*. Also, the *weights* of the ball and rod are *nonimpulsive*. Using the kinematic data shown in Fig. 19–11c, we require

$$(\downarrow+) \qquad (H_A)_1 = (H_A)_2$$

$$m_B(v_B)_1(1.5) = m_B(v_B)_2(1.5) + m_R(v_G)_2(1.5) + I_G\omega_2$$

$$\left(\frac{2}{32.2}\right)(30)(1.5) = \left(\frac{2}{32.2}\right)(v_B)_2(1.5) + \left(\frac{10}{32.2}\right)(v_G)_2(1.5) + \left(\frac{1}{12}\frac{10}{32.2}(3)^2\right)\omega_2$$

Since $(v_G)_2 = 1.5\omega_2$ then

$$2.795 = 0.09317(v_B)_2 + 0.9317\omega_2 \qquad (1)$$

Coefficient of Restitution.   With reference to Fig. 19–11c, we have

$$e = \frac{(v_G)_2 - (v_B)_2}{(v_B)_1 - (v_G)_1}; \qquad 0.4 = \frac{(1.5\omega_2) - (v_B)_2}{30 - 0}$$

$$12.0 = 1.5\omega_2 - (v_B)_2$$

Solving,

$$(v_B)_2 = -6.52 \text{ ft/s} = 6.52 \text{ ft/s} \leftarrow$$

$$\omega = 3.65 \text{ rad/s} \; \text{\rotatebox{0}{$\curvearrowleft$}}$$

Ans.

(b)

(c)

Problems

*19–24. The turntable of a record player has a mass of 0.75 kg and a radius of gyration $k_z = 125$ mm. It is *turning freely* at $\omega_T = 2$ rad/s when a 50-g record (thin disk) falls on it. Determine the final angular velocity of the turntable just after the record stops slipping on the turntable.

Prob. 19–24

19–25. The man sits on the swivel chair holding two 5-lb weights with his arms outstretched. If he is turning at 3 rad/s in this position, determine his angular velocity when the weights are drawn in 0.3 ft from the axis of rotation. Assume he weighs 160 lb and has a radius of gyration $k_z = 0.55$ ft about the z axis. Neglect the mass of his arms and the size of the weights for the calculation.

Prob. 19–25

19–26. The 5-lb rod AB supports the 3-lb disk at its end. If the disk is given an angular velocity $\omega_D = 8$ rad/s while the rod is held stationary and then released, determine the angular velocity of the rod after the disk has stopped spinning relative to the rod due to frictional resistance at the bearing A. Motion is in the *horizontal plane*. Neglect friction at the bearing B.

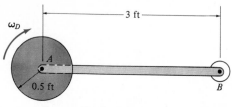

Prob. 19–26

19–27. Two children A and B each having a mass of 30 kg sit at the edge of the merry-go-round which is rotating at $\omega = 2$ rad/s. Excluding the children, the merry-go-round has a mass of 180 kg and a radius of gyration $k_z = 0.6$ m. Determine the angular velocity of the merry-go-round if A jumps off horizontally in the $-n$ direction with a speed of 2 m/s measured with respect to the merry-go-round. What is the merry-go-round's angular velocity if B then jumps off horizontally in the $+t$ direction with a speed of 2 m/s measured with respect to the merry-go-round? Assume each child to be a particle. Neglect friction.

Prob. 19–27

413

*19–28. A horizontal circular platform has a weight of 300 lb and a radius of gyration $k_z = 8$ ft about the z axis passing through its center O. The platform is free to rotate about the z axis and is initially at rest. A man having a weight of 150 lb begins to run along the edge in a circular path of radius 10 ft. If he has a speed of 4 ft/s and maintains this speed relative to the platform, compute the angular velocity of the platform. Neglect friction.

Prob. 19–28

19–29. The space satellite has a mass of 125 kg and a moment of inertia $I_z = 0.940$ kg · m², excluding the four solar panels A, B, C, and D. Each solar panel has a mass of 20 kg and can be approximated as a thin plate. If the satellite is originally spinning about the z axis at a constant rate $\omega_z = 0.5$ rad/s when $\theta = 90°$, determine the rate of spin if all the panels are raised and reach the upward position, $\theta = 0°$, at the same instant.

Prob. 19–29

19–30. For safety reasons, the 20-kg supporting leg of a sign is designed to break away with negligible resistance at B when the leg is subjected to the impact of a car. Assuming that the leg is pin-supported at A and approximates a thin rod, determine the impulse the car bumper exerts on it, if after the impact the leg appears to rotate upward to an angle $\theta_{max} = 150°$ before stopping.

Prob. 19–30

19–31. A thin rod having a mass of 4 kg is balanced vertically as shown. Determine the height h at which it can be struck with a horizontal force $\mathbf{F}$ and not slip on the floor. For the solution this requires that the frictional force at A be essentially zero.

Prob. 19–31

*19–32. A man has a moment of inertia I_z about the z axis. He is originally at rest and standing on a small platform which can turn freely. If he is handed a wheel which is rotating at $\boldsymbol{\omega}$ and has a moment of inertia I about its spinning axis, determine his angular velocity if (a) he holds the wheel upright as shown; (b) turns the wheel out, $\theta = 90°$; and (c) turns the wheel downward, $\theta = 180°$.

Prob. 19–32

19–33. If the man in Prob. 19–32 is given the wheel when it is at *rest* and he starts it spinning with an angular velocity $\boldsymbol{\omega}$, determine his angular velocity if (a) he holds the wheel upright as shown; (b) turns the wheel out, $\theta = 90°$; and (c) turns the wheel downward, $\theta = 180°$.

19–34. Determine the height h at which a billiard ball of mass m must be struck so that no frictional force develops between it and the table at A. Assume that the cue C only exerts a horizontal force **P** on the ball.

Prob. 19–34

19–35. The plank has a weight of 30 lb, center of gravity at G, and it rests on the two sawhorses at A and B. If the end D is raised 2 ft above the top of the sawhorses and is released from rest, determine how high end C will rise from the top of the sawhorses after the plank falls so that it rotates clockwise about A, strikes and pivots on the sawhorse at B, and rotates clockwise off the sawhorse at A.

Prob. 19–35

*19–36. A 7-g bullet having a velocity of 800 m/s is fired into the edge of the 5-kg disk as shown. Determine the angular velocity of the disk just after the bullet becomes embedded in it. Also, calculate how far θ the disk will swing until it stops. The disk is originally at rest.

Prob. 19–36

19–37. The pendulum consists of a slender 2-kg rod AB and 5-kg disk. It is released from rest without rotating. When it falls 0.3 m, the end A strikes the hook S, which provides a permanent connection. Determine the angular velocity of the pendulum after it has rotated 90°. Treat the pendulum's weight during impact as a nonimpulsive force.

Prob. 19–37

19–38. A thin rod of mass m has an angular velocity ω_1 while rotating on a smooth surface. Determine its new angular velocity just after its end strikes and hooks onto the peg and the rod starts to rotate about P without rebounding. Solve the problem (a) using the parameters given; (b) setting $m = 2$ kg, $\omega_1 = 4$ rad/s, $l = 1.5$ m.

Prob. 19–38

19–39. The rod AB weighs 4 lb and is hanging in the vertical position. A 2-lb block, sliding on a smooth horizontal surface with a velocity of 12 ft/s, strikes the rod at its end B. Determine the velocity of the block immediately after the collision. The coefficient of restitution between the block and the rod at B is $e = 0.8$.

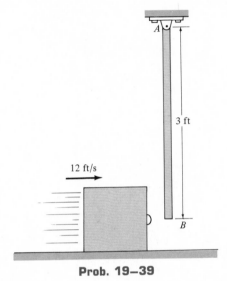

Prob. 19–39

*19–40. The 6-lb slender rod AB is released from rest when it is in the *horizontal position* so that it begins to rotate clockwise. A 1-lb ball is thrown at the rod with a velocity $v = 50$ ft/s. The ball strikes the rod at C at the instant the rod is in the vertical position as shown. Determine the angular velocity of the rod just after the impact. Take $e = 0.7$.

Prob. 19–40

19–41. The disk has a mass of 15 kg. If it is released from rest when $\theta = 30°$, determine the maximum angle θ of rebound after it collides with the wall. The coefficient of restitution between the disk and the wall is $e = 0.6$. When $\theta = 0°$, the disk hangs such that it just touches the wall. Neglect friction at the pin C.

Prob. 19–41

19–42. The pendulum consists of a 10-lb sphere and 4-lb rod. If it is released from rest when $\theta_1 = 90°$, determine the angle θ_2 of rebound after the sphere strikes the floor. Take $e = 0.8$.

Prob. 19–42

19–43. The ball has a weight of 3 lb and rolls without slipping along a horizontal plane with a velocity $v_O = 0.5$ ft/s. Provided it does not slip or rebound, determine the velocity of its center O as it just starts to roll up the inclined plane.

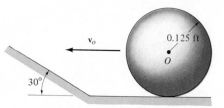

Prob. 19–43

*19–44. The disk has a mass m and radius r. If it strikes a rough step having a height $\frac{1}{8}r$ as shown, determine the smallest angular velocity ω_1 the disk can have and not rebound off the step when it strikes it.

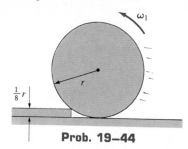

Prob. 19–44

417

Review 2: Planar Kinematics and Kinetics of a Rigid Body

Having presented the various topics in planar kinematics and kinetics in Chapters 16 through 19, we will now summarize these principles and provide an opportunity for applying them to the solution of various types of problems.

Kinematics. Here we are interested in studying the geometry of motion, without concern for the forces which cause the motion. Before solving a planar kinematics problem, it is *first* necessary to *classify the motion* as being either rectilinear or curvilinear translation, rotation about a fixed axis, or general plane motion. In particular, problems involving general plane motion can be solved either with reference to a fixed axis (absolute motion analysis) or using a translating or rotating frame of reference (relative motion analysis). The choice generally depends upon the type of constraints and the problem geometry. In all cases, application of the necessary equations may be clarified by drawing a kinematic diagram. Remember that the *velocity* of a point is always *tangent* to its path of motion, and the *acceleration* of a point can have *components* in the *n-t* directions when the path is *curved*.

Translation. When the body moves with rectilinear or curvilinear translation, *all* the points on the body have the *same motion*.

$$\mathbf{v}_B = \mathbf{v}_A, \qquad \mathbf{a}_B = \mathbf{a}_A$$

Rotation About a Fixed Axis.

Angular Motion

Variable Angular Acceleration. Provided a mathematical relationship is given between *any* two of the *four* variables θ, ω, α, and t, then a *third* variable can be determined by solving one of the following equations which relate all three variables.

$$\omega = \frac{d\theta}{dt} \qquad \alpha = \frac{d\omega}{dt} \qquad \alpha \, d\theta = \omega \, d\omega$$

Constant Angular Acceleration. The following equations apply when it is *absolutely certain* that the angular acceleration is constant.

$$\theta = \theta_0 + \omega_0 t + \tfrac{1}{2}\alpha_c t^2 \qquad \omega = \omega_0 + \alpha t \qquad \omega^2 = \omega_0^2 + 2\alpha_c(\theta - \theta_0)$$

Motion of Point P

$$v = \omega r \qquad\qquad \mathbf{v} = \boldsymbol{\omega} \times \mathbf{r}$$
$$a_t = \alpha r \quad a_n = \omega^2 r \qquad \mathbf{a} = \boldsymbol{\alpha} \times \mathbf{r} + \boldsymbol{\omega} \times (\boldsymbol{\omega} \times \mathbf{r})$$

General Plane Motion—Relative Motion Analysis.
Recall that when *translating axes* are placed at the "base point" A, the *relative motion* of point B with respect to A is simply *circular motion of B about A*. The following equations apply to two points A and B located on the *same* rigid body.

$$\mathbf{v}_B = \mathbf{v}_A + \mathbf{v}_{B/A}$$
$$\mathbf{a}_B = \mathbf{a}_A + \mathbf{a}_{B/A}$$

Rotating and translating axes are often used to analyze the motion of rigid bodies which are connected together by collars or slider blocks.

$$\mathbf{v}_B = \mathbf{v}_A + \boldsymbol{\Omega} \times \mathbf{r}_{B/A} + (\mathbf{v}_{B/A})_{\text{rel}}$$
$$\mathbf{a}_B = \mathbf{a}_A + \dot{\boldsymbol{\Omega}} \times \mathbf{r}_{B/A} + \boldsymbol{\Omega} \times (\boldsymbol{\Omega} \times \mathbf{r}_{B/A}) + 2\boldsymbol{\Omega} \times (\mathbf{v}_{B/A})_{\text{rel}} + (\mathbf{a}_{B/A})_{\text{rel}}$$

Kinetics. To analyze the forces which cause the motion we must use the principles of kinetics. When applying the necessary equations, it is important to first establish the inertial coordinate system and define the positive directions of the axes. The *directions* should be the *same* as those selected when writing any equations of kinematics provided *simultaneous solution* of equations becomes necessary.

Equations of Motion. These equations are used to determine accelerated motions or forces causing the motion. If used to determine position, velocity, or time of motion, then kinematics will have to be considered for part of the solution. Before applying the equations of motion, *always draw a free-body diagram* in order to identify all the forces acting on the body. Also, establish the directions of the acceleration of the mass center and the angular accelera-

rotates about P

if P coincides with G
use
$$\Sigma M_G = I_G \alpha$$

if not
use
$$\Sigma M_P = I_G \alpha + mad$$

tion of the body. (A kinetic diagram may also be drawn in order to graphically represent $m\mathbf{a}_G$ and $I_G\boldsymbol{\alpha}$. This diagram is particularly convenient for resolving $m\mathbf{a}_G$ into components and for identifying the terms in the moment sum $\Sigma(M_k)_P$.)

The three equations of motion are

$$\Sigma F_x = m(a_G)_x$$
$$\Sigma F_y = m(a_G)_y$$
$$\Sigma M_G = I_G\alpha \quad \text{or} \quad \Sigma M_P = \Sigma(M_k)_P$$

In particular, if the body is rotating about a fixed axis, moments may also be summed about a point O on the axis, in which case

$$\Sigma M_O = \Sigma(M_k)_O = I_O\alpha$$

Work and Energy. *The equation of work and energy is used to solve problems involving force, velocity, and displacement.* Before applying this equation, *always draw a free-body diagram* of the body in order to identify the forces which do work. Recall that the kinetic energy of the body consists of that due to translational motion of the mass center, $\mathbf{v}_G$, *and* rotational motion of the body, $\boldsymbol{\omega}$.

$$T_1 + \Sigma U_{1-2} = T_2$$

where

$$T = \tfrac{1}{2}mv_G^2 + \tfrac{1}{2}I_G\omega^2$$

$$U_F = \int F \cos\theta \, ds \quad \text{(variable force)}$$

$$U_{F_c} = F_c \cos\theta \, (s_2 - s_1) \quad \text{(constant force)}$$

$$U_W = W \, \Delta y \quad \text{(weight)}$$

$$U_s = -(\tfrac{1}{2}ks_2^2 - \tfrac{1}{2}ks_1^2) \quad \text{(spring)}$$

$$U_M = M\theta \quad \text{(constant couple)}$$

If the forces acting on the body are conservative forces, then apply the conservation of energy equation. This equation is easier to use than the equation of work and energy, since it applies only at *two points* on the path and *does not* require calculation of the work done by a force as the body moves along the path.

$$T_1 + V_1 = T_2 + V_2$$

where

$$V_g = Wy \quad \text{(gravitational potential energy)}$$

$$V_e = \tfrac{1}{2}ks^2 \quad \text{(elastic potential energy)}$$

Impulse and Momentum. *The principles of linear and angular impulse and momentum are used to solve problems involving force, velocity, and time.* Before applying the equations, *draw a free-body diagram* in order to identify all the forces which cause linear and angular impulses on the body. Also, establish the directions of the velocity of the mass center and the angular velocity of the body just before and just after the impulses are applied. (As an alternative procedure, the impulse and momentum diagrams may accompany the solution in order to graphically account for the terms in the equations. These diagrams are particularly advantageous when computing the angular impulses and angular momenta about a point other than the body's mass center.)

$$m(\mathbf{v}_G)_1 + \Sigma \int \mathbf{F}\, dt = m(\mathbf{v}_G)_2$$

$$(\mathbf{H}_G)_1 + \Sigma \int \mathbf{M}_G\, dt = (\mathbf{H}_G)_2$$

or

$$(\mathbf{H}_O)_1 + \Sigma \int \mathbf{M}_O\, dt = (\mathbf{H}_O)_2$$

Conservation of Momentum. If nonimpulsive forces or no impulsive forces act on the body in a particular direction, or the motions of several bodies are involved in the problem, then consider applying the conservation of linear or angular momentum for the solution. Investigation of the free-body diagram (or the impulse diagram) will aid in determining the directions for which the impulsive forces are zero, or axes for which the impulsive forces cause zero angular momentum. When this occurs, then

$$m(\mathbf{v}_G)_1 = m(\mathbf{v}_G)_2$$
$$(\mathbf{H}_O)_1 = (\mathbf{H}_O)_2$$

The problems that follow involve application of all the above concepts. They are presented in *random order* so that practice may be gained at identifying the various types of problems and developing the skills necessary for their solution.

Problems

Except when stated otherwise, in the following problems assume that the coefficients of static and kinetic friction are equal, i.e., $\mu = \mu_s = \mu_k$.

R2–1. The 40-lb cylinder rests on an inclined surface having a coefficient of friction $\mu = 0.1$. A cord is wrapped around the outer surface of the cylinder. If the cylinder is released from rest, determine the velocity of the center of the cylinder one second after it is released. Neglect the mass of the cord.

Prob. R2–1

R2–2. The double pendulum consists of two rods. Rod AB has a constant angular velocity of 3 rad/s, and rod BC has a constant angular velocity of 2 rad/s, both measured counterclockwise. Determine the velocity and acceleration of point C at the instant shown.

Prob. R2–2

R2–3. If the coefficient of friction between the crate and the platform is $\mu = 0.3$, determine the largest initial angular acceleration α which the links AB and CD can have without causing the crate to slip. The platform is initially at rest when $\theta = 45°$. Assume that the crate does not tip.

Prob. R2–3

*R2–4. At a given instant rod AB has the angular motion shown. If, at this same instant, the *disk* rotates *clockwise* with an angular velocity of 1 rad/s and has an angular acceleration of 2 rad/s², both *measured relative to the rod*, determine the velocity and acceleration of point C on the disk at this instant.

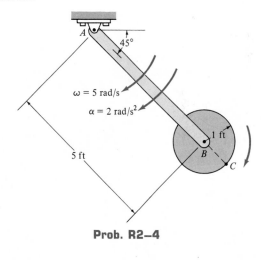

Prob. R2–4

R2–5. The circular plate A has a mass of 60 kg and a radius of gyration $k_G = 275$ mm. If it is placed over the rotor B, which always maintains a constant angular velocity $\omega = 15$ rad/s, determine the angular acceleration of the plate when it is released from rest. The wall at C is smooth, and the coefficient of friction between the plate and the rotor is $\mu = 0.2$.

50 mm 200 mm 400 mm $\omega = 15$ rad/s

Prob. R2–5

R2–6. The 10-lb block is sliding on the smooth surface when the corner D hits a stop block S. Determine the minimum velocity $\mathbf{v}$ the block can have which would allow it to tip over on its side so that it lands in the dotted position shown. Neglect the size of S. *Hint:* During impact consider the weight of the block to be nonimpulsive.

Prob. R2–6

R2–7. Pulley A weighs 15 lb and has a radius of gyration $k_O = 0.8$ ft. If the system is released from rest, determine the velocity of the center O of the pulley after the 10-lb block moves downward 4 ft. Neglect the mass of the cord and the pulley at B.

$\frac{1}{2}$ ft 1 ft 10 lb

Prob. R2–7

*R2–8. Link CD has an angular velocity of 3 rad/s at the instant shown. Determine the angular velocity of link AB at this instant.

3 in. 6 in. 3 rad/s

Prob. R2–8

R2–9. The links AB and BC each have a mass of 6 kg. An elastic cable is attached to A and C and has an unstretched length of 0.2 m. If the links are extended out into the horizontal position, then released from rest ($\theta \approx 0°$) so that they move out of equilibrium, compute the velocity of C at the instant $\theta = 45°$. Neglect the mass of the roller.

Prob. R2–9

R2–10. The hoisting gear A has an initial angular velocity of 60 rad/s and a constant deceleration of 1 rad/s². Determine the velocity and acceleration of the block which is being hoisted by the hub on gear B when $t = 3$ s.

Prob. R2–10

R2–11. A 7-kg automobile tire is released from rest at A on the incline and rolls without slipping to point B, where it then travels in free flight. Determine the maximum height h the tire attains. The radius of gyration of the tire about its mass center is $k_G = 0.3$ m.

Prob. R2–11

***R2–12.** The rod is bent into the shape of a sine curve and is forced to rotate about the y axis by connecting the spindle S to a motor. If the rod starts from rest in the position shown and a motor drives it for a short time with an angular acceleration $\alpha = (1.5e^t)$ rad/s², where t is in seconds, determine the magnitudes of the angular velocity and angular displacement of the rod when $t = 3$ s. Locate the point on the rod which has the greatest velocity and acceleration, and compute the magnitudes of the velocity and acceleration of this point when $t = 3$ s. The curve defining the rod is $z = 0.25 \sin (\pi y)$, where the argument for the sine is given in radians when y is in meters.

Prob. R2–12

R2–13. The device is used to indicate the safe load acting at the end of the boom B when it is in any angular position. It consists of a fixed dial plate D and an indicator arm ACE, which is pinned to the plate at C and to a short link EF. If the boom is pin-connected to the trunk frame at G and is rotating upward at $\omega_B = 3$ rad/s, determine the velocity of the dial pointer A at the instant shown, i.e., when EF and AC are in the vertical position.

Prob. R2–13

R2–14. The pendulum consists of a 30-lb sphere and a 10-lb slender rod. Compute the reaction at the pin O just after the cord AB is cut.

Prob. R2–14

R2–15. The center of the pulley is being lifted vertically with an acceleration of 4 m/s² at the instant it has a velocity of 2 m/s. If the cable does not slip on the pulley's surface, determine the accelerations of the cylinder B and point C on the pulley.

Prob. R2–15

***R2–16.** The cylinder having a mass of 5 kg is initially at rest when it is placed into contact with the wall B and the rotor at A. If the rotor always maintains a constant clockwise angular velocity $\omega = 6$ rad/s, determine the initial angular acceleration of the cylinder. The coefficient of friction at the contacting surfaces B and C is $\mu = 0.2$.

Prob. R2–16

R2–17. The spool has a mass of 20 kg and a radius of gyration $k_O = 0.3$ m. A cord is wrapped around its inner hub and is subjected to a variable force $P = (20t + 50)$ N, where t is in seconds. If the spool rolls without slipping at A, determine the speed of its center O when $t = 4$ s. The spool is initially at rest.

Prob. R2–17

R2–18. Blocks A and B weigh 50 and 100 lb, respectively. If the magnitude of the force P is 100 lb, determine the normal force exerted by block A on block B. Neglect friction and the weights of the pulleys, cord, and bars of the triangular frame.

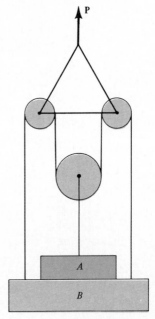

Prob. R2–18

R2–19. The uniform crate C has a mass of 50 kg and it rests on the surface of a truck for which the coefficient of friction is $\mu = 0.30$. If the truck has an initial speed $v_1 = 20$ m/s, determine the shortest distance s in which the truck can stop without causing the crate to move relative to the truck.

Prob. R2–19

*R2–20. The lever ABC has a mass of 4 kg and can be approximated by a slender rod. If a vertical force of 120 N is applied to the lever from an at-rest position $\theta = 30°$, determine the angular velocity of the lever at the instant $\theta = 90°$. The spring has a stiffness $k = 500$ N/m and is unstretched when $\theta = 0°$. Neglect friction and the mass of the slider blocks at A and B.

Prob. R2–20

R2–21. The cylinder C is subjected to rotational motion caused by the driving wheel at A. Roller B maintains contact with the inner surface of the cylinder and is free to rotate. Likewise, the roller at D is free to rotate. If the angular acceleration of A is $\alpha_A = (0.3t^2)$ rad/s^2 when the cylinder is originally at rest, determine the angular velocity of B when $t = 5$ s.

Prob. R2–21

R2–22. The 3-kg ball rests on a horizontal surface and is being pushed forward by moving the piston P with a force of 2 N. If the ball rolls without slipping at C and the coefficient of friction between the piston and the ball at D is $\mu_D = 0.20$, compute the velocity of the ball's mass center G after the piston moves 50 mm. Neglect the mass of the piston and assume that the guides at A and B are smooth. The ball is originally at rest.

Prob. R2–22

R2–23. The disk is driven by a motor in the counterclockwise direction. If the angular position can be described by the relation $\theta = (0.5t + 0.2t^2)$ rad, where t is in seconds, determine the number of revolutions turned, the angular velocity, and angular acceleration of the disk when $t = 20$ s.

75 mm
Motion
θ

Prob. R2–23

*R2–24. The link OA is pinned at O and rotates because of the sliding action of rod R along the horizontal groove. If R starts from rest when $\theta = 0°$ and has a constant acceleration $a_R = 60$ mm/s^2 to the right, determine the angular velocity and angular acceleration of OA when $t = 2$ s.

400 mm
O
ω
α
θ
R
A

Prob. R2–24

R2–25. The uniform connecting rod BC has a mass of 3 kg and is pin-connected at its end points. Determine the vertical forces which the pins exert on the ends B and C of the rod at the instant (a) $\theta = 0°$, and (b) $\theta = 90°$. The crank AB is turning with a constant angular velocity $\omega_{AB} = 5$ rad/s.

700 mm
C
θ
200 mm
B
200 mm
ω_{AB}
A

Prob. R2–25

R2–26. The semicircular disk weighs 10 lb. If it is released from rest in the position shown, determine the velocity of point A when the disk has rotated counterclockwise $90°$. Assume that the disk rolls without slipping on the surface. The moment of inertia of the disk about its mass center is $I_G = 0.0994$ slug-ft^2.

0.424 ft
A
G
2 ft

Prob. R2–26

R2–27. At the instant shown, link AB has a clockwise angular velocity of 0.5 rad/s and a counterclockwise angular deceleration of 2 rad/s^2. Determine the acceleration of point C and the angular acceleration of link CD at this instant.

400 mm
C
300 mm
$\alpha_{AB} = 2$ rad/s^2
A
B
600 mm
$\omega_{AB} = 0.5$ rad/s
500 mm
D

Prob. R2–27

*R2–28. The board rests on the surface of two drums. At the instant shown, it has an acceleration of 0.5 m/s^2 to the right, while at the same instant points on the outer rim of each drum have an acceleration with a magnitude of 3 m/s^2. If the board does not slip on the drums, determine its speed due to the motion.

$a = 0.5$ m/s^2
250 mm
250 mm
Motion
Motion

Prob. R2–28

R2–29. The uniform slender rod AB has a mass of 4 kg and a length of 2 m. If it is suspended horizontally by a spring at A and a cord at B, determine the angular acceleration of the rod and the acceleration of the rod's mass center at the instant the cord at B is cut. *Hint:* The stiffness of the spring is not needed for the calculation.

Prob. R2–29

R2–30. If the slider block C is fixed to the disk that has a constant counterclockwise angular velocity of 4 rad/s, determine the angular velocity and angular acceleration of the slotted arm AB at the instant shown.

Prob. R2–30

R2–31. The center O of the wheel is moving to the right with a speed of 3 m/s. If no slipping occurs at the ground A, determine the velocities of points B and C.

Prob. R2–31

*R2–32. The assembly weighs 10 lb and has a radius of gyration $k_G = 0.6$ ft about its mass center G. The kinetic energy of the assembly is 31 ft-lb when it is in the position shown. If it is rolling counterclockwise on the surface without slipping, determine its linear momentum at this instant.

Prob. R2–32

R2–33. The shaft consists of two cones, each having a mass of 15 kg. If it is acted upon by a moment $M = 3 \text{ N} \cdot \text{m}$ as shown, determine its angular velocity when $t = 2$ s starting from rest. What are the vertical reactions at A and B during the motion?

Prob. R2–33

R2–34. The 20-lb block B is attached to the 10-lb disk pulley A. If a 20-lb force P is applied as shown, determine the acceleration of B and the angular acceleration of A. Neglect the mass of the cord and the pulley at C.

Prob. R2–34

R2–35. The 50-kg cylinder has an angular velocity of 30 rad/s when it is brought into contact with the horizontal surface at C. If the coefficient of friction is $\mu = 0.2$, determine how long it takes for the cylinder to stop spinning. What force is developed at the pin A during this time? The axis of the cylinder is connected to *two* symmetrical links. (Only AB is shown.) For the computation, neglect the weight of the links.

Prob. R2–35

*R2–36. The 3-lb semicircular disk is released from rest at the instant shown. Determine the acceleration of point A the instant it is released. The coefficient of friction is large enough to prevent slipping. The moment of inertia of the disk about its mass center is $I_G = 0.0045$ slug-ft^2.

Prob. R2–36

R2–37. The tire has a mass of 9 kg and a radius of gyration $k_O = 225$ mm. If it is released from rest and rolls down the plane without slipping, determine the speed of its center O when $t = 3$ s.

Prob. R2–37

R2–38. If link BC rotates clockwise with an angular velocity of 12 rad/s, while the outer gear rack rotates counterclockwise with an angular velocity of 3 rad/s, determine the angular velocity of gear A.

Prob. R2–38

R2–39. The assembly consists of a disk A which is pinned at O and weighs 15 lb, a 1-ft rod weighing 2 lb, and 1-ft-diameter sphere weighing 10 lb. The spring is originally stretched 1 ft. If the cord BC is cut, determine the velocity of point B on the sphere when the rod swings down to the vertical position. During the motion, the spring wraps around the outer rim of the disk.

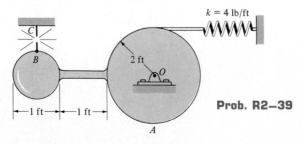

Prob. R2–39

*R2–40. The link AB is turning clockwise with a constant angular velocity of 10 rad/s. Determine the angular acceleration of link DC at the instant shown and the acceleration of point C.

Prob. R2–40

R2–41. Compute the velocity and acceleration of the rod R for any angle θ of the cam C if the cam rotates with a constant angular velocity ω. The pin connection at O does not cause an interference with the motion of A on C.

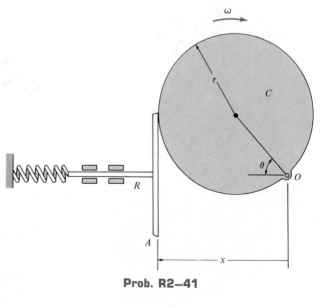

Prob. R2–41

431

R2–42. The motor M is used via the gear arrangement to lift block E. If the motor shaft is turning gear A at a constant rate $\omega_A = 4$ rad/s, determine the velocity of E.

R2–43. The disk A maintains a constant clockwise angular velocity of 30 rad/s. If the 20-kg disk B is initially at rest when it is brought into contact with A, determine the time required for B to attain the same angular velocity as A. The coefficient of friction between the two disks is $\mu = 0.3$. Neglect the mass of bar BC.

$\vdash$150 mm$\dashv$ 250 mm D

200 mm

A 100 mm 400 mm

Top view

400 mm B

200 mm

400 mm

200 mm

C

A

$\omega = 30$ rad/s

Prob. R2–43

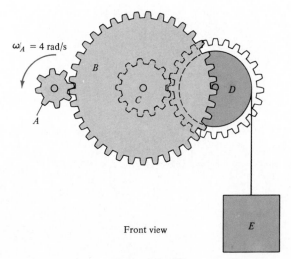

$\omega_A = 4$ rad/s

Front view

Prob. R2–42

*R2–44. If, at a given instant, point B has a downward velocity $v_B = 3$ m/s, determine the velocity of point A at this instant. Notice that for this motion to occur, the wheel must slip on the surface of the horizontal plane.

ω

0.4 m

v_B

B

0.15 m

A

Prob. R2–44

R2–45. The 20-lb link AB is pinned to a moving frame at A and held in the vertical position by means of a string BC which can support a maximum tension of 10 lb. Determine the smallest horizontal acceleration to which the link may be subjected so that the string does not break. Also, what is the corresponding reaction at the pin A?

3 ft

4 ft

3 ft

Prob. R2–45

R2–46. At the instant shown, two forces act on the 30-lb slender rod which is pinned at O. Determine the magnitude of force $\mathbf{F}$ and the initial angular acceleration of the rod so that the horizontal reaction which the *pin exerts on the rod* is 5 lb directed to the right.

3 ft

20 lb

3 ft

F

2 ft

Prob. R2–46

R2–47. Determine the acceleration of point A and the angular acceleration of the gear if at the given instant the gear racks have the velocities and accelerations shown.

$a_C = 5 \text{ m/s}^2$

$v_C = 6 \text{ m/s}$

25 mm

$a_D = 3 \text{ m/s}^2$

$v_D = 4 \text{ m/s}$

Prob. R2–47

***R2–48.** The block and pulley system is released from rest. Using the data shown in the figure, determine the speed which block D attains 10 s later. Assume that the cord does not slip on the pulleys. Neglect the mass of the cord.

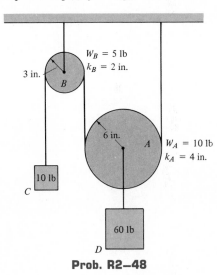

3 in.

$W_B = 5 \text{ lb}$
$k_B = 2 \text{ in.}$

B

6 in.

A

$W_A = 10 \text{ lb}$
$k_A = 4 \text{ in.}$

10 lb

C

60 lb

D

Prob. R2–48

R2–49. The ring has a mass of 10 kg, a mass center at G, and a radius of gyration $k_G = 135$ mm. If its angular velocity is $\omega_1 = 2$ rad/s when it is in the position shown, compute its angular acceleration at this instant and the normal force of the ring on the ground. Assume that slipping does not occur.

Prob. R2–49

R2–50. At a given instant, the slider block A has the velocity and deceleration shown. Determine the acceleration of block B and the angular acceleration of the link at this same instant.

Prob. R2–50

R2–51. Disks A and B weigh 15 and 25 lb, respectively. The connecting link CD weighs 5 lb. Determine the angular velocity of disk A just after the link falls through an angle of $180°$. Assume that slipping does not occur between the two disks and that disk B is *fixed from rotating*.

Prob. R2–51

*R2–52. A horizontal circular platform weighs 180 lb and is free to rotate about a vertical axis passing through its center. Initially the platform is rotating with an angular velocity of 2 rad/s. A 60-lb dog jumps on the platform and runs in the opposite direction of rotation with a velocity of 30 ft/s relative to the platform. Determine the angular velocity of the platform assuming that the dog runs along a circular path having a radius $r = 10$ ft. The centroidal radius of gyration of the platform about its central vertical axis is $k = 4$ ft.

Three-Dimensional Kinematics of a Rigid Body

Three types of planar rigid-body motion have been presented in Chapter 16: translation, rotation about a fixed axis, and general plane motion. In this chapter the three-dimensional motion of a rigid body, which consists of rotation about a fixed point and general motion, will be discussed. This is followed by a general study of the motion of particles and rigid bodies using coordinate systems which both translate and rotate. The analysis of these motions is more complex than planar-motion analysis since the angular acceleration of the body will measure a change in both the *magnitude* and *direction* of the body's angular velocity. In order to simplify the motion's three-dimensional aspects, throughout the chapter we will make complete use of vector analysis.*

Rotation About a Fixed Point 20.1*

When a rigid body rotates about a fixed point, the distance *r* from the point to a particle *P* located in the body is the *same* for *any position* of the body. Thus, the path of motion for the particle lies on the *surface of a sphere* having a radius *r* and centered at the fixed point. Since motion along this path occurs only from a series of rotations made during a finite time interval, it is perhaps wise to first develop a familiarity with some of the properties of rotational displacements.

*A brief review of vector analysis is given in Appendix C.

Euler's Theorem. This theorem states that two "component" rotations about different axes passing through a point are equivalent to a single resultant rotation about an axis passing through the point. If more than two rotations are applied, they can be combined into pairs, and each pair can be further reduced to combine into one rotation.

Finite Rotations. If the component rotations used in Euler's theorem are *finite,* it is important that the *order* in which they are applied be maintained. This is because finite rotations do *not* obey the law of vector addition, and hence they cannot be classified as vector quantities. To show this, consider the two finite rotations $\boldsymbol{\theta}_1 + \boldsymbol{\theta}_2$ applied to the block in Fig. 20–1a. Each rotation has a magnitude of 90° and a direction defined by the right-hand rule, as indicated by the black arrowhead. The resultant orientation of the block is shown at the right. When these two rotations are applied in the order $\boldsymbol{\theta}_2 + \boldsymbol{\theta}_1$, as shown in Fig. 20–1b, the resultant position of the block is *not* the same as it is in Fig. 20–1a. Consequently, *finite rotations* do not obey the commutative law of addition ($\boldsymbol{\theta}_1 + \boldsymbol{\theta}_2 \neq \boldsymbol{\theta}_2 + \boldsymbol{\theta}_1$), and therefore *they cannot be classified as vectors.* If smaller, yet finite, rotations had been used to illustrate this point, e.g., 10° instead of 90°, the *resultant* orientation of the block after each combination of rotations would also be different; however, in this case, only by a small amount.

(a)

(b)

Fig. 20–1

Infinitesimal Rotations. When defining the angular motions of a body subjected to three-dimensional motion, only rotations which are *infinitesimally small* will be considered. *Such rotations may be classified as vectors, since they can be added vectorially in any manner.* To show this, let us for purposes of simplicity consider the rigid body itself to be a sphere which is allowed to rotate about its central fixed point O, Fig. 20–2a. If we impose two infinitesimal rotations $d\boldsymbol{\theta}_1 + d\boldsymbol{\theta}_2$ on the body, it is seen that point P moves along the path $d\boldsymbol{\theta}_1 \times \mathbf{r} + d\boldsymbol{\theta}_2 \times \mathbf{r}$ and ends up at P'. Had the two successive rotations occurred in the order $d\boldsymbol{\theta}_2 + d\boldsymbol{\theta}_1$, then the resultant displacements of P would have been $d\boldsymbol{\theta}_2 \times \mathbf{r} + d\boldsymbol{\theta}_1 \times \mathbf{r}$. Since the vector cross product obeys the distributive law, by comparison $(d\boldsymbol{\theta}_1 + d\boldsymbol{\theta}_2) \times \mathbf{r} = (d\boldsymbol{\theta}_2 + d\boldsymbol{\theta}_1) \times \mathbf{r}$. It is therefore possible to conclude that infinitesimal rotations $d\boldsymbol{\theta}$ are indeed vectors, since these quantities have both a magnitude and direction for which the order of (vector) addition is not important, i.e., $d\boldsymbol{\theta}_1 + d\boldsymbol{\theta}_2 = d\boldsymbol{\theta}_2 + d\boldsymbol{\theta}_1$. Furthermore, as shown in Fig. 20–2a, the two "component" rotations $d\boldsymbol{\theta}_1$ and $d\boldsymbol{\theta}_2$ are equivalent to a single resultant rotation $d\boldsymbol{\theta} = d\boldsymbol{\theta}_1 + d\boldsymbol{\theta}_2$, a consequence of Euler's theorem.

Angular Velocity. If the body is subjected to an angular rotation $d\boldsymbol{\theta}$ about a fixed point, the angular velocity of the body is defined by the time derivative,

$$\boldsymbol{\omega} = \dot{\boldsymbol{\theta}} \tag{20–1}$$

The line specifying the direction of $\boldsymbol{\omega}$, which is collinear with $d\boldsymbol{\theta}$, is referred to as the *instantaneous axis of rotation,* Fig. 20–2b. In general, this axis changes direction during each instant of time. Since $\boldsymbol{\omega}$ is a vector quantity, it follows from vector addition that if the body is subjected to two component angular motions, $\boldsymbol{\omega}_1 = \dot{\boldsymbol{\theta}}_1$ and $\boldsymbol{\omega}_2 = \dot{\boldsymbol{\theta}}_2$, the resultant angular velocity is $\boldsymbol{\omega} = \boldsymbol{\omega}_1 + \boldsymbol{\omega}_2$.

Velocity. Once $\boldsymbol{\omega}$ is specified, the velocity of any point P on a body rotating about a fixed point can be determined using the same methods as for a body rotating about a fixed axis (Sec. 16.3). Hence, by the cross product,

$$\mathbf{v} = \boldsymbol{\omega} \times \mathbf{r} \tag{20–2}$$

Here $\mathbf{r}$ defines the position of P measured from the fixed point O, Fig. 20–3.

Angular Acceleration. The body's angular acceleration is determined from the time derivative of the angular velocity; i.e.,

$$\boldsymbol{\alpha} = \dot{\boldsymbol{\omega}} \tag{20–3}$$

For motion about a fixed point, $\boldsymbol{\alpha}$ must account for a change in *both* the magnitude and direction of $\boldsymbol{\omega}$, so that, in general, $\boldsymbol{\alpha}$ is not directed along the instantaneous axis of rotation, Fig. 20–3.

(a)

(b)

Fig. 20–2

Fig. 20–3

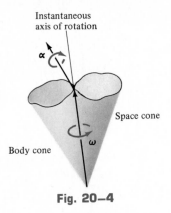

Instantaneous
axis of rotation

Space cone

Body cone

Fig. 20–4

As the direction of the instantaneous axis of rotation (or the line of action of $\boldsymbol{\omega}$) changes in space, the locus of points defined by the axis generates a fixed *space cone*. If the change in this axis is viewed with respect to the rotating body, the locus of the axis generates a *body cone*, Fig. 20–4. At any given instant, these cones are tangent along the instantaneous axis of rotation, and when the body is in motion, the body cone appears to roll either on the inside or the outside surface of the fixed space cone. Provided that the paths defined by the open ends of the cones are described by the tip of the $\boldsymbol{\omega}$ vector, $\boldsymbol{\alpha}$ must act tangent to these paths at any given instant, since the time rate of change of $\boldsymbol{\omega}$ is equal to $\boldsymbol{\alpha}$, Fig. 20–4.

Acceleration. If $\boldsymbol{\omega}$ and $\boldsymbol{\alpha}$ are known at a given instant, the acceleration of any point P on the body can be obtained by time differentiation of Eq. 20–2, which yields

$$\mathbf{a} = \boldsymbol{\alpha} \times \mathbf{r} + \boldsymbol{\omega} \times (\boldsymbol{\omega} \times \mathbf{r}) \qquad (20\text{–}4)$$

The form of this equation is the same as that developed in Sec. 16.3, which defines the acceleration of a point located on a body subjected to rotation about a fixed axis.

$\star$ 20.2 The Time Derivative of a Vector Measured from a Fixed and Translating-Rotating System

Fig. 20–5

In many types of problems involving the motion of a body about a fixed point, the angular velocity $\boldsymbol{\omega}$ is specified in terms of its component angular motions. For example, the disk in Fig. 20–5 spins about the horizontal y axis at $\boldsymbol{\omega}_s$ while it rotates or precesses about the vertical z axis at $\boldsymbol{\omega}_p$. Therefore, its resultant angular velocity is $\boldsymbol{\omega} = \boldsymbol{\omega}_s + \boldsymbol{\omega}_p$. If the angular acceleration $\boldsymbol{\alpha}$ of such a body is to be determined, it is sometimes easier to compute the time derivative of $\boldsymbol{\omega}$, Eq. 20–3, by using a coordinate system which has a *rotation* defined by one or more of the components of $\boldsymbol{\omega}$.[*] For this reason, and for other uses later, an equation will presently be derived that relates the time derivative of any vector $\mathbf{A}$ defined from a translating-rotating reference to its derivative defined from a fixed reference.

Consider the x, y, z axes of the moving frame of reference to have an angular velocity $\boldsymbol{\Omega}$ which is measured from the fixed X, Y, Z axes, Fig. 20–6a. In the following discussion, it will be convenient to express vector $\mathbf{A}$ in terms of its $\mathbf{i}, \mathbf{j}, \mathbf{k}$ components, which define the directions of the moving axes. Hence,

---

[*] In the case of the spinning disk, Fig. 20–5, the x, y, z axes may be given an angular velocity of $\boldsymbol{\omega}_p$.

$$\mathbf{A} = A_x\mathbf{i} + A_y\mathbf{j} + A_z\mathbf{k}$$

In general, the time derivative of **A** must account for the change in both the vector's magnitude and direction. However, if this derivative is taken *with respect to the moving frame of reference,* only a change in the magnitudes of the components of **A** must be accounted for, since the directions of the components do not change with respect to the moving reference. Hence,

$$(\dot{\mathbf{A}})_{xyz} = \dot{A}_x\mathbf{i} + \dot{A}_y\mathbf{j} + \dot{A}_z\mathbf{k} \tag{20-5}$$

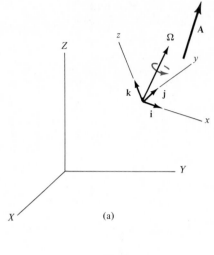

When the time derivative of **A** is taken *with respect to the fixed frame of reference,* the *directions* of **i, j,** and **k** change only on account of the rotation $\mathbf{\Omega}$ of the axes and not their translation. Hence, in general,

$$(\dot{\mathbf{A}})_{XYZ} = \dot{A}_x\mathbf{i} + \dot{A}_y\mathbf{j} + \dot{A}_z\mathbf{k} + A_x\dot{\mathbf{i}} + A_y\dot{\mathbf{j}} + A_z\dot{\mathbf{k}}$$

The time derivatives of the unit vectors will now be considered. For example, $\dot{\mathbf{i}} = d\mathbf{i}/dt$ represents only a change in the direction of **i** with respect to time, since **i** has a fixed magnitude of 1 unit. As shown in Fig. 20–6*b*, the change, $d\mathbf{i}$, is *tangent to the path* described by the tip of **i** as **i** moves due to the rotation $\mathbf{\Omega}$. Accounting for both the magnitude and direction of $d\mathbf{i}$, we can therefore define $\dot{\mathbf{i}}$ using the cross product, $\dot{\mathbf{i}} = \mathbf{\Omega} \times \mathbf{i}$. In general,

$$\dot{\mathbf{i}} = \mathbf{\Omega} \times \mathbf{i} \qquad \dot{\mathbf{j}} = \mathbf{\Omega} \times \mathbf{j} \qquad \dot{\mathbf{k}} = \mathbf{\Omega} \times \mathbf{k}$$

(a)

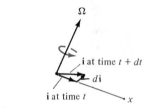

(b)

Fig. 20–6

These formulations were also developed in Sec. 16.10, regarding planar motion of the axes. Substituting the results into the above equation and using Eq. 20–5 yields

$$(\dot{\mathbf{A}})_{XYZ} = (\dot{\mathbf{A}})_{xyz} + \mathbf{\Omega} \times \mathbf{A} \tag{20-6}$$

This result is rather important and it will be used throughout Sec. 20.4 and Chapter 21. In words, it states that *the time derivative of* **A** *as observed from the fixed X, Y, Z frame of reference is equal to the time rate of change of* **A** *as observed from the x, y, z translating-rotating frame of reference, Eq. 20–5, plus* $\mathbf{\Omega} \times \mathbf{A}$, *the change of* **A** *caused by the rotation of the x, y, z frame.*

The following two example problems numerically illustrate the use of this equation for obtaining the angular acceleration of a body rotating about a fixed point.

Example 20–1

The disk shown in Fig. 20–7 is spinning about its horizontal axis with a constant angular velocity $\omega_s = 3$ rad/s, while the horizontal platform upon which the disk is mounted is rotating about the vertical axis at a constant rate $\omega_p = 1$ rad/s. Determine the angular acceleration of the disk and the velocity and acceleration of point A on the disk when it is in the position shown.

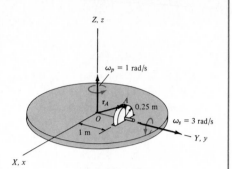

Fig. 20–7

Solution

Point O represents a fixed point of rotation for the disk if one considers a hypothetical extension of the disk to this point. To determine the velocity and acceleration of point A, it is first necessary to determine the resultant angular velocity $\boldsymbol{\omega}$ and angular acceleration $\boldsymbol{\alpha}$ of the disk, since these vectors are used in Eqs. 20–2 and 20–4.

Angular Velocity. The angular velocity is simply the vector addition of the two component motions. Thus,

$$\boldsymbol{\omega} = \boldsymbol{\omega}_s + \boldsymbol{\omega}_p = \{3\mathbf{j} + 1\mathbf{k}\} \text{ rad/s}$$

Angular Acceleration. Since the magnitude of $\boldsymbol{\omega}$ is constant, only a change in its direction, as seen from a fixed reference, creates the angular acceleration $\boldsymbol{\alpha}$ of the disk. One way to obtain $\boldsymbol{\alpha}$ is to compute the time derivative of *each of the two components* of $\boldsymbol{\omega}$ using Eq. 20–6. At the instant shown in Fig. 20–7, imagine the fixed X, Y, Z and a rotating x, y, z frame to be in coincidence. If the rotating x, y, z frame is chosen to have an angular velocity of $\boldsymbol{\Omega} = \boldsymbol{\omega}_p = \{1\mathbf{k}\}$ rad/s, then $\boldsymbol{\omega}_s$ will *always* be directed along the y (not Y) axis, and the time rate of change of $\boldsymbol{\omega}_s$ as seen from x, y, z is *zero*; i.e., $(\dot{\boldsymbol{\omega}}_s)_{xyz} = \mathbf{0}$ (the magnitude of $\boldsymbol{\omega}_s$ is constant). Consequently,

$$(\dot{\boldsymbol{\omega}}_s)_{XYZ} = (\dot{\boldsymbol{\omega}}_s)_{xyz} + \boldsymbol{\omega}_p \times \boldsymbol{\omega}_s = \mathbf{0} + (1\mathbf{k}) \times (3\mathbf{j}) = \{-3\mathbf{i}\} \text{ rad/s}^2$$

By the same choice of axes rotation, $\boldsymbol{\Omega} = \boldsymbol{\omega}_p$, or even with $\boldsymbol{\Omega} = \mathbf{0}$, the time derivative $(\dot{\boldsymbol{\omega}}_p)_{xyz} = \mathbf{0}$, since $\boldsymbol{\omega}_p$ is *always* directed along the z (or Z) axis and has a constant magnitude. Hence,

$$(\dot{\boldsymbol{\omega}}_p)_{XYZ} = (\dot{\boldsymbol{\omega}}_p)_{xyz} + \boldsymbol{\omega}_p \times \boldsymbol{\omega}_p = \mathbf{0} + \mathbf{0} = \mathbf{0}$$

The angular acceleration is therefore

$$\boldsymbol{\alpha} = (\dot{\boldsymbol{\omega}})_{XYZ} = (\dot{\boldsymbol{\omega}}_s)_{XYZ} + (\dot{\boldsymbol{\omega}}_p)_{XYZ} = \{-3\mathbf{i}\} \text{ rad/s}^2 \qquad \textit{Ans.}$$

Velocity and Acceleration. Since $\boldsymbol{\omega}$ and $\boldsymbol{\alpha}$ have been determined, the velocity and acceleration of point A can be computed using Eqs. 20–2 and 20–4. Realizing that $\mathbf{r}_A = \{1\mathbf{j} + 0.25\mathbf{k}\}$ m, Fig. 20–7, we have

$$\mathbf{v}_A = \boldsymbol{\omega} \times \mathbf{r}_A = (3\mathbf{j} + 1\mathbf{k}) \times (1\mathbf{j} + 0.25\mathbf{k}) = \{-0.25\mathbf{i}\} \text{ m/s} \qquad \textit{Ans.}$$

$$\mathbf{a}_A = \boldsymbol{\alpha} \times \mathbf{r}_A + \boldsymbol{\omega} \times (\boldsymbol{\omega} \times \mathbf{r}_A)$$

$$= (-3\mathbf{i}) \times (1\mathbf{j} + 0.25\mathbf{k}) + (3\mathbf{j} + 1\mathbf{k}) \times [(3\mathbf{j} + 1\mathbf{k}) \times (1\mathbf{j} + 0.25\mathbf{k})]$$

$$= \{0.50\mathbf{j} - 2.25\mathbf{k}\} \text{ m/s}^2 \qquad \textit{Ans.}$$

Example 20–2

At the instant $\theta = 60°$, the gyrotop in Fig. 20–8 has three components of angular motion directed as shown and having magnitudes which are defined as follows:

> spin: $\omega_s = 10$ rad/s, increasing at the rate of 6 rad/s^2
> nutation: $\omega_n = 3$ rad/s, increasing at the rate of 2 rad/s^2
> precession: $\omega_p = 5$ rad/s, increasing at the rate of 4 rad/s^2

Determine the angular velocity and angular acceleration of the top.

Fig. 20–8

Solution

Angular Velocity. The top is rotating about the fixed point O. If the fixed and rotating frames are considered to be coincident at the instant shown, then the angular velocity can be expressed in terms of $\mathbf{i}, \mathbf{j}, \mathbf{k}$ components, appropriate to the x, y, z frame; i.e.,

$$\boldsymbol{\omega} = -\omega_n\mathbf{i} + \omega_s \sin \theta\mathbf{j} + (\omega_p + \omega_s \cos \theta)\mathbf{k}$$
$$= -3\mathbf{i} + 10 \sin 60°\mathbf{j} + (5 + 10 \cos 60°)\mathbf{k}$$
$$= \{-3\mathbf{i} + 8.66\mathbf{j} + 10\mathbf{k}\} \text{ rad/s} \qquad \textit{Ans.}$$

Angular Acceleration. As in the solution of Example 20–1, the angular acceleration $\boldsymbol{\alpha}$ will be determined by investigating separately the time rate of change of *each of the angular velocity components* as observed from the fixed reference X, Y, Z. This is done by using Eq. 20–6 and choosing an $\boldsymbol{\Omega}$ for the x, y, z reference so that the component of $\boldsymbol{\omega}$ which is being considered is viewed as having a constant direction when observed from x, y, z.

Careful examination of the motion of the top reveals that $\boldsymbol{\omega}_s$ has a *constant direction* relative to x, y, z if these axes rotate at $\boldsymbol{\Omega} = \boldsymbol{\omega}_p + \boldsymbol{\omega}_n$. Since $\boldsymbol{\omega}_n$ *always* lies in the fixed X-Y plane, this vector has a *constant direction* if the motion is viewed from axes x, y, z having a rotation of $\boldsymbol{\Omega} = \boldsymbol{\omega}_p$ (not $\boldsymbol{\Omega} = \boldsymbol{\omega}_s + \boldsymbol{\omega}_p$). Finally, the component $\boldsymbol{\omega}_p$ is *always directed* along the Z axis so that here it is not necessary to think of x, y, z as rotating, i.e., $\boldsymbol{\Omega} = 0$. Expressing the data in terms of the $\mathbf{i}, \mathbf{j}, \mathbf{k}$ components, we therefore have

For $\boldsymbol{\omega}_s$, $\boldsymbol{\Omega} = \boldsymbol{\omega}_n + \boldsymbol{\omega}_p$;

$$(\dot{\boldsymbol{\omega}}_s)_{XYZ} = (\dot{\boldsymbol{\omega}}_s)_{xyz} + (\boldsymbol{\omega}_n + \boldsymbol{\omega}_p) \times \boldsymbol{\omega}_s$$
$$= (6 \sin 60°\mathbf{j} + 6 \cos 60°\mathbf{k}) + (-3\mathbf{i} + 5\mathbf{k}) \times (10 \sin 60°\mathbf{j} + 10 \cos 60°\mathbf{k})$$
$$= \{-43.30\mathbf{i} + 20.20\mathbf{j} - 22.98\mathbf{k}\} \text{ rad/s}^2$$

For $\boldsymbol{\omega}_n$, $\boldsymbol{\Omega} = \boldsymbol{\omega}_p$;

$$(\dot{\boldsymbol{\omega}}_n)_{XYZ} = (\dot{\boldsymbol{\omega}}_n)_{xyz} + \boldsymbol{\omega}_p \times \boldsymbol{\omega}_n = -2\mathbf{i} + (5\mathbf{k}) \times (-3\mathbf{i}) = \{-2\mathbf{i} - 15\mathbf{j}\} \text{ rad/s}^2$$

For $\boldsymbol{\omega}_p$, $\boldsymbol{\Omega} = 0$;

$$(\dot{\boldsymbol{\omega}}_p)_{XYZ} = (\dot{\boldsymbol{\omega}}_p)_{xyz} + 0 \times \boldsymbol{\omega}_p = \{4\mathbf{k}\} \text{ rad/s}^2$$

Thus, the angular acceleration of the top is

$$\boldsymbol{\alpha} = (\dot{\boldsymbol{\omega}}_s)_{XYZ} + (\dot{\boldsymbol{\omega}}_n)_{XYZ} + (\dot{\boldsymbol{\omega}}_p)_{XYZ}$$
$$= \{-45.3\mathbf{i} + 5.20\mathbf{j} - 19.0\mathbf{k}\} \text{ rad/s}^2 \qquad \textit{Ans.}$$

★20.3 General Motion

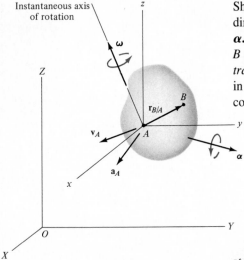

Instantaneous axis of rotation

Fig. 20–9

Shown in Fig. 20–9 is a rigid body subjected to general motion in three dimensions for which the angular velocity is $\boldsymbol{\omega}$ and the angular acceleration is $\boldsymbol{\alpha}$. If point A has a known motion of $\mathbf{v}_A$ and $\mathbf{a}_A$, the motion of any other point B may be determined by using a relative-motion analysis. In this section a *translating coordinate system* will be used to define the relative motion, and in the next section a reference that is both rotating and translating will be considered.

If the origin of the translating coordinate system x, y, z ($\boldsymbol{\Omega} = \mathbf{0}$) is located at the "base point" A, then, at the instant shown, the motion of the body may be regarded as the sum of an instantaneous translation of the body having a motion of $\mathbf{v}_A$ and $\mathbf{a}_A$ and a rotation of the body about an instantaneous axis passing through the base point. Since the body is rigid, the motion of point B measured by an observer located at A is the same as *motion of the body about a fixed point*. This relative motion occurs about the instantaneous axis of rotation and is defined by $\mathbf{v}_{B/A} = \boldsymbol{\omega} \times \mathbf{r}_{B/A}$, Eq. 20–2, and $\mathbf{a}_{B/A} = \boldsymbol{\alpha} \times \mathbf{r}_{B/A} + \boldsymbol{\omega} \times (\boldsymbol{\omega} \times \mathbf{r}_{B/A})$, Eq. 20–4. For translating axes the relative motions are related to absolute motions by $\mathbf{v}_B = \mathbf{v}_A + \mathbf{v}_{B/A}$ and $\mathbf{a}_B = \mathbf{a}_A + \mathbf{a}_{B/A}$, Eqs. 16–13 and 16–15, so that the absolute velocity and acceleration of point B can be determined from the equations

$$\mathbf{v}_B = \mathbf{v}_A + (\boldsymbol{\omega} \times \mathbf{r}_{B/A}) \qquad (20\text{–}7)$$

and

$$\mathbf{a}_B = \mathbf{a}_A + (\boldsymbol{\alpha} \times \mathbf{r}_{B/A}) + \boldsymbol{\omega} \times (\boldsymbol{\omega} \times \mathbf{r}_{B/A}) \qquad (20\text{–}8)$$

These two equations are identical to those describing the general plane motion of a rigid body, Eqs. 16–14 and 16–16. However, difficulty in application arises for three-dimensional motion, because $\boldsymbol{\alpha}$ measures the change in *both* the magnitude and direction of $\boldsymbol{\omega}$. (Recall that for general plane motion $\boldsymbol{\alpha}$ and $\boldsymbol{\omega}$ are always parallel or perpendicular to the plane of motion, and therefore $\boldsymbol{\alpha}$ measures only a change in the magnitude of $\boldsymbol{\omega}$.) In some problems the constraints or connections of a body will require that the directions of the angular motions or displacement paths of points on the body be defined. As illustrated in the following example, this information is useful for obtaining some of the terms in the above equations.

Example 20–3

One end of the rigid bar CD shown in Fig. 20–10a slides along the grooved wall slot AB, and the other end slides along the vertical member EF. If the collar at C is moving toward B at a speed of 3 m/s, determine the velocity of the collar at D and the angular velocity of the bar. The bar is connected to the collars at its end points by ball-and-socket joints.

Solution

Bar CD is subjected to general motion. Why? The velocity of point D on the bar may be related to the velocity of point C by the equation

$$\mathbf{v}_D = \mathbf{v}_C + (\boldsymbol{\omega} \times \mathbf{r}_{D/C})$$

The fixed and translating frames of reference are assumed to coincide at the instant considered, Fig. 20–10b. We have

$$\mathbf{v}_D = -v_D\mathbf{k} \qquad \mathbf{v}_C = \{3\mathbf{j}\} \text{ m/s}$$
$$\mathbf{r}_{D/C} = \{1\mathbf{i} + 2\mathbf{j} - 0.5\mathbf{k}\} \text{ m} \qquad \boldsymbol{\omega} = \omega_x\mathbf{i} + \omega_y\mathbf{j} + \omega_z\mathbf{k}$$

Substituting these quantities into the above equation gives

$$-v_D\mathbf{k} = 3\mathbf{j} + \begin{vmatrix} \mathbf{i} & \mathbf{j} & \mathbf{k} \\ \omega_x & \omega_y & \omega_z \\ 1 & 2 & -0.5 \end{vmatrix}$$

Expanding and equating the respective $\mathbf{i}$, $\mathbf{j}$, $\mathbf{k}$ components yields

$$-0.5\omega_y - 2\omega_z = 0 \qquad (1)$$
$$0.5\omega_x + 1\omega_z + 3 = 0 \qquad (2)$$
$$2\omega_x - 1\omega_y + v_D = 0 \qquad (3)$$

These equations contain four unknowns.* A fourth equation can be written if the direction of $\boldsymbol{\omega}$ is specified. In particular, any component of $\boldsymbol{\omega}$ acting along the bar's axis has no effect on moving the collars. This is because the bar is *free to rotate* about its axis. Therefore, if $\boldsymbol{\omega}$ is specified as acting *perpendicular* to the axis of the bar, then $\boldsymbol{\omega}$ must have a unique magnitude to satisfy the above equations. Perpendicularity is guaranteed provided the dot product of $\boldsymbol{\omega}$ and $\mathbf{r}_{D/C}$ is zero (see Eq. C–14). Hence,

$$\boldsymbol{\omega} \cdot \mathbf{r}_{D/C} = (\omega_x\mathbf{i} + \omega_y\mathbf{j} + \omega_z\mathbf{k}) \cdot (1\mathbf{i} + 2\mathbf{j} - 0.5\mathbf{k}) = 0$$
$$1\omega_x + 2\omega_y - 0.5\omega_z = 0 \qquad (4)$$

Solving Eqs. (1) to (4) simultaneously yields

$$\omega_x = -4.86 \text{ rad/s} \quad \omega_y = 2.29 \text{ rad/s} \quad \omega_z = -0.571 \text{ rad/s} \qquad \textit{Ans.}$$
$$v_D = 12.0 \text{ m/s} \downarrow \qquad \textit{Ans.}$$

---

*Although this is the case the magnitude of $\mathbf{v}_D$ can be obtained. For example, solve Eqs. (1) and (2) for ω_y and ω_x in terms of ω_z and substitute into Eq. (3). It will be noted that ω_z will *cancel out*, which will allow a solution for v_D.

(a)

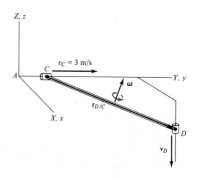

(b)

Fig. 20–10

Problems

20–1. The fan is mounted on a swivel support such that at the instant shown it is rotating about the z axis at $\omega_1 = 0.8$ rad/s, which is increasing at 12 rad/s^2. The blade is spinning at $\omega_2 = 16$ rad/s, which is decreasing at 2 rad/s^2. Determine the angular velocity and angular acceleration of the blade at this instant.

Prob. 20–1

20–2. Gear A is *fixed* while gear B is free to rotate on the shaft S. If the shaft is turning about the z axis at $\omega_z = 5$ rad/s, which is increasing at 2 rad/s^2, determine the velocity and acceleration of point P at the instant shown. The face of gear B lies in a vertical plane.

Prob. 20–2

20–3. The antenna is following the motion of a jet plane. At the instant $\theta = 25°$ and $\phi = 75°$, the constant angular rates of change are $\dot{\theta} = 0.4$ rad/s and $\dot{\phi} = 0.6$ rad/s. Compute the velocity and acceleration of the signal horn A at this instant. The distance OA is 0.8 m.

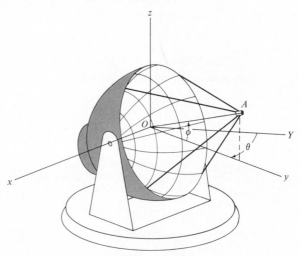

Prob. 20–3

*20–4. The ladder of the fire truck rotates about the z axis with an angular velocity $\omega_1 = 0.15$ rad/s, which is increasing at 0.8 rad/s^2. At the same instant it is rotating upward at a constant rate $\omega_2 = 0.6$ rad/s. Compute the velocity and acceleration of point A located at the top of the ladder at this instant.

Prob. 20–4

20–5. The keg is in the shape of a truncated cone and rotates about the z axis at a constant rate $\omega_z = 0.4$ rad/s without slipping on the horizontal plane. Determine the magnitudes of the velocity and acceleration of point A on the keg.

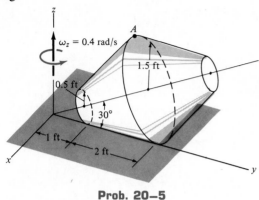

Prob. 20–5

20–6. The drill pipe P turns at a constant angular rate $\omega_P = 4$ rad/s. Determine the angular velocity and angular acceleration of the conical rock bit, which rolls without slipping. Also, what are the velocity and acceleration of point A?

Prob. 20–6

20–7. The bevel gear A rolls on the fixed gear B. If the shaft to which A is attached is rotating at 4 rad/s and has an angular acceleration of 3 rad/s^2 as shown, determine the angular velocity and angular acceleration of gear A.

Prob. 20–7

*20–8. The tower crane is rotating about the z axis with a constant rate $\omega_1 = 0.25$ rad/s, while the boom OA is rotating downward with a constant rate $\omega_2 = 0.4$ rad/s. Compute the velocity and acceleration of point A located at the top of the boom at the instant shown.

Probs. 20–8 / 20–9

20–9. At the instant shown, the tower crane is rotating about the z axis with an angular velocity $\omega_1 = 0.25$ rad/s, which is increasing at 0.6 rad/s^2. The boom OA is rotating downward with an angular velocity $\omega_2 = 0.4$ rad/s, which is increasing at 0.8 rad/s^2. Compute the velocity and acceleration of point A located at the top of the boom at this instant.

445

20–10. Gear A is fixed to the crankshaft S, while gear C is fixed and gear B is free to rotate. If the crankshaft is turning at $\omega_1 = 80$ rad/s about its axis, determine the magnitudes of the angular velocity of the propeller and the angular acceleration of gear B. $r_A = 0.4$ ft and $r_B = 0.1$ ft.

Prob. 20–10

20–11. The rod is attached to smooth collars at its end points using ball-and-socket joints. Determine the speed of B if A is moving downward at 8 ft/s. Also, determine the angular velocity of the rod if it is directed perpendicular to the axis of the rod.

Prob. 20–11

*20–12. The differential of an automobile allows the two rear wheels to rotate at different speeds when the automobile travels along a curve. For operation, the rear axles are attached to the wheels at one end and have beveled gears A and B on their other ends. The differential case D is placed over the left axle, but can rotate about C, independent of the axle. The case supports a pinion gear E on a shaft, which meshes with gears A and B. Finally, a ring gear G is *fixed* to the differential case so that the case rotates with the ring gear when the latter is driven by the drive pinion H. This gear, like the differential case, is free to rotate about the left wheel axle. If the drive pinion is turning at $\omega_H = 100$ rad/s and the pinion gear E is spinning about its shaft at $\omega_E = 30$ rad/s, compute the angular velocity, ω_A and ω_B, of each axle.

Prob. 20–12

20–13. The rod AB is attached to collars at its ends by ball-and-socket joints. If collar A has a velocity $v_A = 15$ ft/s at the instant shown, determine the velocity of collar B.

Prob. 20–13

20–14. Rod *AB* is attached to the rotating arm using ball-and-socket joints. If *AC* is rotating with a constant angular velocity of 8 rad/s about the pin at *C*, determine the angular velocity of link *BD* at the instant shown.

Prob. 20–14

20–15. Disk *A* is rotating at a constant angular velocity of 10 rad/s. If rod *BC* is joined to the disk and a collar by ball-and-socket joints, determine the velocity of the collar *B* at the instant shown, and the rod's angular velocity if ω_{BC} is directed perpendicular to the axis of the rod.

Probs. 20–15 / 20–16

***20–16.** If the disk in Prob. 20–15 has an angular acceleration $\alpha = 5$ rad/s², determine the acceleration of the collar *B* at the instant shown, when $\omega = 10$ rad/s. Also, what is the rod's angular acceleration α_{BC} if it is directed perpendicular to the axis of the rod?

20–17. The rod *AB* is attached to collars at its ends by ball-and-socket joints. If collar *A* moves upward with a velocity of 8 ft/s, determine the angular velocity of the rod and the velocity of collar *B* at the instant shown. Assume that the rod's angular velocity is directed perpendicular to the rod.

Probs. 20–17 / 20–18

20–18. The rod *AB* is attached to collars at its ends by ball-and socket joints. If collar *A* moves upward with a velocity of 8 ft/s and an acceleration of 4 ft/s², determine the angular acceleration of the rod and the acceleration of the collar *B* at the instant shown. Assume that the rod's angular velocity and angular acceleration are directed perpendicular to the rod.

447

20–19. The rod AB is attached to the collars at its ends by ball-and-socket joints. If collar A has a velocity $v_A = 5$ ft/s, determine the angular velocity of the rod and the velocity of collar B at the instant shown. Assume the angular velocity of the rod is directed perpendicular to the rod.

20–22. Solve Prob. 20–15 if the connection at B consists of a pin as shown in the figure below, rather than a ball-and-socket joint. *Hint:* The constraint allows rotation of the rod both along bar DE ($\mathbf{j}$ direction) and along the axis of the pin ($\mathbf{n}$ direction). Since there is no rotational component in the $\mathbf{u}$ direction, i.e., perpendicular to $\mathbf{n}$ and $\mathbf{j}$ where $\mathbf{u} = \mathbf{j} \times \mathbf{n}$, an additional equation for solution can be obtained from $\boldsymbol{\omega} \cdot \mathbf{u} = 0$. The vector $\mathbf{n}$ is in the same direction as $\mathbf{r}_{B/C} \times \mathbf{r}_{D/C}$.

Probs. 20–19 / 20–20

Prob. 20–22

*20–20. The rod AB is attached to the collars at its ends by ball-and-socket joints. If collar A has an acceleration $\mathbf{a}_A = \{6\mathbf{i}\}$ ft/s^2, and a velocity $\mathbf{v}_A = \{5\mathbf{i}\}$ ft/s, determine the angular acceleration of the rod and the acceleration of collar B at the instant shown. Assume the angular velocity and angular acceleration of the rod are both directed perpendicular to the rod.

20–21. The rod ABC is supported at A by a ball-and-socket and at point C by the y-z plane. The segment AB lies in the x-y plane. If at the instant when $\theta = 60°$ and $\dot{\theta} = 2$ rad/s point C has the coordinates shown, determine the angular velocity of the rod and the velocity of point C at this instant.

20–23. The rod assembly is supported at B by a ball-and-socket joint and at A by a clevis connection. If the collar at B moves in the x-z plane with a speed $v_B = 5$ ft/s, determine the velocity of points A and C on the rod assembly. *Hint:* See Prob. 20–22.

Prob. 20–23

Prob. 20–21

Relative-Motion Analysis Using Translating and Rotating Axes $\overset{\star}{\textbf{20.4}}$

The most general way to analyze the three-dimensional motion of a rigid body requires the use of a system of x, y, z axes which both translate and rotate relative to a second frame X, Y, Z. This analysis also provides a means for determining the motions of two points located on separate members of a mechanism, and for determining the relative motion of one particle with respect to another when one or both particles are moving along *rotating paths*. In this section two equations will be developed which relate the velocities and accelerations of two points A and B, of which one point moves relative to a frame of reference subjected to both translation and rotation. Because of the generality in the derivation, A and B may represent either two particles moving independently of one another or two points located in the same or different rigid bodies.

As shown in Fig. 20–11, the locations of points A and B are specified relative to the X, Y, Z frame of reference by position vectors $\mathbf{r}_A$ and $\mathbf{r}_B$. The base point A represents the origin of the x, y, z coordinate system, which is translating and rotating with respect to X, Y, Z. At the instant considered, the velocity and acceleration of point A are $\mathbf{v}_A$ and $\mathbf{a}_A$, respectively, and the angular velocity and angular acceleration of the x, y, z axes are $\mathbf{\Omega}$ and $\dot{\mathbf{\Omega}} = d\mathbf{\Omega}/dt$, respectively. All these vectors are *measured* with respect to the X, Y, Z frame of reference, although they may be expressed in Cartesian component form along either set of axes.

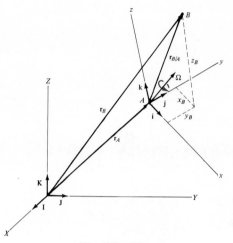

Fig. 20–11

Position. If the position of "B with respect to A" is specified by the *relative-position vector* $\mathbf{r}_{B/A}$, Fig. 20–11, then, by vector addition,

$$\mathbf{r}_B = \mathbf{r}_A + \mathbf{r}_{B/A} \qquad (20\text{–}9)$$

where

$\mathbf{r}_B$ = position of B

$\mathbf{r}_A$ = position of the origin A

$\mathbf{r}_{B/A}$ = relative position of "B with respect to A"

Velocity. The velocity of point B measured from X, Y, Z is determined by taking the time derivative of Eq. 20–9, which yields

$$(\dot{\mathbf{r}}_B)_{XYZ} = (\dot{\mathbf{r}}_A)_{XYZ} + (\dot{\mathbf{r}}_{B/A})_{XYZ}$$

The first two terms represent $\mathbf{v}_B$ and $\mathbf{v}_A$. The last term is evaluated by applying Eq. 20–6, since $\mathbf{r}_{B/A}$ is measured between two points in a rotating reference. Hence,

$$(\dot{\mathbf{r}}_{B/A})_{XYZ} = (\dot{\mathbf{r}}_{B/A})_{xyz} + \mathbf{\Omega} \times \mathbf{r}_{B/A} = (\mathbf{v}_{B/A})_{\text{rel}} + \mathbf{\Omega} \times \mathbf{r}_{B/A} \qquad (20\text{–}10)$$

Here $(\mathbf{v}_{B/A})_{\text{rel}}$ is the relative velocity of B with respect to A measured from x, y, z. Thus,

$$\mathbf{v}_B = \mathbf{v}_A + \boldsymbol{\Omega} \times \mathbf{r}_{B/A} + (\mathbf{v}_{B/A})_{\text{rel}} \qquad (20\text{–}11)$$

where

$\mathbf{v}_B$ = velocity of B

$\mathbf{v}_A$ = velocity of the origin A of the x, y, z frame of reference

$(\mathbf{v}_{B/A})_{\text{rel}}$ = relative velocity of "B with respect to A" as measured by an observer attached to the rotating x, y, z frame of reference

$\boldsymbol{\Omega}$ = angular velocity of the x, y, z frame of reference

$\mathbf{r}_{A/B}$ = relative position of "B with respect to A"

Acceleration. The acceleration of point B measured from X, Y, Z is determined by taking the time derivative of Eq. 20–11, which yields

$$(\dot{\mathbf{v}}_B)_{XYZ} = (\dot{\mathbf{v}}_A)_{XYZ} + (\dot{\boldsymbol{\Omega}})_{XYZ} \times \mathbf{r}_{B/A} + \boldsymbol{\Omega} \times (\dot{\mathbf{r}}_{B/A})_{XYZ} + (\dot{\mathbf{v}}_{B/A})_{XYZ}$$

The time derivatives defined in the first and second terms represent $\mathbf{a}_B$ and $\mathbf{a}_A$ respectively. The fourth term is evaluated using Eq. 20–10, and the last term is evaluated by applying Eq. 20–6, which yields

$$(\dot{\mathbf{v}}_{B/A})_{XYZ} = (\dot{\mathbf{v}}_{B/A})_{xyz} + \boldsymbol{\Omega} \times (\mathbf{v}_{B/A})_{\text{rel}}$$
$$= (\mathbf{a}_{B/A})_{\text{rel}} + \boldsymbol{\Omega} \times (\mathbf{v}_{B/A})_{\text{rel}}$$

Here $(\mathbf{a}_{B/A})_{\text{rel}}$ is the relative acceleration of B with respect to A measured from x, y, z. Substituting this result and Eq. 20–10 into the above equation and simplifying, we have

$$\mathbf{a}_B = \mathbf{a}_A + \dot{\boldsymbol{\Omega}} \times \mathbf{r}_{B/A} + \boldsymbol{\Omega} \times (\boldsymbol{\Omega} \times \mathbf{r}_{B/A}) + 2\boldsymbol{\Omega} \times (\mathbf{v}_{B/A})_{\text{rel}} + (\mathbf{a}_{B/A})_{\text{rel}}$$

$$(20\text{–}12)$$

where

$\mathbf{a}_B$ = acceleration of B

$\mathbf{a}_A$ = acceleration of the origin A of the x, y, z frame of reference

$(\mathbf{a}_{B/A})_{\text{rel}}, (\mathbf{v}_{B/A})_{\text{rel}}$ = relative acceleration and relative velocity of "B with respect to A" as measured by an observer attached to the rotating x, y, z frame of reference

$\dot{\boldsymbol{\Omega}}, \boldsymbol{\Omega}$ = angular acceleration and angular velocity of the x, y, z frame of reference

$\mathbf{r}_{B/A}$ = relative position of "B with respect to A"

Equations 20–11 and 20–12 are identical to those used in Sec. 16.10 for analyzing relative plane motion.* In that case, however, application was simplified since $\boldsymbol{\Omega}$ and $\dot{\boldsymbol{\Omega}}$ have a *constant direction* which is always perpendicular to the plane of motion. For three-dimensional motion, $\dot{\boldsymbol{\Omega}}$ must be com-

*Refer to Sec. 16.10 for an interpretation of the terms.

puted by using Eq. 20–6, since it depends upon the change in both the magnitude and direction of $\mathbf{\Omega}$. Furthermore, in some problems, calculation of $\mathbf{v}_A$, $\mathbf{a}_A$ and $(\mathbf{v}_{B/A})_{\text{rel}}$, $(\mathbf{a}_{B/A})_{\text{rel}}$ must also be performed by using Eq. 20–6, since these quantities depend upon the angular rate at which $\mathbf{r}_A$ and $\mathbf{r}_{B/A}$ are "swinging" as measured from their respective frames of reference.

PROCEDURE FOR ANALYSIS

The following procedure provides a method for applying Eqs. 20–11 and 20–12 to solve problems involving the three-dimensional motion of particles or rigid bodies.

Coordinate Axes. Define the location and orientation of the X, Y, Z and x, y, z coordinate axes. Most often solutions are easily obtained if at the instant considered: (1) the origins are *coincident*, (2) the axes are collinear, and/or (3) the axes are parallel. Since several components of angular velocity may be involved in a problem, the calculations will be reduced if the x, y, z axes are selected such that only one component of angular velocity is observed in this frame ($\mathbf{\Omega}_{\text{rel}}$) and the frame rotates with $\mathbf{\Omega}$ defined by the other components of angular velocity.

Kinematic Equations. After the origin of the moving reference, A, is defined and the moving point B is specified, Eqs. 20–11 and 20–12 should be written in symbolic form as

$$\mathbf{v}_B = \mathbf{v}_A + \mathbf{\Omega} \times \mathbf{r}_{B/A} + (\mathbf{v}_{B/A})_{\text{rel}}$$

$$\mathbf{a}_B = \mathbf{a}_A + \dot{\mathbf{\Omega}} \times \mathbf{r}_{B/A} + \mathbf{\Omega} \times (\mathbf{\Omega} \times \mathbf{r}_{B/A}) + 2\mathbf{\Omega} \times (\mathbf{v}_{B/A})_{\text{rel}} + (\mathbf{a}_{B/A})_{\text{rel}}$$

Vectors $\mathbf{\Omega}$, $\mathbf{r}_A$, and $\mathbf{r}_{B/A}$ should be defined from the problem data and represented in Cartesian form. *Motion of the moving reference* ($\mathbf{v}_A$, $\mathbf{a}_A$, and $\dot{\mathbf{\Omega}}$) is determined by applying Eq. 20–6 to compute the time derivatives of $\mathbf{r}_A$ and $\mathbf{\Omega}$. In a similar manner, if $\mathbf{r}_{B/A}$ has an angular motion $\mathbf{\Omega}_{\text{rel}}$ when observed from the moving reference, then the *motion of B with respect to the moving reference*, $(\mathbf{v}_{B/A})_{\text{rel}}$ and $(\mathbf{a}_{B/A})_{\text{rel}}$, must be determined by applying Eq. 20–6 to compute the time derivatives of $\mathbf{r}_{B/A}$. In all cases, the kinematic quantities which are involved in computing the time derivatives of $\mathbf{r}_A$, $\mathbf{r}_{B/A}$, and $\mathbf{\Omega}$ should be treated as *variables*. After the final forms of $\mathbf{v}_A$, $\mathbf{a}_A$, $(\mathbf{v}_{B/A})_{\text{rel}}$, $(\mathbf{a}_{B/A})_{\text{rel}}$ and $\mathbf{\Omega}$ are obtained, numerical problem data may be substituted and the kinematic terms evaluated. The components of all these vectors may be selected either along the X, Y, Z axes or along x, y, z. The choice is arbitrary, provided a consistent set of unit vectors is used.

Finally, substitute the data into the kinematic equations and perform the vector operations.

The following two examples numerically illustrate this procedure.

Example 20–4

A motor M and attached rod AB have the angular motion shown in Fig. 20–12. A collar C on the rod is located 0.25 m from A, and is moving downward along the rod with a velocity of 3 m/s and an acceleration of 2 m/s^2. Determine the velocity and acceleration of C at this instant.

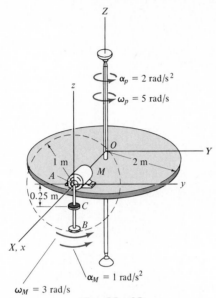

Fig. 20–12

Solution

Coordinate Axes. The origin of the fixed X, Y, Z reference is chosen at the center of the platform, and the origin of the moving x, y, z frame at point A, Fig. 20–12. Since the collar is subjected to two components of angular motion, $\boldsymbol{\omega}_p$ and $\boldsymbol{\omega}_M$, it is viewed as having an angular velocity of $\boldsymbol{\Omega}_{\text{rel}} = \boldsymbol{\omega}_M$ in x, y, z and the x, y, z axes are attached to the platform so that $\boldsymbol{\Omega} = \boldsymbol{\omega}_p$.

Kinematic Equations. Equations 20–11 and 20–12, applied to points C and A, become

$$\mathbf{v}_C = \mathbf{v}_A + \boldsymbol{\Omega} \times \mathbf{r}_{C/A} + (\mathbf{v}_{C/A})_{\text{rel}} \tag{1}$$

$$\mathbf{a}_C = \mathbf{a}_A + \dot{\boldsymbol{\Omega}} \times \mathbf{r}_{C/A} + \boldsymbol{\Omega} \times (\boldsymbol{\Omega} \times \mathbf{r}_{C/A}) + 2\boldsymbol{\Omega} \times (\mathbf{v}_{C/A})_{\text{rel}} + (\mathbf{a}_{C/A})_{\text{rel}} \tag{2}$$

Motion of Moving Reference

$$\boldsymbol{\Omega} = \boldsymbol{\omega}_p = 5\mathbf{k}$$

$$\dot{\boldsymbol{\Omega}} = (\dot{\boldsymbol{\omega}}_p)_{xyz} + \boldsymbol{\omega}_p \times \boldsymbol{\omega}_p = 2\mathbf{k} + 5\mathbf{k} \times 5\mathbf{k} = 2\mathbf{k}$$

$$\mathbf{r}_A = 2\mathbf{i}$$

$$\mathbf{v}_A = \dot{\mathbf{r}}_A = (\dot{\mathbf{r}}_A)_{xyz} + \boldsymbol{\omega}_p \times \mathbf{r}_A = 0 + (5\mathbf{k} \times 2\mathbf{i}) = 10\mathbf{j}$$

$$\mathbf{a}_A = \ddot{\mathbf{r}}_A = [(\ddot{\mathbf{r}}_A)_{xyz} + \boldsymbol{\omega}_p \times (\dot{\mathbf{r}}_A)_{xyz}] + \dot{\boldsymbol{\omega}}_p \times \mathbf{r}_A + \boldsymbol{\omega}_p \times \dot{\mathbf{r}}_A$$

$$= 0 + 0 + (2\mathbf{k} \times 2\mathbf{i}) + (5\mathbf{k} \times 10\mathbf{j}) = -50\mathbf{i} + 4\mathbf{j}$$

Since point A moves in a circular path lying in the X-Y plane we can also apply Eqs. 16–14 and 16–16 to obtain these same results.

Motion of C with Respect to Moving Reference

$$\boldsymbol{\Omega}_{\text{rel}} = \boldsymbol{\omega}_M = 3\mathbf{i} \qquad \dot{\boldsymbol{\Omega}}_{\text{rel}} = \dot{\boldsymbol{\omega}}_M = 1\mathbf{i}$$

Note that $\boldsymbol{\omega}_M$ does not change direction relative to x, y, z.

$$\mathbf{r}_{C/A} = -0.25\mathbf{k}$$

$$(\mathbf{v}_{C/A})_{\text{rel}} = \dot{\mathbf{r}}_{C/A} = (\dot{\mathbf{r}}_{C/A})_{xyz} + \boldsymbol{\omega}_M \times \mathbf{r}_{C/A}$$

$$= -3\mathbf{k} + [3\mathbf{i} \times (-0.25\mathbf{k})] = 0.75\mathbf{j} - 3\mathbf{k}$$

$$(\mathbf{a}_{C/A})_{\text{rel}} = \ddot{\mathbf{r}}_{C/A} = [(\ddot{\mathbf{r}}_{C/A})_{xyz} + \boldsymbol{\omega}_M \times (\dot{\mathbf{r}}_{C/A})_{xyz}] + \dot{\boldsymbol{\omega}}_M \times \mathbf{r}_{C/A} + \boldsymbol{\omega}_M \times \dot{\mathbf{r}}_{C/A}$$

$$= [-2\mathbf{k} + 3\mathbf{i} \times (-3\mathbf{k})] + [(1\mathbf{i}) \times (-0.25\mathbf{k})] + [(3\mathbf{i}) \times (0.75\mathbf{j} - 3\mathbf{k})]$$

$$= 18.25\mathbf{j} + 0.25\mathbf{k}$$

Try to obtain these same results by applying Eqs. 20–11 and 20–12 between the x, y, z axes and coincident (rotating) x', y', z' axes having $\boldsymbol{\Omega} = 3\mathbf{i}$, $\dot{\boldsymbol{\Omega}} = 1\mathbf{i}$.

Substituting the data into Eqs. (1) and (2) yields

$$\mathbf{v}_C = \mathbf{v}_A + \boldsymbol{\Omega} \times \mathbf{r}_{C/A} + (\mathbf{v}_{C/A})_{\text{rel}}$$

$$= 10\mathbf{j} + [5\mathbf{k} \times (-0.25\mathbf{k})] + (0.75\mathbf{j} - 3\mathbf{k})$$

$$= \{10.8\mathbf{j} - 3\mathbf{k}\} \text{ m/s} \qquad\qquad Ans.$$

$$\mathbf{a}_C = \mathbf{a}_A + \dot{\boldsymbol{\Omega}} \times \mathbf{r}_{C/A} + \boldsymbol{\Omega} \times (\boldsymbol{\Omega} \times \mathbf{r}_{C/A}) + 2\boldsymbol{\Omega} \times (\mathbf{v}_{C/A})_{\text{rel}} + (\mathbf{a}_{C/A})_{\text{rel}}$$

$$= (-50\mathbf{i} + 4\mathbf{j}) + [2\mathbf{k} \times (-0.25\mathbf{k})] + 5\mathbf{k} \times [5\mathbf{k} \times (-0.25\mathbf{k})]$$

$$+ 2[5\mathbf{k} \times (0.75\mathbf{j} - 3\mathbf{k})] + (18.25\mathbf{j} + 0.25\mathbf{k})$$

$$= \{-57.5\mathbf{i} + 22.2\mathbf{j} + 0.25\mathbf{k}\} \text{ m/s}^2 \qquad\qquad Ans.$$

Example 20–5

The pendulum shown in Fig. 20–13 consists of two rods. AB is pin-supported at A and swings only in the Y-Z plane, whereas a bearing at B allows the attached rod BD to spin about rod AB. At a given instant, the rods have the angular motions shown. If a collar C, located 0.2 m from B, has a velocity of 3 m/s and an acceleration of 2 m/s² along the rod, determine the velocity and acceleration of the collar at this instant.

Fig. 20–13

Solution I $(\boldsymbol{\Omega} = \boldsymbol{\omega}_1,\ \boldsymbol{\Omega}_{\text{rel}} = \boldsymbol{\omega}_2)$

Coordinate Axes. Since the rod is rotating about the fixed point A, the origin of the X, Y, Z frame will be chosen at A. Motion of the collar is conveniently observed from B, so the origin of the x, y, z frame is located at this point.

Kinematic Equations

$$\mathbf{v}_C = \mathbf{v}_B + \boldsymbol{\Omega} \times \mathbf{r}_{C/B} + (\mathbf{v}_{C/B})_{\text{rel}} \tag{1}$$

$$\mathbf{a}_C = \mathbf{a}_B + \dot{\boldsymbol{\Omega}} \times \mathbf{r}_{C/B} + \boldsymbol{\Omega} \times (\boldsymbol{\Omega} \times \mathbf{r}_{C/B}) + 2\boldsymbol{\Omega} \times (\mathbf{v}_{C/B})_{\text{rel}} + (\mathbf{a}_{C/B})_{\text{rel}} \tag{2}$$

Motion of Moving Reference

$$\boldsymbol{\Omega} = \boldsymbol{\omega}_1 = 4\mathbf{i}, \quad \dot{\boldsymbol{\Omega}} = (\dot{\boldsymbol{\omega}}_1)_{xyz} + \boldsymbol{\omega}_1 \times \boldsymbol{\omega}_1 = 1.5\mathbf{i} + 0 = 1.5\mathbf{i}$$

$$\mathbf{r}_B = -0.5\mathbf{k}$$

$$\mathbf{v}_B = \dot{\mathbf{r}}_B = (\dot{\mathbf{r}}_B)_{xyz} + \boldsymbol{\omega}_1 \times \mathbf{r}_B = 0 + 4\mathbf{i} \times (-0.5\mathbf{k}) = 2\mathbf{j} \left.\begin{array}{l}\\ \\ \\ \\ \end{array}\right.$$

$$\mathbf{a}_B = \ddot{\mathbf{r}}_B = [(\ddot{\mathbf{r}}_B)_{xyz} + \boldsymbol{\omega}_1 \times (\dot{\mathbf{r}}_B)_{xyz}] + \dot{\boldsymbol{\omega}}_1 \times \mathbf{r}_B + \boldsymbol{\omega}_1 \times \dot{\mathbf{r}}_B$$

$$= [0 + 0] + [1.5\mathbf{i} \times (-0.5\mathbf{k})] + (4\mathbf{i} \times 2\mathbf{j}) = 0.75\mathbf{j} + 8\mathbf{k}$$

This is the same as applying Eqs. 16–14 and 16–16 since B has circular motion in the X-Y plane.

Motion of C with Respect to Moving Reference

$$\boldsymbol{\Omega}_{\text{rel}} = \boldsymbol{\omega}_2 = 5\mathbf{k}, \quad \dot{\boldsymbol{\Omega}}_{\text{rel}} = \dot{\boldsymbol{\omega}}_2 = -6\mathbf{k}\}$$

Note that $\boldsymbol{\omega}_2$ does not change direction relative to x, y, z.

$$\mathbf{r}_{C/B} = 0.2\mathbf{j}$$

$$(\mathbf{v}_{C/B})_{\text{rel}} = \dot{\mathbf{r}}_{C/B} = (\dot{\mathbf{r}}_{C/B})_{xyz} + \boldsymbol{\omega}_2 \times \mathbf{r}_{C/B} = 3\mathbf{j} + (5\mathbf{k} \times 0.2\mathbf{j}) = -1\mathbf{i} + 3\mathbf{j}$$

$$(\mathbf{a}_{C/B})_{\text{rel}} = \ddot{\mathbf{r}}_{C/B} = [(\ddot{\mathbf{r}}_{C/B})_{xyz} + \boldsymbol{\omega}_2 \times (\dot{\mathbf{r}}_{C/B})_{xyz}] + \dot{\boldsymbol{\omega}}_2 \times \mathbf{r}_{C/B} + \boldsymbol{\omega}_2 \times \dot{\mathbf{r}}_{C/B}$$

$$= [2\mathbf{j} + (5\mathbf{k} \times 3\mathbf{j})] + (-6\mathbf{k} \times 0.2\mathbf{j}) + [5\mathbf{k} \times (-1\mathbf{i} + 3\mathbf{j})]$$

$$= -28.8\mathbf{i} - 3\mathbf{j}$$

The same results can be obtained by applying Eqs. 20–11 and 20–12 between the x, y, z axes and coincident (rotating) x', y', z' axes having $\boldsymbol{\Omega} = 5\mathbf{k}$, $\dot{\boldsymbol{\Omega}} = -6\mathbf{k}$.

Substituting the data into Eqs. (1) and (2) yields

$$\mathbf{v}_C = \mathbf{v}_B + \boldsymbol{\Omega} \times \mathbf{r}_{C/B} + (\mathbf{v}_{C/B})_{\text{rel}} = 2\mathbf{j} + (4\mathbf{i} \times 0.2\mathbf{j}) + (-1\mathbf{i} + 3\mathbf{j})$$

$$= \{-1\mathbf{i} + 5\mathbf{j} + 0.8\mathbf{k}\} \text{ m/s} \qquad \qquad \textit{Ans.}$$

$$\mathbf{a}_C = \mathbf{a}_B + \dot{\boldsymbol{\Omega}} \times \mathbf{r}_{C/B} + \boldsymbol{\Omega} \times (\boldsymbol{\Omega} \times \mathbf{r}_{C/B}) + 2\boldsymbol{\Omega} \times (\mathbf{v}_{C/B})_{\text{rel}} + (\mathbf{a}_{C/B})_{\text{rel}}$$

$$= (0.75\mathbf{j} + 8\mathbf{k}) + (1.5\mathbf{i} \times 0.2\mathbf{j}) + [4\mathbf{i} \times (4\mathbf{i} \times 0.2\mathbf{j})]$$

$$\qquad\qquad + 2[4\mathbf{i} \times (-1\mathbf{i} + 3\mathbf{j})] + (-28.8\mathbf{i} - 3\mathbf{j})$$

$$= \{-28.8\mathbf{i} - 5.45\mathbf{j} + 32.3\mathbf{k}\} \text{ m/s}^2 \qquad \qquad \textit{Ans.}$$

Solution II $(\Omega = \omega_1 + \omega_2, \ \Omega_{\text{rel}} = 0)$
Motion of Moving Reference

$$\Omega = \omega_1 + \omega_2 = 4\mathbf{i} + 5\mathbf{k}$$

From the constraints of the problem ω_1 does not change direction; however, the direction of ω_2 is changed by ω_1. Thus, a simple way of obtaining $\dot{\Omega}$ is to consider x', y', z' axes coincident with the X, Y, Z axes at A, such that the primed axes have an angular velocity ω_1. Then

$$\dot{\Omega} = \dot{\omega}_1 + \dot{\omega}_2 = [(\dot{\omega}_1)_{x'y'z'} + \omega_1 \times \omega_1] + [(\dot{\omega}_2)_{x'y'z'} + \omega_1 \times \omega_2]$$

$$= [1.5\mathbf{i} + 0] + [-6\mathbf{k} + (4\mathbf{i} \times 5\mathbf{k})] = 1.5\mathbf{i} - 20\mathbf{j} - 6\mathbf{k}$$

Also,
$$\mathbf{r}_B = -0.5\mathbf{k}$$

Here again only ω_1 changes the direction of $\mathbf{r}_B$ so that the time derivatives of $\mathbf{r}_B$ can be computed from the primed axes used above, which are rotating at ω_1. Hence,

$$\mathbf{v}_B = \dot{\mathbf{r}}_B = (\dot{\mathbf{r}}_B)_{x'y'z'} + \omega_1 \times \mathbf{r}_B$$

$$= 0 + 4\mathbf{i} \times (-0.5\mathbf{k}) = 2\mathbf{j}$$

$$\mathbf{a}_B = \ddot{\mathbf{r}}_B = [(\ddot{\mathbf{r}}_B)_{x'y'z'} + \omega_1 \times (\dot{\mathbf{r}}_B)_{x'y'z'}] + \dot{\omega}_1 \times \mathbf{r}_B + \omega_1 \times \dot{\mathbf{r}}_B$$

$$= [0 + 0] + [1.5\mathbf{i} \times (-0.5\mathbf{k})] + (4\mathbf{i} \times 2\mathbf{j}) = 0.75\mathbf{j} + 8\mathbf{k}$$

Motion of C with Respect to Moving Reference

$$\Omega_{\text{rel}} = 0$$

$$\dot{\Omega}_{\text{rel}} = 0$$

$$\mathbf{r}_{C/B} = 0.2\mathbf{j}$$

$$(\mathbf{v}_{C/B})_{\text{rel}} = \dot{\mathbf{r}}_{C/B} = (\dot{\mathbf{r}}_{C/B})_{xyz} + \Omega_{\text{rel}} \times \mathbf{r}_{C/B}$$

$$= 3\mathbf{j} + 0 = 3\mathbf{j}$$

$$(\mathbf{a}_{C/B})_{\text{rel}} = \ddot{\mathbf{r}}_{C/B} = [(\ddot{\mathbf{r}}_{C/B})_{xyz} + \Omega_{\text{rel}} \times (\dot{\mathbf{r}}_{C/B})_{xyz}] + \dot{\Omega}_{C/B} \times \mathbf{r}_{C/B} + \Omega_{C/B} \times \dot{\mathbf{r}}_{C/B}$$

$$= [2\mathbf{j} + 0] + 0 + 0 = 2\mathbf{j}$$

Substituting the data into Eqs. (1) and (2) yields

$$\mathbf{v}_C = \mathbf{v}_B + \Omega \times \mathbf{r}_{C/B} + (\mathbf{v}_{C/B})_{\text{rel}}$$

$$= 2\mathbf{j} + [(4\mathbf{i} + 5\mathbf{k}) \times (0.2\mathbf{j})] + 3\mathbf{j}$$

$$= \{-1\mathbf{i} + 5\mathbf{j} + 0.8\mathbf{k}\} \text{ m/s} \qquad\qquad\qquad Ans.$$

$$\mathbf{a}_C = \mathbf{a}_B + \dot{\Omega} \times \mathbf{r}_{C/B} + \Omega \times (\Omega \times \mathbf{r}_{C/B}) + 2\Omega \times (\mathbf{v}_{C/B})_{\text{rel}} + (\mathbf{a}_{C/B})_{\text{rel}}$$

$$= (0.75\mathbf{j} + 8\mathbf{k}) + [(1.5\mathbf{i} - 20\mathbf{j} - 6\mathbf{k}) \times (0.2\mathbf{j})]$$

$$+ (4\mathbf{i} + 5\mathbf{k}) \times [(4\mathbf{i} + 5\mathbf{k}) \times 0.2\mathbf{j}] + 2[(4\mathbf{i} + 5\mathbf{k}) \times 3\mathbf{j}] + 2\mathbf{j}$$

$$= \{-28.8\mathbf{i} - 5.45\mathbf{j} + 32.3\mathbf{k}\} \text{ m/s}^2 \qquad\qquad Ans.$$

Problems

*20–24. Solve Example 20–5 by fixing the x, y, z axes to rod BD so that $\Omega = \omega_1 + \omega_2$. In this case the collar appears only to move radially outward along BD, hence $\Omega_{C/B} = 0$.

20–25. Solve Example 20–5 such that the x, y, z axes move with curvilinear translation, $\Omega = 0$, in which case the collar appears to have both an angular velocity $\Omega_{C/B} = \omega_1 + \omega_2$ and radial motion.

20–26. The boom AB of the locomotive crane is rotating about the Z axis with an angular velocity $\omega_1 = 0.5$ rad/s and has a rate of increase $\dot\omega_1 = 3$ rad/s². At the same instant, $\theta = 30°$ and the boom is rotating upward at a constant rate $\dot\theta = 3$ rad/s. Determine the velocity and acceleration of the tip B of the boom at this instant.

Prob. 20–26

20–27. At the instant shown, the telescopic boom is rotating about the z axis with an angular velocity $\omega_1 = 1.5$ rad/s and angular acceleration $\dot\omega_1 = 2$ rad/s². At this same instant the boom OA is extending at $\dot r = 4$ ft/s, which is increasing at $\ddot r = 2.5$ ft/s². Also, the bucket is rotating at a constant rate $\omega_2 = 8$ rad/s, measured relative to the boom. Determine the velocity and acceleration of point P on the bucket at this instant.

Prob. 20–27

*20–28. At a given instant, rod BD is rotating about the vertical axis with an angular velocity $\omega_{BD} = 7$ rad/s and an angular acceleration $\alpha_{BD} = 4$ rad/s². Also, $\theta = 60°$ and link AC is rotating downward such that $\dot\theta = 2$ rad/s and $\ddot\theta = 3$ rad/s². Determine the velocity and acceleration of point A on the link at this instant.

Prob. 20–28

20–29. At the instant shown, the helicopter is moving upward with a velocity $v_H = 4$ ft/s and has an acceleration $a_H = 2$ m/s^2. At the same instant the frame H, *not* the horizontal blade, is rotating about a vertical axis with an angular velocity $\omega_H = 0.9$ rad/s. If the tail blade B is rotating with an angular velocity $\omega_{B/H} = 180$ rad/s, measured relative to H, determine the velocity and acceleration of point P located on the tip of the blade at the instant the blade is in the vertical position.

Prob. 20–29

20–31. At a given instant the crane is moving along the wall with a velocity $v_{CD} = 8$ m/s and acceleration of 9 m/s^2. Simultaneously, it has the angular motions shown. If the trolley T is moving outwards along the boom AB with a relative speed of 3 ft/s and relative acceleration of 5 ft/s^2, determine the velocity and acceleration of the trolley.

Prob. 20–31

20–30. At a given instant, the antenna has an angular motion $\omega_1 = 3$ rad/s and $\dot{\omega}_1 = 2$ rad/s^2 about the Z axis. At this same instant $\theta = 30°$, the angular motion about the X axis is $\omega_2 = 1.5$ rad/s, and $\dot{\omega}_2 = 4$ rad/s^2. Determine the velocity and acceleration of the signal horn A at this instant. The distance from O to A is $d = 3$ ft.

Prob. 20–30

*20–32. At a given instant, the rod has the angular motions shown, while the collar C is moving down *relative* to the rod with a velocity of 6 ft/s and an acceleration of 2 ft/s^2. Determine the collar's velocity and acceleration at this instant.

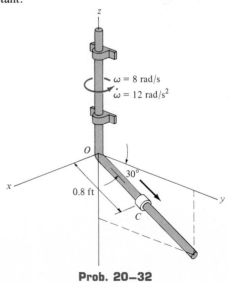

Prob. 20–32

20–33. At the instant shown, the arm AB is rotating about the fixed pin A with an angular velocity $\omega_1 = 4$ rad/s and angular acceleration $\dot{\omega}_1 = 3$ rad/s^2. At this same instant, rod BD is rotating relative to rod AB with an angular velocity $\omega_2 = 5$ rad/s and an angular acceleration $\dot{\omega}_2 = 7$ rad/s^2. Also, the collar C is moving along rod BD with a velocity of 3 m/s and an acceleration of 2 m/s^2, both measured relative to the rod. Determine the velocity and acceleration of the collar at this instant.

Prob. 20–33

20–34. At the instant shown, the arm AB is rotating about the fixed bearing with an angular velocity $\omega_1 = 2$ rad/s and angular acceleration $\dot{\omega}_1 = 6$ rad/s^2. At the same instant, rod BD is rotating relative to rod AB at $\omega_2 = 7$ rad/s, which is increasing at $\dot{\omega}_2 = 1$ rad/s^2. Also, the collar C is moving along rod BD with a velocity $\dot{r} = 2$ ft/s and a *deceleration* $\ddot{r} = -0.5$ ft/s^2, both measured relative to the rod. Determine the velocity and acceleration of the collar at this instant.

Prob. 20–34

20–35. At the given instant, the rod is spinning about the z axis with an angular velocity $\omega_1 = 8$ rad/s and angular acceleration $\dot{\omega}_1 = 2$ rad/s^2. At this same instant, the disk is spinning at a constant rate $\omega_2 = 4$ rad/s, measured *relative* to the rod. Determine the velocity and acceleration of point P on the disk at this instant.

Prob. 20–35

*20–36. At the instant shown, the frame $ABDE$ is rotating about the y axis with an angular velocity $\omega_1 = 6$ rad/s and an angular acceleration $\dot{\omega}_1 = 2$ rad/s^2. At this same instant, the rotating rod FG has an angular motion *relative to the frame* as shown in the figure. If the collar C is moving along the rod toward F with a velocity of 0.5 m/s and acceleration of 0.7 m/s^2, both measured relative to rod FG, determine the collar's velocity and acceleration at this instant.

Prob. 20–36

20–37. At the instant shown, the arm OA of the conveyor belt is rotating about the z axis with a constant angular velocity $\omega_1 = 6$ rad/s, while at the same instant the arm is rotating upward at a constant rate $\omega_2 = 4$ rad/s. If the conveyor is running with a constant rate $\dot{r} = 5$ ft/s, determine the velocity and acceleration of the package P at the instant shown. Neglect the size of the package.

20–39. At a given instant the portal crane is moving forward with a velocity $v_C = 6$ ft/s and an acceleration $a_C = 2$ ft/s^2. Simultaneously, it has the angular motions shown at $\theta = 60°$. Determine the velocity and acceleration of point A at the top of boom AB.

Probs. 20–37 / 20–38

20–38. At the instant shown, the arm OA of the conveyor belt is rotating about the z axis with a constant angular velocity $\omega_1 = 6$ rad/s, while at the same instant the arm is rotating upward at a constant rate $\omega_2 = 4$ rad/s. If the conveyor is running at a rate $\dot{r} = 5$ ft/s, which is increasing at $\ddot{r} = 8$ ft/s^2, determine the velocity and acceleration of the package P at the instant shown. Neglect the size of the package.

Prob. 20–39

*20–40. At the instant shown, the rod AB is rotating about the z axis with an angular velocity $\omega_1 = 4$ rad/s and an angular acceleration $\dot{\omega}_1 = 3$ rad/s^2. At this same instant, the circular rod has an angular motion relative to the rod as shown in the figure. If the collar C is moving down around the circular rod with a speed of 3 in./s, which is increasing at 8 in./s^2, both measured relative to the rod, determine the collar's velocity and acceleration at this instant.

20–41. At the instant shown, the industrial manipulator is rotating about the z axis at $\omega_1 = 5$ rad/s, and about joint B at $\omega_2 = 2$ rad/s. Determine the velocity and acceleration of the grip A at this instant when $\phi = 30°$, $\theta = 45°$, and $r = 1.6$ m.

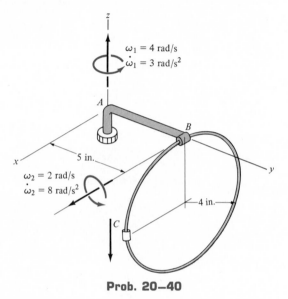

Prob. 20–40

Probs. 20–41 / 20–42

20–42. At the instant shown, the industrial manipulator is rotating about the z axis at $\omega_1 = 5$ rad/s, which is increasing at the rate $\dot{\omega}_1 = 7$ rad/s^2. Simultaneously, the arm BA is rotating about joint B at $\omega_2 = 2$ rad/s, which is increasing at the rate $\dot{\omega}_2 = 6$ rad/s^2, and the arm BA is telescoping outward (elongating) at $\dot{r} = 0.5$ m/s, which is increasing at $\ddot{r} = 1.3$ m/s^2. Determine the velocity and acceleration of the grip A at this instant when $\phi = 30°$, $\theta = 45°$, and $r = 1.6$ m.

Three-Dimensional Kinetics of a Rigid Body

In general, as a body moves through space it has a simultaneous translation and rotation at a given instant. The kinetic aspects of the *translation* have been discussed in Chapter 17, where it was shown that a system of external forces acting on the body may be related to the acceleration of the body's mass center by the equation $\Sigma \mathbf{F} = m\mathbf{a}_G$. In this chapter emphasis is placed primarily upon the *rotational* aspects of rigid-body motion, since motion of the body's mass center, defined by $\Sigma \mathbf{F} = m\mathbf{a}_G$, is treated in the same manner as particle motion.

The rotational equations of motion relate the body's components of angular motion to the moment components created by the external forces about some point located either on or off the body. To apply these equations, it is first necessary to formulate the moments and products of inertia of the body and to compute the body's angular momentum. Afterward, the principles of impulse and momentum and work and energy will be at our disposal for solving problems in three dimensions. The rotational equations of motion are developed in Sec. 21.4. Of special interest are problems involving the motion of an unsymmetrical body about a fixed axis; motion of a gyroscope, Sec. 21.5; and torque-free motion, Sec. 21.6.

Moments and Products of Inertia $\mathbf{21.1}^{*}$

When studying the planar kinetics of a body, it was necessary to introduce the moment of inertia I_G, which was computed about an axis perpendicular to the plane of motion and passing through the mass center G. For the kinetic analysis of three-dimensional motion it will sometimes be necessary to calculate six

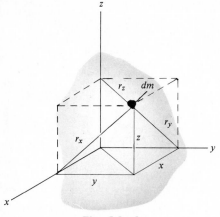

Fig. 21–1

inertial quantities. These terms, called the moments and products of inertia, describe in a particular way the distribution of mass for a body relative to a given coordinate system that has a specified orientation and point of origin.

Moment of Inertia. Consider the rigid body shown in Fig. 21–1. The *moment of inertia* for a differential element dm of the body about any one of the three coordinate axes is defined as the product of the mass of the element and the square of the shortest distance from the axis to the element. For example, as noted in the figure, $r_x = \sqrt{y^2 + z^2}$, so that the mass moment of inertia of dm about the x axis is

$$dI_{xx} = r_x^2 \, dm = (y^2 + z^2) \, dm$$

The moment of inertia I_{xx} for the entire body is determined by integrating this expression over the entire mass of the body. Hence, for each of the axes, we may write

$$I_{xx} = \int_m r_x^2 \, dm = \int_m (y^2 + z^2) \, dm$$

$$I_{yy} = \int_m r_y^2 \, dm = \int_m (x^2 + z^2) \, dm \qquad (21\text{–}1)$$

$$I_{zz} = \int_m r_z^2 \, dm = \int_m (x^2 + y^2) \, dm$$

Here it is seen that the moment of inertia is *always a positive quantity,* since it is the summation of the product of the mass dm, which is always positive, and distances squared.

Product of Inertia. The *product of inertia* for a differential element dm is defined with respect to a set of *two orthogonal planes* as the product of the mass of the element and the perpendicular (or shortest) distances from the planes to the element. For example, with respect to the x-z and y-z planes, the product of inertia dI_{xy} for the element dm shown in Fig. 21–1 is

$$dI_{xy} = xy \, dm$$

Note also that $dI_{yx} = dI_{xy}$. By integrating over the entire mass, the product of inertia of the body for each combination of planes may be expressed as

$$I_{xy} = I_{yx} = \int_m xy \, dm$$

$$I_{yz} = I_{zy} = \int_m yz \, dm \qquad (21\text{–}2)$$

$$I_{xz} = I_{zx} = \int_m xz \, dm$$

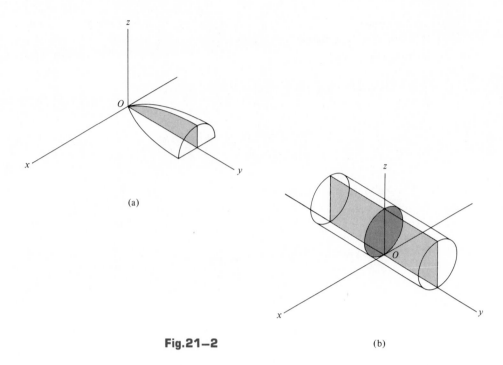

Fig.21–2

(a)

(b)

Unlike the moment of inertia, which is always positive, the product of inertia may be positive, negative, or zero. The result depends upon the signs of the two defining coordinates, which vary independently from one another. In particular, if either one or both of the orthogonal planes are *planes of symmetry* for the mass, the *product of inertia* with respect to these planes will be *zero*. In such cases, elements of mass will occur in *pairs,* located on each side of the plane of symmetry. On one side of the plane the product of inertia for the element will be positive, while on the other side the product of inertia for the corresponding element will be negative, the sum therefore yielding zero. Examples of this are shown in Fig. 21–2. In the first case, Fig. 21–2*a*, the *y-z* plane is a plane of symmetry, and hence for point O, $I_{xz} = I_{xy} = 0$. Computation for I_{yz} will yield a *positive* result, since all elements of mass are located using only positive y and z coordinates. For the cylinder, with the coordinate axes located as shown in Fig. 21–2*b*, the *x-z* and *y-z* planes are both planes of symmetry. Thus, for point O, $I_{zx} = I_{yz} = I_{xy} = 0$.

Parallel-Axis and Parallel-Plane Theorems. The techniques of integration which are used to determine the moment of inertia of a body were described in Sec. 17.1. Also discussed were methods to determine the moment of inertia of a composite body, i.e., a body that is composed of simpler segments, as tabulated on the inside back cover of the book. In both of these cases the *parallel-axis theorem* is often used for the calculations. This theorem, which

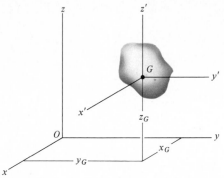

Fig. 21-3

was proven in Sec. 17.1, is used to transfer the moment of inertia of a body from an axis passing through its mass center G to a parallel axis passing through some other point. In this regard, if G has coordinates x_G, y_G, z_G defined from the x, y, z axes, Fig. 21-3, then the parallel-axis theorems used to calculate the moments of inertia about the x, y, z axes are

$$\begin{aligned} I_{xx} &= (I_{x'x'})_G + m(y_G^2 + z_G^2) \\ I_{yy} &= (I_{y'y'})_G + m(x_G^2 + z_G^2) \\ I_{zz} &= (I_{z'z'})_G + m(x_G^2 + y_G^2) \end{aligned} \qquad (21\text{-}3)$$

The products of inertia of a body or a composite are computed in the same manner as the body's moments of inertia. Here, however, the *parallel-plane theorems* are important. These theorems are used to transfer the products of inertia of the body from a set of three orthogonal planes passing through the body's mass center to a corresponding set of three parallel planes passing through some other point O. Defining the perpendicular distances between the planes as x_G, y_G, and z_G, Fig. 21-3, the parallel-plane theorems can be written as

$$\begin{aligned} I_{xy} &= (I_{x'y'})_G + mx_Gy_G \\ I_{yz} &= (I_{y'z'})_G + my_Gz_G \\ I_{zx} &= (I_{z'x'})_G + mz_Gx_G \end{aligned} \qquad (21\text{-}4)$$

The derivation of these formulas is similar to that given for the parallel-axis theorem, Sec. 17.1.

Inertia Tensor. The inertial properties of a body are completely characterized by nine terms, six of which are independent of one another. This set of terms is defined using Eqs. 21-1 and 21-2 and can be written as

$$\begin{pmatrix} I_{xx} & -I_{xy} & -I_{xz} \\ -I_{yx} & I_{yy} & -I_{yz} \\ -I_{zx} & -I_{zy} & I_{zz} \end{pmatrix}$$

This array is called an *inertia tensor*. It has a unique set of values for a body when it is computed for each location of the origin O and orientation of the coordinate axes.

For point O we can specify a unique axis inclination for which the products of inertia for the body are zero when computed with respect to these axes. When this is done, the inertia tensor is said to be "diagonalized" and may be written in the simplified form

$$\begin{pmatrix} I_x & 0 & 0 \\ 0 & I_y & 0 \\ 0 & 0 & I_z \end{pmatrix}$$

Here $I_x = I_{xx}$, $I_y = I_{yy}$, and $I_z = I_{zz}$ are termed the *principal moments of iner-*

tia for the body, which are computed from the *principal axes of inertia*. Of these three principal moments of inertia, one will be a maximum and another a minimum of the body's moment of inertia.

Mathematical determination of the directions of principal axes of inertia will not be discussed here (see Prob. 21–18). There are many cases, however, in which the principal axes may be determined by inspection. From the previous discussion it was noted that if the coordinate axes are oriented such that *two* of the three orthogonal planes containing the axes are planes of *symmetry* for the body, then all the products of inertia for the body are zero with respect to the coordinate planes, and hence the coordinate axes are principal axes of inertia. For example, the *x, y, z* axes shown in Fig. 21–2*b* represent the principal axes of inertia for the cylinder at point *O*.

Moment of Inertia About an Arbitrary Axis. Consider the body shown in Fig. 21–4, where the nine elements of the inertia tensor have been computed for the *x, y, z* axes having an origin at *O*. Here we wish to compute the moment of inertia of the body about the *Oa* axis, for which the direction is defined by the unit vector $\mathbf{u}_a$. By definition $I_{aa} = \int b^2 \, dm$, where *b* is the *perpendicular distance* from *dm* to *Oa*. If the position of *dm* is located using $\mathbf{r}$, then $b = r \sin \theta$, which represents the *magnitude* of the cross product $\mathbf{u}_a \times \mathbf{r}$. Hence, the moment of inertia can be expressed as

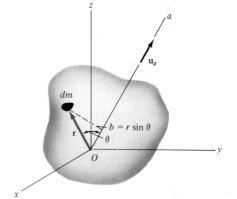

Fig. 21–4

$$I_{aa} = \int_m |(\mathbf{u}_a \times \mathbf{r})|^2 \, dm = \int_m (\mathbf{u}_a \times \mathbf{r}) \cdot (\mathbf{u}_a \times \mathbf{r}) \, dm$$

Provided $\mathbf{u}_a = u_x \mathbf{i} + u_y \mathbf{j} + u_z \mathbf{k}$ and $\mathbf{r} = x\mathbf{i} + y\mathbf{j} + z\mathbf{k}$, so that $\mathbf{u}_a \times \mathbf{r} = (u_y z - u_z y)\mathbf{i} + (u_z x - u_x z)\mathbf{j} + (u_x y - u_y x)\mathbf{k}$, then, after substituting and performing the dot-product operation, we can write the moment of inertia as

$$I_{aa} = \int_m [(u_y z - u_z y)^2 + (u_z x - u_x z)^2 + (u_x y - u_y x)^2] \, dm$$

$$= u_x^2 \int_m (y^2 + z^2) \, dm + u_y^2 \int_m (z^2 + x^2) \, dm + u_z^2 \int_m (x^2 + y^2) \, dm$$

$$- 2u_x u_y \int_m xy \, dm - 2u_y u_z \int_m yz \, dm - 2u_z u_x \int_m zx \, dm$$

Recognizing the integrals to be the moments and products of inertia of the body, Eqs. 21–1 and 21–2, we have

$$I_{aa} = I_{xx} u_x^2 + I_{yy} u_y^2 + I_{zz} u_z^2 - 2I_{xy} u_x u_y - 2I_{yz} u_y u_z - 2I_{zx} u_z u_x \quad (21\text{–}5)$$

Thus, if the inertia tensor is specified for the *x, y, z* axes, the moment of inertia of the body about the inclined *Oa* axis can be computed by using Eq. 21–5. For the calculation the direction cosines u_x, u_y, u_z of the axes must be determined. These terms specify the cosines of the coordinate direction angles α, β, γ made between the *Oa* axis and the *x, y, z* axes, respectively (see Appendix C).

Example 21–1

Determine the moment of inertia of the bent rod shown in Fig. 21–5a about the *Aa* axis. The mass of each of the three segments is shown in the figure.

(a)

Solution

The moment of inertia I_{Aa} can be computed by using Eq. 21–5. It is first necessary, however, to determine the moments and products of inertia of the rod about the x, y, z axes. This is done using the formula for the moment of inertia of a slender rod, $I = \frac{1}{12}ml^2$, and the parallel-axis and parallel-plane theorems, Eqs. 21–3 and 21–4. Dividing the rod into three parts and locating the mass center of each segment, Fig. 21–5b, we have

Fig. 21–5

$$I_{xx} = \left[\frac{1}{12}(2)(0.2)^2 + 2(0.1)^2\right] + [0 + 2(0.2)^2]$$
$$+ \left[\frac{1}{12}(4)(0.4)^2 + 4((0.2)^2 + (0.2)^2)\right] = 0.480 \text{ kg} \cdot \text{m}^2$$

$$I_{yy} = \left[\frac{1}{12}(2)(0.2)^2 + 2(0.1)^2\right] + \left[\frac{1}{12}(2)(0.2)^2 + 2((-0.1)^2 + (0.2)^2)\right]$$
$$+ [0 + 4((-0.2)^2 + (0.2)^2)] = 0.453 \text{ kg} \cdot \text{m}^2$$

$$I_{zz} = [0 + 0] + \left[\frac{1}{12}(2)(0.2)^2 + 2(0.1)^2\right]$$
$$+ \left[\frac{1}{12}(4)(0.4)^2 + 4((-0.2)^2 + (0.2)^2)\right] = 0.400 \text{ kg} \cdot \text{m}^2$$

$$I_{xy} = [0 + 0] + [0 + 0] + [0 + 4(-0.2)(0.2)] = -0.160 \text{ kg} \cdot \text{m}^2$$
$$I_{yz} = [0 + 0] + [0 + 0] + [0 + 4(0.2)(0.2)] = 0.160 \text{ kg} \cdot \text{m}^2$$
$$I_{zx} = [0 + 0] + [0 + 2(0.2)(-0.1)] + [0 + 4(0.2)(-0.2)]$$
$$= -0.200 \text{ kg} \cdot \text{m}^2$$

The *Aa* axis is defined by the unit vector

$$\mathbf{u}_{Aa} = \frac{\mathbf{r}_D}{r_D} = \frac{-0.2\mathbf{i} + 0.4\mathbf{j} + 0.2\mathbf{k}}{\sqrt{(-0.2)^2 + (0.4)^2 + (0.2)^2}}$$
$$= -0.408\mathbf{i} + 0.816\mathbf{j} + 0.408\mathbf{k}$$

Thus,

(b)

$$u_x = -0.408, \qquad u_y = 0.816, \qquad u_z = 0.408$$

Substituting the computed data into Eq. 21–5 yields

$$I_{Aa} = I_{xx}u_x^2 + I_{yy}u_y^2 + I_{zz}u_z^2 - 2I_{xy}u_xu_y - 2I_{yz}u_yu_z - 2I_{zx}u_zu_x$$
$$= 0.480(-0.408)^2 + (0.453)(0.816)^2 + 0.400(0.408)^2$$
$$- 2(-0.160)(-0.408)(0.816) - 2(0.160)(0.816)(0.408)$$
$$- 2(-0.200)(0.408)(-0.408)$$

$$= 0.168 \text{ kg} \cdot \text{m}^2 \qquad\qquad \text{Ans.}$$

Problems

21–1. Show that the sum of the moments of inertia of a body, $I_{xx} + I_{yy} + I_{zz}$, is independent of the orientation of the x, y, z axes and thus depends only on the location of the origin.

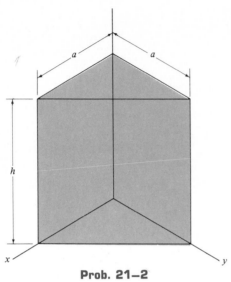

Prob. 21–2

21–2. Determine by direct integration the product of inertia I_{yz} for the homogeneous prism. The density of the material is ρ. Express the result in terms of the total mass m of the prism.

21–3. Determine by direct integration the product of inertia I_{xy} of the homogeneous prism in Prob. 21–2. The density of the material is ρ. Express the result in terms of the total mass m of the prism.

*21–4. Determine by direct integration the product of inertia I_{yz} for the homogeneous tetrahedron. The mass density of the material is ρ. Express the result in terms of the total mass m of the solid.

Prob. 21–4

21–5. Determine by direct integration the product of inertia I_{xy} for the homogeneous tetrahedron. The density of the material is ρ. Express the result in terms of the total mass m of the solid. *Suggestion:* Use a triangular element of thickness dz and then express dI_{xy} in terms of the size and mass of the element using the result of Prob. 21–3.

Prob. 21–5

21–6. Determine the inertia tensor for the homogeneous cylinder of mass m with respect to the x', y', z' coordinate axes. Is x', y', z' a set of principal axes of inertia?

Prob. 21–6

21–7. Compute the moment of inertia of the disk about the axis of shaft AB. The disk has a total mass of 15 kg.

Prob. 21–7

*21–8. The bent rod has a weight of 1.5 lb/ft. Locate the center of gravity $G(\bar{x}, \bar{y})$ and determine the principal moments of inertia $I_{x'}$, $I_{y'}$, and $I_{z'}$ of the rod with respect to the x', y', z' axes.

Prob. 21–8

21–9. Determine the moments of inertia about the x, y, z axes of the rod assembly. The rods have a mass of 0.75 kg/m.

Prob. 21–9

21–10. The bent rod has a mass per unit length of 4 kg/m. Determine the moment of inertia of the rod about the Oa axis.

Prob. 21–10

21–11. Determine the moment of inertia I_{xx} of the composite plate assembly. The plates have a specific weight of 6 lb/ft^2.

Probs. 21–11 / 21–12

*21–12. Determine the product of inertia I_{yz} of the composite plate assembly. The plates have a specific weight of 6 lb/ft^2.

21–13. The thin plate has a weight of 5 lb and each of the four rods weighs 3 lb. Determine the moment of inertia of the assembly about the z axis.

Prob. 21–13

21–14. The assembly consists of a 15-lb plate A, 40-lb plate B, and four 7-lb rods. Determine the moments of inertia with respect to the principal axes x, y, z.

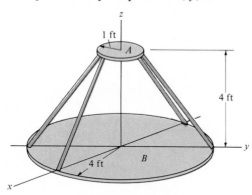

Prob. 21–14

21–15. Determine the moment of inertia of both the 1.5-kg rod and 4-kg disk about the z' axis.

Prob. 21–15

*21–16. Rod AB has a weight of 6 lb and each sphere has a weight of 8 lb. Determine the moment of inertia of the assembly about the x axis.

Prob. 21–16

21–17. The assembly consists of two square plates A and B which have a mass of 3 kg each and a rectangular plate C which has a mass of 4.5 kg. Determine the moments of inertia with respect to the principal x, y, z axes.

Prob. 21–17

*21.2 Angular Momentum

In this section we will develop the necessary equations used to determine the angular momentum of a rigid body about an arbitrary point. This formulation will provide the necessary means for developing both the principle of impulse and momentum and the equations of rotational motion for the rigid body.

Consider the rigid body in Fig. 21–6, which has a total mass m and center of mass located at G. The X, Y, Z coordinate system represents an inertial frame of reference, and hence, its axes are fixed or translate with a constant velocity. The angular momentum as measured from this reference will be computed relative to the arbitrary point A. The position vectors $\mathbf{r}_A$ and $\boldsymbol{\rho}_A$ are drawn from the origin of coordinates to point A and from A to the ith particle of the body. If the particle's mass is Δm_i, the angular momentum about point A is

$$\Delta \mathbf{H}_A = \boldsymbol{\rho}_A \times \Delta m_i \mathbf{v}_i$$

where $\mathbf{v}_i$ represents the particle's velocity as measured from the X, Y, Z coordinate system. If the body has an angular velocity $\boldsymbol{\omega}$ at the instant considered, $\mathbf{v}_i$ may be related to the velocity of A by the kinematic equation

$$\mathbf{v}_i = \mathbf{v}_A + \boldsymbol{\omega} \times \boldsymbol{\rho}_A$$

Thus,

$$\Delta \mathbf{H}_A = \boldsymbol{\rho}_A \times \Delta m_i (\mathbf{v}_A + \boldsymbol{\omega} \times \boldsymbol{\rho}_A)$$
$$= (\boldsymbol{\rho}_A \, \Delta m_i) \times \mathbf{v}_A + \boldsymbol{\rho}_A \times (\boldsymbol{\omega} \times \boldsymbol{\rho}_A) \, \Delta m_i$$

For the entire body, summing all n particles,

$$\mathbf{H}_A = (\Sigma \boldsymbol{\rho}_A \, \Delta m_i) \times \mathbf{v}_A + \Sigma \boldsymbol{\rho}_A \times (\boldsymbol{\omega} \times \boldsymbol{\rho}_A) \, \Delta m_i$$

As $n \to \infty$, $\Delta m_i \to dm$ and the summations become integrals:

$$\mathbf{H}_A = \left(\int_m \boldsymbol{\rho}_A \, dm \right) \times \mathbf{v}_A + \int_m \boldsymbol{\rho}_A \times (\boldsymbol{\omega} \times \boldsymbol{\rho}_A) \, dm \qquad (21\text{–}6)$$

Fig. 21–6

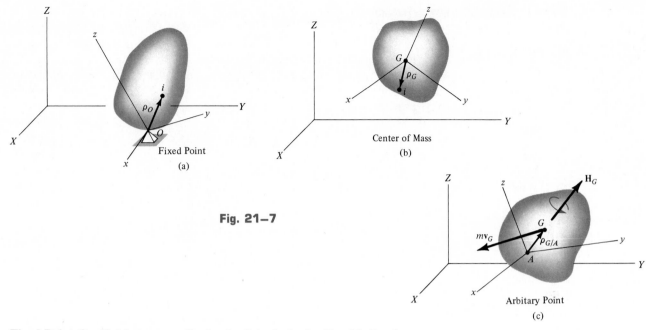

Fig. 21–7

Fixed Point O. If A becomes a *fixed point* O in the body, Fig. 21–7a, then $\mathbf{v}_A = \mathbf{0}$ and Eq. 21–6 reduces to

$$\mathbf{H}_O = \int_m \boldsymbol{\rho}_O \times (\boldsymbol{\omega} \times \boldsymbol{\rho}_O)\, dm \qquad (21\text{–}7)$$

Center of Mass G. If A is located at the *center of mass G* of the body, Fig. 21–7b, then $\int_m \boldsymbol{\rho}_A\, dm = \mathbf{0}$ and

$$\mathbf{H}_G = \int_m \boldsymbol{\rho}_G \times (\boldsymbol{\omega} \times \boldsymbol{\rho}_G)\, dm \qquad (21\text{–}8)$$

Arbitrary Point A. In general, A may be some point other than O or G, Fig. 21–7c, in which case Eq. 21–6 may nevertheless be simplified to the following form (see Prob. 21–19).

$$\mathbf{H}_A = (\boldsymbol{\rho}_{G/A} \times m\mathbf{v}_G) + \mathbf{H}_G \qquad (21\text{–}9)$$

Here it can be seen that the angular momentum consists of two parts—the moment of the linear momentum $m\mathbf{v}_G$ of the body about point A added (vectorially) to the angular momentum $\mathbf{H}_G$. Equation 21–9 may also be used for computing the angular momentum of the body about a fixed point O in the body; the results, of course, will be the same as those computed using the more convenient Eq. 21–7.

Rectangular Components of H. To make practical use of Eqs. 21–7 to 21–9, the angular momentum must be expressed in terms of its scalar components. For this purpose, it is convenient to choose a second set of x, y, z axes having an arbitrary orientation relative to the X, Y, Z axes, Fig. 21–7. For a general explanation, note that Eqs. 21–7 to 21–9 all contain the form

$$\mathbf{H} = \int_m \boldsymbol{\rho} \times (\boldsymbol{\omega} \times \boldsymbol{\rho}) \, dm$$

Expressing $\mathbf{H}$, $\boldsymbol{\rho}$, and $\boldsymbol{\omega}$ in terms of x, y, and z components, we have

$$H_x\mathbf{i} + H_y\mathbf{j} + H_z\mathbf{k} = \int_m (x\mathbf{i} + y\mathbf{j} + z\mathbf{k}) \times [(\omega_x\mathbf{i} + \omega_y\mathbf{j} + \omega_z\mathbf{k})$$
$$\times (x\mathbf{i} + y\mathbf{j} + z\mathbf{k})] \, dm$$

Expanding the cross products and combining terms yields

$$H_x\mathbf{i} + H_y\mathbf{j} + H_z\mathbf{k} = \left[\omega_x \int_m (y^2 + z^2) \, dm - \omega_y \int_m xy \, dm - \omega_z \int_m xz \, dm \right]\mathbf{i}$$
$$+ \left[-\omega_x \int_m xy \, dm + \omega_y \int_m (x^2 + z^2) \, dm - \omega_z \int_m yz \, dm \right]\mathbf{j}$$
$$+ \left[-\omega_x \int_m zx \, dm - \omega_y \int_m yz \, dm + \omega_z \int_m (x^2 + y^2) \, dm \right]\mathbf{k}$$

Equating the respective $\mathbf{i}$, $\mathbf{j}$, $\mathbf{k}$ components and recognizing that the integrals represent the moments and products of inertia, we obtain

$$
\begin{aligned}
H_x &= I_{xx}\omega_x - I_{xy}\omega_y - I_{xz}\omega_z \\
H_y &= -I_{yx}\omega_x + I_{yy}\omega_y - I_{yz}\omega_z \\
H_z &= -I_{zx}\omega_x - I_{zy}\omega_y + I_{zz}\omega_z
\end{aligned}
\tag{21–10}
$$

These three equations represent the scalar form of the $\mathbf{i}$, $\mathbf{j}$, and $\mathbf{k}$ components of $\mathbf{H}_O$ or $\mathbf{H}_G$ (given in vector form by Eqs. 21–7 and 21–8). The angular momentum of the body about the arbitrary point A, other than the fixed point O or the center of mass G, may also be expressed in scalar form. Here it is necessary to use Eq. 21–9 and to represent $\boldsymbol{\rho}_{G/A}$ and $\mathbf{v}_G$ in Cartesian component form, carry out the cross-product operation, and substitute the components, Eqs. 21–10, for $\mathbf{H}_G$.

Equations 21–10 may be simplified further if the x, y, z coordinate axes are oriented such that they become *principal axes of inertia* for the body at the point. When these axes are used, computation of the products of inertia $I_{xy} = I_{yz} = I_{zx} = 0$, and if the principal moments of inertia about the x, y, z axes are represented as $I_x = I_{xx}$, $I_y = I_{yy}$, and $I_z = I_{zz}$, the three components of angular momentum become

$$
H_x = I_x\omega_x \qquad H_y = I_y\omega_y \qquad H_z = I_z\omega_z
\tag{21–11}
$$

Principle of Impulse and Momentum. Now that the means for computing the angular momentum for a body have been presented, the *principle of impulse and momentum,* as discussed in Sec. 19.2, may be used to solve kinetics problems which involve *force, velocity, and time.* For this case, the following two vector equations are available:

$$m(\mathbf{v}_G)_1 + \Sigma \int_{t_1}^{t_2} \mathbf{F}\, dt = m(\mathbf{v}_G)_2 \qquad (21\text{--}12)$$

$$\mathbf{H}_1 + \Sigma \int_{t_1}^{t_2} \mathbf{M}_O\, dt = \mathbf{H}_2 \qquad (21\text{--}13)$$

In three dimensions each vector term can be represented by three scalar components, and therefore a total of *six scalar equations* can be written. Three equations relate the linear impulse and momentum in the x, y, z directions, and three equations relate the body's angular impulse and momentum about the x, y, z axes. Before applying Eqs. 21–12 and 21–13 to the solution of problems, the material in Secs. 19.2 and 19.3 should be reviewed.

Kinetic Energy 21.3*

In order to apply the principle of work and energy to the solution of problems involving general rigid-body motion, it is first necessary to formulate expressions for the kinetic energy of the body. In this regard, consider the rigid body shown in Fig. 21–8, which has a total mass m and center of mass located at G. The kinetic energy of the ith particle of the body having a mass Δm_i and velocity $\mathbf{v}_i$, measured relative to the inertial X, Y, Z frame of reference, is

$$\Delta T_i = \tfrac{1}{2}\Delta m_i\, v_i^2 = \tfrac{1}{2}\Delta m_i\, (\mathbf{v}_i \cdot \mathbf{v}_i)$$

Provided the velocity of an arbitrary point A in the body is known, $\mathbf{v}_i$ may be related to $\mathbf{v}_A$ by the equation $\mathbf{v}_i = \mathbf{v}_A + \boldsymbol{\omega} \times \boldsymbol{\rho}_A$, where $\boldsymbol{\omega}$ is the instantaneous angular velocity of the body, measured from the X, Y, Z coordinate system, and $\boldsymbol{\rho}_A$ is a position vector drawn from A to i. Using this expression for $\mathbf{v}_i$, the kinetic energy for the particle may be written as

$$\Delta T_i = \tfrac{1}{2}\Delta m_i\, (\mathbf{v}_A + \boldsymbol{\omega} \times \boldsymbol{\rho}_A) \cdot (\mathbf{v}_A + \boldsymbol{\omega} \times \boldsymbol{\rho}_A)$$
$$= \tfrac{1}{2}(\mathbf{v}_A \cdot \mathbf{v}_A)\, \Delta m_i + \mathbf{v}_A \cdot (\boldsymbol{\omega} \times \boldsymbol{\rho}_A)\, \Delta m_i + \tfrac{1}{2}(\boldsymbol{\omega} \times \boldsymbol{\rho}_A) \cdot (\boldsymbol{\omega} \times \boldsymbol{\rho}_A)\, \Delta m_i$$

The kinetic energy for the entire body is obtained by summing the kinetic energies of all n particles of the body. Thus,

$$T = \tfrac{1}{2}(\mathbf{v}_A \cdot \mathbf{v}_A) \Sigma\, \Delta m_i + \mathbf{v}_A \cdot (\boldsymbol{\omega} \times \Sigma \boldsymbol{\rho}_A\, \Delta m_i) + \tfrac{1}{2}\Sigma(\boldsymbol{\omega} \times \boldsymbol{\rho}_A) \cdot (\boldsymbol{\omega} \times \boldsymbol{\rho}_A)\, \Delta m_i$$

If $n \to \infty$, $\Delta m_i \to dm$, and the summations become integrals, so that

$$T = \tfrac{1}{2}m(\mathbf{v}_A \cdot \mathbf{v}_A) + \mathbf{v}_A \cdot \left(\boldsymbol{\omega} \times \int_m \boldsymbol{\rho}_A\, dm\right) + \tfrac{1}{2}\int_m (\boldsymbol{\omega} \times \boldsymbol{\rho}_A) \cdot (\boldsymbol{\omega} \times \boldsymbol{\rho}_A)\, dm$$

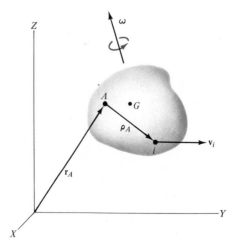

Fig. 21–8

The last term on the right may be rewritten using the vector identity $\mathbf{a} \times \mathbf{b} \cdot \mathbf{c} = \mathbf{a} \cdot \mathbf{b} \times \mathbf{c}$, where $\mathbf{a} = \boldsymbol{\omega}$, $\mathbf{b} = \boldsymbol{\rho}_A$, and $\mathbf{c} = \boldsymbol{\omega} \times \boldsymbol{\rho}_A$. The final result is therefore,

$$T = \tfrac{1}{2}m(\mathbf{v}_A \cdot \mathbf{v}_A) + \mathbf{v}_A \cdot \left(\boldsymbol{\omega} \times \int_m \boldsymbol{\rho}_A \, dm \right) + \tfrac{1}{2}\boldsymbol{\omega} \cdot \int_m \boldsymbol{\rho}_A \times (\boldsymbol{\omega} \times \boldsymbol{\rho}_A) \, dm$$

$$(21\text{--}14)$$

This equation is rarely used because of the computations involving the integrals. Simplification occurs, however, if the reference point A is either a fixed point O or the center of mass G.

Fixed Point O. If A is a *fixed point* O in the body, Fig. 21–7a, then $\mathbf{v}_A = \mathbf{0}$, and using Eq. 21–7, we can express Eq. 21–14 as

$$T = \tfrac{1}{2}\boldsymbol{\omega} \cdot \mathbf{H}_O$$

If the x, y, z axes represent the principal axes of inertia for the body, then $\boldsymbol{\omega} = \omega_x\mathbf{i} + \omega_y\mathbf{j} + \omega_z\mathbf{k}$ and $\mathbf{H}_O = I_x\omega_x\mathbf{i} + I_y\omega_y\mathbf{j} + I_z\omega_z\mathbf{k}$. Substituting into the above equation and performing the dot product operations yields

$$T = \tfrac{1}{2}I_x\omega_x^2 + \tfrac{1}{2}I_y\omega_y^2 + \tfrac{1}{2}I_z\omega_z^2 \qquad (21\text{--}15)$$

Center of Mass G. If A is located at the *center of mass G* of the body, Fig. 21–7b, then $\int_m \boldsymbol{\rho}_A \, dm = \mathbf{0}$ and, using Eq. 21–8, we can write Eq. 21–14 as

$$T = \tfrac{1}{2}mv_G^2 + \tfrac{1}{2}\boldsymbol{\omega} \cdot \mathbf{H}_G$$

In a manner similar to that for a fixed point, the last term on the right side may be represented in scalar form, in which case

$$T = \tfrac{1}{2}mv_G^2 + \tfrac{1}{2}I_x\omega_x^2 + \tfrac{1}{2}I_y\omega_y^2 + \tfrac{1}{2}I_z\omega_z^2 \qquad (21\text{--}16)$$

Here it is seen that the kinetic energy consists of two parts; namely, the translational kinetic energy of the mass center, $\tfrac{1}{2}mv_G^2$, and the body's rotational kinetic energy.

Principle of Work and Energy. Using one of the above expressions for computing the kinetic energy of a body, the *principle of work and energy* may be applied to solve kinetics problems which involve *force, velocity, and displacement*. For this case only one scalar equation can be written for each body, i.e.,

$$T_1 + \Sigma U_{1-2} = T_2 \qquad (21\text{--}17)$$

Before applying this equation to the solution of problems the material in Chapter 18 should be reviewed.

Example 21–2

The rod in Fig. 21–9a has a weight of 1.5 lb/ft. Determine its angular velocity just after the end A falls onto the hook at E. The hook provides a permanent connection for the rod because of the spring-lock mechanism S. Just before striking the hook the rod is falling downward with a speed $(v_G)_1 = 10$ ft/s.

Solution

The principle of impulse and momentum is used since impact occurs.

Impulse and Momentum Diagrams. Figure 21–9b. During the short time Δt, the impulsive force **F** acting at A changes the momentum of the rod. (The impulse created by the rod's weight **W** during this time is small compared to $\int \mathbf{F}\,dt$, so that it is neglected, i.e., the weight is a nonimpulsive force.) Hence, the angular momentum of the rod is *conserved* about point A since the moment of $\int \mathbf{F}\,dt$ about A is zero.

Conservation of Angular Momentum. Equation 21–9 must be used for computing the angular momentum of the rod, since A does not become a *fixed point* until *after* the impulsive interaction with the hook. Thus, with reference to Fig. 21–9b, $(\mathbf{H}_A)_1 = (\mathbf{H}_A)_2$, or

$$\mathbf{r}_{G/A} \times m(\mathbf{v}_G)_1 = \mathbf{r}_{G/A} \times m(\mathbf{v}_G)_2 + (\mathbf{H}_G)_2 \tag{1}$$

From Fig. 21–9a, $\mathbf{r}_{G/A} = \{-0.667\mathbf{i} + 0.5\mathbf{j}\}$ ft. Furthermore, the primed axes are principal axes of inertia for the rod because $I_{x'y'} = I_{x'z'} = I_{z'y'} = 0$. Hence, from Eqs. 21–11, $(\mathbf{H}_G)_2 = I_{x'}\omega_x\mathbf{i} + I_{y'}\omega_y\mathbf{j} + I_{z'}\omega_z\mathbf{k}$. The principal moments of inertia are $I_{x'} = I_{y'} = 0.0155$ slug · ft², $I_{z'} = 0.0401$ slug · ft² (see Prob. 21–8). Substituting into Eq. (1), we have

$$(-0.667\mathbf{i} + 0.5\mathbf{j}) \times \left[\left(\frac{4.5}{32.2}\right)(-10\mathbf{k})\right]$$

$$= (-0.667\mathbf{i} + 0.5\mathbf{j}) \times \left[\left(\frac{4.5}{32.2}\right)(-v_G)_2\mathbf{k}\right]$$

$$+ 0.0155\omega_x\mathbf{i} + 0.0155\omega_y\mathbf{j} + 0.0401\omega_z\mathbf{k}$$

Expanding and equating the respective **i**, **j**, and **k** components yields

$$-0.699 = -0.0699v_G + 0.0155\omega_x \tag{2}$$
$$-0.932 = -0.0932v_G + 0.0155\omega_y \tag{3}$$
$$0 = 0.0401\omega_z \tag{4}$$

Kinematics. There are four unknowns in the above equations; however, another equation may be obtained by relating $\boldsymbol{\omega}$ to $(\mathbf{v}_G)_2$ using *kinematics*. Since $\omega_z = 0$ (Eq. 4) and after impact the rod rotates about the fixed point A, Eq. 20–2 may be applied, in which case $\mathbf{v}_G = \boldsymbol{\omega} \times \mathbf{r}_{G/A}$, or

$$-(v_G)_2\mathbf{k} = (\omega_x\mathbf{i} + \omega_y\mathbf{j}) \times (-0.667\mathbf{i} + 0.5\mathbf{j})$$
$$-(v_G)_2 = 0.5\omega_x + 0.667\omega_y \tag{5}$$

Solving Eqs. (2) to (5) simultaneously yields

$$(v_G)_2 = \{-8.62\mathbf{k}\} \text{ ft/s} \qquad \boldsymbol{\omega} = \{-6.21\mathbf{i} - 8.29\mathbf{j}\} \text{ rad/s} \qquad Ans.$$

(a)

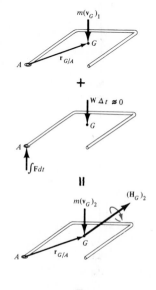

(b)

Fig. 21–9

Example 21–3

A 5-N · m torque is applied to the vertical shaft *CD* shown in Fig. 21–10a, which allows the 10-kg gear *A* to turn freely about *CE*. Assuming that gear *A* starts from rest, determine the angular velocity of *CD* after it has turned two revolutions. Neglect the mass of shaft *CD* and axle *CE* and assume that gear *A* is approximated by a thin disk. Gear *B* is fixed.

(a)

Solution

The principle of work and energy may be applied to solve this problem. Why?

Work. If shaft *CD*, the axle *CE*, and gear *A* are considered as a system of connected bodies, only the applied torque **M** does work. For two revolutions of *CD*, this work is $\Sigma U_{1-2} = (5 \text{ N} \cdot \text{m})(4\pi \text{ rad}) = 62.83 \text{ J}$.

Kinetic Energy. Since the gear is initially at rest, the initial kinetic energy of the system is zero. A kinematic diagram for the gear is shown in Fig. 21–10b. If the angular velocity of *CD* is taken as $\boldsymbol{\omega}_{CD}$, then the angular velocity of gear *A* is $\boldsymbol{\omega}_A = \boldsymbol{\omega}_{CD} + \boldsymbol{\omega}_{CE}$. The gear may be imagined as a portion of a hypothetical massless extended body which is rotating about the *fixed point C*. The instantaneous axis of rotation for this body is along line *CH*, because both points *C* and *H* on the body (gear) have zero velocity and must therefore lie on this axis. This requires that the components ω_{CD} and ω_{CE} be related by the equation $\omega_{CD}/0.1 \text{ m} = \omega_{CE}/0.3 \text{ m}$ or $\omega_{CE} = 3\omega_{CD}$. The angular velocity of the gear at the instant considered is therefore

(b)

Fig. 21–10

$$\boldsymbol{\omega}_A = -\omega_{CE}\mathbf{i} + \omega_{CD}\mathbf{k} = -3\omega_{CD}\mathbf{i} + \omega_{CD}\mathbf{k} \qquad (1)$$

The *x*, *y*, *z* axes in Fig. 21–10a represent *principal axes of inertia at C* for the gear. Since point *C* is a fixed point of rotation, Eq. 21–15 may be applied to determine the kinetic energy, i.e.,

$$T = \tfrac{1}{2}I_x\omega_x^2 + \tfrac{1}{2}I_y\omega_y^2 + \tfrac{1}{2}I_z\omega_z^2 \qquad (2)$$

Using the parallel-axis theorem, the moments of inertia of the gear about point *C* are as follows:

$$I_x = \frac{1}{2}(10)(0.1)^2 = 0.05 \text{ kg} \cdot \text{m}^2$$

$$I_y = I_z = \frac{1}{4}(10)(0.1)^2 + 10(0.3)^2 = 0.925 \text{ kg} \cdot \text{m}^2$$

Since $\omega_x = -3\omega_{CD}$, $\omega_y = 0$, $\omega_z = \omega_{CD}$, Eq. (2) becomes

$$T_A = \frac{1}{2}(0.05)(-3\omega_{CD})^2 + 0 + \frac{1}{2}(0.925)(\omega_{CD})^2 = 0.6875\omega_{CD}^2$$

Principle of Work and Energy. Applying the principle of work and energy, we obtain

$$T_1 + \Sigma U_{1-2} = T_2 \qquad 0 + 62.83 = 0.6875\omega_{CD}^2$$

$$\omega_{CD} = 9.56 \text{ rad/s} \qquad \qquad \textit{Ans.}$$

Problems

21–18. If a body contains *no planes of symmetry*, the principal moments of inertia can be computed mathematically. To show how this is done, consider the rigid body which is spinning with an angular velocity $\boldsymbol{\omega}$ directed along one of its principal axes of inertia. If the principal moment of inertia about this axis is I, the angular momentum can be expressed as $\mathbf{H} = I\boldsymbol{\omega} = I\omega_x\mathbf{i} + I\omega_y\mathbf{j} + I\omega_z\mathbf{k}$. The components of $\mathbf{H}$ may also be expressed by Eqs. 21–10, where the inertia tensor is assumed to be known. Equate the $\mathbf{i}$, $\mathbf{j}$, and $\mathbf{k}$ components of both expressions for $\mathbf{H}$ and consider ω_x, ω_y, and ω_z to be unknown. The solution of these three equations is obtained provided the determinant of the coefficients is zero. Show that this determinant, when expanded, yields the cubic equation

$$I^3 - (I_{xx} + I_{yy} + I_{zz})I^2 + (I_{xx}I_{yy} + I_{yy}I_{zz} + I_{zz}I_{xx}$$
$$- I_{xy}^2 - I_{yz}^2 - I_{zx}^2)I - (I_{xx}I_{yy}I_{zz} - 2I_{xy}I_{yz}I_{zx}$$
$$- I_{xx}I_{yz}^2 - I_{yy}I_{zx}^2 - I_{zz}I_{xy}^2) = 0$$

The three positive roots of I, obtained from the solution of this equation, represent the principal moments of inertia I_x, I_y, and I_z.

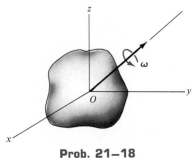

Prob. 21–18

21–19. Show that if the angular momentum of a body is computed with respect to an arbitrary point A, then $\mathbf{H}_A$ can be expressed by Eq. 21–9. This requires substituting $\boldsymbol{\rho}_A = \boldsymbol{\rho}_G + \boldsymbol{\rho}_{G/A}$ into Eq. 21–6 and expanding, noting that $\int \boldsymbol{\rho}_G \, dm = \mathbf{0}$ by definition of the mass center and $\mathbf{v}_G = \mathbf{v}_A + \boldsymbol{\omega} \times \boldsymbol{\rho}_{G/A}$.

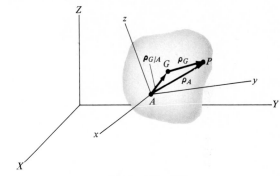

Prob. 21–19

*21–20. Determine the kinetic energy of the 7-kg disk and 1.5-kg rod when the assembly is rotating about the z axis at $\omega = 5$ rad/s.

Probs. 21–20 / 21–21

21–21. Determine the magnitude of the angular momentum of the 7-kg disk and 1.5-kg rod about point C. The shaft BA is rotating at $\omega = 5$ rad/s.

21–22. Each of the two disks has a weight of 10 lb. The axle AB weighs 3 lb. If the assembly is rotating about the z axis at $\omega_z = 6$ rad/s, determine its angular momentum about the z axis and its kinetic energy. The disks roll without slipping.

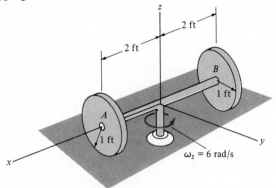

Prob. 21–22

21–23. The 15-kg rectangular plate is free to rotate about the y axis because of the bearing supports at A and B. When the plate is balanced in the vertical plane, a 3-g bullet is fired into it, perpendicular to its surface, with a velocity $\mathbf{v} = \{-2000\mathbf{i}\}$ m/s. Compute the angular velocity of the plate at the instant it has rotated 180°. If the bullet strikes corner D with the same velocity $\mathbf{v}$, instead of at C, does the angular velocity remain the same? Why or why not?

Prob. 21–23

*21–24. The circular disk has a weight of 15 lb and is mounted on the shaft AB at an angle of 45° with the horizontal. Determine the angular velocity of the shaft when $t = 2$ s if a torque $M = (4e^{0.1t})$ N · m, where t is in seconds, is applied to the shaft. The shaft is originally spinning at $\omega_1 = 8$ rad/s when the torque is applied.

Prob. 21–24

21–25. The 5-kg thin plate is suspended at O using a ball-and-socket joint. It is rotating with a constant angular velocity $\boldsymbol{\omega} = \{2\mathbf{k}\}$ rad/s when the corner at A strikes the hook at S, which provides a permanent connection. Determine the angular velocity of the plate immediately after impact.

Prob. 21–25

21–26. The 15-lb plate is subjected to a force $F = 8$ lb which is always directed perpendicular to the face of the plate. If the plate is originally at rest, determine its angular velocity after it has rotated one revolution (360°). The plate is supported by ball-and-socket joints at A and B.

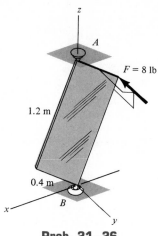

Prob. 21–26

21–27. The circular plate has a weight of 19 lb and a diameter $d = 1.5$ ft. If it is released from rest and falls horizontally, $h = 2.5$ ft, onto the hook at S, which provides a permanent connection, determine the velocity of the mass center of the plate just after the connection with the hook is made.

Prob. 21–27

*21–28. The 4-lb rod AB is attached to the 1-lb collar at A and a 2-lb link BC using ball-and-socket joints. If the rod is released from rest in the position shown, determine the angular velocity of the link after it has rotated 180°.

Prob. 21–28

21–29. Rod AB has a weight of 6 lb and is attached to two smooth collars at its ends by ball-and-socket joints. If collar A is moving downward with a speed of 8 ft/s when $z = 3$ ft, determine the speed of collar A at the instant $z = 0$. The spring has an unstretched length of 2 ft. Neglect the mass of the collars. Assume the angular velocity of rod AB is perpendicular to its axis.

Prob. 21–29

479

21–30. The assembly consists of a 4-kg rod AB which is connected to link OA and the collar at B by ball-and-socket joints. When $\theta = 0°$, the system is at rest, the spring is unstretched, and a 7-N · m couple moment is applied to the link at O. Determine the angular velocity of the link at the instant $\theta = 90°$.

Prob. 21–30

21–31. The rod assembly is supported at G by a ball-and-socket joint. Each rod has a mass of 0.5 kg/m. If the assembly is originally at rest and an impulse of $\mathbf{I} = \{-8\mathbf{k}\}$ N · s is applied at D, determine the angular velocity of the assembly just after the impact.

Prob. 21–31

***21–32.** The space capsule has a mass of 3.5 Mg and the radii of gyration are $k_x = k_z = 0.8$ m and $k_y = 0.5$ m. If it is initially coasting with a velocity $\mathbf{v}_G = \{600\mathbf{j}\}$ m/s, compute its angular velocity just after it is struck by a meteoroid having a mass of 0.60 kg and a speed $v_m = 1.5$ km/s directed as shown. Assume that the meteoroid embeds itself into the capsule at $A(x_A = 0,\ y_A = 0.5$ m, $z_A = 0.75$ m) and that the capsule initially has no angular velocity.

Prob. 21–23

Equations of Motion 21.4*

Having become familiar with the techniques used to describe both the inertia properties and the angular momentum of a body, we can now write the equations which describe the motion of the body in their most useful forms.

Equations of Translational Motion. The *translational motion* of a body is defined in terms of the acceleration of the body's mass center, which is measured from an inertial X, Y, Z reference. The equation of translational motion for the body can be written in vector form as

$$\Sigma \mathbf{F} = m\mathbf{a}_G \qquad (21\text{--}18)$$

or by the three scalar equations

$$\Sigma F_x = m(a_G)_x$$
$$\Sigma F_y = m(a_G)_y \qquad (21\text{--}19)$$
$$\Sigma F_z = m(a_G)_z$$

Here, $\Sigma \mathbf{F} = \Sigma F_x \mathbf{i} + \Sigma F_y \mathbf{j} + \Sigma F_z \mathbf{k}$ represents the sum of all the external forces acting on the body.

Equations of Rotational Motion. In Sec. 15.6, we developed Eq. 15–17, namely,

$$\Sigma \mathbf{M}_O = \dot{\mathbf{H}}_O \qquad (21\text{--}20)$$

which states that the sum of the moments about a fixed point O of all the external forces acting on a system of particles (contained in a rigid body) is equal to the time rate of change of the total angular momentum of the body about point O. When moments of the external forces acting on the particles are summed about the system's *mass center G,* one again obtains the same simple form of Eq. 21–20, relating the moment summation $\Sigma \mathbf{M}_G$ to the angular momentum $\mathbf{H}_G$. To show this, consider the system of particles in Fig. 21–11, where x, y, z represents an inertial frame of reference and the x', y', z' axes, with origin at G, *translate* with respect to this frame. In general, G is *accelerating,* so by definition the translating frame is *not* an inertial reference. The angular momentum of the ith particle with respect to this frame is, however,

$$(\mathbf{H}_i)_G = \mathbf{r}_{i/G} \times m_i \mathbf{v}_{i/G}$$

where $\mathbf{r}_{i/G}$ and $\mathbf{v}_{i/G}$ represent the relative position and relative velocity of the ith particle with respect to G. Taking the time derivative gives

$$(\dot{\mathbf{H}}_i)_G = \dot{\mathbf{r}}_{i/G} \times m_i \mathbf{v}_{i/G} + \mathbf{r}_{i/G} \times m_i \dot{\mathbf{v}}_{i/G}$$

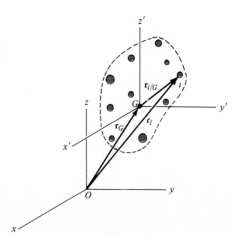

Fig. 21–11

By definition, $\mathbf{v}_{i/G} = \dot{\mathbf{r}}_{i/G}$. Thus, the first term on the right side is zero since the cross product of equal vectors is zero. Also, $\mathbf{a}_{i/G} = \dot{\mathbf{v}}_{i/G}$, so that

$$(\dot{\mathbf{H}}_i)_G = \mathbf{r}_{i/G} \times m_i \mathbf{a}_{i/G}$$

Similar expressions can be written for the other particles of the system. When the results are summed for all the particles, we get

$$\dot{\mathbf{H}}_G = \Sigma(\mathbf{r}_{i/G} \times m_i \mathbf{a}_{i/G})$$

Here $\dot{\mathbf{H}}_G$ is the time rate of change of the total angular momentum of the body computed relative to point G.

The relative acceleration for the ith particle is defined by the equation $\mathbf{a}_{i/G} = \mathbf{a}_i - \mathbf{a}_G$, where $\mathbf{a}_i$ and $\mathbf{a}_G$ represent, respectively, the accelerations of the ith particle and point G measured with respect to the *inertial frame of reference*. Substituting and expanding, using the distributive property of the vector cross product, yields

$$\dot{\mathbf{H}}_G = \Sigma(\mathbf{r}_{i/G} \times m_i \mathbf{a}_i) - (\Sigma m_i \mathbf{r}_{i/G}) \times \mathbf{a}_G$$

By definition of the mass center, the sum $(\Sigma m_i \mathbf{r}_{i/G}) = (\Sigma m_i)\bar{\mathbf{r}}$ is equal to zero, since the position vector $\bar{\mathbf{r}}$ relative to G is zero. Hence, the last term in the above equation is zero. Using the equation of motion, the product $m_i \mathbf{a}_i$ may be replaced by the resultant *external force* $\mathbf{F}_i$ acting on the ith particle. Denoting $\Sigma \mathbf{M}_G = \Sigma \mathbf{r}_{i/G} \times \mathbf{F}_i$, the final result may be written as

$$\Sigma \mathbf{M}_G = \dot{\mathbf{H}}_G \tag{21-21}$$

The rotational equation of motion for the body will now be developed from either Eq. 21–20 or Eq. 21–21. In this regard, the scalar components of the angular momentum $\mathbf{H}_O$ or $\mathbf{H}_G$ are defined by Eqs. 21–10 or, if principal axes of inertia are used either at point O or G, by Eqs. 21–11. If these components are computed about x, y, z axes that are *rotating* with an angular velocity $\boldsymbol{\Omega}$, which may be *different* from the body's angular velocity $\boldsymbol{\omega}$, then the time derivative $\dot{\mathbf{H}} = d\mathbf{H}/dt$, as used in Eqs. 21–20 and 21–21, must account for the rotation of the x, y, z axes as measured from the inertial X, Y, Z axes. Hence, the time derivative of $\mathbf{H}$ must be determined from Eq. 20–6, in which case Eqs. 21–20 and 21–21 become

$$\Sigma \mathbf{M}_O = (\dot{\mathbf{H}}_O)_{xyz} + \boldsymbol{\Omega} \times \mathbf{H}_O$$
$$\Sigma \mathbf{M}_G = (\dot{\mathbf{H}}_G)_{xyz} + \boldsymbol{\Omega} \times \mathbf{H}_G \tag{21-22}$$

Here $(\dot{\mathbf{H}})_{xyz}$ is the time rate of change of $\mathbf{H}$ measured from the x, y, z reference.

There are three ways in which one can define the motion of the x, y, z axes. Obviously, motion of this reference should be chosen to yield the simplest set of moment equations for the solution of a particular problem.

x, y, z Axes Having Motion $\boldsymbol{\Omega} = \mathbf{0}.$ If the body has general motion, the x, y, z axes may be chosen with origin at G, such that the axes only *translate* relative to the inertial X, Y, Z frame of reference. Doing this would certainly simplify Eq. 21–22, since $\boldsymbol{\Omega} = \mathbf{0}.$ However, the body may have a rotation $\boldsymbol{\omega}$ about these axes, and therefore the moments and products of inertia of the body would have to be expressed as *functions of time*. In most cases this would be a difficult task, so that such a choice of axes has restricted value.

x, y, z Axes Having Motion $\boldsymbol{\Omega} = \boldsymbol{\omega}.$ The x, y, z axes may be chosen such that they are *fixed in and move with the body*. The moments and products of inertia of the body relative to these axes will be *constant* during the motion. Since $\boldsymbol{\Omega} = \boldsymbol{\omega},$ Eqs. 21–22 become

$$\Sigma \mathbf{M}_O = (\dot{\mathbf{H}}_O)_{xyz} + \boldsymbol{\omega} \times \mathbf{H}_O$$
$$\Sigma \mathbf{M}_G = (\dot{\mathbf{H}}_G)_{xyz} + \boldsymbol{\omega} \times \mathbf{H}_G \qquad (21\text{–}23)$$

We may express each of these vector equations as three scalar equations using Eqs. 21–10. Neglecting the subscripts O and G yields

$$\Sigma M_x = I_{xx}\dot{\omega}_x - (I_{yy} - I_{zz})\omega_y\omega_z - I_{xy}(\dot{\omega}_y - \omega_z\omega_x) - I_{yz}(\omega_y^2 - \omega_z^2)$$
$$- I_{zx}(\dot{\omega}_z + \omega_x\omega_y)$$
$$\Sigma M_y = I_{yy}\dot{\omega}_y - (I_{zz} - I_{xx})\omega_z\omega_x - I_{yz}(\dot{\omega}_z - \omega_x\omega_y) - I_{zx}(\omega_z^2 - \omega_x^2)$$
$$- I_{xy}(\dot{\omega}_x + \omega_y\omega_z) \qquad (21\text{–}24)$$
$$\Sigma M_z = I_{zz}\dot{\omega}_z - (I_{xx} - I_{yy})\omega_x\omega_y - I_{zx}(\dot{\omega}_x - \omega_y\omega_z) - I_{xy}(\omega_x^2 - \omega_y^2)$$
$$- I_{yz}(\dot{\omega}_y + \omega_z\omega_x)$$

Notice that for a rigid body symmetric with respect to the xy reference plane, and undergoing general plane motion in this plane, $I_{xz} = I_{yz} = 0$, and $\omega_x = \omega_y = d\omega_x/dt = d\omega_y/dt = 0$. Equations 21–24 reduce to the form $\Sigma M_x = \Sigma M_y = 0$, and $\Sigma M_z = I_{zz}\alpha_z$ (where $\alpha_z = \dot{\omega}_z$), which is essentially the third of Eqs. 17–11 or 17–16 depending upon the choice of point G or O for summing moments.

If the $x, y,$ and z axes are chosen as *principal axes of inertia*, the products of inertia are zero, $I_{xx} = I_x$, etc., and Eqs. 21–24 reduce to the form

$$\Sigma M_x = I_x\dot{\omega}_x - (I_y - I_z)\omega_y\omega_z$$
$$\Sigma M_y = I_y\dot{\omega}_y - (I_z - I_x)\omega_z\omega_x \qquad (21\text{–}25)$$
$$\Sigma M_z = I_z\dot{\omega}_z - (I_x - I_y)\omega_x\omega_y$$

This set of equations is known historically as the *Euler equations of motion,* named after the Swiss mathematician Leonhard Euler (1707–1783). They *only* apply for moments summed either about point O or G.

When applying these equations it should be realized that $\dot{\omega}_x$, $\dot{\omega}_y$, and $\dot{\omega}_z$ represent the time derivatives of the magnitudes of the x, y, z components of

$\boldsymbol{\omega}$. Since the x, y, z axes are rotating at $\boldsymbol{\Omega} = \boldsymbol{\omega}$, then, from Eq. 20–6, it may be noted that $(\dot{\boldsymbol{\omega}})_{XYZ} = (\dot{\boldsymbol{\omega}})_{xyz} + \boldsymbol{\omega} \times \boldsymbol{\omega}$. Since $\boldsymbol{\omega} \times \boldsymbol{\omega} = \mathbf{0}$, $(\dot{\boldsymbol{\omega}})_{XYZ} = (\dot{\boldsymbol{\omega}})_{xyz}$. This important result indicates that the required time derivative of $\boldsymbol{\omega}$ can be obtained either by first finding the components of $\boldsymbol{\omega}$ along the x, y, z axes and then taking the time derivative of the magnitudes of these components, i.e., $(\dot{\boldsymbol{\omega}})_{xyz}$, or by first finding the time derivative of $\boldsymbol{\omega}$ with respect to the X, Y, Z axes, i.e., $(\dot{\boldsymbol{\omega}})_{XYZ}$, and then determining the components $\dot{\omega}_x$, $\dot{\omega}_y$, and $\dot{\omega}_z$. In practice, it is generally easier to compute $\dot{\omega}_x$, $\dot{\omega}_y$, and $\dot{\omega}_z$ on the basis of finding $(\dot{\boldsymbol{\omega}})_{XYZ}$.

x, y, z Axes Having Motion $\boldsymbol{\Omega} \neq \boldsymbol{\omega}$. To simplify the calculations for the time derivative of $\boldsymbol{\omega}$, it is often convenient to choose the x, y, z axes having an angular velocity $\boldsymbol{\Omega}$ which is different from the angular velocity $\boldsymbol{\omega}$ of the body. This is particularly suitable for the analysis of spinning tops and gyroscopes which are *symmetrical* about the spinning axis.* When this is the case, the moments and products of inertia remain constant during the motion.

Equations 21–22 are applicable for such a set of chosen axes. Each of these two vector equations may be reduced to a set of three scalar equations which are derived in a manner similar to Eqs. 21–25,† i.e.,

$$\Sigma M_x = I_x \dot{\omega}_x - I_y \Omega_z \omega_y + I_z \Omega_y \omega_z$$
$$\Sigma M_y = I_y \dot{\omega}_y - I_z \Omega_x \omega_z + I_x \Omega_z \omega_x \tag{21–26}$$
$$\Sigma M_z = I_z \dot{\omega}_z - I_x \Omega_y \omega_x + I_y \Omega_x \omega_y$$

Here Ω_x, Ω_y, Ω_z represent the x, y, z components of $\boldsymbol{\Omega}$, measured from the inertial frame of reference.

Any one of these sets of moment equations, Eqs. 21–24, 21–25, or 21–26, represents a series of three first-order nonlinear differential equations. These equations are "coupled," since the angular-velocity components are present in all the terms. Success in determining the solution for a particular problem therefore depends upon what is unknown in these equations. Difficulty certainly arises when one attempts to solve for the unknown components of $\boldsymbol{\omega}$, given the external moments as functions of time. Further complications can arise if the moment equations are coupled to the three scalar equations of translation, Eqs. 21–19. This can happen because of the existence of kinematic constraints which relate the rotation of the body to the translation of its mass center, as in the case of a hoop which rolls without slipping. Problems necessitating the simultaneous solution of differential equations generally require application of numerical methods with the aid of a computer. In many engineering problems, however, one is required to determine the applied moments acting on the body, given information about the motion of the body. Fortunately, many of these types of problems have direct solutions, so that there is no need to resort to computer techniques.

---

*A detailed discussion of such devices is given in Sec. 21.5.
†See Prob. 21–35.

PROCEDURE FOR ANALYSIS

The following procedure provides a method for solving problems involving the three-dimensional motion of a rigid body.

Free-Body Diagram. Draw a *free-body diagram* of the body at the instant considered and specify the x, y, z coordinate system. The origin of this reference must be located either at the body's mass center G, or at point O, considered fixed in an inertial reference frame and located either in the body or on a hypothetical massless extension of the body. Depending upon the nature of the problem, a decision should be made as to what type of rotational motion Ω this coordinate system should have, i.e., $\Omega = 0$, $\Omega = \omega$, or $\Omega \neq \omega$. When deciding, one should keep in mind that the moment equations are simplified when the axes move in such a manner that they represent principal axes of inertia for the body at all times.

Kinematics. Compute the necessary moments and products of inertia and the components of the body's angular velocity and angular acceleration. In some cases a *kinematic diagram* might be helpful, since it provides a graphical aid for determining the acceleration of the body's mass center and for computing the angular acceleration. The angular velocity can usually be determined from the constraints of the body or from the given problem data. If this vector is changing direction, Eq. 20–6 [$(\dot{\omega})_{XYZ} = (\dot{\omega})_{xyz} + \Omega \times \omega$] must be used to determine the components of angular acceleration. In particular, note that if $\Omega = \omega$ then $(\dot{\omega})_{XYZ} = (\dot{\omega})_{xyz}$.

Equations of Motion. Apply either the two vector equations 21–18 and 21–22, or the six scalar component equations appropriate for the x, y, z coordinate axes chosen for the problem.

The following example problems numerically illustrate application of this procedure.

Example 21–4

The gear shown in Fig. 21–12a has a mass of 10 kg and is mounted at an angle of 10° with a rotating shaft having negligible mass. If $I_z = 0.1$ kg · m², $I_x = I_y = 0.05$ kg · m², and the shaft is rotating with a constant angular velocity of $\omega_{AB} = 30$ rad/s, determine the reactions that the bearing supports A and B exert on the shaft at the instant shown.

(a)

Solution

Free-Body Diagram (Figure 21–12b). The origin of the x, y, z coordinate system is located at the gear's center of mass G, which is also a fixed point. The axes are fixed in and rotate with the gear; hence $\mathbf{\Omega} = \mathbf{\omega}$.

Kinematics. As shown in Fig. 21–12c, the angular velocity $\mathbf{\omega}$ of the gear is constant in magnitude and is always directed along the axis of the shaft AB. Since this vector is measured from the X, Y, Z inertial frame of reference, for any position of the x, y, z axes $\mathbf{\omega}$ has x, y, z components of

(b)

$$\omega_x = 0, \qquad \omega_y = -30 \sin 10°, \qquad \omega_z = 30 \cos 10°$$

Since $\mathbf{\omega}$ has a constant direction and magnitude when observed from X, Y, Z, $(\dot{\mathbf{\omega}})_{XYZ} = (\dot{\mathbf{\omega}})_{xyz} = \mathbf{0}$. Hence, $\dot{\omega}_x = \dot{\omega}_y = \dot{\omega}_z = 0$. Also, since G is a fixed point, $(a_G)_x = (a_G)_y = (a_G)_z = 0$.

Equations of Motion. Applying Eqs. 21–25 yields

$$\Sigma M_x = I_x \dot{\omega}_x - (I_y - I_z)\omega_y\omega_z$$
$$-(A_Y)(0.2) + (B_Y)(0.25) = 0 - (0.05 - 0.1)(-30 \sin 10°)(30 \cos 10°)$$
$$-0.2A_Y + 0.25B_Y = -7.70 \qquad (1)$$

$$\Sigma M_y = I_y \dot{\omega}_y - (I_z - I_x)\omega_z\omega_x$$
$$A_X(0.2) \cos 10° - B_X(0.25) \cos 10° = 0 + 0$$
$$A_X = 1.25B_X \qquad (2)$$

(c)

Fig. 21–12

$$\Sigma M_z = I_z \dot{\omega}_z - (I_x - I_y)\omega_x\omega_y$$
$$A_X(0.2) \sin 10° - B_X(0.25) \sin 10° = 0$$
$$A_X = 1.25B_X$$

Applying Eqs. 21–19, we have

$$\Sigma F_X = m(a_G)_X; \qquad A_X + B_X = 0 \qquad (3)$$
$$\Sigma F_Y = m(a_G)_Y; \qquad A_Y + B_Y - 98.1 = 0 \qquad (4)$$
$$\Sigma F_Z = m(a_G)_Z; \qquad 0 = 0$$

Solving Eqs. (1) to (4) simultaneously gives

$$A_X = B_X = 0, \ A_Y = 71.6 \text{ N}, \ B_Y = 26.4 \text{ N} \qquad \textit{Ans.}$$

Example 21-5

The airplane shown in Fig. 21–13a is in the process of making a steady *horizontal* turn at the rate of ω_p. During this motion, the airplane's propeller is spinning at the rate of ω_s. If the propeller has two blades, determine the moments which the propeller shaft exerts on the propeller when the blades are in the vertical position. For simplicity, assume the blades to be a uniform slender bar having a moment of inertia I about an axis perpendicular to the blades and passing through their center, and having zero moment of inertia about a longitudinal axis.

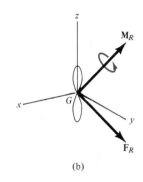

(a)

Solution

Free-Body Diagram (Figure 21–13b).   The effect of the connecting shaft on the propeller is indicated by the resultants $\mathbf{F}_R$ and $\mathbf{M}_R$. (The propeller's weight is assumed to be negligible.) The x, y, z axes are taken to be *fixed to the propeller* so that $\mathbf{\Omega} = \mathbf{\omega} = \mathbf{\omega}_s + \mathbf{\omega}_p$. Furthermore, the axes so chosen represent the principal axes of inertia for the propeller.

Kinematics.   The moments of inertia I_x and I_y are equal ($I_x = I_y = I$) and $I_z = 0$. The angular velocity at the instant shown, as measured from a fixed X, Y, Z frame coincident with x, y, z, Fig. 21–13c, is $\mathbf{\omega} = \mathbf{\omega}_s + \mathbf{\omega}_p = \omega_s\mathbf{i} + \omega_p\mathbf{k}$, so that the components of $\mathbf{\omega}$ are

$$\omega_x = \omega_s, \qquad \omega_y = 0, \qquad \omega_z = \omega_p$$

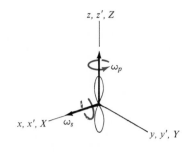

(b)

Since $\mathbf{\omega}$ is *changing direction,* the time derivative of $\mathbf{\omega}$ will be computed with respect to the fixed X, Y, Z axes and *then* $(\dot{\mathbf{\omega}})_{XYZ}$ will be resolved into components along the moving x, y, z axes. To simplify this calculation, a third coordinate system x', y', z' will be chosen which has an angular velocity $\mathbf{\Omega}' = \mathbf{\omega}_p$ and is coincident with the X, Y, Z axes at the instant shown. Thus

$$(\dot{\mathbf{\omega}})_{XYZ} = (\dot{\mathbf{\omega}})_{x'y'z'} + \mathbf{\Omega}' \times \mathbf{\omega} \tag{1}$$

$$= (\dot{\mathbf{\omega}}_s)_{x'y'z'} + (\dot{\mathbf{\omega}}_p)_{x'y'z'} + \mathbf{\omega}_p \times (\mathbf{\omega}_s + \mathbf{\omega}_p)$$

$$= \mathbf{0} + \mathbf{0} + \mathbf{\omega}_p \times \mathbf{\omega}_s + \mathbf{\omega}_p \times \mathbf{\omega}_p$$

$$= \mathbf{0} + \mathbf{0} + \omega_p\mathbf{k} \times \omega_s\mathbf{i} + \mathbf{0} = \omega_p\omega_s\mathbf{j}$$

Therefore, since $\mathbf{\Omega} = \mathbf{\omega}$, $(\dot{\mathbf{\omega}})_{XYZ} = (\dot{\mathbf{\omega}})_{xyz}$ and

$$\dot{\omega}_x = 0, \qquad \dot{\omega}_y = \omega_p\omega_s, \qquad \dot{\omega}_z = 0$$

z, z', Z

x, x', X

y, y', Y

(c)

Fig. 21–13

Equations of Motion.   Using Eqs. 21–25, we have

$$\Sigma M_x = I_x\dot{\omega}_x - (I_y - I_z)\omega_y\omega_z = I(0) - (I - 0)(0)\omega_p$$

$$M_x = 0 \hspace{4cm} \textit{Ans.}$$

$$\Sigma M_y = I_y\dot{\omega}_y - (I_z - I_x)\omega_z\omega_x = I(\omega_p\omega_s) - (0 - I)\omega_p\omega_s$$

$$M_y = 2I\omega_p\omega_s \hspace{3cm} \textit{Ans.}$$

$$\Sigma M_z = I_z\dot{\omega}_z - (I_x - I_y)\omega_x\omega_y = 0(0) - (I - I)(\omega_s)(0)$$

$$M_z = 0 \hspace{4cm} \textit{Ans.}$$

Problems

21–33. Derive the scalar form of the rotational equation of motion along the *x* axis, when $\Omega \neq \omega$ and the moments and products of inertia of the body are *not constant* with respect to time.

21–34. Derive the scalar form of the rotational equation of motion along the *x* axis, when $\Omega \neq \omega$ and the moments and the products of inertia of the body are *constant* with respect to time.

21–35. Derive the Euler equations of motion for $\Omega \neq \omega$, i.e., Eqs. 21–26.

*21–36. The 40-kg flywheel (disk) is mounted 20 mm off its true center at *G*. If the shaft is rotating at a constant speed $\omega = 8$ rad/s, determine the maximum reactions exerted on the journal bearings at *A* and *B*.

Prob. 21–36

21–37. The *thin rod* has a mass of 0.8 kg and a total length of 150 mm. It is rotating about its midpoint at a constant rate of $\dot{\theta} = 6$ rad/s, while the table to which its axle *A* is fastened is rotating at 2 rad/s. Determine the *x*, *y*, *z* moment components which the axle transmits to the rod when the rod is in any position θ.

Prob. 21–37

21–38. The rod assembly has a mass of 1.5 kg/m. A resultant frictional moment **M**, developed in the journal bearings *A* and *B*, causes the shaft to decelerate at 1 rad/s² when it has an angular velocity $\omega = 6$ rad/s. Determine the reactions on the bearings at the instant shown. What is the frictional moment **M** causing the deceleration?

Prob. 21–38

21–39. The 5-kg rod *AB* is supported by a rotating arm. The support at *A* is a journal bearing, which develops reactions normal to the rod. The support at *B* is a thrust bearing, which develops reactions both normal to the rod and along the axis of the rod. Neglecting friction, determine the *x*, *y*, *z* components of reaction at these supports when the frame rotates with a constant angular velocity of $\omega = 10$ rad/s.

Prob. 21–39

*21–40. The uniform hatch door, having a mass of 15 kg and a mass center at G, is supported in the horizontal plane by bearings at A and B. If a vertical force $F = 300$ N is applied to the door as shown, determine the components of reaction at the bearings and the angular acceleration of the door. The bearing at A will resist a component of force in the y direction, whereas the bearing at B will not. For the calculation, assume the door to be a thin plate and neglect the size of the bearing. The door is originally at rest. Set $a = 100$ mm, $b = 200$ mm, $c = 150$ mm, and $d = 30$ mm.

Prob. 21–40

21–41. The cylinder has a mass of 30 kg and is mounted on an axle which is supported by bearings at A and B. If the axle is turning at a constant rate of 40 rad/s, determine the components of force acting on the bearings at the instant shown.

Prob. 21–41

21–42. The rod assembly has a weight of 5 lb/ft. It is supported at B by a smooth journal bearing, which develops x and y force reactions, and at A by a smooth thrust bearing, which develops x, y, and z force reactions. If a 50-lb · ft torque is applied along rod AB, determine the components of reaction at the bearing at the instant the assembly has an angular velocity $\omega = 10$ rad/s.

Prob. 21–42

21–43. The 15-lb cylinder is rotating about shaft AB with a constant angular speed $\omega = 4$ rad/s. If the supporting shaft at C, initially at rest, is given an angular acceleration $\alpha_C = 12$ rad/s², determine the components of reaction at the bearings A and B. The bearing at A cannot support a force component along the x axis, whereas the bearing at B does.

Prob. 21–43

489

*21–44. The man sits on a swivel chair which is rotating with a constant angular velocity of 3 rad/s. He holds the uniform 5-lb rod AB horizontal. He suddenly gives it an angular acceleration of 2 rad/s², measured relative to him, as shown. Determine the required force and moment components at the grip, A, necessary to do this.

Prob. 21–44

21–45. The 4-kg slender rod AB is pinned at A and held at B by a cord. The axle CD is supported at the ends by ball-and-socket joints and is rotating with a constant angular velocity of 2 rad/s. Determine the tension developed in the cord and the magnitude of force developed at the pin A.

Probs. 21–45 / 21–46

21–46. Determine the tension in the cord and the magnitude of force at the pin A in Prob. 21–45 if the supporting axle has an angular acceleration $\alpha = 1.5$ rad/s² at the instant $\omega = 2$ rad/s. Assume that α acts in the same direction as ω shown in Prob. 21–45.

21–47. The rod assembly is supported on bearings at A and B. The mass of each rod is 1.5 kg/m. If the shaft AB is rotating with a speed $\omega = 5$ rad/s, determine the reactions at the journal bearings when the assembly is in the position shown. Also, what is the shaft's angular acceleration?

Prob. 21–47

*21–48. The rod assembly is supported by a ball-and-socket joint at C and a journal bearing support at D, which develops only x and y force reactions. The rods have a mass of 0.75 kg/m. Determine the angular acceleration of the rods and the components of reaction at the supports at the instant $\omega = 8$ rad/s as shown.

Prob. 21–48

21–49. The bent uniform rod ACD has a weight of 5 lb/ft and is supported at A by a pin and at B by a cord. If the vertical shaft rotates with a constant angular velocity of $\omega = 20$ rad/s, determine the x, y, z components of force and moment developed at A and the tension in the cord.

Prob. 21–49

21–50. Two uniform rods, each having a weight of 10 lb, are pin-connected to the edge of a rotating disk. If the disk has a constant angular velocity $\omega_D = 4$ rad/s, determine the angle θ made by each rod during the motion, and the components of the force and moment developed at the pin A. *Suggestion:* Use the x, y, z axes oriented as shown.

Prob. 21–50

491

*21.5 Gyroscopic Motion

In this section the equations used for analyzing the motion of a body (or top) which is symmetrical with respect to an axis and moving about a fixed point lying on the axis will be developed. These equations will then be applied to study the motion of a particularly interesting device, the gyroscope.

The body's motion will be analyzed using *Euler angles* ϕ, θ, ψ (phi, theta, psi). To illustrate how these angles are formed, reference is made to the top shown in Fig. 21–14a. The top is attached to point O and has an orientation relative to the fixed X, Y, Z axes at some instant of time as shown in Fig. 21–14d. To define this final position, a second set of x, y, z axes will be needed. For purposes of discussion, assume that this reference is fixed in the top. Starting with the X, Y, Z and x, y, z axes in coincidence, Fig. 21–14a, the final position of the top is determined using the following three orderly steps:

1. Rotate the top about the Z (or z) axis through an angle ϕ $(0 \leq \phi < 2\pi)$, Fig. 21–14b.
2. Rotate the top about the x axis through an angle θ $(0 \leq \theta \leq \pi)$, Fig. 21–14c.
3. Rotate the top about the z axis through an angle ψ $(0 \leq \psi < 2\pi)$ to obtain the final position, Fig. 20–14d.

When the top is in motion, in general each of the Euler angles is changing with time. Although the three finite rotations defined above are not vectors, the infinitesimal rotations $d\phi$, $d\theta$, and $d\psi$ are vectors, and thus the angular

(a)

Precession

(b)

Nutation

(c)

Spin

(d)

Fig. 21–14

velocity $\boldsymbol{\omega}$ of the top can be expressed in terms of the time derivatives of the Euler angles. The angular-velocity components $\dot{\phi}$, $\dot{\theta}$, and $\dot{\psi}$ are known as the *precession, nutation,* and *spin,* respectively. Their positive directions are shown in Fig. 21–14. It is seen that these vectors are not all perpendicular to one another; however, $\boldsymbol{\omega}$ of the top can still be expressed in terms of these three components.

In our case the body (top) is symmetric with respect to the z or spin axis. If we consider the top in Fig. 21–15, for which the x axis is oriented such that at the instant considered the spin angle $\psi = 0$ and *the x, y, z axes follow the motion of the body only in nutation and precession,* i.e., $\boldsymbol{\Omega} = \boldsymbol{\omega}_p + \boldsymbol{\omega}_n$, then the nutation and spin are always directed along the x and z axes, respectively. Hence, the angular velocity of the body is specified only in terms of the Euler angle θ, i.e.,

$$\boldsymbol{\omega} = \omega_x\mathbf{i} + \omega_y\mathbf{j} + \omega_z\mathbf{k}$$
$$= \dot{\theta}\mathbf{i} + (\dot{\phi}\sin\theta)\mathbf{j} + (\dot{\phi}\cos\theta + \dot{\psi})\mathbf{k} \qquad (21\text{-}27)$$

Since motion of the axes is not affected by the spin component,

$$\boldsymbol{\Omega} = \Omega_x\mathbf{i} + \Omega_y\mathbf{j} + \Omega_z\mathbf{k}$$
$$= \dot{\theta}\mathbf{i} + (\dot{\phi}\sin\theta)\mathbf{j} + (\dot{\phi}\cos\theta)\mathbf{k} \qquad (21\text{-}28)$$

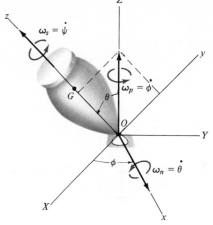

Fig. 21–15

The x, y, z axes in Fig. 21–15 represent *principal axes of inertia* of the body for *any* spin of the body about these axes. Hence, the moments of inertia are constant and will be represented as $I_{xx} = I_{yy} = I$ and $I_{zz} = I_z$. Since $\boldsymbol{\Omega} \neq \boldsymbol{\omega}$, the Euler equations 21–26 are used to establish the rotational equations of motion. Substituting into these equations the respective angular-velocity components defined by Eqs. 21–27 and 21–28, their corresponding time derivatives, and the moment of inertia components yields

$$\Sigma M_x = I(\ddot{\theta} - \dot{\phi}^2\sin\theta\cos\theta) + I_z\dot{\phi}\sin\theta(\dot{\phi}\cos\theta + \dot{\psi})$$
$$\Sigma M_y = I(\ddot{\phi}\sin\theta + 2\dot{\phi}\dot{\theta}\cos\theta) - I_z\dot{\theta}(\dot{\phi}\cos\theta + \dot{\psi}) \qquad (21\text{-}29)$$
$$\Sigma M_z = I_z(\ddot{\psi} + \ddot{\phi}\cos\theta - \dot{\phi}\dot{\theta}\sin\theta)$$

Each moment summation applies only at the fixed point O or the center of mass G of the body. Since the equations represent a coupled set of nonlinear second-order differential equations, in general a closed-form solution may not be obtained. Instead, the Euler angles ϕ, θ, and ψ may be obtained graphically as functions of time using numerical analysis and computer techniques.

A special case, however, does exist for which simplification of Eqs. 21–29 is possible. Commonly referred to as *steady precession,* it occurs when the nutation angle θ, precession $\dot{\phi}$, and spin $\dot{\psi}$ all remain *constant.* Equations 21–29 then reduce to the form

$$\Sigma M_x = -I\dot{\phi}^2\sin\theta\cos\theta + I_z\dot{\phi}\sin\theta(\dot{\phi}\cos\theta + \dot{\psi}) \qquad (21\text{-}30)$$

$$\Sigma M_y = 0$$
$$\Sigma M_z = 0$$

Equation 21–30 may be further simplified by noting that, from Eq. 21–27, $\omega_z = \dot{\phi} \cos \theta + \dot{\psi}$, so that

$$\Sigma M_x = -I\dot{\phi}^2 \sin \theta \cos \theta + I_z \dot{\phi}(\sin \theta)\omega_z$$

or

$$\Sigma M_x = \dot{\phi} \sin \theta(I_z\omega_z - I\dot{\phi} \cos \theta) \qquad (21\text{–}31)$$

It is interesting to note what effects the spin $\dot{\psi}$ has on the moment about the x axis. In this regard consider the spinning rotor shown in Fig. 21–16. Here $\theta = 90°$, in which case Eq. 21–30 reduces to the form

$$\Sigma M_x = I_z \dot{\phi} \dot{\psi}$$

or

$$\Sigma M_x = I_z \Omega_y \omega_z \qquad (21\text{–}32)$$

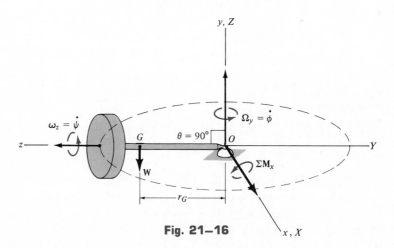

Fig. 21–16

From the figure it is seen that the vectors $\Sigma \mathbf{M}_x$, $\mathbf{\Omega}_y$, and $\boldsymbol{\omega}_z$ all act along their respective *positive axes* and therefore are mutually perpendicular. Instinctively, one would expect the rotor to fall down under the influence of gravity! However, this is not the case at all provided the product $I_z\Omega_y\omega_z$ is correctly chosen to counterbalance the moment $\Sigma M_x = Wr_G$ of the rotor's weight about O. This unusual phenomenon of rigid-body motion is often referred to as the *gyroscopic effect*.

Perhaps a more intriguing demonstration of the gyroscopic effect comes from studying the action of a *gyroscope*, frequently referred to as a *gyro*. A gyro is a rotor which spins at a very high rate about its axis of symmetry. This rate of spin is considerably greater than its precessional rate of rotation about the vertical axis. Hence, for all practical purposes, the angular momentum of the gyro can be assumed directed along its axis of spin. Thus, for the gyro rotor shown in Fig. 21–17, $\omega_z \gg \Omega_y$, and the magnitude of the angular

Fig. 21–17

momentum about point O, as computed by Eqs. 21–11, reduces to the form $H_O = I_z\omega_z$. Since both the magnitude and direction of $\mathbf{H}_O$ are constant as observed from x, y, z, direct application of Eq. 21–22 yields

$$\Sigma\mathbf{M}_x = \mathbf{\Omega}_y \times \mathbf{H}_O \qquad (21\text{–}33)$$

Using the right-hand rule applied to the cross product, it is seen that $\mathbf{\Omega}_y$ always swings $\mathbf{H}_O$ (or $\boldsymbol{\omega}_z$) toward the sense of $\Sigma\mathbf{M}_x$. In effect, the *change in direction* of the gyro's angular momentum, $d\mathbf{H}_O$, causes an angular impulse about O, i.e., $d\mathbf{H}_O = \Sigma\mathbf{M}_x\,dt$, Eq. 21–20. Note that since $H_O = I_z\omega_z$ and ΣM_x, Ω_y, and H_O are mutually perpendicular, Eq. 21–33 reduces to Eq. 21–32.

When a gyro is mounted in gimbal rings, Fig. 21–18, it becomes *free* of external moments applied to its base. Thus, in theory, its angular momentum **H** will never precess but, instead, maintain its same fixed orientation along the axis of spin when the base is rotated. This type of gyroscope is called a *free gyro* and is useful as a gyrocompass when the spin axis of the gyro is directed north. In reality, the gimbal mechanism is never completely free of friction, so that such a device is useful only for the local navigation of ships and aircraft. The gyroscopic effect is also useful as a means of stabilizing both the rolling motion of ships at sea and the trajectories of missiles and projectiles. Furthermore, this effect is of significant importance in the design of shafts and bearings for rotors which are subjected to forced precessions.

Fig. 21–18

Example 21–6

The top shown in Fig. 21–19a has a mass of 0.5 kg and is precessing about the vertical axis at a constant angle of $\theta = 60°$. If it spins with an angular velocity $\omega_s = 100$ rad/s, determine the precessional velocity ω_p. Assume that the axial and transverse moments of inertia of the top are $4.5(10^{-4})$ kg · m² and $12.0(10^{-4})$ kg · m², respectively, measured with respect to the fixed point O.

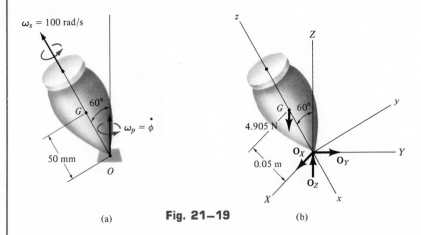

Fig. 21–19

(a)   (b)

Solution

Equation 21–30 will be used for the solution since the motion is a *steady precession*. As shown on the free-body diagram, Fig. 21–19b, the coordinate axes are established in the usual manner, that is, with the positive z axis in the direction of spin, the positive Z axis in the direction of precession, and the positive x axis in the direction of the moment $\Sigma\mathbf{M}_x$ (refer to Fig. 21–15). Thus,

$$\Sigma M_x = -I\dot\phi^2 \sin\theta \cos\theta + I_z\dot\phi \sin\theta(\dot\phi \cos\theta + \dot\psi)$$

$$4.905(0.05) \sin 60° = -12.0(10^{-4})\dot\phi^2 \sin 60° \cos 60°$$
$$+ 4.5(10^{-4})\dot\phi \sin 60°(\dot\phi \cos 60° + 100)$$

or

$$\dot\phi^2 - 120.0\dot\phi + 654.0 = 0 \qquad (1)$$

Solving this quadratic equation for the precession gives

$$\dot\phi = 114 \text{ rad/s} \qquad \text{(high precession)} \qquad Ans.$$

and

$$\dot\phi = 5.72 \text{ rad/s} \qquad \text{(low precession)} \qquad Ans.$$

In reality low precession of the top would generally be observed since high precession would require a large kinetic energy.

Example 21-7

The 1-kg disk shown in Fig. 21–20a is spinning about its axis with a constant angular velocity $\omega_D = 70$ rad/s. The block at B has a mass of 2 kg, and by adjusting its position s one can change the precession of the disk about its supporting pivot at O. Compute the position s which will enable the disk to have a constant precessional velocity $\omega_p = 0.5$ rad/s about the pivot. Neglect the weight of the shaft.

Fig. 21–20

Solution

The free-body diagram of the disk is shown in Fig. 21–20b, where **F** represents the force reaction of the shaft on the disk. The origin for both the x, y, z and X, Y, Z coordinate systems is located at point O, which represents a *fixed point* for the disk. (Although point O does not lie on the disk, imagine a massless extension of the disk to this point.) In the conventional sense, the Z axis is chosen along the axis of precession, and the z axis is along the axis of spin, so that $\theta = 90°$. Since the precession is *steady*, Eq. 21–31 may be used for the solution. This equation reduces to

$$\Sigma M_x = \dot{\phi} I_z \omega_z$$

which is the same as Eq. 21–32. Substituting the required data gives

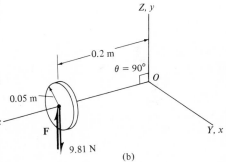

$$9.81(0.2) - F(0.2) = 0.5\left[\frac{1}{2}(1)(0.05)^2\right](-70)$$

$$F = 10.0 \text{ N}$$

As shown on the free-body diagram of the shaft and block B, Fig. 21–20c, summing moments about the x axis requires

$$(19.62)s = (10.0)(0.20)$$
$$= 0.102 \text{ m} = 102 \text{ mm} \qquad \textbf{\textit{Ans.}}$$

497

*21.6 Torque-Free Motion

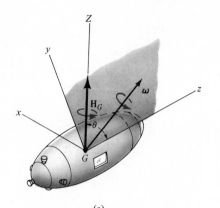

(a)

Fig. 21–21a

When the only external force acting on a body is caused by gravitation, the general motion of the body is referred to as *torque-free motion*. This type of motion is characteristic of planets, artificial satellites, and projectiles—provided the effects of air friction are neglected.

In order to describe the characteristics of this motion, the distribution of the body's mass will be assumed *axisymmetric*. The satellite shown in Fig. 21–21a is an example of such a body, where the z axis represents an axis of symmetry. The origin of the x, y, z coordinates is located at the mass center G, such that $I_{zz} = I_z$ and $I_{xx} = I_{yy} = I$ for the body. If it is assumed that gravitation is the only external force present, the summation of moments about the mass center is zero. From Eq. 21–22, this requires the angular momentum of the body to be constant, i.e.,

$$\mathbf{H}_G = \text{const}$$

At the instant considered, it will be assumed that the inertial frame of reference is oriented such that the positive Z axis is directed along $\mathbf{H}_G$ and the y axis lies in the plane formed by the z and Z axes, Fig. 21–21a. The Euler angle formed between Z and z is θ, and therefore, with this choice of axes the angular momentum may be expressed as

$$\mathbf{H}_G = H_G \sin \theta \mathbf{j} + H_G \cos \theta \mathbf{k}$$

Furthermore, using Eqs. 21–11, we have

$$\mathbf{H}_G = I\omega_x \mathbf{i} + I\omega_y \mathbf{j} + I_z \omega_z \mathbf{k}$$

where ω_x, ω_y, ω_z represent the x, y, z components of the body's angular velocity. Equating the respective **i, j,** and **k** components of the above two equations yields

$$\omega_x = 0, \qquad \omega_y = \frac{H_G \sin \theta}{I}, \qquad \omega_z = \frac{H_G \cos \theta}{I_z} \qquad (21\text{–}34)$$

or

$$\boldsymbol{\omega} = \frac{H_G \sin \theta}{I}\mathbf{j} + \frac{H_G \cos \theta}{I_z}\mathbf{k} \qquad (21\text{–}35)$$

In a similar manner, equating the respective **i, j,** and **k** components of Eq. 21–27 to those of Eq. 21–34, we obtain

$$\dot{\theta} = 0$$

$$\dot{\phi} \sin \theta = \frac{H_G \sin \theta}{I}$$

$$\dot{\phi} \cos \theta + \dot{\psi} = \frac{H_G \cos \theta}{I_z}$$

Solving, we get

$$\theta = \text{const}$$

$$\dot{\phi} = \frac{H_G}{I}$$

$$\dot{\psi} = \frac{I - I_z}{II_z} H_G \cos \theta$$

(21–36)

Thus, for torque-free motion of an axisymmetrical body, the angle θ formed between the angular-momentum vector and the spin of the body remains constant. Furthermore, the angular momentum $\mathbf{H}_G$, precession $\dot{\phi}$, and spin $\dot{\psi}$ for the body remain constant at all times during the motion. Eliminating H_G from the second and third of Eqs. 21–36 yields the following relationship between the spin and precession:

$$\dot{\psi} = \frac{I - I_z}{I_z} (\dot{\phi}) \cos \theta$$

(21–37)

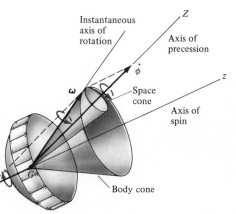

(b)

Fig. 21–21b

As shown in Fig. 21–21b, the body precesses about the Z axis, which is fixed in direction, while it spins about the z axis. These two components of angular motion may be analyzed by using a simple cone model, introduced in Sec. 20.1. The *space cone* defining the precession is fixed from rotating, since the precession has a fixed direction, while the *body cone* rotates around the space cone's outer surface without slipping. On this basis, an attempt should be made to imagine the motion. The interior angle of each cone is chosen such that the resultant angular velocity of the body is directed along the line of contact of the two cones. This line of contact represents the instantaneous axis of rotation for the body cone, and hence the angular velocity of both the body cone and the body must be directed along this line. Since the spin is a function of the moments of inertia I and I_z of the body, Eq. 21–36, the cone model in Fig. 21–21b is satisfactory for describing the motion, provided $I > I_z$. Torque-free motion which meets these requirements is called *regular precession*. If $I < I_z$, the spin is negative and the precession positive. This motion is represented by the satellite motion shown in Fig. 21–22 ($I < I_z$). The cone model may again be used to represent the motion; however, to preserve the correct vector addition of spin and precession to obtain the angular velocity $\boldsymbol{\omega}$, the inside surface of the body cone must roll on the outside surface of the (fixed) space cone. This motion is referred to as *retrograde precession*.

Fig. 21–22

Example 21–8

The motion of a football is observed using a slow-motion projector. From the film, the spin of the football is seen to be directed 30° from the horizontal, as shown in Fig. 21–23a. Also, the football is precessing about the vertical axis at a rate $\dot{\phi} = 3$ rad/s. If the ratio of the axial to transverse moments of inertia of the football is $\frac{1}{3}$, measured with respect to the center of gravity, determine the magnitude of the football's spin and its angular velocity. Neglect the effect of air resistance.

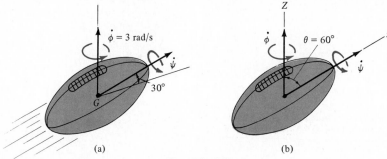

Fig. 21–23

Solution

Since the weight of the football is the only force acting, the motion is torque-free. In the conventional sense, if the z axis is established along the axis of spin and the Z axis along the precession axis, as shown in Fig. 21–23b, then the angle $\theta = 60°$. Applying Eq. 21–37 gives

$$\dot{\psi} = \frac{I - I_z}{I_z}(\dot{\phi}) \cos \theta = \frac{I - \frac{1}{3}I}{\frac{1}{3}I}(3) \cos 60°$$

$$= 3 \text{ rad/s} \qquad\qquad \textit{Ans.}$$

Using Eqs. 21–34, where $H_G = \dot{\phi}I$ (Eq. 21–36), we have

$$\omega_x = 0$$

$$\omega_y = \frac{H_G \sin \theta}{I} = \frac{3I \sin 60°}{I} = 2.60 \text{ rad/s}$$

$$\omega_z = \frac{H_G \cos \theta}{I_z} = \frac{3I \cos 60°}{\frac{1}{3}I} = 4.50 \text{ rad/s}$$

Thus,

$$\omega = \sqrt{(\omega_x)^2 + (\omega_y)^2 + (\omega_z)^2}$$

$$= \sqrt{(0)^2 + (2.60)^2 + (4.50)^2}$$

$$= 5.20 \text{ rad/s} \qquad\qquad \textit{Ans.}$$

Problems

21–51. The propeller on a single-engine airplane has a mass of 15 kg and a centroidal radius of gyration of 0.3 m computed about the axis of spin. When viewed from the front of the airplane, the propeller is turning clockwise at 350 rad/s about the spin axis. If the airplane enters a vertical curve having a radius $\rho = 80$ m and is traveling at 200 km/h, determine the gyroscopic bending moment which the propeller exerts on the bearings of the engine when the airplane is in its lowest position as shown.

Prob. 21–51

*__*21–52.__ The turbine on a ship has a mass of 400 kg and is mounted on bearings A and B as shown. Its center of mass is at G, its radius of gyration is $k_z = 0.3$ m, and $k_x = k_y = 0.5$ m. If it is spinning at 200 rad/s, determine the vertical reactions at the bearings when the ship undergoes the following motions: (a) rolling, $\omega_1 = 0.2$ rad/s, (b) turning, $\omega_2 = 0.8$ rad/s, (c) pitching, $\omega_3 = 1.4$ rad/s.

Prob. 21–52

21–53. The conical top has a mass of 0.8 kg, and the moments of inertia are $I_x = I_y = 3.5(10^{-3})$ kg·m^2 and $I_z = 0.8(10^{-3})$ kg·m^2. If it spins freely in the ball-and-socket joint at A with an angular velocity $\omega_s = 750$ rad/s, compute the precession of the top about the axis of the shaft AB.

Prob. 21–53

21–54. The top consists of a thin disk that has a weight of 8 lb and a radius of 0.3 ft. The rod has a negligible mass and a length of 0.5 ft. If the top is spinning with an angular velocity $\omega = 300$ rad/s, determine the steady-state precessional angular velocity ω_p of the rod when $\theta = 40°$.

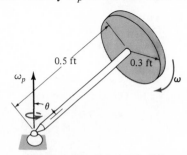

Probs. 21–54 / 21–55

21–55. Solve Prob. 21–54 when $\theta = 90°$.

*21–56. The top has a mass of 90 g, a mass center at G, and a radius of gyration $k = 18$ mm about its axis of symmetry. About any transverse axis acting through point O the radius of gyration is $k_t = 35$ mm. If the top is pinned at O and the precession is $\omega_p = 0.5$ rad/s, determine the spin $\boldsymbol{\omega}_s$.

Prob. 21–56

21–57. The rotor assembly on the engine of a jet airplane consists of the turbine, drive shaft, and compressor. The total mass is 700 kg, the radius of gyration about the shaft axis is $k_{AB} = 0.35$ m, and the mass center is at G. If the rotor has an angular velocity $\omega_{AB} = 1000$ rad/s, and the plane is pulling out of a vertical curve while traveling at 250 m/s, determine the components of reaction at the bearings A and B due to the gyroscopic effect.

Prob. 21–57

21–58. The toy gyroscope consists of a rotor R which is attached to the frame of negligible mass. If the rotor is spinning about its axle at an angular speed $\omega_R = 150$ rad/s, determine the constant angular velocity ω_p at which the frame is precessing about the pivot point at O. OA moves in the horizontal plane. The rotor has a mass of 200 g and a radius of gyration $k_{OA} = 20$ mm about OA.

Prob. 21–58

21–59. The 30-lb wheel is rolling without slipping. If it has a radius of gyration $k_{AB} = 1.2$ ft about its axle AB, and the vertical drive shaft is turning at 8 rad/s, determine the normal reaction the wheel exerts on the ground at C.

Prob. 21–59

*21–60. An airplane descends at a steep angle and then levels off horizontally to land. If the propeller is turning clockwise when observed from the rear of the plane, determine the direction in which the plane tends to turn as caused by the gyroscopic effect.

21–61. The car is traveling at $v_C = 100$ km/h around the horizontal curve having a radius of curvature of 80 m. If each wheel has a mass of 16 kg, a radius of gyration $k_G = 300$ mm about its spinning axis, and a diameter of 400 mm, determine the difference in the normal force between the rear wheels caused by the gyroscopic effect. The distance between the wheels is 1.30 m.

1.30 m

80 m

$v_C = 100$ km/h

Prob. 21–61

21–62. The 4-kg disk is thrown in the air with a spin $\omega_z = 6$ rad/s. If the angle θ is measured as 160°, determine the precession about the Z axis.

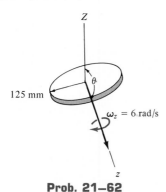

Z

θ

125 mm

$\omega_z = 6$ rad/s

z

Prob. 21–62

21–63. The satellite has a mass of 1.8 Mg, and about axes passing through the mass center G the axial and transverse radii of gyration are $k_z = 0.8$ m and $k_t = 1.2$ m, respectively. If it is spinning at $\omega_s = 6$ rad/s when it is launched, determine its angular momentum. Precession occurs about the Z axis.

ω_s

z

Z

5°

G

Prob. 21–63

21–64. The rocket has a mass of 4 Mg and radii of gyration $k_z = 0.85$ m and $k_y = 2.3$ m. It is initially spinning about the z axis at $\omega_z = 0.05$ rad/s when a meteoroid M strikes it at A and creates an impulse $\mathbf{I} = \{300\mathbf{i}\}$ kN · s. Determine the axis of precession after the impact.

y

x

G

A

ω_z

z

3 m

M

Prob. 21–64

21–65. The projectile shown is subjected to torque-free motion. The transverse and axial moments of inertia are I and I_z, respectively. If θ represents the angle between the precessional axis Z and the axis of symmetry z, and β is the angle between the angular velocity ω and the z axis, show that β and θ are related by the equation $\tan \theta = (I/I_z) \tan \beta$.

Prob. 21–65

21–66. The radius of gyration about an axis passing through the axis of symmetry of the 2.5-Mg satellite is $k_z = 2.3$ m, and about any transverse axis passing through the center of mass G, $k_t = 3.4$ m. If the satellite has a steady-state precession of two revolutions per hour about the Z axis, determine the rate of spin about the z axis.

Prob. 21–66

21–67. While the rocket is in free flight, it has a spin of 3 rad/s and precesses about an axis measured 10° from the axis of spin. If the ratio of the axial to transverse moments of inertia of the rocket is 1/15, computed about axes which pass through the mass center G, determine the angle which the resultant angular velocity makes with the spin axis. Construct the body and space cones used to describe the motion. Is the precession regular or retrograde?

Prob. 21–67

Vibrations

A *vibration* is the periodic motion of a body or system of connected bodies displaced from a position of equilibrium. In general, there are two types of vibration, free and forced. *Free vibration* occurs when the motion is maintained by gravitational or elastic restoring forces, such as the swinging motion of a pendulum or the vibration of an elastic rod. *Forced vibration* is caused by an external periodic or intermittent force applied to the system. Both of these types of vibration may be either damped or undamped. *Undamped* vibrations can continue indefinitely because frictional effects are neglected in the analysis. Since in reality both internal and external frictional forces are present, the motion of all vibrating bodies is actually *damped*.

In this chapter we will study the characteristics of the above types of vibrating motion. The analysis will apply to those bodies which are constrained to move only in one direction. These single-degree-of-freedom systems require only one coordinate to specify completely the position of the system at any time. The analysis of multi-degree-of-freedom systems is based on this simplified case and is thoroughly treated in textbooks devoted to vibrational theory.

⋆**22.1** Undamped Free Vibration

Equilibrium
position

(a)

(b)

Fig. 22–1

The simplest type of vibrating motion is undamped free vibration, represented by the model shown in Fig. 22–1a. The block has a mass m and is attached to a spring having a stiffness k. Vibrating motion is provided by displacing the block a distance x from its equilibrium position and allowing the spring to restore it to its original position. As the spring pulls on the block, the block will attain a velocity such that it will proceed to move out of equilibrium when $x = 0$. Provided the supporting surface is smooth, oscillation will continue indefinitely.

The time-dependent path of motion of the block may be determined by applying the equation of motion to the block when it is in the displaced position x. The free-body diagram is shown in Fig. 22–1b. The elastic restoring force $F = kx$ is always directed toward the equilibrium position, whereas the acceleration **a** acts in the direction of *positive displacement*. Noting that $a = d^2x/dt^2 = \ddot{x}$, we have

$$\xrightarrow{+}\Sigma F_x = ma_x; \qquad\qquad -kx = m\ddot{x}$$

Here it is seen that the acceleration is proportional to the position. Motion described in this manner is called *simple harmonic motion*. Rearranging the terms into a "standard form" gives

$$\ddot{x} + p^2 x = 0 \qquad\qquad (22\text{–}1)$$

The constant p is called the *circular frequency,* expressed in rad/s, and in this case

$$p = \sqrt{\frac{k}{m}} \qquad\qquad (22\text{–}2)$$

Equation 22–1 may also be obtained by considering the block to be suspended, as shown in Fig. 22–2a, and measuring the displacement y from the block's *equilibrium position*. The free-body diagram is shown in Fig. 22–2b. When the block is in equilibrium, the spring exerts an upward force of $F = W = mg$ on the block. Hence, when the block is displaced a distance y downward from this position, the magnitude of the spring force is $F = W + ky$.

(a)　　　**Fig.22–2**　　　(b)

Applying the equation of motion gives

$$+ \downarrow \Sigma F_y = ma_y; \qquad\qquad -W - ky + W = m\ddot{y}$$

or

$$\ddot{y} + p^2 y = 0$$

which is the same form as Eq. 22–1, where p is defined by Eq. 22–2.

Equation 22–1 is a homogeneous, second-order, linear, differential equation with constant coefficients. It can be shown, using the methods of differential equations, that the general solution of this equation is

$$x = A \sin pt + B \cos pt \qquad\qquad (22\text{–}3)$$

where A and B represent two constants of integration. The block's velocity and acceleration are determined by taking successive time derivatives, which yields

$$v = \dot{x} = Ap \cos pt - Bp \sin pt \qquad\qquad (22\text{–}4)$$
$$a = \ddot{x} = -Ap^2 \sin pt - Bp^2 \cos pt \qquad\qquad (22\text{–}5)$$

When Eqs. 22–3 and 22–5 are substituted into Eq. 22–1, the differential equation is indeed satisfied, and therefore Eq. 22–3 represents the true solution to Eq. 22–1.

The integration constants A and B in Eq. 22–3 are generally determined from the initial conditions of the problem. For example, suppose that the block in Fig. 22–1a has been displaced a distance x_1 to the right from its equilibrium position and given an initial (positive) velocity $\mathbf{v}_1$ directed to the right. Substituting $x = x_1$ at $t = 0$ into Eq. 22–3 yields $B = x_1$. Since $v = v_1$ at $t = 0$, using Eq. 22–4 we obtain $A = v_1/p$. If these values are substituted into Eq. 22–3, the equation describing the motion becomes

$$x = \frac{v_1}{p} \sin pt + x_1 \cos pt \qquad\qquad (22\text{–}6)$$

Equation 22–3 may also be expressed in terms of simple sinusoidal motion. Let

$$A = C \cos \phi \qquad\qquad (22\text{–}7)$$

and

$$B = C \sin \phi \qquad\qquad (22\text{–}8)$$

where C and ϕ are new constants to be determined in place of A and B. Substituting into Eq. 22–3 yields

$$x = C \cos \phi \sin pt + C \sin \phi \cos pt$$

Since $\sin (\theta + \phi) = \sin \theta \cos \phi + \cos \theta \sin \phi$, then

$$x = C \sin (pt + \phi) \qquad\qquad (22\text{–}9)$$

507

If this equation is plotted on an x versus pt axis, the graph shown in Fig. 22–3 is obtained. The maximum displacement of the block from its equilibrium position is defined as the *amplitude* of vibration. From either the figure or Eq. 22–9 the amplitude is C. The angle ϕ is called the *phase angle* since it represents the amount by which the curve is displaced from the origin when $t = 0$. The constants C and ϕ are related to A and B by Eqs. 22–7 and 22–8. Squaring and adding these two equations, the amplitude becomes

$$C = \sqrt{A^2 + B^2} \tag{22–10}$$

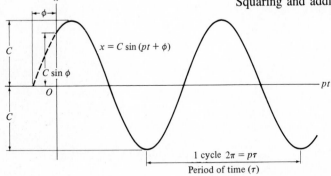

Fig. 22–3

If Eq. 22–8 is divided by Eq. 22–7, the phase angle is

$$\phi = \tan^{-1}\frac{B}{A} \tag{22–11}$$

Note that the sine curve, Eq. 22–9, completes one *cycle*, and hence the cyclic motion of the block is repeated in time $t = \tau$ (tau), so that $p\tau = 2\pi$ or

$$\tau = \frac{2\pi}{p} \tag{22–12}$$

This length of time is called a *period*, Fig. 22–3. Using Eq. 22–2, the period may also be represented as

$$\tau = 2\pi\sqrt{\frac{m}{k}} \tag{22–13}$$

The *frequency* f is defined as the number of cycles completed per unit of time, which is the reciprocal of the period:

$$f = \frac{1}{\tau} = \frac{p}{2\pi} \tag{22–14}$$

or

$$f = \frac{1}{2\pi}\sqrt{\frac{k}{m}} \tag{22–15}$$

The frequency is expressed in cycles/s. This ratio of units is called a *hertz* (Hz), where 1 Hz = 1 cycle/s = 2π rad/s.

When a body or system of connected bodies is given an initial displacement from its equilibrium position and released, it will vibrate with a definite frequency known as the *natural frequency*. This type of vibration is called *free vibration*, since no external forces except gravitational or elastic forces act on the body during the motion. Provided the *amplitude* of vibration remains *constant*, the motion is said to be *undamped*.

The undamped free vibration of a body having a single degree of freedom has the same characteristics as simple harmonic motion of the block and spring discussed above. Consequently, the body's motion is described by a differential equation of the *same form* as Eq. 22–1, i.e.,

$$\ddot{x} + p^2 x = 0 \qquad (22\text{–}16)$$

Hence, if the circular frequency p of the body is known, the period of vibration τ, natural frequency f, and other vibrating characteristics of the body can be established using the above equations.

PROCEDURE FOR ANALYSIS

As in the case of the block and spring, the circular frequency p of a rigid body or system of connected rigid bodies having a single degree of freedom can be determined using the following procedure:

Free-Body Diagram. Draw the free-body diagram of the body when the body is displaced by a *small amount* from its equilibrium position. Locate the body with respect to its equilibrium position by using an appropriate *inertial coordinate q*. The acceleration of the body's mass center $\mathbf{a}_G$ or the body's angular acceleration $\boldsymbol{\alpha}$ should have a sense which is in the positive direction of the position coordinate. These vectors may be sketched near, *but not on,* the free-body diagram. If it is decided that the rotational equation of motion $\Sigma M_P = \Sigma(M_k)_P$ is to be used, then it may be beneficial to also draw the kinetic diagram since it graphically accounts for the components $m(\mathbf{a}_G)_x$, $m(\mathbf{a}_G)_y$, and $I_G\boldsymbol{\alpha}$, and thereby makes it convenient for visualizing the terms needed in the moment sum $\Sigma(M_k)_P$.

Equation of Motion. Apply the equation of motion to relate the elastic or gravitational restoring forces and couples acting on the body to the body's accelerated motion.

Kinematics. Using kinematics, express the body's accelerated motion in terms of the second time derivative of the position coordinate, $\ddot{q}$. Substitute this result into the equation of motion and determine p by rearranging the terms so that the resulting equation is of the form $\ddot{q} + p^2 q = 0$.

The following examples illustrate this procedure.

Example 22–1

Determine the period of vibration for the simple pendulum shown in Fig. 22–4a. The bob has a mass m and is attached to a cord of length l. Neglect the size of the bob.

Solution

Free-Body Diagram. Motion of the system will be related to the position coordinate $(q =)\theta$, Fig. 22–4b. When the bob is displaced by an angle θ, the *restoring force* acting on the bob is created by the *weight component* $mg \sin \theta$. Furthermore, note that $\mathbf{a}_t$ is shown acting in the direction of *increasing* θ.

Fig. 22–4

Equation of Motion. Applying the equation of motion in the *tangential direction*, since it involves the restoring force, yields

$$+\nearrow \Sigma F_t = ma_t; \qquad -mg \sin \theta = ma_t \qquad (1)$$

Kinematics. $a_t = d^2s/dt^2 = \ddot{s}$. Furthermore, s may be related to θ by the equation $s = l\theta$, so that $a_t = l\ddot{\theta}$. Hence, Eq. (1) reduces to the form

$$\ddot{\theta} + \frac{g}{l} \sin \theta = 0 \qquad (2)$$

The solution of this equation involves the use of an elliptic integral. For *small displacements,* however, $\sin \theta \approx \theta$, in which case

$$\ddot{\theta} + \frac{g}{l} \theta = 0 \qquad (3)$$

Comparing this equation with Eq. 22–16 ($\ddot{x} + p^2x = 0$), which is the "standard form" for simple harmonic motion, it is seen that $p = \sqrt{g/l}$. From Eq. 22–12, the period of time required for the bob to make one complete swing is therefore

$$\tau = \frac{2\pi}{p} = 2\pi \sqrt{\frac{l}{g}} \qquad \textit{Ans.}$$

This interesting result, originally discovered by Galileo Galilei through experiment, indicates that the period depends only on the length of the cord and not on the mass of the pendulum bob.

The solution of Eq. (3) is given by Eq. 22–3, where $p = \sqrt{g/l}$ and θ is substituted for x. Like the block and spring, the constants A and B in this problem may be determined if, for example, one knows the displacement and velocity of the bob at a given instant.

Example 22–2

The 10-kg rectangular plate shown in Fig. 22–5*a* is suspended at its center from a rod having a torsional stiffness $k = 1.5$ N · m/rad. Determine the natural period of vibration of the plate when it is given a small angular displacement θ in the plane of the plate.

(a)   **Fig. 22–5**   (b)

Solution

Free-Body Diagram (Figure 22–5*b*).   Since the plate is displaced in its own plane, the torsional *restoring* moment created by the rod is $M = k\theta$. This moment acts in the direction opposite to the angular displacement θ. The angular acceleration $\ddot{\theta}$ acts in the direction of *positive* θ.

Equation of Motion

$$\Sigma M_O = I_O \alpha; \qquad\qquad -k\theta = I_O \ddot{\theta}$$

or

$$\ddot{\theta} + \frac{k}{I_O}\theta = 0$$

Since this equation is in "standard form," the circular frequency is $p = \sqrt{k/I_O}$.

The moment of inertia of the plate about an axis coincident with the rod is $I_O = \frac{1}{12}m(a^2 + b^2)$. Hence,

$$I_O = \frac{1}{12}(10)[(0.2)^2 + (0.3)^2] = 0.108 \text{ kg} \cdot \text{m}^2$$

The natural period of vibration is, therefore,

$$\tau = \frac{2\pi}{p} = 2\pi\sqrt{\frac{I_O}{k}} = 2\pi\sqrt{\frac{0.108}{1.5}} = 1.69 \text{ s} \qquad\qquad \textit{Ans.}$$

Example 22–3

The bent rod shown in Fig. 22–6*a* has a negligible mass and supports a 5-kg collar at its end. Determine the natural period of vibration for the system.

(a)

Solution
Free-Body and Kinetic Diagrams (Figure 22–6*b*).   Here the rod is displaced by a small amount θ from the equilibrium position. Since the spring is subjected to an initial compression of x_{st} for statical equilibrium, for a displacement $x > x_{st}$ the spring exerts a force of $F_s = kx - kx_{st}$ on the rod. To obtain the standard form, Eq. 22–16, $5\mathbf{a}_y$ acts *upward*, which is in accordance with positive θ displacement.

(b)

Equation of Motion.   Moments will be summed about point B to eliminate the unknown reaction at this point. Since θ is small,

$$\zeta +\Sigma M_B = \Sigma(M_k)_B; \quad kx(0.1) - kx_{st}(0.1) + 49.05(0.2) = -5a_y(0.2)$$

The second term on the left side, $-kx_{st}(0.1)$, represents the moment created by the spring force which is necessary to hold the collar in *equilibrium*, i.e., at $x = 0$. Since this moment is equal and opposite to the moment $49.05(0.2)$ created by the weight of the collar, these two terms cancel in the above equation, so that

$$kx(0.1) = -5a_y(0.2) \qquad (1)$$

Kinematics.   The positions of the spring and the collar may be related to the angle θ, Fig. 22–6*c*. Since θ is small, $x = 0.1\theta$ and $y = 0.2\theta$. Therefore, $a_y = \ddot{y} = 0.2\ddot{\theta}$. Substituting into Eq. (1) yields

$$400(0.1\theta)0.1 = -5(0.2\ddot{\theta})0.2$$

Rewriting this equation in standard form gives

$$\ddot{\theta} + 20\theta = 0$$

Comparing with $\ddot{x} + p^2x = 0$ (Eq. 22–16), we have

$$p^2 = 20, \quad p = 4.47 \text{ rad/s}$$

The natural period of vibration is therefore,

$$\tau = \frac{2\pi}{p} = \frac{2\pi}{4.47} = 1.40 \text{ s} \qquad \qquad \textit{Ans.}$$

(c)

Fig. 22–6

Example 22-4

A 10-lb block is suspended from a cord that passes over a 15-lb disk, as shown in Fig. 22-7a. The spring has a stiffness $k = 200$ lb/ft. Determine the natural period of vibration for the system.

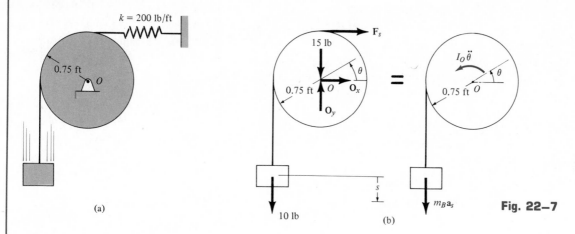

(a)

(b)

Fig. 22-7

Solution

Free-Body and Kinetic Diagrams (Figure 22-7b). The *system* consists of the disk, which undergoes a rotation defined by the angle θ, and the block, which translates by an amount s. The vector $I_O\ddot{\theta}$ acts in the direction of *positive* θ, and consequently $m_B\mathbf{a}_s$ acts downward in the direction of *positive* s.

Equation of Motion. Summing moments about point O to eliminate the reactions $\mathbf{O}_x$ and $\mathbf{O}_y$, realizing that $I_O = \frac{1}{2}mr^2$, yields

$$\zeta + \Sigma M_O = \Sigma(M_k)_O;$$

$$10(0.75) - F_s(0.75) = \frac{1}{2}\left(\frac{15}{32.2}\right)(0.75)^2\ddot{\theta} + \left(\frac{10}{32.2}\right)a_s(0.75) \quad (1)$$

Kinematics. As shown on the kinematic diagram in Fig. 22-7c, a small positive displacement θ of the disk causes the block to lower by an amount $s = 0.75\theta$; hence, $a_s = \ddot{s} = 0.75\ddot{\theta}$. When $\theta = 0°$, the spring force required for *equilibrium* of the disk is 10 lb, acting to the right. For position θ, the spring force is $F_s = (200 \text{ lb/ft})(0.75\theta \text{ ft}) + 10$ lb. Substituting these results into Eq. (1) and simplifying yields

$$\ddot{\theta} + 368\theta = 0$$

Hence,

$$p^2 = 368 \qquad p = 19.2 \text{ rad/s}$$

Therefore, the natural period of vibration is

$$\tau = \frac{2\pi}{p} = \frac{2\pi}{19.2} = 0.328 \text{ s} \qquad \qquad \textit{Ans.}$$

(c)

Problems

22–1. A block having a weight of 8 lb is suspended from a spring having a stiffness $k = 40$ lb/ft. If the block is pushed $y = 0.2$ ft upward from its equilibrium position and then released from rest, determine the equation which describes the motion. What are the amplitude and the natural frequency of the vibration? Assume that positive displacement is measured downward.

22–2. An 8-kg block is suspended from a spring having a stiffness $k = 80$ N/m. If the block is given an upward velocity of 0.4 m/s when it is 90 mm above its equilibrium position, determine the equation which describes the motion and the maximum upward displacement of the block measured from the equilibrium position. Assume that positive displacement is measured downward.

22–3. When a 3-kg block is suspended from a spring, the spring is stretched a distance of 60 mm. Determine the natural frequency and the period of vibration for a 0.2-kg block attached to the same spring.

*22–4. A spring has a stiffness of 600 N/m. If a 4-kg block is attached to the spring, pushed 50 mm above its equilibrium position, and released from rest, determine the equation which describes the block's motion. Assume that positive displacement is measured downward.

■22–5. A spring is stretched 175 mm by an 8-kg block. If the block is displaced 100 mm downward from its equilibrium position and given a downward velocity of 1.50 m/s, determine the differential equation which describes the motion. Assume that positive displacement is measured downward. Use the Runge-Kutta method to determine the position of the block, measured from its unstretched position, when $t = 0.22$ s. Use a time increment of $\Delta t = 0.02$ s.

22–6. If the block in Prob. 22–5 is given an upward velocity of 4 m/s when it is displaced downward a distance of 60 mm from its equilibrium position, determine the equation which describes the motion. What is the amplitude of the motion? Assume that positive displacement is measured downward.

22–7. An 800-lb elevator is suspended from a single cable which has a stiffness $k = 241(10^3)$ lb/in. If the elevator is moving downward at 20 ft/s when the motor controlling the cable suddenly jams (stops), determine the maximum tension in the cable and the frequency of vibration of the elevator. Neglect the mass of the cable.

*22–8. A pendulum has a 0.4-m-long cord and is given a tangential velocity of 0.2 m/s toward the vertical from a position $\theta = 0.3$ rad. Determine the equation which describes the angular motion.

Probs. 22–8 / 22–9

22–9. Determine to the nearest degree the maximum angular displacement of the bob in Prob. 22–8 if it is initially displaced $\theta = 0.2$ rad from the vertical and given a tangential velocity of 0.4 m/s away from the vertical.

22–10. Determine the frequency of vibration for the block-and-spring mechanisms.

(a)

(b)

Prob. 22–10

22–11. The circular disk has a mass m and is pinned at O. Determine the period of oscillation if it is displaced a small amount and released.

Prob. 22–11

*22–12. The body of arbitrary shape has a mass m, mass center at G, and a radius of gyration about G of k_G. If it is displaced by a slight amount θ from its equilibrium position and released, determine the period of vibration.

Prob. 22–12

22–13. The connecting rod is supported by a knife edge at A and the period of vibration is measured as $\tau_A = 3.38$ s. It is then removed and rotated 180° so that it is supported by the knife edge at B. In this case the period of vibration is measured as $\tau_B = 3.96$ s. Determine the location d of the center of gravity G, and compute the radius of gyration k_G. *Hint:* See Prob. 22–12.

Prob. 22–13

22–14. The square plate has a mass m and is suspended at its corner by the pin O. Determine the period of oscillation if it is displaced a small amount and released.

Prob. 22–14

515

22–15. The bell has a mass of 375 kg, a center of mass at G, and a radius of gyration about point D of $k_D = 0.4$ m. The tongue consists of a slender rod attached to the inside of the bell at C. If a 1.8-kg mass is attached to the end of the rod, determine the length l of the rod so that the bell will "ring silent," i.e., so that the period of vibration of the tongue is the same as that of the bell. For the calculation, neglect the small distance between C and D and neglect the mass of the rod.

Prob. 22–15

*22–16.** While standing in an elevator, the man holds a pendulum which consists of an 18-in. cord and a 0.5-lb bob. If the elevator is descending with an acceleration $a = 4$ ft/s², determine the period of vibration for small amplitudes of swing.

Prob. 22–16

22–17. The bar has a length l and mass m. It is supported at its ends by rollers of negligible mass. If it is given a small displacement and released, determine the frequency of vibration.

Prob. 22–17

22–18. The slender rod has a mass of 0.2 kg and is supported at O by a pin and at its end A by two springs, each having a stiffness $k = 4$ N/m. The period of vibration of the rod can be set by fixing the 0.5-kg collar C to the rod at an appropriate location along its length. If the springs are originally unstretched when the rod is vertical, determine the position y of the collar so that the period of vibration becomes $\tau = 1$ s.

Prob. 22–18

22–19. The 50-lb wheel has a radius of gyration about its mass center G of $k_G = 0.7$ ft. Determine the frequency of vibration if it is displaced slightly from the equilibrium position and released. Assume no slipping.

Prob. 22–19

*22–20. The pointer on a metronome supports a 0.4-lb slider A, which is positioned at a fixed distance from the pivot O of the pointer. When the pointer is displaced, a torsional spring at O exerts a restoring moment on the pointer having a magnitude $M = (1.2\theta)$ lb · ft, where θ represents the angle of displacement from the vertical, measured in radians. Determine the period of vibration when the pointer is displaced a small amount θ and released. Neglect the mass of the pointer.

Prob. 22–20

22–21. The uniform beam is supported at its ends by two springs A and B, each having the same stiffness k. When nothing is supported on the beam, it has a period of vertical vibration of 0.83 s. If a 50-kg mass is placed at its center, the period of vertical vibration is 1.52 s. Compute the stiffness of each spring and the mass of the beam.

Prob. 22–21

22–22. The platform AB when empty has a mass of 400 kg, mass center at G_1, and period of oscillation $\tau_1 = 2.38$ s. If a car having a mass of 1.2 Mg and center of mass at G_2 is placed on the platform, the period of oscillation becomes $\tau_2 = 3.16$ s. Determine the moment of inertia of the car about an axis passing through G_2.

Prob. 22–22

22–23. A uniform board is supported on two wheels which rotate in opposite directions at a constant angular speed. If the coefficient of friction between the wheels and board is μ, determine the frequency of vibration of the board if it is displaced slightly, a distance x from the midpoint between the wheels, and released.

Prob. 22–23

*22–24. The cylinder of radius r and mass m is displaced a small amount along the curved surface. If it rolls without slipping, determine the frequency of oscillation when it is released.

Prob. 22–24

22–25. The plate of mass m is supported by three symmetrically placed cords of length l as shown. If the plate is given a slight rotation about a vertical axis through its center and released, determine the period of oscillation.

Prob. 22–25

★**22.2** Energy Methods

Equilibrium position

Fig. 22–8

The simple harmonic motion of a body, discussed in the first section, is due only to gravitational and elastic restoring forces acting on the body. Since these types of forces are *conservative,* it is possible to use the conservation of energy equation to obtain the body's natural frequency or period of vibration. To show how to do this, consider the block and spring in Fig. 22–8. When the block is displaced an arbitrary amount x from the equilibrium position, the kinetic energy is $T = \frac{1}{2}mv^2 = \frac{1}{2}m\dot{x}^2$ and the potential energy is $V = \frac{1}{2}kx^2$. By the conservation of energy equation, Eq. 14–22, it is necessary that

$$T + V = \text{const}$$
$$\tfrac{1}{2}m\dot{x}^2 + \tfrac{1}{2}kx^2 = \text{const} \qquad (22\text{–}17)$$

The differential equation describing the *accelerated motion* of the block can be obtained by *differentiating* this equation with respect to time, i.e.,

$$m\dot{x}\ddot{x} + kx\dot{x} = 0$$
$$\dot{x}(m\ddot{x} + kx) = 0$$

Since the velocity $\dot{x}$ is not *always* zero in a vibrating system,

$$\ddot{x} + p^2x = 0, \qquad p = \sqrt{k/m}$$

which is the same as Eq. 22–1.

For a system of connected bodies, determination of the natural frequency or the equation of motion by time differentiation of the energy equation is advantageous, since it is *not necessary* to dismember the system to account for reactive and connective forces which do no work. There is also an added advantage in that the circular frequency p may be obtained *directly*. For example, consider the total mechanical energy of the block and spring in Fig. 22–8 when the block is at its *maximum displacement*. In this position the block is temporarily at rest, so that the kinetic energy is zero and the potential energy, stored in the spring, is a maximum. Therefore, Eq. 22–17 becomes $\frac{1}{2}kx_{max}^2$ = const. At the instant the block passes the equilibrium position, the kinetic energy of the block is a maximum and the potential energy of the spring is zero. Hence, Eq. 22–17 becomes $\frac{1}{2}m(\dot{x})_{max}^2$ = const. Since the vibrating motion of the block is *harmonic*, the *solution* for the displacement and velocity may be written in the form of Eq. 22–9 and its time derivative, i.e.,

$$x = C \sin (pt + \phi), \qquad \dot{x} = Cp \cos (pt + \phi)$$

so that

$$x_{max} = C, \qquad \dot{x}_{max} = Cp$$

Applying the conservation of energy equation ($T + V$ = const) yields

$$V_{max} = T_{max}; \qquad \frac{1}{2}kx_{max}^2 = \frac{1}{2}m\dot{x}_{max}^2 = \text{const}$$

or

$$kC^2 = mC^2p^2$$

Solving for p yields

$$p = \sqrt{\frac{k}{m}}$$

which is identical to Eq. 22–2.

PROCEDURE FOR ANALYSIS

The following procedure provides a method for determining the circular frequency p of a body or system of connected bodies using the conservation of energy equation.

Energy Equation. Draw the body when it is displaced by a *small amount* from its equilibrium position and define the location of the body from its equilibrium position by an appropriate position coordinate q. Formulate the equation of energy for the body, $T + V$ = const, in terms of the position coordinate.* Recall that, in general, the kinetic energy must account for both the body's translational and rotational motion, $T = \frac{1}{2}mv_G^2 + \frac{1}{2}I_G\omega^2$, Eq. 18–2, and that the potential energy is the sum of the gravitational and elastic potential energies of the body, $V = V_g + V_e$, Eq. 18–13. In particular, V_g should be measured from a datum for which $q = 0$ (equilibrium position).

Time Derivative. Take the time derivative of the energy equation and factor out the common terms. The resultant differential equation represents the equation of motion for the system. The value of p is obtained after rearranging the terms in the standard form $\ddot{q} + p^2q = 0$.

The following examples illustrate this procedure.

*It is suggested that the material in Sec. 18.5 be reviewed.

Example 22–5

The thin hoop shown in Fig. 22–9a is supported by a peg at O. Determine the period of oscillation for small amplitudes of swing. The hoop has a mass m.

(a)

Solution

Energy Equation. A diagram of the hoop when it is displaced a small amount $(q =)\theta$ from the equilibrium position is shown in Fig. 22–9b. Using the table on the inside back cover of the book and the parallel-axis theorem to determine I_O, we can express the kinetic energy as

$$T = \tfrac{1}{2}I_O\omega^2 = \tfrac{1}{2}[mr^2 + mr^2]\dot{\theta}^2 = mr^2\dot{\theta}^2$$

Datum

(b)

Fig. 22–9

If a horizontal datum is placed through the center of gravity of the hoop when $\theta = 0$, then the center of gravity moves upward $r(1 - \cos\theta)$ in the displaced position. For *small angles*, the $\cos\theta$ may be replaced by the first two terms of its series expansion, $\cos\theta = 1 - \theta^2/2 + \cdots$. Therefore, the potential energy is

$$V = mgr\left[1 - \left(1 - \frac{\theta^2}{2}\right)\right] = mgr\frac{\theta^2}{2}$$

The total energy in the system is

$$T + V = mr^2\dot{\theta}^2 + mgr\frac{\theta^2}{2}$$

Time Derivative

$$mr^2 2\dot{\theta}\ddot{\theta} + mgr\theta\dot{\theta} = 0$$
$$mr\dot{\theta}(2r\ddot{\theta} + g\theta) = 0$$

Since $\dot{\theta}$ is not always equal to zero, from the terms in parentheses

$$\ddot{\theta} + \frac{g}{2r}\theta = 0$$

Hence,

$$p = \sqrt{\frac{g}{2r}}$$

so that

$$\tau = \frac{2\pi}{p} = 2\pi\sqrt{\frac{2r}{g}} \qquad\qquad \textit{Ans.}$$

Example 22–6

A 10-kg block is suspended from a cord wrapped around a 5-kg disk, as shown in Fig. 22–10a. If the spring has a stiffness $k = 200$ N/m, determine the natural period of vibration for the system.

Solution

Energy Equation. A diagram of the block and disk when they are displaced by respective amounts s and θ from the equilibrium position is shown in Fig. 22–10b. Since $s = 0.15\theta$, the kinetic energy of the system is

$$T = \tfrac{1}{2}m_b v_b^2 + \tfrac{1}{2}I_O \omega_d^2$$

$$= \frac{1}{2}(10)(0.15\dot\theta)^2 + \frac{1}{2}\left[\frac{1}{2}(5)(0.15)^2\right](\dot\theta)^2$$

$$= 0.141(\dot\theta)^2$$

(a)

Establishing the datum at the equilibrium position of the block and realizing that the spring stretches s_{st} for equilibrium, we can write the potential energy as

$$V = \tfrac{1}{2}k(s_{st} + s)^2 - Ws$$

$$= \frac{1}{2}(200)(s_{st} + 0.15\theta)^2 - 98.1(0.15\theta)$$

The total energy for the system is, therefore,

$$T + V = 0.141(\dot\theta)^2 + 100(s_{st} + 0.15\theta)^2 - 14.72\theta$$

Time Derivative

$$0.282(\dot\theta)\ddot\theta + 200(s_{st} + 0.15\theta)0.15\dot\theta - 14.72\dot\theta = 0$$

Since $s_{st} = 98.1/200 = 0.4905$ m, the above equation reduces to the standard form

$$\ddot\theta + 16\theta = 0$$

so that

$$p = \sqrt{16} = 4 \text{ rad/s}$$

Thus,

$$\tau = \frac{2\pi}{p} = \frac{2\pi}{4} = 1.57 \text{ s} \qquad\qquad Ans.$$

(b)

Fig. 22–10

Problems

22–26. Solve Prob. 22–11 using energy methods.

22–27. Solve Prob. 22–12 using energy methods.

*22–28. Solve Prob. 22–14 using energy methods.

22–29. Solve Prob. 22–19 using energy methods.

22–30. The thin hoop is supported by a knife edge. Determine the period of oscillation for small amplitudes of swing.

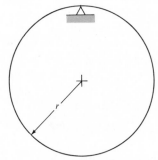

Prob. 22–30

22–31. The machine has a mass m and is uniformly supported by four springs each having a stiffness k. Determine the period of vertical vibration.

Prob. 22–31

*22–32. Determine the period of vibration of the pendulum. Consider the two rods to be slender, each having a weight $W = 8$ lb/ft.

Prob. 22–32

22–33. If the disk has a mass of 8 kg, determine the natural frequency of vibration. The springs are originally unstretched.

Prob. 22–33

22–34. The 7-kg disk is pin-connected at its midpoint. Determine the period of vibration of the disk if the springs have sufficient tension in them to prevent the belt from slipping on the disk as it oscillates. *Hint:* Assume that the initial stretch in each spring is δ_O. This term will cancel out after taking the time derivative of the energy equation.

Prob. 22–34

22–35. Determine the differential equation of motion of the block of mass m when it is displaced slightly and released. Motion occurs in the horizontal plane. The springs are originally unstretched and each is attached to the block.

Prob. 22–35

*22–36. The bar has a mass of 8 kg and is suspended from two springs such that when it is in equilibrium, the springs make an angle of 45° with the horizontal as shown. Determine the period of vibration if the bar is pulled down a short distance and released.

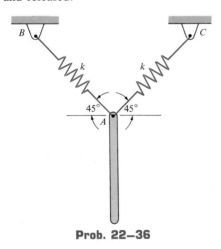

Prob. 22–36

22–37. Using energy methods, determine the differential equation of motion of the 12-kg spool. Assume that it does not slip at the surface of contact as it oscillates. The radius of gyration of the spool about its center of mass is $k_G = 125$ mm.

Prob. 22–37

22–38. The slender rod has a weight $W = 4$ lb/ft. If it is supported in the horizontal plane by a ball-and-socket joint at A and a cable at B, determine the natural frequency of vibration when the end B is given a small horizontal displacement and then released.

Prob. 22–38

★22.3 Undamped Forced Vibration

Undamped forced vibration is considered to be one of the most important types of vibrating motion in engineering work. The principles which describe the nature of this motion may be applied to the analysis of forces which cause vibration in various types of machines and structures.

Periodic Force. The block and spring shown in Fig. 22–11*a* provide a convenient "model" which represents the vibrational characteristics of a system subjected to a periodic force $F = F_O \sin \omega t$. This force has a maximum magnitude of F_O and a *forcing frequency* ω. The free-body diagram for the block when it is displaced a distance x is shown in Fig. 22–11*b*. Applying the equation of motion yields

$$\xrightarrow{+} \Sigma F_x = ma_x; \qquad F_O \sin \omega t - kx = m\ddot{x}$$

or

$$\ddot{x} + \frac{k}{m}x = \frac{F_O}{m} \sin \omega t \qquad (22\text{--}18)$$

This equation is referred to as a nonhomogeneous second-order differential equation. The general solution consists of a complementary solution, x_c, *plus* a particular solution, x_p.

The *complementary solution* is determined by setting the term on the right side of Eq. 22–18 equal to zero and solving the resulting homogeneous equation, which is equivalent to Eq. 22–1. The solution is defined by Eq. 22–3, i.e.,

$$x_c = A \sin pt + B \cos pt \qquad (22\text{--}19)$$

where p is the circular frequency, $p = \sqrt{k/m}$, Eq. 22–2.

(a)

(b)

Fig. 22–11

Since the motion is periodic, the *particular solution* of Eq. 22–18 may be determined by assuming a solution of the form

$$x_p = C \sin \omega t \qquad (22\text{--}20)$$

where C is a constant. Taking the second time derivative and substituting into Eq. 22–18 yields

$$-C\omega^2 \sin \omega t + \frac{k}{m}(C \sin \omega t) = \frac{F_O}{m} \sin \omega t$$

Factoring out $\sin \omega t$ and solving for C gives

$$C = \frac{F_O/m}{\dfrac{k}{m} - \omega^2} = \frac{F_O/k}{1 - \left(\dfrac{\omega}{p}\right)^2} \qquad (22\text{--}21)$$

Substituting into Eq. 22–20, we obtain the particular solution

$$x_p = \frac{F_O/k}{1 - \left(\dfrac{\omega}{p}\right)^2} \sin \omega t \qquad (22\text{--}22)$$

The *general solution* is therefore

$$x = x_c + x_p = A \sin pt + B \cos pt + \frac{F_O/k}{1 - \left(\dfrac{\omega}{p}\right)^2} \sin \omega t \qquad (22\text{--}23)$$

Here x describes two types of vibrating motion of the block. The *complementary solution* x_c defines the *free vibration*, which depends upon the circular frequency $p = \sqrt{k/m}$ and the constants A and B, Fig. 22–12a. Specific values for A and B are obtained by evaluating Eq. 22–23 at a given instant when the displacement and velocity are known. The *particular solution* x_p describes the *forced vibration* of the block caused by the applied force $F = F_O \sin \omega t$, Fig. 22–12b. The resultant vibration x is shown in Fig. 22–12c. Since all vibrating systems are subject to *friction*, the free vibration, x_c, will in time dampen out. For this reason the free vibration is referred to as *transient*, and the forced vibration is called *steady state*, since it is the only vibration that remains, Fig. 22–12d.

From Eq. 22–21 it is seen that the *amplitude* of forced vibration depends upon the *frequency ratio* ω/p. If the *magnification factor* MF is defined as the ratio of the amplitude of steady-state vibration, $(x_p)_{max}$, to the static deflection F_O/k, which is caused by the amplitude of the periodic force F_O, then, from Eq. 22–21,

$$\text{MF} = \frac{(x_p)_{max}}{F_O/k} = \frac{1}{1 - \left(\dfrac{\omega}{p}\right)^2} \qquad (22\text{--}24)$$

(a)

(b)

(c)

(d)

Fig. 22–12

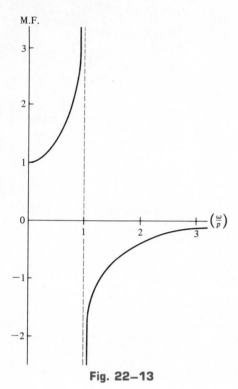

Fig. 22–13

This equation is graphed in Fig. 22–13, where it is seen that for $\omega \approx 0$, the MF ≈ 1. In this case, because of the low frequency $\omega < p$, the vibration of the block will be in phase with the applied force **F**. If the force or displacement is applied with a frequency close to the natural frequency of the system, i.e., $\omega/p \approx 1$, the amplitude of vibration of the block becomes extremely large. This condition is called *resonance,* and in practice, resonating vibrations can cause tremendous stress and rapid failure of parts. When the cyclic force $F_O \sin \omega t$ is applied at high frequencies ($\omega > p$), the value of the MF becomes negative, indicating that the motion of the block is out of phase with the force. Under these conditions, as the block is displaced to the right, the force acts to the left, and vice versa. For extremely high frequencies ($\omega \gg p$) the block remains almost stationary, and hence the MF is approximately zero.

Periodic Support Displacement. Forced vibrations can also arise from the periodic excitation of the support of a system. The model shown in Fig. 22–14a represents the periodic vibration of a block which is caused by harmonic movement $\delta = \delta_O \sin \omega t$ of the support. The free-body diagram for the block in this case is shown in Fig. 22–14b. The coordinate x is measured from the point of zero displacement of the support, i.e., when the radius vector OA coincides with OB, Fig. 22–14a. Therefore, general displacement of the spring is $(x - \delta_O \sin \omega t)$. Applying the equation of motion yields

$$\xrightarrow{+}\Sigma F_x = ma_x; \qquad -k(x - \delta_O \sin \omega t) = m\ddot{x}$$

or

$$\ddot{x} + \frac{k}{m}x = \frac{k\delta_O}{m} \sin \omega t \qquad (22\text{–}25)$$

By comparison, this equation is identical to the form of Eq. 22–18, *provided* F_O is *replaced* by $k\delta_O$. If this substitution is made into the solutions defined by Eqs. 22–21 to 22–23, the results are appropriate for describing the motion of the block when subjected to the support displacement $\delta = \delta_O \sin \omega t$.

(a)

(b)

Fig. 22–14

Example 22–7

The instrument shown in Fig. 22–15 is rigidly attached to a platform P, which in turn is supported by *four* springs, each having a stiffness $k = 800$ N/m. Initially the platform is at rest when the floor is subjected to a displacement $\delta = 10 \sin (8t)$ mm, where t is in seconds. If the instrument is constrained to move vertically, and the total mass of the instrument and platform is 20 kg, determine the vertical displacement y of the platform, measured from the equilibrium position, as a function of time. What would be the floor vibration to cause resonance?

Fig. 22–15

Solution

Since the induced vibration is caused by the displacement of the supports, the motion is described by Eq. 22–23, with F_O replaced by $k\delta_O$, i.e.,

$$y = A \sin pt + B \cos pt + \frac{\delta_O}{1 - \left(\dfrac{\omega}{p}\right)^2} \sin \omega t \qquad (1)$$

Here $\delta = \delta_O \sin \omega t = 10 \sin (8t)$ mm, so that

$$\delta_O = 10 \text{ mm}, \qquad \omega = 8 \text{ rad/s}$$

$$p = \sqrt{\frac{k}{m}} = \sqrt{\frac{4(800)}{20}} = 12.6 \text{ rad/s}$$

From Eq. 22–22, with $k\delta_O$ replacing F_O, the amplitude of vibration caused by the floor displacement is

$$(y_p)_{max} = \frac{\delta_O}{1 - \left(\dfrac{\omega}{p}\right)^2} = \frac{10}{1 - \left(\dfrac{8}{12.6}\right)^2} = 16.7 \text{ mm} \qquad (2)$$

Hence, Eq. (1) and its time derivative become

$$y = A \sin (12.6t) + B \cos (12.6t) + 16.7 \sin (8t)$$
$$\dot{y} = A(12.6) \cos (12.6t) - B(12.6) \sin (12.6t) + 133.3 \cos (8t)$$

The constants A and B are evaluated from these equations. Since $y = \dot{y} = 0$ at $t = 0$, then

$$0 = 0 + B + 0; \qquad \qquad B = 0$$
$$0 = A(12.6) - 0 + 133.3; \qquad A = -10.6$$

The vibrating motion is therefore described by the equation

$$y = -10.6 \sin (12.6t) + 16.7 \sin (8t) \qquad \qquad Ans.$$

Resonance will occur when the amplitude of vibration caused by the floor displacement approaches infinity. From Eq. (2), this requires that

$$\omega = p = 12.6 \text{ rad/s} \qquad \qquad Ans.$$

***22.4** Viscous Damped Free Vibration

The vibration analysis considered thus far has not included the effects of friction or damping in the system, and as a result, the solutions obtained are only in close agreement with the actual motion. Since all vibrations die out in time, the presence of damping forces should be included in the analysis.

In many cases damping is attributed to the resistance created by the substance, such as water, oil, or air, in which the system vibrates. Provided the body moves slowly through this substance, the resistance to motion is directly proportional to the body's speed. The type of force developed under these conditions is called a *viscous damping force*. The magnitude of this force may be expressed by an equation of the form

$$F = c\dot{x} \qquad (22\text{–}26)$$

where the constant c is called the *coefficient of viscous damping* and has units of $N \cdot s/m$ or $lb \cdot s/ft$.

The vibrating motion of a body or system having viscous damping may be characterized by the block and spring shown in Fig. 22–16a. The effect of damping is provided by the *dashpot* connected to the block on the right side. Damping occurs when the piston P moves to the right or left within the enclosed cylinder. The cylinder contains a fluid, and the motion of the piston is retarded since the fluid must flow around or through a small hole in the piston. The dashpot is assumed to have a coefficient of viscous damping c.

If the block is displaced a distance x from its equilibrium position, the resulting free-body diagram is shown in Fig. 22–16b. Both the spring force kx and the damping force $c\dot{x}$ oppose the forward motion of the block, so that applying the equation of motion yields

$$\overset{+}{\rightarrow}\Sigma F_x = ma_x; \qquad\qquad -kx - c\dot{x} = m\ddot{x}$$

or

$$m\ddot{x} + c\dot{x} + kx = 0 \qquad (22\text{–}27)$$

(a)   (b)

Fig. 22–16

This linear, second-order, homogeneous, differential equation has solutions of the form

$$x = e^{\lambda t}$$

where e is the base of the natural logarithm and λ is a constant. The value of λ (lambda) may be obtained by substituting this solution into Eq. 22–27, which yields

$$m\lambda^2 e^{\lambda t} + c\lambda e^{\lambda t} + k e^{\lambda t} = 0$$

or

$$e^{\lambda t}(m\lambda^2 + c\lambda + k) = 0$$

Since $e^{\lambda t}$ is always positive, a solution is possible provided

$$m\lambda^2 + c\lambda + k = 0$$

Hence, by the quadratic formula, the two values of λ are

$$\lambda_1 = -\frac{c}{2m} + \sqrt{\left(\frac{c}{2m}\right)^2 - \frac{k}{m}}$$

$$\lambda_2 = -\frac{c}{2m} - \sqrt{\left(\frac{c}{2m}\right)^2 - \frac{k}{m}}$$

(22–28)

The general solution of Eq. 22–27 is therefore a linear combination of exponentials which involves both of these roots. There are three possible combinations of λ_1 and λ_2 which must be considered for the general solution. Before discussing these combinations, however, it is first necessary to consider the definition of the *critical damping coefficient* c_c as the value of c which makes the radical in Eqs. 22–28 equal to zero; i.e.,

$$\left(\frac{c_c}{2m}\right)^2 - \frac{k}{m} = 0$$

or

$$c_c = 2m\sqrt{\frac{k}{m}} = 2mp$$

(22–29)

Here the value of p is the circular frequency $p = \sqrt{k/m}$, Eq. 22–2.

Overdamped System. When $c > c_c$, the roots λ_1 and λ_2 are both real. The general solution of Eq. 22–27 may then be written as

$$x = Ae^{\lambda_1 t} + Be^{\lambda_2 t}$$

(22–30)

Motion corresponding to this solution is *nonvibrating*. The effect of damping is so strong that when the block is displaced and released, it simply creeps back to its original position without oscillating. The system is said to be *overdamped*.

Critically Damped System. If $c = c_c$, then $\lambda_1 = \lambda_2 = -c_c/2m = -p$. This situation is known as *critical damping*, since it represents a condition where c has the smallest value necessary to cause the system to be nonvibrating. Using the methods of differential equations, it may be shown that the solution to Eq. 22–27 for critical damping is

$$x = (A + Bt)e^{-pt} \qquad (22\text{–}31)$$

Underdamped System. Most often $c < c_c$, in which case the system is referred to as *underdamped*. In this case the roots λ_1 and λ_2 are complex numbers and it may be shown that the general solution of Eq. 22–27 can be written as

$$x = D[e^{-(c/2m)t} \sin (p_d t + \phi)] \qquad (22\text{–}32)$$

where D and ϕ are constants generally determined from the initial conditions of the problem. The constant p_d is called the *damped natural frequency* of the system. It has a value of

$$p_d = \sqrt{\frac{k}{m} - \left(\frac{c}{2m}\right)^2} = p\sqrt{1 - \left(\frac{c}{c_c}\right)^2} \qquad (22\text{–}33)$$

where the ratio c/c_c is called the *damping factor*.

The graph of Eq. 22–32 is shown in Fig. 22–17. The initial limit of motion, D, diminishes with each cycle of vibration, since motion is confined within the bounds of the exponential curve. Using the damped natural frequency p_d, the period of damped vibration may be written as

$$\tau_d = \frac{2\pi}{p_d} \qquad (22\text{–}34)$$

Since $p_d < p$, Eq. 22–33, the period of damped vibration, τ_d, will be greater than that of free vibration, $\tau = 2\pi/p$.

$De^{-(\frac{c}{2m})t}$

$-De^{-(\frac{c}{2m})t}$

$x = D[e^{-(\frac{c}{2m})t} \sin (p_d t + \phi)]$

Fig. 22–17

Viscous Damped Forced Vibration **22.5**[*]

The most general case of single-degree-of-freedom vibrating motion occurs when the system includes the effects of forced motion and induced damping. The analysis of this particular type of vibration is of practical value when applied to systems having significant damping characteristics.

If a dashpot is attached to the block and spring shown in Fig. 22–11a, the differential equation which describes the motion becomes

$$m\ddot{x} + c\dot{x} + kx = F_O \sin \omega t \qquad (22\text{--}35)$$

A similar equation may be written for a block and spring having a periodic support displacement, Fig. 22–14a, which includes the effects of damping. In that case, however, F_O is replaced by $k\delta_O$. Since Eq. 22–35 is nonhomogeneous, the general solution is the sum of a complementary solution, x_c, and a particular solution, x_p. The complementary solution is determined by setting the right side of Eq. 22–35 equal to zero and solving the homogeneous equation, which is equivalent to Eq. 22–27. The solution is therefore given by Eq. 22–30, 22–31, or 22–32, depending upon the values of λ_1 and λ_2. Because all systems contain friction, however, this solution will dampen out with time. Only the particular solution, which describes the *steady-state vibration* of the system, will remain. Since the applied forcing function is harmonic, the steady-state motion will also be harmonic. Consequently, the particular solution will be of the form

$$x_p = A' \sin \omega t + B' \cos \omega t \qquad (22\text{--}36)$$

The constants A' and B' are determined by taking the necessary time derivatives and substituting them into Eq. 22–35, which after simplification yields

$$[-A'm\omega^2 - cB'\omega + kA'] \sin \omega t + [-B'm\omega^2 + cA'\omega + kB'] \cos \omega t = F_O \sin \omega t$$

Since this equation holds for all time, the constant coefficients of $\sin \omega t$ and $\cos \omega t$ may be equated; i.e.,

$$-A'm\omega^2 - cB'\omega + kA' = F_O$$
$$-B'm\omega^2 + cA'\omega + kB' = 0$$

Solving for A' and B', realizing that $p^2 = k/m$, yields

$$A' = \frac{\left(\dfrac{F_O}{m}\right)(p^2 - \omega^2)}{(p^2 - \omega^2)^2 + \left(\dfrac{c\omega}{m}\right)^2}$$

$$B' = \frac{-F_O \dfrac{c\omega}{m^2}}{(p^2 - \omega^2)^2 + \left(\dfrac{c\omega}{m}\right)^2} \qquad (22\text{--}37)$$

531

It is also possible to express Eq. 22–36 in a form similar to Eq. 22–9,

$$x_p = C' \sin(\omega t - \phi') \tag{22–38}$$

in which case the constants C' and ϕ' are

$$C' = \frac{F_O/k}{\sqrt{\left[1 - \left(\dfrac{\omega}{p}\right)^2\right]^2 + \left(2\dfrac{c}{c_c}\dfrac{\omega}{p}\right)^2}}$$

$$\tag{22–39}$$

$$\phi' = \tan^{-1}\left(\frac{c\omega/k}{1 - \left(\dfrac{\omega}{p}\right)^2}\right)$$

The angle ϕ' represents the phase difference between the applied force and the resulting steady-state vibration of the damped system.

The *magnification factor* MF has been defined in Sec. 22.3 as the ratio of the amplitude of deflection caused by the forced vibration to the deflection caused by a static force $\mathbf{F}_O$. From Eq. 22–38, the forced vibration has an amplitude of C'; thus,

$$MF = \frac{C'}{F_O/k} = \frac{1}{\sqrt{\left[1 - \left(\dfrac{\omega}{p}\right)^2\right]^2 + \left(2\dfrac{c}{c_c}\dfrac{\omega}{p}\right)^2}} \tag{22–40}$$

The MF is plotted in Fig. 22–18 versus the frequency ratio ω/p for various values of the damping factor c/c_c. It can be seen from this graph that the magnification of the amplitude increases as the damping factor decreases. Resonance obviously occurs only when the damping is zero and the frequency ratio equals 1.

Fig. 22–18

Example 22–8

The 30-kg electric motor shown in Fig. 22–19 is supported by *four* springs, each spring having a stiffness of 200 N/m. If the rotor R is unbalanced such that its effect is equivalent to a 4-kg mass located 60 mm from the axis of rotation, determine the amplitude of vibration when the rotor is turning at $\omega = 10$ rad/s. The damping factor is $c/c_c = 0.15$.

Fig. 22–19

Solution

The periodic force which causes the motor to vibrate is the centrifugal force due to the unbalanced rotor. This force has a constant magnitude of

$$F_O = ma_n = mr\omega^2 = 4 \text{ kg}(0.06 \text{ m})(10 \text{ rad/s})^2 = 24 \text{ N}$$

Oscillation in the vertical direction may be expressed in the periodic form $F = F_O \sin \omega t$, where $\omega = 10$ rad/s. Thus,

$$F = 24 \sin 10t$$

The stiffness of the entire system of four springs is $k = 4(200) = 800$ N/m. Therefore, the circular frequency of vibration is

$$p = \sqrt{\frac{k}{m}} = \sqrt{\frac{800}{30}} = 5.16 \text{ rad/s}$$

Since the damping factor is known, the steady-state amplitude may be determined from the first of Eqs. 22–39, i.e.,

$$
\begin{aligned}
C' &= \frac{F_O/k}{\sqrt{\left[1 - \left(\dfrac{\omega}{p}\right)^2\right]^2 + \left(2\dfrac{c}{c_c}\dfrac{\omega}{p}\right)^2}} \\[2ex]
&= \frac{24/800}{\sqrt{\left[1 - \left(\dfrac{10}{5.16}\right)^2\right]^2 + \left[2(0.15)\dfrac{10}{5.16}\right]^2}} \\[2ex]
&= 0.0107 \text{ m} = 10.7 \text{ mm} \qquad\qquad\qquad \textit{Ans.}
\end{aligned}
$$

*22.6 Electrical Circuit Analogues

(a)

(b)

Fig. 22–20

The characteristics of a vibrating mechanical system may be represented by an electric circuit. Consider the circuit shown in Fig. 22–20a, which consists of an inductor L, a resistor R, and a capacitor C. When a voltage $E(t)$ is applied, it causes a current of magnitude i to flow through the circuit. As the current flows past the inductor the voltage drop is $L(di/dt)$, when it flows across the resistor the drop is Ri, and when it arrives at the capacitor the drop is $(1/C) \int i \, dt$. Since current cannot flow past a capacitor, it is only possible to measure the charge q acting on the capacitor. The charge may, however, be related to the current by the equation $i = dq/dt$. Thus, the voltage drops which occur across the inductor, resistor, and capacitor may be written as $L \, d^2q/dt^2$, $R \, dq/dt$, and q/C, respectively. According to Kirchhoff's voltage law, the applied voltage balances the sum of the voltage drops around the circuit. Therefore,

$$L\frac{d^2q}{dt^2} + R\frac{dq}{dt} + \frac{1}{C}q = E(t) \qquad (22\text{–}41)$$

Consider now the model of a single-degree-of-freedom mechanical system, Fig. 22–20b, which is subjected to both a general forcing function $F(t)$ and damping. The equation of motion for this system was established in the previous section and can be written as

$$m\frac{d^2x}{dt^2} + c\frac{dx}{dt} + kx = F(t) \qquad (22\text{–}42)$$

By comparison, it is seen that Eqs. 22–41 and 22–42 have the same form, hence mathematically the problem of analyzing an electric circuit is the same as that of analyzing a vibrating mechanical system. The analogues between the two equations are given in Table 22–1.

This analogy has important application to experimental work, for it is much easier to simulate the vibration of a complex mechanical system using an electric circuit, which can be constructed on an analogue computer, than to make an equivalent mechanical spring and dashpot model.

Table 22–1 Electrical-Mechanical Analogues

Electrical		*Mechanical*	
Electric charge	q	Displacement	x
Electric current	i	Velocity	dx/dt
Voltage	$E(t)$	Applied force	$F(t)$
Inductance	L	Mass	m
Resistance	R	Viscous damping coefficient	c
Reciprocal of capacitance	$1/C$	Spring stiffness	k

Problems

22–39. If the block is subjected to the impressed force $F = F_O \cos \omega t$, show that the differential equation of motion is $\ddot{y} + (k/m)y = (F_O/m) \cos \omega t$, where y is measured from the equilibrium position of the block. What is the general solution of this equation?

$F = F_O \cos \omega t$

Prob. 22–39

*22–40. A 5-kg block is suspended from a spring having a stiffness of 300 N/m. If the block is acted upon by a vertical force $F = (7 \sin 8t)$ N, where t is in seconds, determine the equation which describes the motion of the block when it is pulled down 100 mm from the equilibrium position and released from rest at $t = 0$. Assume that positive displacement is measured downward.

$k = 300$ N/m

$F = 7 \sin 8t$

Prob. 22–40

22–41. A 7-kg block is suspended from a spring that has a stiffness $k = 350$ N/m. The block is drawn downward 70 mm from the equilibrium position and released from rest at $t = 0$. If the support moves with an impressed displacement $\delta = (20 \sin 4t)$ mm, where t is in seconds, determine the equation which describes the vertical motion of the block. Assume that positive displacement is measured downward.

22–42. The fan has a mass of 25 kg and is fixed to the end of a horizontal beam that has a negligible mass. The fan blade is mounted eccentrically on the shaft such that it is equivalent to an unbalanced 3.5-kg mass located 100 mm from the axis of rotation. If the static deflection of the beam is 50 mm as a result of the weight of the fan, determine the angular speed of the fan at which resonance will occur. *Hint:* See the first part of Example 22–8.

ω

Probs. 22–42 / 22–43 / 22–44

22–43. What will be the amplitude of steady-state vibration of the fan in Prob. 22–42 if the angular speed of the fan is 18 rad/s? *Hint:* See the first part of Example 22–8.

*22–44. What is the amplitude of steady-state vibration of the fan in Prob. 22–42 when the angular speed of the fan is 10 rad/s? *Hint:* See the first part of Example 22–8.

22–45. The engine is mounted on a foundation block which is spring-supported. Describe the steady-state vibration of the system if the block and engine have a total weight $W = 1500$ lb and the engine, when running, creates an impressed force $F = (50 \sin 2t)$ lb, where t is in seconds. Assume that the system vibrates only in the vertical direction, with the positive displacement measured downward, and that the total stiffness of the springs can be represented as $k = 2000$ lb/ft.

Probs. 22–45 / 22–46

22–46. Determine the rotational speed ω of the engine in Prob. 22–45 which will cause resonance.

22–47. The 80-lb block is attached to a spring, the end of which is subjected to a periodic support displacement $\delta_A = (0.5 \sin 8t)$ ft, where t is in seconds. Determine the amplitude of the steady-state motion.

Prob. 22–47

*22–48. The electric motor has a mass of 50 kg and is supported by *four springs,* each spring having a stiffness of 100 N/m. If the motor turns a disk D which is mounted eccentrically, 20 mm from the disk's center, determine the angular rotation ω at which resonance occurs. Assume that the motor only vibrates in the vertical direction. *Hint:* See the first part of Example 22–8.

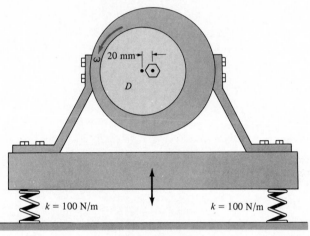

Prob. 22–48

22–49. The instrument is centered uniformly on a platform P, which in turn is supported by *four* springs, each spring having a stiffness $k = 130$ N/m. If the floor is subjected to a vibration $\omega = 7$ Hz, having a vertical displacement amplitude $\delta_O = 0.17$ ft, determine the vertical displacement amplitude of the platform and instrument. The instrument and the platform have a total weight of 18 lb.

Prob. 22–49

22–50. The light elastic rod supports a 4-kg sphere. When an 18-N vertical force is applied to the sphere, the rod deflects 14 mm. If the wall oscillates with a harmonic frequency of 2 Hz and has an amplitude of 15 mm, determine the amplitude of vibration for the sphere.

0.75 m

Prob. 22–50

22–51. The small block at A has a mass of 4 kg and is mounted on the bent rod having negligible mass. If the rotor at B causes a harmonic movement $\delta_B = (0.1 \cos 15t)$ m, where t is in seconds, determine the amplitude of vibration of the block.

0.6 m

A O

1.2 m

$k = 15$ N/m ω

B

Prob. 22–51

*22–52. The 450-kg trailer is pulled with a constant speed over the surface of a bumpy road, which may be approximated by a cosine curve having an amplitude of 50 mm and wave length of 4 m. If the two springs s which support the trailer each have a stiffness of 800 N/m, determine the speed v which will cause the greatest vibration (resonance) of the trailer. Neglect the weight of the wheels.

v

s

100 mm

2 m   2 m

Probs. 22–52 / 22–53

22–53. Determine the amplitude of vibration of the trailer in Prob. 22–52 if the speed $v = 15$ km/h.

22–54. A block having a weight of 7 lb is suspended from a spring having a stiffness $k = 75$ lb/ft. The support to which the spring is attached is given a simple harmonic motion which may be expressed by $\delta = (0.15 \sin 2t)$ ft, where t is in seconds. If the damping factor is $c/c_c = 0.8$, determine the phase angle ϕ of forced vibration.

22–55. Determine the magnification factor of the block, spring, and dashpot combination in Prob. 22–54.

*22–56. A block having a mass of 0.8 kg is suspended from a spring having a stiffness of 120 N/m. If a dashpot provides a damping force of 2.5 N when the speed of the block is 0.2 m/s, determine the period of free vibration.

22–57. The 20-kg block is subjected to the action of the harmonic force $F = (90 \cos 6t)$ N, where t is in seconds. Write the equation which describes the steady-state motion.

$k = 400$ N/m

$F = 90 \cos 6t$

20 kg

$k = 400$ N/m

$c = 125$ N · s/m

Prob. 22–57

22–58. The barrel of a cannon has a mass of 700 kg, and after firing it recoils a distance of 0.64 m. If it returns to its original position by means of a single recuperator having a damping coefficient of 2 kN · s/m, determine the required stiffness of each of the two springs fixed to the base and attached to the barrel so that the barrel recuperates without vibration.

22–59. A block having a mass of 7 kg is suspended from a spring that has a stiffness $k = 600$ N/m. If it is given an upward velocity of 0.6 m/s from its equilibrium position at $t = 0$, determine the position of the block as a function of time. Assume that positive displacement of the block is downward, and that motion takes place in a medium which furnishes a damping force $F = (50|v|)$ N, where v is measured in m/s.

*****22–60.** The damping factor, c/c_c, may be determined experimentally by measuring the successive amplitudes of vibrating motion of a system. If two of these maximum displacements can be approximated by x_1 and x_2, as shown in Fig. 22–17, show that the ratio $\ln x_1/x_2 = 2\pi(c/c_c)/\sqrt{1 - (c/c_c)^2}$. The quantity $\ln x_1/x_2$ is called the *logarithmic decrement*.

22–61. The block shown in Fig. 22–16 has a mass of 20 kg and the spring has a stiffness $k = 600$ N/m. When the block is displaced and released, two successive amplitudes are measured as $x_1 = 150$ mm and $x_2 = 87$ mm. Determine the coefficient of viscous damping, c. *Hint:* See Prob. 22–60.

22–62. The 48-kg electric motor is fastened to the midpoint of a pin-supported beam having negligible mass. It is found that the beam deflects 35 mm when the motor is not running. If the motor turns an eccentric flywheel which is equivalent to an unbalanced mass of 4 kg located 100 mm from the axis of rotation, determine the amplitude of steady-state vibration when the flywheel is turning at 17 rad/s. The damping factor is $c/c_c = 0.20$.

Prob. 22–62

22–63. The bar has a weight of 6 lb. If the stiffness of the spring is $k = 8$ lb/ft and the dashpot has a damping coefficient $c = 60$ lb · s/ft, determine the differential equation which describes the motion in terms of the angle θ of the bar's rotation. Also, what should the damping coefficient of the dashpot be if the bar is to be critically damped?

Prob. 22–63

*****22–64.** The 4-kg circular disk is attached to three springs, each spring having a stiffness of $k = 180$ N/m. If the disk is immersed in a fluid and given a downward velocity of 0.3 m/s at the equilibrium position, determine the equation which describes the motion. Assume that positive displacement is measured downward, and that fluid resistance acting on the disk furnishes a damping force having a magnitude of $F = (60|v|)$ N, where v is in m/s.

Prob. 22–64

22–65. Draw the electrical circuit that is equivalent to the mechanical system shown. What is the differential equation which describes the motion of the current in the circuit?

Prob. 22–65

22–67. Determine the mechanical analogue for the electrical circuit. What differential equations describe the mechanical and electrical systems?

$E = E_O \sin \omega t$

Prob. 22–67

22–66. Draw the electrical circuit that is equivalent to the mechanical system shown. What is the differential equation which describes the motion of the current in the circuit?

Prob. 22–66

*22–68.** Draw the electric circuit that is analogous to the mechanical system shown. Determine the differential equation which describes the motion of the current in the circuit.

$F = F_O \cos \omega t$

Prob. 22–68

22–69. Determine the differential equation of motion for the damped vibratory system shown. What type of motion occurs?

$k = 100 \text{ N/m}$

25 kg

$c = 200 \text{ N} \cdot \text{s/m}$ $c = 200 \text{ N} \cdot \text{s/m}$

Prob. 22–69

539

Mathematical Expressions

Quadratic Formula

If $ax^2 + bx + c = 0$, then $x = \dfrac{-b \pm \sqrt{b^2 - 4ac}}{2a}$

Hyperbolic Functions

$$\sinh x = \frac{e^x - e^{-x}}{2}, \quad \cosh x = \frac{e^x + e^{-x}}{2}, \quad \tanh x = \frac{\sinh x}{\cosh x}$$

Trigonometric Identities

$\sin^2 \theta + \cos^2 \theta = 1$

$\sin (\theta \pm \phi) = \sin \theta \cos \phi \pm \cos \theta \sin \phi$

$\sin 2\theta = 2 \sin \theta \cos \theta$

$\cos (\theta \pm \phi) = \cos \theta \cos \phi \mp \sin \theta \sin \phi$

$\cos 2\theta = \cos^2 \theta - \sin^2 \theta$

$\cos \theta = \pm \sqrt{\dfrac{1 + \cos 2\theta}{2}}, \quad \sin \theta = \pm \sqrt{\dfrac{1 - \cos 2\theta}{2}}$

$\tan \theta = \dfrac{\sin \theta}{\cos \theta}$

$1 + \tan^2 \theta = \sec^2 \theta$

$1 + \text{ctn}^2 \theta = \csc^2 \theta$

Power-Series Expansions

$$\sin x = x - \frac{x^3}{3!} + \frac{x^5}{5!} - \frac{x^7}{7!} + \cdots$$

$$\cos x = 1 - \frac{x^2}{2!} + \frac{x^4}{4!} - \frac{x^6}{6!} + \cdots$$

$$\sinh x = x + \frac{x^3}{3!} + \frac{x^5}{5!} + \cdots$$

$$\cosh x = 1 + \frac{x^2}{2!} + \frac{x^4}{4!} + \cdots$$

Derivatives

$$\frac{d}{dx}(u^n) = nu^{n-1}\frac{du}{dx}$$

$$\frac{d}{dx}(uv) = u\frac{dv}{dx} + v\frac{du}{dx}$$

$$\frac{d}{dx}\left(\frac{u}{v}\right) = \frac{v\frac{du}{dx} - u\frac{dv}{dx}}{v^2}$$

$$\frac{d}{dx}(\sin u) = \cos u\frac{du}{dx}$$

$$\frac{d}{dx}(\cos u) = -\sin u\frac{du}{dx}$$

$$\frac{d}{dx}(\tan u) = \sec^2 u\frac{du}{dx}$$

$$\frac{d}{dx}(\cot u) = -\csc^2 u\frac{du}{dx}$$

$$\frac{d}{dx}(\sec u) = \tan u \sec u\frac{du}{dx}$$

$$\frac{d}{dx}(\csc u) = -\csc u \cot u\frac{du}{dx}$$

$$\frac{d}{dx}(\sinh u) = \cosh u\frac{du}{dx}$$

$$\frac{d}{dx}(\cosh u) = \sinh u\frac{du}{dx}$$

Integrals

$$\int x^n \, dx = \frac{x^{n+1}}{n+1} + C, \ n \neq -1$$

$$\int \frac{dx}{a+bx} = \frac{1}{b}\ln(a+bx) + C$$

$$\int \sqrt{a+bx}\, dx = \frac{2}{3b}\sqrt{(a+bx)^3} + C$$

$$\int x\sqrt{a+bx}\, dx = \frac{-2(2a-3bx)\sqrt{(a+bx)^3}}{15b^2} + C$$

$$\int x^2\sqrt{a+bx}\, dx = \frac{2(8a^2 - 12abx + 15b^2x^2)\sqrt{(a+bx)^3}}{105b^3} + C$$

$$\int \sqrt{a^2 - x^2}\, dx = \frac{1}{2}\left[x\sqrt{a^2-x^2} + a^2\sin^{-1}\frac{x}{a}\right] + C, \ a > 0$$

$$\int x\sqrt{a^2 - x^2}\, dx = -\frac{1}{3}\sqrt{(a^2-x^2)^3} + C$$

$$\int x^2\sqrt{a^2 - x^2}\, dx = -\frac{x}{4}\sqrt{(a^2-x^2)^3} + \frac{a^2}{8}\left(x\sqrt{a^2-x^2} + a^2\sin^{-1}\frac{x}{a}\right) + C, \ a > 0$$

$$\int \sqrt{x^2 \pm a^2}\, dx = \frac{1}{2}\left[x\sqrt{x^2+a^2} \pm a^2\ln(x + \sqrt{x^2 \pm a^2})\right] + C$$

$$\int x\sqrt{x^2 \pm a^2}\, dx = \frac{1}{3}\sqrt{(x^2 \pm a^2)^3} + C$$

$$\int \frac{x\, dx}{\sqrt{x^2 \pm a^2}} = \sqrt{x^2 \pm a^2} + C$$

$$\int \frac{dx}{\sqrt{a+bx+cx^2}} = \frac{1}{\sqrt{c}}\ln\left[\sqrt{a+bx+cx^2} + x\sqrt{c} + \frac{b}{2\sqrt{c}}\right] + C, \ c > 0$$

$$= \frac{1}{\sqrt{-c}}\sin^{-1}\left(\frac{-2cx-b}{\sqrt{b^2-4ac}}\right) + C, \ c < 0$$

$$\int \sin x\,dx = -\cos x + C$$

$$\int \cos x\,dx = \sin x + C$$

$$\int x\cos(ax)\,dx = \frac{1}{a^2}\cos(ax) + \frac{x}{a}\sin(ax) + C$$

$$\int x^2\cos(ax)\,dx = \frac{2x}{a^2}\cos(ax) + \frac{a^2x^2 - 2}{a^3}\sin(ax) + C$$

$$\int e^{ax}\,dx = \frac{1}{a}e^{ax} + C$$

$$\int \sinh x\,dx = \cosh x + C$$

$$\int \cosh x\,dx = \sinh x + C$$

B Numerical and Computer Analysis

Occasionally the application of the laws of mechanics will lead to a system of equations for which a closed-form solution is difficult or impossible to obtain. When confronted with this situation, engineers will often use a numerical method which in most cases can be programmed on a microcomputer or "programmable" pocket calculator. Here we will briefly present a computer program for solving a set of linear algebraic equations and three numerical methods which can be used to solve an algebraic or transcendental equation, evaluate a definite integral, and solve an ordinary differential equation. Application of each method will be explained by example, and an associated computer program written in Microsoft BASIC, which is designed to run on most personal computers, is provided.* A text on numerical analysis should be consulted for further discussion regarding a check of the accuracy of each method and the inherent errors that can develop.

Linear Algebraic Equations. Application of the equations of static equilibrium or the equations of motion sometimes requires solving a set of linear algebraic equations. The computer program listed in Fig. B–1 can be used for this purpose. It is based on the method of a Gauss elimination and can solve at most ten equations with ten unknowns. To do so, the equations should first be written in the following general format:

$$A_{11}x_1 + A_{12}x_2 + \cdots + A_{1n}x_n = B_1$$
$$A_{21}x_1 + A_{22}x_2 + \cdots + A_{2n}x_n = B_2$$
$$\cdot$$
$$\cdot$$
$$\cdot$$
$$A_{n1}x_1 + A_{n2}x_2 + \cdots + A_{nn}x_n = B_n$$

The "A" and "B" coefficients are "called" for when running the program. The output presents the unknowns $x_1, \ldots, x_n$.

*Similar types of programs can be written or purchased for programmable pocket calculators.

```
1 PRINT"Linear system of equations":PRINT
2 DIM A(10,11)
3 INPUT"Input number of equations : ",N
4 PRINT
5 PRINT"A  coefficients"
6 FOR I = 1 TO N
7 FOR J = 1 TO N
8 PRINT "A(";I;",";J;
9 INPUT")=",A(I,J)
10 NEXT J
11 NEXT I
12 PRINT
13 PRINT"B  coefficients"
14 FOR I = 1 TO N
15 PRINT "B(";I;
16 INPUT")=",A(I,N+1)
17 NEXT I
18 GOSUB 25
19 PRINT
20 PRINT"Unknowns"
21 FOR I = 1 TO N
22 PRINT "X(";I;")=";A(I,N+1)
23 NEXT I
24 END
25 REM Subroutine Guassian
26 FOR M=1 TO N
27 NP=M
28 BG=ABS(A(M,M))
29 FOR I = M TO N
30 IF ABS(A(I,M))<=BG THEN 33
31 BG=ABS(A(I,M))
32 NP=I
33 NEXT I
34 IF NP=M THEN 40
35 FOR I = M TO N+1
36 TE=A(M,I)
37 A(M,I)=A(NP,I)
38 A(NP,I)=TE
39 NEXT I
40 FOR I = M+1 TO N
41 FC=A(I,M)/A(M,M)
42 FOR J = M+1 TO N+1
43 A(I,J)=A(I,J)-FC*A(M,J)
44 NEXT J
45 NEXT I
46 NEXT M
47 A(N,N+1)=A(N,N+1)/A(N,N)
48 FOR I = N-1 TO 1 STEP -1
49 SM=0
50 FOR J=I+1 TO N
51 SM=SM+A(I,J)*A(J,N+1)
52 NEXT J
53 A(I,N+1)=(A(I,N+1)-SM)/A(I,I)
54 NEXT I
55 RETURN
```

Fig. B–1

Example B–1

Solve the two equations

$$3x_1 + x_2 = 4$$
$$2x_1 - x_2 = 10$$

Solution

When the program begins to run, it first calls for the number of equations (2), then the A coefficients in the following sequence: $A_{11} = 3$, $A_{12} = 1$, $A_{21} = 2$, $A_{22} = -1$ and finally the B coefficients $B_1 = 4$, $B_2 = 10$. The output appears as:

Unknowns
$X(1) = 2.8$ *Ans.*
$X(2) = -4.4$ *Ans.*

Simpson's Rule. This numerical method can be used to determine the area under a curve given as a graph or as an explicit function $y = f(x)$. Likewise it can be used to compute the value of a definite integral which involves the function $y = f(x)$. To do so, the area must be subdivided into an *even number* of strips or intervals having a width h. The curve between three consecutive ordinates is approximated by a parabola, and the entire area or definite integral is then determined from the formula

$$\int_{x_0}^{x_n} f(x)\,dx \simeq \frac{h}{3}[y_0 + 4(y_1 + y_3 + \cdots + y_{n-1})$$
$$+ 2(y_2 + y_4 + \cdots + y_{n-2}) + y_n] \quad (B-1)$$

The computer program for this equation is given in Fig. B–2. For its use, one must first enter the function (line 6), then the upper and lower limits of the integral (lines 7 and 8), and finally the number of intervals (line 9). The value of the integral is then given as the output.

```
1 PRINT"Simpson's rule":PRINT
2 PRINT" To execute this program :":PRINT
3 PRINT"   1- Modify right-hand side of the equation given below,
4 PRINT"      then press RETURN key"
5 PRINT"   2- Type   RUN 6":PRINT:EDIT 6
6 DEF FNF(X)=LOG(X)
7 PRINT:INPUT" Enter Lower Limit = ",A
8 INPUT" Enter Upper Limit = ",B
9 INPUT" Enter Number (even) of Intervals = ",N%
10 H=(B-A)/N%:AR=FNF(A):X=A+H
11 FOR J%=2 TO N%
12 K=2*(2-J%+2*INT(J%/2))
13 AR=AR+K*FNF(X)
14 X=X+H:NEXT J%
15 AR=H*(AR+FNF(B))/3
16 PRINT" Integral = ",AR
17 END
```

Fig. B–2

Example B–2

Evaluate the definite integral $\int_2^5 \ln x \, dx$

Solution

The interval $x_0 = 2$ to $x_6 = 5$ will be divided into six equal parts ($n = 6$), each having a width $h = (5 - 2)/6 = 0.5$. We then compute $y = f(x) = \ln x$ at each point of subdivision.

n	x_n	y_n
0	2	0.693
1	2.5	0.916
2	3	1.099
3	3.5	1.253
4	4	1.386
5	4.5	1.504
6	5	1.609

Thus, Eq. B–1 becomes

$$\int_2^5 \ln x \, dx \simeq \frac{0.5}{3}[0.693 + 4(0.916 + 1.253 + 1.504)$$
$$+ 2(1.099 + 1.386) + 1.609]$$
$$\simeq 3.66 \qquad\qquad Ans.$$

This answer is equivalent to the exact answer to three significant figures. Obviously, accuracy to a larger number of significant figures can be improved by selecting a smaller interval h (or larger n).

Using the computer program, we enter the function $\ln x$, line 6 in Fig. B–2, the upper and lower limits 2 and 5, and the number of intervals $n = 6$. The output appears as

$$\text{Integral} = 3.66082 \qquad\qquad Ans.$$

The Secant Method. This numerical method is used to find the real roots of an algebraic or transcendental equation $f(x) = 0$. The method derives its name from the fact that the formula used is established from the slope of the secant line to the graph $y = f(x)$. This slope is $(f(x_n) - f(x_{n-1}))/(x_n - x_{n-1})$ and the secant formula is

$$x_{n+1} = x_n - f(x_n) \left(\frac{x_n - x_{n-1}}{f(x_n) - f(x_{n-1})} \right) \qquad \text{(B–2)}$$

For application it is necessary to provide two initial guesses, x_0 and x_1, and thereby evaluate x_2 from Eq. B–2 ($n = 1$). One then proceeds to reapply Eq. B–2 with x_1 and the calculated value of x_2 and obtain x_3 ($n = 2$), etc., until the value $x_{n+1} \simeq x_n$. One can see this will occur if x_n is approaching the root of the function $f(x) = 0$ since the correction term on the right of Eq. B–2 will tend toward zero. In particular, the larger the slope the smaller the correction to x_n and the faster the root will be found. Likewise, if the slope is very small in the neighborhood of the root, the method leads to large corrections for x_n and convergence to the root is slow and may even lead to a failure to find it. In such cases other numerical techniques must be used for solution.

A computer program based on Eq. B–2 is listed in Fig. B–3. The user must supply the function (line 7) along with two initial guesses x_0 and x_1 used to approximate the solution. The output specifies the value of the root. If it cannot be determined it is so stated.

```
1 PRINT"Secant method":PRINT
2 PRINT" To execute this program :":PRINT
3 PRINT"    1) Modify right hand side of the equation given below,"
4 PRINT"       then press RETURN key."
5 PRINT"    2) Type  RUN 7"
6 PRINT:EDIT 7
7 DEF FNF(X)=.5*SIN(X)-2*COS(X)+1.3
8 INPUT"Enter point #1 =",X
9 INPUT"Enter point #2 =",X1
10 IF X=X1 THEN 14
11 EP=.00001:TL=2E-20
12 FP=(FNF(X1)-FNF(X))/(X1-X)
13 IF ABS(FP)>TL THEN 15
14 PRINT"Root can not be found.":END
15 DX=FNF(X1)/FP
16 IF ABS(DX)>EP THEN 19
17 PRINT "Root = ";X1;"      Function evaluated at this root = ";FNF(X1)
18 END
19 X=X1:X1=X1-DX
20 GOTO 12
```

Fig. B–3

Example B–3

Determine the root of the equation

$$f(x) = 0.5 \sin x - 2 \cos x + 1.30 = 0$$

Solution

Guesses of the initial roots will be $x_0 = 45°$ and $x_1 = 30°$. Applying Eq. B–2,

$$x_2 = 30° - (-0.1821)\frac{(30° - 45°)}{(-0.1821 - 0.2393)} = 36.48°$$

Using this value in Eq. B–2, along with $x_1 = 30°$, we have

$$x_3 = 36.48° - (-0.0108)\frac{36.48° - 30°}{(-0.0108 + 0.1821)} = 36.89°$$

Repeating the process with this value and $x_2 = 36.48°$ yields

$$x_4 = 36.89° - (0.0005)\left(\frac{36.89° - 36.48°}{(0.0005 + 0.0108)}\right) = 36.87°$$

Thus $x = 36.9°$ is appropriate to three significant figures.

If the problem is solved using the computer program, we enter the function, line 7 in Fig. B–3. The first and second guesses must be in radians. Choosing these to be 0.8 rad, and 0.5 rad, the result appears as Root = 0.6435022.

Function evaluated at this root = 1.66893E–06.

This result converted from radians to degrees is therefore $x = 36.9°$.

Runge-Kutta Method. This numerical method is used to solve an ordinary differential equation. It consists of applying a set of formulas which are used to find specific values of y for corresponding incremental values h in x. The formulas given in general form are as follows:

First-Order Equation: To integrate $\dot{x} = f(t, x)$ step-by-step use

$$x_{i+1} = x_i + \frac{1}{6}(k_1 + 2k_2 + 2k_3 + k_4) \tag{B-3}$$

where

$$k_1 = hf(t_i, x_i)$$

$$k_2 = hf\left(t_i + \frac{h}{2}, x_i + \frac{k_1}{2}\right)$$

$$k_3 = hf\left(t_i + \frac{h}{2}, x_i + \frac{k_2}{2}\right) \tag{B-4}$$

$$k_4 = hf(t_i + h, x_i + k_3)$$

Second-Order Equation. To integrate $\ddot{x} = f(t, x, \dot{x})$

$$x_{i+1} = x_i + h\left[\dot{x}_i + \frac{1}{6}(k_1 + k_2 + k_3)\right]$$

$$\dot{x}_{i+1} = \dot{x}_i + \frac{1}{6}(k_1 + 2k_2 + 2k_3 + k_4) \tag{B-5}$$

where

$$k_1 = hf(t_i, x_i, \dot{x}_i)$$

$$k_2 = hf\left(t_i + \frac{h}{2}, x_i + \frac{h}{2}\dot{x}_i, \dot{x}_i + \frac{k_1}{2}\right)$$

$$k_3 = hf\left(t_i + \frac{h}{2}, x_i + \frac{h}{2}\dot{x}_i + \frac{h}{4}k_1, \dot{x}_i + \frac{k_2}{2}\right) \tag{B-6}$$

$$k_4 = hf\left(t_i + h, x_i + h\dot{x}_i + \frac{h}{2}k_2, \dot{x}_i + k_3\right)$$

To use these equations, one starts with initial values $t_i = t_0$, $x_i = x_0$ and $\dot{x}_i = \dot{x}_0$ (for the second-order equation). Choosing an increment h for t_0 the four constants k are computed and these results are substituted into Eq. B–3 or B–5 in order to compute $x_{i+1} = x_1$, $\dot{x}_{i+1} = x_1$, corresponding to $t_{i+1} = t_1 = t_0 + h$. Repeating this process using t_1, x_1, $\dot{x}_1$ and h, values for x_2, $\dot{x}_2$ and $t_2 = t_1 + h$ are then computed, etc.

Computer programs which solve first and second-order differential equations by this method are listed in Figs. B–4 and B–5, respectively. In order to use these programs, the operator specifies the function $\dot{x} = f(t, x)$ or $\ddot{x} = f(t, x, \dot{x})$ (line 7), the initial values t_0, x_0, $\dot{x}_0$ (for second-order equation), the final time t_n, and the step size h. The output gives the values of t, x, and $\dot{x}$ for each time increment until t_n is reached.

```
1 PRINT"Runge-Kutta Method for 1-st order Differential Equation":PRINT
2 PRINT" To execute this program :":PRINT
3 PRINT"   1) Modify right hand side of the equation given below,"
4 PRINT"      then Press RETURN key"
5 PRINT"   2) Type  RUN 7"
6 PRINT:EDIT 7
7 DEF FNF(T,X)=5*T+X
8 CLS:PRINT" Initial Conditions":PRINT
9 INPUT"Input  t   = ",T
10 INPUT"       x   = ",X
11 INPUT"Final  t   = ",T1
12 INPUT"step size  = ",H:PRINT
13 PRINT"        t            x"
14 IF T>=T1+H THEN 23
15 PRINT USING"######.#####";T;X
16 K1=H*FNF(T,X)
17 K2=H*FNF(T+.5*H,X+.5*K1)
18 K3=H*FNF(T+.5*H,X+.5*K2)
19 K4=H*FNF(T+H,X+K3)
20 T=T+H
21 X=X+(K1+K2+K2+K3+K3+K4)/6
22 GOTO 14
23 END
```

Fig. B—4

```
1 PRINT"Runge-Kutta Method for 2-nd order Differential Equation":PRINT
2 PRINT" To execute this program :":PRINT
3 PRINT"   1) Modify right hand side of the equation given below,"
4 PRINT"      then Press RETURN key"
5 PRINT"   2) Type  RUN 7"
6 PRINT:EDIT 7
7 DEF FNF(T,X,XD)=
8 INPUT"Input  t   = ",T
9 INPUT"       x   = ",X
10 INPUT"      dx/dt = ",XD
11 INPUT"Final  t   = ",T1
12 INPUT"step size  = ",H:PRINT
13 PRINT"        t          x          dx/dt"
14 IF T>=T1+H THEN 24
15 PRINT USING"######.#####";T;X;XD
16 K1=H*FNF(T,X,XD)
17 K2=H*FNF(T+.5*H,X+.5*H*XD,XD+.5*K1)
18 K3=H*FNF(T+.5*H,X+(.5*H)*(XD+.5*K1),XD+.5*K2)
19 K4=H*FNF(T+H,X+H*XD+.5*H*K2,XD+K3)
20 T=T+H
21 X=X+H*XD+H*(K1+K2+K3)/6
22 XD=XD+(K1+K2+K2+K3+K3+K4)/6
23 GOTO 14
24 END
```

Fig. B—5

Example B–4

Solve the differential equation $\dot{x} = 5t + x$. Obtain the results for two steps using time increments of $h = 0.02$ s. At $t_0 = 0$, $x_0 = 0$.

Solution

This is a first-order equation, so Eqs. B–3 and B–4 apply. Thus, for $t_0 = 0$, $x_0 = 0$, $h = 0.02$, we have

$$k_1 = 0.02[0 + 0] = 0$$
$$k_2 = 0.02[5(0.1) + 0] = 0.001$$
$$k_3 = 0.02[5(0.01) + 0.0005] = 0.00101$$
$$k_4 = 0.02[5(0.02) + 0.00101] = 0.00202$$

$$x_1 = 0 + \frac{1}{6}[0 + 2(0.001) + 2(0.00101) + 0.00202] = 0.00101$$

Using the values $t_1 = 0 + 0.02 = 0.02$ and $x_1 = 0.00101$ with $h = 0.02$, the value for x_2 is now computed from Eqs. B–3 and B–4.

$$k_1 = 0.02[5(0.02) + 0.00101] = 0.00202$$
$$k_2 = 0.02[5(0.03) + 0.00202] = 0.00304$$
$$k_3 = 0.02[5(0.03) + 0.00253] = 0.00305$$
$$k_4 = 0.02[5(0.04) + 0.00406] = 0.00408$$

$$x_2 = 0.001 + \frac{1}{6}[0.00202 + 2(0.00304) + 2(0.00305) + 0.00408]$$

$$x_2 = 0.00405 \qquad\qquad\qquad\qquad Ans.$$

To solve this problem using the computer program in Fig. B–4, the function is first entered on line 7, then the data $t_0 = 0$, $x_0 = 0$, $t_n = 0.04$, and $h = 0.02$ is specified. The results appear as

t	x
0.00000	0.00000
0.02000	0.00101
0.04000	0.00405

Ans.

Vector Analysis

The following discussion provides a brief review of the vector analysis used in statics. A more detailed treatment of these topics is given in *Engineering Mechanics: Statics*.

Vector. A vector, **A**, is a quantity which has a magnitude and direction, and adds according to the parallelogram law. As shown in Fig. C–1, $\mathbf{A} = \mathbf{B} + \mathbf{C}$, where **A** is the *resultant vector* and **B** and **C** are *component vectors*.

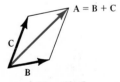

Fig. C–1

Unit Vector. A unit vector, $\mathbf{u}_A$, has a magnitude of one "dimensionless" unit and acts in the same direction as **A**. It is determined by dividing **A** by its magnitude A, i.e.,

$$\mathbf{u}_A = \frac{\mathbf{A}}{A} \qquad \text{(C–1)}$$

Cartesian Vector Notation. The directions of the positive x, y, z axes are defined by the Cartesian unit vectors, **i**, **j**, **k**, respectively.

As shown in Fig. C–2, vector **A** is formulated by the addition of its x, y, z components as

$$\mathbf{A} = A_x\mathbf{i} + A_y\mathbf{j} + A_z\mathbf{k} \qquad \text{(C–2)}$$

The *magnitude* of **A** is determined from

$$A = \sqrt{A_x^2 + A_y^2 + A_z^2} \qquad \text{(C–3)}$$

Fig. C–2

Fig. C–3

The *direction* of **A** is defined in terms of its *coordinate direction angles*, α, β, γ, measured from the *tail* of **A** to the *positive x, y, z* axes, Fig. C–3. These angles are determined from the *direction cosines* which represent the **i**, **j**, **k** components of the unit vector $\mathbf{u}_A$; i.e., from Eqs. C–1 and C–2,

$$\mathbf{u}_A = \frac{A_x}{A}\mathbf{i} + \frac{A_y}{A}\mathbf{j} + \frac{A_z}{A}\mathbf{k} \qquad \text{(C–4)}$$

so that the direction cosines are

$$\cos\alpha = \frac{A_x}{A}, \qquad \cos\beta = \frac{A_y}{A}, \qquad \cos\gamma = \frac{A_z}{A} \qquad \text{(C–5)}$$

Hence $\mathbf{u}_A = \cos\alpha\,\mathbf{i} + \cos\beta\,\mathbf{j} + \cos\gamma\,\mathbf{k}$, and using Eq. C–3, it is seen that

$$\cos^2\alpha + \cos^2\beta + \cos^2\gamma = 1 \qquad \text{(C–6)}$$

The Cross Product. The cross product of two vectors **A** and **B**, which yields the resultant vector **C**, is written as

$$\mathbf{C} = \mathbf{A} \times \mathbf{B} \qquad \text{(C–7)}$$

and reads **C** equals **A** "cross" **B**. The *magnitude* of **C** is

$$C = AB\sin\theta \qquad \text{(C–8)}$$

Fig. C–4

where θ is the angle made between the *tails* of **A** and **B** ($0° \leq \theta \leq 180°$). The *direction* of **C** is determined by the right-hand rule, whereby the fingers of the right hand are curled *from* **A** *to* **B** and the thumb points in the direction of **C**, Fig. C–4. This vector is perpendicular to the plane containing vectors **A** and **B**.

The vector cross product is *not* commutative, i.e., $\mathbf{A} \times \mathbf{B} \neq \mathbf{B} \times \mathbf{A}$. Rather,

$$\mathbf{A} \times \mathbf{B} = -\mathbf{B} \times \mathbf{A} \qquad \text{(C–9)}$$

The distributive law is valid, i.e.,

$$\mathbf{A} \times (\mathbf{B} + \mathbf{D}) = \mathbf{A} \times \mathbf{B} + \mathbf{A} \times \mathbf{D} \qquad \text{(C–10)}$$

And, the cross product may be multiplied by a scalar m in any manner, i.e.,

$$m(\mathbf{A} \times \mathbf{B}) = (m\mathbf{A}) \times \mathbf{B} = \mathbf{A} \times (m\mathbf{B}) = (\mathbf{A} \times \mathbf{B})m \qquad \text{(C–11)}$$

If **A** and **B** are expressed in Cartesian component form, then the cross product, Eq. C–7, may be evaluated by expanding the determinant

$$\mathbf{C} = \mathbf{A} \times \mathbf{B} = \begin{vmatrix} \mathbf{i} & \mathbf{j} & \mathbf{k} \\ A_x & A_y & A_z \\ B_x & B_y & B_z \end{vmatrix} \qquad \text{(C–12)}$$

which yields

$$\mathbf{C} = (A_y B_z - A_z B_y)\mathbf{i} - (A_x B_z - A_z B_x)\mathbf{j} + (A_x B_y - A_y B_x)\mathbf{k}$$

Recall that the cross product is used in statics to define the moment of a force $\mathbf{F}$ about point O, in which case

$$\mathbf{M}_O = \mathbf{r} \times \mathbf{F} \qquad \text{(C–13)}$$

where $\mathbf{r}$ is a position vector directed from point O to *any point* on the line of action of force $\mathbf{F}$.

The Dot Product. The dot product of two vectors $\mathbf{A}$ and $\mathbf{B}$, which yields a scalar, is defined as

$$\mathbf{A} \cdot \mathbf{B} = AB \cos \theta \qquad \text{(C–14)}$$

and reads $\mathbf{A}$ ''dot'' $\mathbf{B}$. The angle θ is formed between the *tails* of $\mathbf{A}$ and $\mathbf{B}$ ($0° \leq \theta \leq 180°$).

The dot product is commutative, i.e.,

$$\mathbf{A} \cdot \mathbf{B} = \mathbf{B} \cdot \mathbf{A} \qquad \text{(C–15)}$$

The distributive law is valid, i.e.,

$$\mathbf{A} \cdot (\mathbf{B} + \mathbf{D}) = \mathbf{A} \cdot \mathbf{B} + \mathbf{A} \cdot \mathbf{D} \qquad \text{(C–16)}$$

And, scalar multiplication can be performed in any manner, i.e.,

$$m(\mathbf{A} \cdot \mathbf{B}) = (m\mathbf{A}) \cdot \mathbf{B} = \mathbf{A} \cdot (m\mathbf{B}) = (\mathbf{A} \cdot \mathbf{B})m \qquad \text{(C–17)}$$

If $\mathbf{A}$ and $\mathbf{B}$ are expressed in Cartesian component form, then the dot product, Eq. C–14, can be determined from

$$\mathbf{A} \cdot \mathbf{B} = A_x B_x + A_y B_y + A_z B_z \qquad \text{(C–18)}$$

The dot product may be used to determine the *angle θ formed between two vectors*. From Eq. C–14,

$$\theta = \cos^{-1}\left(\frac{\mathbf{A} \cdot \mathbf{B}}{AB}\right) \qquad \text{(C–19)}$$

It is also possible to find the *component of a vector in a given direction* using the dot product. For example, the magnitude of the component (or projection) of vector $\mathbf{A}$ in the direction of $\mathbf{B}$, Fig. C–5, is defined by $A \cos \theta$. From Eq. C–14, this magnitude is

Fig. C–5

$$A \cos \theta = \mathbf{A} \cdot \frac{\mathbf{B}}{B} = \mathbf{A} \cdot \mathbf{u}_B \qquad \text{(C–20)}$$

where $\mathbf{u}_B$ represents a unit vector acting in the direction of $\mathbf{B}$, Fig. C–5.

Differentiation and Integration of Vector Functions. The rules for differentiation and integration of the sums and products of scalar functions apply as well to vector functions. Consider, for example, the two vector functions $\mathbf{A}(s)$ and $\mathbf{B}(s)$. Provided these functions are smooth and continuous for all s, then

$$\frac{d}{ds}(\mathbf{A} + \mathbf{B}) = \frac{d\mathbf{A}}{ds} + \frac{d\mathbf{B}}{ds} \tag{C–21}$$

$$\int (\mathbf{A} + \mathbf{B})\, ds = \int \mathbf{A}\, ds + \int \mathbf{B}\, ds \tag{C–22}$$

For the cross product,

$$\frac{d}{ds}(\mathbf{A} \times \mathbf{B}) = \left(\frac{d\mathbf{A}}{ds} \times \mathbf{B}\right) + \left(\mathbf{A} \times \frac{d\mathbf{B}}{ds}\right) \tag{C–23}$$

Similarly, for the dot product,

$$\frac{d}{ds}(\mathbf{A} \cdot \mathbf{B}) = \frac{d\mathbf{A}}{ds} \cdot \mathbf{B} + \mathbf{A} \cdot \frac{d\mathbf{B}}{ds} \tag{C–24}$$

Answers*

Chapter 12

12–1. $a_c = 6.40$ ft/s^2, $t = 12.5$ s.

12–2. $a_c = 1.74$ m/s^2, $t = 4.80$ s.

12–3. $t = 30$ s, $s = 792$ m.

12–5. $d = 517$ ft sober driver, $d = 616$ ft drunk driver.

12–6. $v_{Avg} = 0.333$ m/s, $v_{sp} = 1$ m/s.

12–7. $s = -18$ ft, $d = 46$ ft.

12–9. $v = 32$ m/s, $s = 67$ m, $d = 66$ m.

12–10. 0.603 ft/s^2.

12–11. 7.87 m.

12–13. 16.9 ft.

12–14. $k = -10.2$ s^{-2}, $s = 1.56$ ft.

12–15. $s = 11.9$ m, $v = 0.250$ m/s.

12–17. $v = 322$ m/s, $t = 19.3$ s.

12–18. $t = 77.6$ s, $d_m = 3.67(10^3)$ ft.

12–19. (a) $a_A = -50$ mm/s^2, (b) $s = 100$ mm, (c) $t \rightarrow \infty$.

12–21. 48.3 s.

12–22. $s = 12.5$, $t = 2.00$ s.

12–23. 11.2 km/s.

12–25. $s = 3600$ ft, $v_{Avg} = 30$ ft/s.

12–26. $a = 0.4$, $0 \leq t < 30$; $a = 0$, $30 < t \leq 50$,
$s = 0.2t^2$, $0 \leq t \leq 30$, $s = 12t - 180$, $30 \leq t \leq 50$.

12–27. 87.5 m.

12–29. 144 m.

12–30. $v_{max} = 0.894$ m/s, $t' = 0.447$ s.

12–31. 114 m.

12–33. 17.9 s.

12–34. $s = 200$, $v = 20$; $s = 400$, $v = 34.6$.

12–35. $t = 15$, $v = 112.5$, $s = 562.5$; $t = 40$, $v = 612.5$,
$s = 9625$.

12–37. $v_{max} = 120$ m/s, $t = 41.9$ s.

12–38. $v = 27.4$ ft/s, $v = 37.4$ ft/s.

12–39. 13.0 s.

12–41. 4.06 s.

12–42. 10.1 s.

12–43. $y = (\frac{1}{8}x)^{3/2} + 5$, $v = 55.1$ m/s, $a = 24.1$ m/s^2.

12–45. $v = 73.1$ m/s, $\alpha = 80.6°$, $\beta = 86.9°$, $\gamma = 9.96°$,
$a = 72.2$ m/s^2, $\alpha = 85.2°$, $\beta = 90°$, $\gamma = 4.76°$.

12–46. $a = 5.31$ m/s^2, $\alpha = 53.0°$, $\beta = 37.0°$, $\gamma = 90°$.

12–47. $\mathbf{v} = \{8.82\mathbf{i} + 494\mathbf{j}\}$ ft/s, $\mathbf{a} = \{432\mathbf{i} - 95.1\mathbf{j}\}$ ft/s^2.

12–49. $y^2 + x^2 - 4y = 0$, $v = 2$ ft/s, $a = 2$ ft/s^2,
$r = 3.37$ ft/s.

12–50. $(4$ ft, 3 ft$)$, $v_B = 8.06$ ft/s, $v_A = 4.47$ ft/s.

12–51. $t = 2$ s, $(6$ m, $0)$, $v_A = 18.3$ m/s, $v_B = 6.71$ m/s.

12–53. $\mathbf{a}_{AB} = \{0.404\mathbf{i} + 7.07\mathbf{j}\}$ m/s^2, $\mathbf{a}_{AC} = \{2.50\mathbf{i}\}$ m/s^2.

*Note: Answers to every fourth problem are omitted.

12–54. $\mathbf{v}_{AD} = \{3.33\mathbf{i} + 1.67\mathbf{j}\}$ m/s.

12–55. 4.28 m/s.

12–57. $v = 201$ m/s, $a = 405$ m/s^2.

12–58. $d = 204$ m, $v = 41.8$ m/s, $a = 4.66$ m/s^2.

12–59. $t = 0.545$ s, $h = 4.79$ ft.

12–61. 87.4 m/s.

12–62. $R = 19.0$ m, $t = 2.48$ s.

12–63. $(v_A)_{min} = 4.32$ m/s, $(v_A)_{max} = 5.85$ m/s.

12–65. $(v_A)_{min} = 0.838$ m/s, $(v_A)_{max} = 1.76$ m/s.

12–66. 41.3°.

12–67. $x = 5.15$ ft, $y = 1.33$ ft.

12–69. $\theta_D = 14.7°$, $\theta_C = 75.3°$, $\Delta t = 1.45$ s.

12–70. 3.13 m.

12–71. $x = 10.8$ ft, $y = 3.73$ ft.

12–73. 8.83 m.

12–74. 9.50 m/s^2.

12–75. 51.5 m/s^2.

12–77. 2.82 ft/s^2.

12–78. 2.63 s.

12–79. $a = 25.5$ ft/s^2, $\theta = 45°$ $\measuredangle\theta$.

12–81. $v = 4.52$ m/s, $a = 6.88$ m/s^2.

12–82. $v = 5.66$ m/s, $a = 1.02$ m/s^2.

12–83. $x = 0$, $y = -4$ m, $(a)_{max} = 50$ m/s^2.

12–85. $v = 19.9$ ft/s, $a = 16.0$ ft/s^2.

12–86. $a_n = 28.2$ m/s^2, $a_t = 40$ m/s^2.

12–87. 3.39 s.

12–89. $v = 115$ ft/s, $a = 58.4$ ft/s^2.

12–90. $\Delta s_{AB} = 14.4$ m, $a_B = 12.8$ m/s^2, $a_A = 190$ m/s^2.

12–91. 2.51 s, $a_B = 65.1$ m/s^2, $a_A = 22.2$ m/s^2.

12–93. $\dot{\mathbf{a}} = (\dddot{r} - 3\dot{r}\dot{\theta}^2 - 3r\dot{\theta}\ddot{\theta})\mathbf{u}_r + (3\dot{r}\ddot{\theta} + r\dddot{\theta} - r\dot{\theta}^3)\mathbf{u}_\theta + (\dddot{z})\mathbf{u}_z$.

12–94. 12.0 ft/s, 0.646 ft/s^2.

12–95. $\dot{\theta} = 0.333$ rad/s, $a = 6.67$ m/s^2.

12–97. $v_r = -0.212$ m/s, $v_\theta = 0.812$ m/s, $a_r = -0.703$ m/s^2, $a_\theta = 0.0839$ m/s^2.

12–98. $v_r = 0$, $v_\theta = 0.8$ m/s, $v_z = -0.0932$ m/s, $a_r = -0.16$ m/s^2, $a_\theta = 0$, $a_z = -7.25(10^{-3})$ m/s^2.

12–99. $v_r = 12$ m/s, $v_\theta = 12.6$ m/s, $a_r = -37.7$ m/s^2, $a_\theta = 72.0$ m/s^2.

12–101. $v = 8.00$ m/s, $a = 160$ m/s^2.

12–102. $v = 4.24$ m/s, $a = 17.6$ m/s^2.

12–103. $v_r = -16.9$ m/s, $v_\theta = 4.87$ m/s, $a_r = -89.4$ m/s^2, $a_\theta = -53.7$ m/s^2.

12–105. $v_r = -0.306$ km/s, $v_\theta = 0.177$ km/s, $a_r = -0.128$ km/s^2, $a_\theta = 0.0676$ km/s^2.

12–106. $v_r = -2.80$ m/s, $v_\theta = 19.8$ m/s.

12–107. $v = 640$ m/s, $a = 657$ m/s^2.

12–109. $v_r = 0.242$ m/s, $v_\theta = 0.943$ m/s, $a_r = -2.33$ m/s^2, $a_\theta = 1.74$ m/s^2.

12–110. 0.75 rad/s.

12–111. $v = 0.242$ m/s, $a = 0.169$ m/s^2, $(v_A)_x = 0.162$ m/s, $(v_A)_y = 0.180$ m/s.

12–113. $\mathbf{u}_b = -0.349\mathbf{i} - 0.0127\mathbf{j} - 0.937\mathbf{k}$, $\mathbf{u}_b' = 0.349\mathbf{i} + 0.0127\mathbf{j} + 0.937\mathbf{k}$.

12–114. $\alpha = 139°$, $\beta = 49.6°$, $\gamma = 95.2°$, $\alpha = 40.9°$, $\beta = 130°$, $\gamma = 84.8°$.

12–115. 3 ft/s.

12–117. **(a)** $v_A = 2.4$ m/s $\uparrow$, $v_B = 0.6$ m/s $\downarrow$, **(b)** $h = 4.5$ m.

12–118. 32 m/s $\downarrow$.

12–119. 1.33 ft.

12–121. 0.667 m/s $\uparrow$.

12–122. 2.22 m.

12–123. $v_B = 4$ ft/s $\uparrow$, $v_{B/A} = 12$ ft/s $\uparrow$.

12–125. $t = 1.07$ s, $v_{A/B} = 5.93$ m/s $\rightarrow$.

12–126. $v_A = -v_B\left(1 + \left(\dfrac{h}{s_A}\right)^2\right)^{1/2}$,

$a_A = -a_B\left(1 + \left(\dfrac{h}{s_A}\right)^2\right)^{1/2} + \dfrac{v_A v_B h^2}{s_A^3}\left(1 + \left(\dfrac{h}{s_A}\right)^2\right)^{-1/2}$.

12–127. $\dot{y} = v_B = \dfrac{-3(y^2 + 16)^{1/2}}{y}$ m/s.

12–129. 0.671 m/s $\uparrow$.

12–130. 1.41 m/s $\uparrow$.

12–131. 6.08 ft/s $\leftarrow$.

12–133. $v_B = 15.8$ ft/s, $\theta = 18.4°$ $\nwarrow^\theta$, $d = 53.3$ ft.

12–134. $v_{r/m} = 12.2$ km/h, $\theta = 35.0°$ $\searrow\theta$.

12–135. 30°.

12–137. 56.2°.

12–138. $v_{B/A} = 26.5$ mi/h, $\theta = 40.9°$ $\measuredangle\theta$, $a_{B/A} = 1.79(10^3)$ mi/h^2, $\theta = 72.0°$ $\measuredangle\theta$.

12–139. $v_{A/B} = 6.69$ m/s, $\theta = 53.3°$, $\theta\nearrow$, $a_{B/A} = 1.52$ m/s^2, $\theta = 41.9°$ $\theta\nearrow$.

12–141. $v_{B/A} = 875$ km/h, $\theta = 41.5°$ $\theta\searrow$.

12–142. $v_{A/B} = 13.5$ ft/s, $t = 1.85$ min.

12–143. $v_B = 13.0$ ft/s, $v_{B/A} = 37.0$ ft/s.

12–145. $t = 0.977$ s, $v_0 = 15.7$ ft/s, $v_{e/B} = 24.1$ ft/s.

12–146. $a_{A/B} = 4.47$ m/s², $\theta = 87.4°$.

Chapter 13

13–1. 1.07 μN.

13–2. 199(10^{18}) N.

13–3. 75.9 N, 20.4 kg, 3.72 m/s².

13–5. $v = 74.0$ m/s, $\theta = 68.7°$ $\theta\searrow$, $s = 185$ m.

13–6. 1.09 s, 1.54 s, 2.32 s.

13–7. 2.04 s.

13–9. (a) $T = 40$ lb, (b) $T = 59.9$ lb.

13–10. 14.8 ft/s².

13–11. 3.31 m/s $\uparrow$.

13–13. 36.8 s.

13–14. 26.8 ft/s $\downarrow$.

13–15. 9.50 m/s².

13–17. 4.52 m/s.

13–18. (a) $s = 0$, (b) $s = 0.955$ m.

13–19. 8.91 m/s.

13–21. 2.30 ft/s² $\uparrow$.

13–22. $F = m(a_B + g)\sqrt{4y^2 + d^2}/(4y)$.

13–23. 2.49 ft.

13–25. $k = 0.0557 \dfrac{\text{lb} \cdot \text{s}^2}{\text{ft}^2}$.

13–26. 68.2 s.

13–27. $s = 93.2$ ft.

13–29. $a_A = 1.51$ m/s² $\uparrow$, $a_B = 6.04$ m/s² $\downarrow$, $T_B = 22.6$ N, $T_A = 90.6$ N.

13–30. 78.7°.

13–31. 4.39 m/s.

13–33. 2.89 m/s².

13–34. 574 N.

13–35. 2.94 m/s².

13–37. 1.99(10^{30}) kg.

13–38. $v = 251$ m/s, $N = 6.18$ kN.

13–39. $\theta = 22.2°$, $F_L = 42.4$ kN, $\theta = 42.5°$.

13–41. 9.90 m/s.

13–42. $N_x = 217$ N, $N_y = 571$ N.

13–43. 40.1 ft/s.

13–45. 2.09 m/s.

13–46. $v = 6.30$ m/s, $F_n = 283$ N, $F_t = 0$, $F_b = 490$ N.

13–49. $T = 9.09$ kN, $\theta = 43.3°$.

13–50. $N_T = 522$ N, $a_t = 8.32$ m/s².

13–51. $a_t = 0$, $N = 8.65$ kip.

13–53. $F_C = 1.11$ kN, $N_C = 6.73$ kN.

13–54. 1.5 ft.

13–55. $N_B = 4.16$ kN, $a_t = 1.92$ m/s².

13–57. $F_s = 117$ N.

13–58. 383 mm.

13–59. $N_C = 75.6$ N, $a = 4.18$ m/s².

13–61. 1.48 m/s.

13–62. 16.1 Mm/h.

13–63. 20.9°.

13–65. $F_r = -480$ N, $F_\theta = 132$ N, $F_z = 23.4$ N.

13–66. $F_r = -29.4$ N, $F_\theta = 0$, $F_z = 392$ N.

13–67. 1.98 lb.

13–69. $N_C = 20.7$ N, $F_{OA} = 0$.

13–70. 5.66 N.

13–71. 0.225 rad/s.

13–73. 0.143 lb.

13–74. 0.432 lb.

13–75. 98.4 lb.

13–77. $N_P = 3.72$ N, $F = 7.44$ N.

13–78. 7.67 N.

13–79. $P = 4.22$ lb, $N = 21.1$ lb.

13–81. $F_r = -126$ N, $F_\theta = 0$.

13–82. $N_C = 1.10$ N, $F_C = 4.91$ N.

13–83. $r = 42.2$ Mm, $v = 3.07$ km/s.

13–85. $v_p = 7.44$ km/s, $v_a = 5.95$ km/s, $e = 0.111$.

13–86. at perigee, $\Delta v = \sqrt{\dfrac{GM_e}{r_0}}(\sqrt{2} - \sqrt{e + 1})$.

13–87. 7.82 km/s, 1.97 h.

13–89. $r_p = 146$ Gm, $v_p = 30.4$ km/s, $v_a = 29.4$ km/s.

13–90. 10.8(10^9) km.

13–91. $v_A = 11.5$ Mm/h, 27.3 Mm.

13–93. $d = 4.70$ Mm, crash.

13–94. (a) $v_p = 14.1$ km/s, (b) $r_a = 775$ Gm.

13–95. 9.60 km/s, $r = 6.50$ Mm.

13–97. $v_A = 6.11$ km/s, $\Delta v_B = 2.37$ km/s.

13–98. $v_A = 6.20$ km/s, $T/2 = 2.64$ h, $\Delta v = 479$ m/s.

Chapter 14

14–1. 13.5 kN.

14–2. 17.7 ft/s.

14–3. 1.53(10⁶) ft · lb.

14–5. 744 lb, 18.0(10³) ft · lb.

14–6. $\Delta T = 7.49(10^3)$ ft · lb, 1.90 kip.

14–7. 1.67 m/s.

14–9. 7.79 m/s.

14–10. 0.782 ft.

14–11. 232 mm.

14–13. $k = 642$ lb/ft, 18.0 ft/s.

14–14. 3.52 ft/s.

14–15. 5.44 ft.

14–17. 2.34 kJ.

14–18. 1.37 m/s.

14–19. 1.77 m/s.

14–21. 5.80 ft/s.

14–22. $v_B = 30.0$ m/s, $s = 130$ m.

14–23. 43.9 N.

14–25. $U_{1-2} = GM_e m \left(\dfrac{1}{r_2} - \dfrac{1}{r_1} \right).$

14–26. 2.45 m/s.

14–27. 1.70 kN.

14–29. 2.77 ft.

14–30. 200 kW.

14–31. 46.2 min.

14–33. 102 kW.

14–34. 1.76 hp.

14–35. 4.44 s.

14–37. 35.4 kW.

14–38. 181 ft.

14–39. $5.33t$ MW.

14–41. 4.36 hp.

14–42. 50.7 kW.

14–43. 0.231 hp.

14–45. 7.79 m/s.

14–46. 1.77 m/s.

14–47. 2.45 m/s.

14–49. $v_B = 7.58$ m/s, $T = 1.56$ kN, $T = 2.90$ kN.

14–50. $h = 24.5$ m, $N_B = 0$, $N_C = 16.8$ kN.

14–51. 1.59 ft.

14–53. 34.8 Mm/h.

14–54. 0.260 m.

14–55. 13.5 m/s, 149 kN.

14–57. $v_B = 98.4$ ft/s, $N_P = 601$ lb.

14–58. $\theta = 27.8°$, $s = 4.63$ m.

14–59. 15.1 m/s, $h = 5.35$ m.

14–61. 11.0 m/s.

14–62. $x_1 = 549$ mm, $x_2 = 366$ mm.

14–63. $x_1 = 184$ mm, $x_2 = 123$ mm.

14–65. $x_1 = 1.02$ m, $x_2 = 0.638$ m.

Chapter 15

15–1. 0.311 lb · s.

15–2. 37.3 s.

15–3. 63.4 N · s ←.

15–5. 0.706 N · s $\measuredangle$40°.

15–6. 79.0 kN · s.

15–7. 70.2 lb · s.

15–9. $F = 19.4$ kN, $T = 12.5$ kN.

15–10. hole cavity: $I = 46$ lb · s, starred cavity: $I = 54$ lb · s.

15–11. 24.7 N · s, $\theta = 30.5°$ $\measuredangle$.

15–13. 104 m/s.

15–14. 0.849 m/s.

15–15. (a) 15 kN · s, (b) 15 kN · s.

15–17. 3.01 m/s ↑.

15–18. $v_2 = 2ct'/\pi m$, $s = ct'^2/\pi m$.

15–19. 7.65 m/s.

15–21. 10.1 ft/s.

15–22. 1.92 m/s.

15–23. 0.486 s.

15–25. 3.5 ft/s.

15–26. $v = 1.55$ m/s ←, $t = 0.493$ s.

15–27. $v_c = 1.88$ ft/s →, $s = 3.75$ ft.

15–29. $v_A = 1.86$ m/s ←, $v_C = 0.781$ m/s →.

15–30. 1.66 m.

15–31. 0.242 m.

15–33. 1.55 ft →.

15–34. 1.55 ft →.

15–35. (a) $v_r = 1.31$ ft/s ←, (b) $v_r = 1.17$ ft/s ←.

15–37. 8.93 ft/s →.

15–38. $(v_A)_2 = 0.353$ m/s, $(v_B)_2 = 2.35$ m/s.

15–39. $(v_A)_2 = 1.64$ m/s →, $(v_B)_2 = 0.364$ m/s ←.

15–42. $v_B = 3.24$ m/s, $\theta = 43.9°$ $\measuredangle\theta$.

15–43. $\cos \phi = 1 - \left(\dfrac{1+e}{2} \right)^4 (1 - \cos \theta).$

15–45. $(v_B)_2 = 1.06$ m/s $\leftarrow$, $(v_A)_2 = 0.968$ m/s, $(\theta_A)_2 = 5.11°$ $\measuredangle$.

15–46. $(v_A)_2 = 4.06$ ft/s, $(v_B)_2 = 6.24$ ft/s.

15–47. 0.456 ft.

15–49. $(v_B)_1 = 3.16$ m/s, $(v_A)_1 = 0.699$ m/s.

15–50. $(v_B)_2 = 3.50$ m/s, $(v_A)_2 = 6.47$ m/s.

15–51. $(v_A)_2 = 6.88$ m/s, $(v_B)_2 = 75.6$ m/s.

15–53. (a) $v_B = 8.81$ m/s, $\theta_1 = 10.5°$ $\measuredangle$, (b) $(v_B)_2 = 4.62$ m/s, $\theta_2 = 20.3°$ $\searrow$, (c) $s = 3.96$ m.

15–54. 1.68 m.

15–55. 90°.

15–57. $(v_B)_2 = \dfrac{\sqrt{3}\, v(1 + e)}{4}$, $\Delta T = \frac{3}{16}mv^2(e^2 - 1)$.

15–58. $\mathbf{H}_A = \{-70\mathbf{i} - 105\mathbf{j} - 28\mathbf{k}\}$ kg $\cdot$ m^2/s.

15–59. (a) $(H_O)_A = 22.3$ kg $\cdot$ m^2/s $\uparrow$, $(H_O)_B = 7.18$ kg $\cdot$ m^2/s $\downarrow$, $(H_O)_C = 21.6$ kg $\cdot$ m^2/s $\downarrow$, (b) $\mathbf{H}_O = \{-16.8\mathbf{i} + 14.9\mathbf{j} - 23.6\mathbf{k}\}$ slug $\cdot$ ft^2/s.

15–61. 4.60 ft/s.

15–62. 13.4 m/s.

15–63. $v_B = 10.8$ ft/s, $U_{AB} = 113$ ft $\cdot$ lb.

15–65. $(v_G)_2 = 4.31$ m/s, $\theta = 33.2°$.

15–66. $v_B = 2.77$ m/s, $v_A = 3.23$ m/s.

15–67. 3.41 s.

15–69. 13.8 ft/s.

15–70. $v_B = 10.2$ km/s, $r_B = 13.8$ Mm.

15–71. 9.41 ft.

15–73. $A_x = 1.80$ kN $\rightarrow$, $A_y = 7.20$ kN $\uparrow$, $B_x = 5.40$ kN $\rightarrow$.

15–74. 11.0 N.

15–75. 302 lb.

15–77. $F_x = 9.87$ lb $\leftarrow$, $F_y = 4.93$ lb $\uparrow$.

15–78. $F_x = 0.143$ lb $\leftarrow$, $F_y = 1.43$ lb $\uparrow$.

15–79. $A_x = 3.98$ kN $\leftarrow$, $A_y = 3.81$ kN $\downarrow$, $M_A = 1.99$ kN $\cdot$ m $\uparrow$.

15–81. At the instant of firing $a = 2.11$ m/s^2, just before fuel runs out $a = 2.40$ m/s^2.

15–82. 32.2 slug/s.

15–83. 11.5 kN.

15–85. 104 mm/s^2.

15–86. 112 mm/s^2.

15–87. 16.1 m/s.

15–89. $R = (20t + 2.48)$ lb.

15–90. $F = (7.85t + 0.32)$ N.

15–91. $\dfrac{dm}{dt} = m_0\left(\dfrac{a_0 + g}{v_{e/r}}\right)e^{[(a_0 + g)/v_{e/r}]t}$ ($v_{e/r}$ is negative).

Review 1

R1–1. 7.86 ft/s $\leftarrow$.

R1–2. $\mathbf{v} = \{5\mathbf{u}_r + 4t\mathbf{u}_z\}$ m/s, $v = 7.81$ m/s, $\mathbf{a} = \{4\mathbf{u}_z\}$ m/s^2, $a = 4$ m/s^2.

R1–3. 38.5 ft/s.

R1–5. 83.1 km/h.

R1–6. 210 N.

R1–7. $4.26(10^3)$ ft/s.

R1–9. $t = 4.58$ s, $x = 1.25$ m.

R1–10. 0.933 m.

R1–11. 2.06 s.

R1–13. $a_B = 0.671$ m/s^2, $v_{B/C} = 1.44$ m/s $\measuredangle$ 56.3°, $a_{B/C} = 0.671$ m/s^2 $\measuredangle$ 63.4°.

R1–14. $\theta = 14.6°$, $\phi = 14.0°$.

R1–15. 10.4 m/s.

R1–17. $a = 322$ mm/s^2, $\theta = 26.6°$ $\nwarrow$.

R1–18. $v_2 = 3.84$ m/s, $T = 5.89$ N.

R1–19. 5.38 ft/s.

R1–21. $(v_A)_2 = -1$ ft/s, $(v_B)_2 = 2.5$ ft/s, $\Delta T = -0.815$ ft $\cdot$ lb.

R1–22. 0.901.

R1–23. 13.8 ft/s.

R1–25. 158 N.

R1–26. $T = 11.8$ N, $a = 5.89$ m/s^2.

R1–27. $t = 12$ s, $v = 18$ m/s, $s = 178$ m.

R1–29. $v_{A/B} = 1.80$ m/s, $s_{B/A} = 1.40$ m.

R1–30. (a) $\Delta s = 256$ mm, (b) $v = 748$ mm/s, $a = 300$ mm/s^2.

R1–31. $a_A = 1.96$ m/s^2, $a_B = 3.92$ m/s^2, $T_A = 54.9$ N, $T_B = 27.5$ N.

R1–33. 834 km/h.

R1–34. $v_B = 55.2$ ft/s, $v_A = 27.6$ ft/s.

R1–35. 4.38 m/s.

R1–37. 24.8 m/s.

R1–38. 24.6 ft/s.

R1–39. 3.58 m/s.

R1–41. 2.67 kN.

R1–42. (a) $a_B = a_A = 2.76$ ft/s^2, (b) $a_A = 67.6$ ft/s^2, $a_B = 5.15$ ft/s^2.

R1–43. 19.4 ft/s.

R1–45. $v_B = 24.0$ ft/s, $N_B = 7.18$ lb, $v_C = 16.0$ ft/s, $N_C = 1.18$ lb.

R1–46. (a) 0.627 hp, (b) 1.25 hp.

R1–47. $\theta_1 = 5.86°$, $\theta_2 = 79.4°$.

R1–49. 852 kW.

R1–50. $F = 68$ N, $N_C = 153$ N.

R1–51. $(v_B)_2 = 0.575$ ft/s, $\theta = 71.5°$↖, $\Delta T = -0.192$ ft · lb.

Chapter 16

16–1. $\theta_2 = 3.32$ rev, $t = 1.67$ s.

16–2. $a_t = 0.562$ ft/s^2, $a_n = 3600$ ft/s^2, $s = 3000$ ft.

16–3. (a) $\alpha = -3.33$ rad/s^2, (b) $\theta = 59.7$ rev.

16–5. 126 rad/s.

16–6. $v_A = v_B = 28.8$ in/s, $a_B = 207$ rad/s, $a_A = 4.8$ in/s^2.

16–7. 2.49 in.

16–9. $\alpha_D = 2.09$ rad/s^2, $\theta = 0.667$ rev.

16–10. $\omega_B = 180$ rad/s, $\omega_C = 360$ rad/s.

16–11. 211 rad/s.

16–13. $v_B = 14.7$ m/s.

16–14. $v_C = 125$ mm/s ↑ , $a_C = 37.5$ mm/s^2 ↓ .

16–15. $v_C = 184$ mm/s ↑ , $a_C = 2.52$ m/s^2 ↓ .

16–17. 22.3 rad/s.

16–18. $v_P = 12.6$ ft/s, $a_P = 160$ ft/s^2, $t = 3.47$ s.

16–19. $v = 3.70$ ft/s, $a = 0.456$ ft/s^2.

16–21. $a = \dfrac{s}{2\pi}\omega^2$.

16–22. 14.4 rad/s ↓ .

16–23. $v = \ell\omega \cos\theta$, $a = -\ell\omega^2 \sin\theta$.

16–25. $v = 750 \cos\theta$ mm/s, $a = -3750 \sin\theta$ mm/s^2.

16–26. $v = 650$ mm/s, $a = -1615$ mm/s^2.

16–27. $v = 0.480 \cos\theta$ m/s, $a = -1.92 \sin\theta$ m/s^2.

16–29. 4.62 rad/s ↑, 4.62 m/s ↓ .

16–30. 18.5 ft/s ← .

16–31. $\dot{x} = -\omega d\left(\sin\theta + \dfrac{d \sin 2\theta}{2((R + r)^2 - d^2 \sin^2\theta)^{1/2}}\right)$.

16–33. $\omega = \dfrac{-rv_A}{y\sqrt{y^2 - r^2}}$, $\alpha = \dfrac{rv_A^2(2y^2 - r^2)}{y^2(y^2 - r^2)^{3/2}}$.

16–34. $v = \ell r\omega\left(\dfrac{r + d \cos\theta}{(d + r \cos\theta)^2}\right)$,

$a = \ell r\omega^2 \sin\theta\left(\dfrac{2r^2 - d^2 + dr \cos\theta}{(d + r \cos\theta)^3}\right)$.

16–35. $\omega_A = 2\left(\dfrac{(\sqrt{2}\cos\theta - 1)}{3 - 2\sqrt{2}\cos\theta}\right)$.

16–37. $v_C = 2.45$ m/s, $_{45°}$↗, $\omega_{CB} = 7.81$ rad/s ↑.

16–38. 5.18 m/s → .

16–39. 2.40 ft/s ↑ .

16–41. 9 m/s ← .

16–42. 2.5 ft/s ← .

16–43. $\omega_{CD} = 4.03$ rad/s ↑, $\omega_{BC} = 1.65$ rad/s ↓.

16–45. 2 rad/s ↑.

16–46. $v_A = 4$ ft/s $_{30°}$↗, $v_B = 7.21$ ft/s $\angle 76.1°$.

16–47. 2.22 rad/s ↓.

16–49. 4.42 in./s $_{30°}$↖.

16–50. $\omega_{CD} = 0.0510$ rad/s ↑, $\omega_{BC} = 1.98$ rad/s ↓.

16–51. $\omega_S = 57.5$ rad/s ↑, $\omega_{OA} = 10.6$ rad/s ↑.

16–53. $\omega_B = 5$ rad/s ↓, $\omega_C = 3.46$ rad/s ↓.

16–54. $\omega_B = 90$ rad/s ↓, $\omega_A = 180$ rad/s ↑.

16–55. $v_C = 0.776$ m/s ↗$_{45°}$, $v_D = 1.06$ m/s ↖$_{45°}$.

16–57. 1.34 m/s ← .

16–58. $\omega = 7.81$ rad/s ↑, $v_C = 2.45$ m/s $_{45°}$↗.

16–59. 2.5 ft/s ← .

16–61. $\omega_{BC} = 1.65$ rad/s ↓, $\omega_{CD} = 4.03$ rad/s ↑.

16–62. $v_H = 0.536$ m/s $\angle$ 53.6°,
$v_F = 0.992$ m/s $_{70.9°}$↖.

16–63. 50 m/s ↓ .

16–65. 11.4 rad/s ↓.

16–66. $\omega = 4$ rad/s ↓, $v_P = 4.88$ m/s ← .

16–67. 1.64 m/s ← .

16–69. $\omega_C = 26.7$ rad/s ↑, $\omega_B = 28.7$ rad/s ↓,
$\omega_A = 14.0$ rad/s ↑.

16–70. $v_B = 8$ ft/s ↑ , $v_C = 2.93$ ft/s ↓ .

16–71. $\omega = 58.8$ rad/s↓.

16–73. 14.4 km/s^2.

16–74. $\omega = 2$ rad/s ↓, $\alpha = 7.68$ rad/s^2 ↓.

16–75. 66.5 ft/s^2.

16–77. 13.9 ft/s^2 $\angle$ 8.80°.

16–78. 12.5 m/s^2 ← .

16–79. $a_A = 73.5$ in./s^2, $\theta = 78.2°$ $\angle\theta$, $a_B = 113$ in./s^2 → .

16–81. $\alpha_{CB} = 7.52$ rad/s^2 ↑, $a_C = 4.22$ m/s^2, $\theta = 60.8°$ ↘$_\theta$.

16–82. $\omega = 0.25$ rad/s ↓, $v_B = 5$ ft/s ↓ , $\alpha = 0.875$ rad/s^2 ↑,
$a_B = 1.51$ ft/s^2, $\theta = 85.2°$ ↘$_\theta$.

16–83. $\alpha = 10.7$ rad/s^2 ↑, $a_D = 14.1$ in./s^2, $\theta = 18.4°$ $\angle\theta$.

16–85. $\omega_{AB} = 7.17$ rad/s ↓, $\alpha_{AB} = 23.1$ rad/s^2 ↑.

16–86. $\mathbf{a}_B = \{-1.60\mathbf{i} - 16.0\mathbf{j}\}$ m/s^2.

16–87. $\mathbf{a}_B = \{1.40\mathbf{i} - 16.2\mathbf{j}\}$ m/s^2.

16–89. (a) $\mathbf{a}_D = \{-2.5\mathbf{i}\}$ ft/s^2, (b) $\mathbf{a}_B = \{-4.17\mathbf{i}\}$ ft/s^2.

16–90. $v_C = 8$ ft/s, $a_C = 32$ ft/s^2.

16–91. $\Omega = 2$ rad/s ↓, $\dot{\Omega} = 1.0$ rad/s^2 ↓.

16–93. $\omega_{CD} = 0.866$ rad/s ↑, $\alpha_{CD} = 3.23$ rad/s^2 ↑.

16–94. $\mathbf{v}_A = \{-19.3\mathbf{i} - 6.5\mathbf{j}\}$ m/s,
$\mathbf{a}_A = \{-56.6\mathbf{i} + 99.2\mathbf{j}\}$ m/s^2.

16–95. $\mathbf{v}_A = \{-2.63\mathbf{i} - 4.50\mathbf{j}\}$ m/s,
$\mathbf{a}_A = \{6.75\mathbf{i} - 9.19\mathbf{j}\}$ m/s^2.

16–97. $(\mathbf{v}_{B/A})_{rel} = \{31.0\mathbf{i}\}$ m/s,
$(\mathbf{a}_{B/A})_{rel} = \{-14.0\mathbf{i} - 206\mathbf{j}\}$ m/s^2.

16–98. $\Omega = 0.720$ rad/s γ, $\dot{\Omega} = 2.02$ rad/s^2 γ.

Chapter 17

17–1. $I_y = \frac{1}{3}m\ell^2$.

17–2. $I_x = \frac{3}{10}mr^2$.

17–3. 57.7 mm.

17–5. $I_x = \frac{2}{5}mr^2$.

17–6. $I_x = \frac{2}{5}mb^2$.

17–7. $I_y = \frac{1}{6}m(a^2 + h^2)$.

17–9. $I_z = \frac{1}{10}ma^2$.

17–10. 2.25 slug $\cdot$ ft^2.

17–11. 7.67 kg $\cdot$ m^2.

17–13. 1.78 m, 4.45 kg $\cdot$ m^2.

17–14. 6.23 kg $\cdot$ m^2.

17–15. 0.203 m, 0.230 kg $\cdot$ m^2.

17–17. 34.2 kg $\cdot$ m^2.

17–18. 0.393 m, 0.0885 kg $\cdot$ m^2.

17–19. $N_B = 95.0$ lb, $N_A = 105$ lb, $s = 9.66$ ft.

17–21. 6.04 ft/s^2.

17–22. 16.1 ft/s^2.

17–23. 16.1 ft/s^2.

17–25. $a_G = 0.3$ m/s^2, $N_A = 5.43$ kN, $N_B = 7.09$ kN.

17–26. $a_G = 0.0157$ m/s^2, $N_A = 72.1$ kN, $N_B = 71.9$ kN.

17–27. 11.2 s.

17–29. $a_B = 2.26$ m/s^2, $N_A = 559$ N, $N_B = 226$ N;
equilibrium, $N_A = 330$ N, $N_B = 454$ N.

17–30. $\mu = 0.458$.

17–31. not possible, $a = 9.81$ m/s^2.

17–33. $a_T = 2.33$ m/s^2.

17–34. $A_x = 9.87$ lb $\rightarrow$, $V_y = 4.86$ lb $\downarrow$, $M_E = 7.29$ lb $\cdot$ ft γ.

17–35. 23.9 lb.

17–37. 3.12 ft.

17–38. $\alpha = 0.587$ rad/s^2.

17–39. $T = 2.68$ kN, $a_C = 1.52$ m/s^2.

17–41. $N_B = 9.40$ kN, $N_C = 4.62$ kN, $N_D^* = 7.56$ kN.

17–42. 9.90 s.

17–43. 8.10 s.

17–45. $\omega = 56.2$ rad/s, $A_x = 0$, $A_y = 98.1$ N.

17–46. $T_B = 100$ N, $T_A = 102$ N.

17–47. $A_P = 7.38$ N, $V_P = 3.75$ N, $M_P = 2.02$ N $\cdot$ m.

17–49. 2.97 rad/s $\downarrow$.

17–50. 4.88 ft/s.

17–51. 6.71 s.

17–53. $O_x = 142$ lb, $O_y = 4.29$ lb.

17–54. 2.71 rad/s.

17–55. $\omega = 4.83$ rad/s γ, $\theta = 92.2°$.

17–57. $A_x = 4.50$ lb, $A_y = 6.50$ lb.

17–58. $A_x = 89.2$ N, $A_y = 66.9$ N, $t = 1.26$ s.

17–59. point A, $\Delta v = 0.267\sqrt{g\ell}$.

17–62. 482 mm.

17–63. 891 rad/s^2.

17–65. $a = \frac{8}{25}(2 - \mu)g$.

17–66. $T_1 = 1.73$ lb, $T_2 = 2.31$ lb, $v_C = 6.68$ ft/s.

17–67. $F = \frac{3}{2}mg \sin \theta(\frac{3}{2}\cos \theta - 1)$,
$N = \frac{1}{4}mg(1 - 3 \cos \theta)^2$.

17–70. $a_G = 12.9$ ft/s^2, $\alpha = 2.58$ rad/s^2.

17–71. $\alpha = 23.4$ rad/s^2, $B_y = 9.62$ lb.

17–73. 193 ft/s^2.

17–74. 167 ft/s^2.

17–75. $\alpha = 1.25$ rad/s^2, $T = 2.32$ kN.

17–77. 5.01 rad/s^2.

17–78. $F = 40.2$ lb, $N_T = 329$ lb.

17–79. 4.06 ft.

17–81. $\alpha_D = 21.9$ rad/s^2 $\downarrow$, $\alpha_C = 11.3$ rad/s^2 $\downarrow$.

17–82. $\alpha = 4.18$ rad/s^2, $T = 43.3$ lb.

17–83. 29.8 rad/s, 0.965 s.

17–85. 38.6 ft/s $\downarrow$.

17–86. 10 N, $\alpha = 16.7$ rad/s^2, $\alpha' = 16.7$ rad/s^2.

17–87. $N_B = 2.35$ kN, $N_C = 2.48$ kN, $v = 8.03$ m/s.

17–89. $N_B = 297$ N $\diagdown$ 30°, $A_x = 1.63$ N $\rightarrow$, $A_y = 344$ N $\uparrow$.

17–90. 4.01 rad/s^2 $\downarrow$.

17–91. 743 N $\cdot$ m $\downarrow$.

17–93. 33.5°.

Chapter 18

18–2. 0.304 ft.

18–3. 3.89 rad/s.

18–5. 9.07 m/s $\uparrow$.

18–6. 9.43 rad/s $\downarrow$.

18–7. 1.84 rad/s.

18–9. $A_x = 148$ N, $A_y = 224$ N, $\theta = 1.34$ rev.

18–10. 7.08 rad/s $\upharpoonleft$.

18–11. 5.05 ft/s $\leftarrow$.

18–13. 8.59 rad/s.

18–14. 0.891 rev.

18–15. 13.3 rad/s $\downarrow$.

18–17. 0.934 rev.

18–18. 4.60 rad/s $\downarrow$.

18–19. 4.70 m/s $\uparrow$.

18–21. 8.37 ft/s $\uparrow$.

18–22. 0.304 ft.

18–23. 13.3 rad/s.

18–25. 0.934 rev.

18–26. 7.34 rad/s.

18–27. 5.93 m/s.

18–29. 0.753 rad/s $\downarrow$.

18–30. 12.7 ft/s.

18–31. 11.9 ft/s.

18–33. 13.3 ft/s.

18–34. 14.2 ft/s.

18–35. 7.37 rad/s $\downarrow$.

18–37. 41.1 ft/s $\leftarrow$.

18–38. 4.00 rad/s $\downarrow$.

18–39. 5.17 rad/s $\downarrow$.

18–41. 1.60 rad/s $\downarrow$.

18–42. 11.4 ft/s $\rightarrow$.

Chapter 19

19–5. 53.7 rad/s.

19–6. $\omega = (2gt/3r) \sin \theta$ (disk), $\omega = (5gt/7r) \sin \theta$ (sphere).

19–7. $v_G = 2$ m/s $_{60°}\searrow$, $\omega = 3.90$ rad/s $\upharpoonleft$.

19–9. $v_A = 1.66$ m/s $\downarrow$, $v_B = 1.30$ m/s $\uparrow$.

19–10. $\omega_A = 1.70$ rad/s, $\omega_B = 5.10$ rad/s.

19–11. 1.32 s.

19–13. $\theta = \tan^{-1}[(1 + r^2/k_O^2)\mu]$.

19–14. 24.1 m/s $\uparrow$.

19–15. 12.7 rad/s.

19–17. 51.5 rad/s $\downarrow$.

19–18. $\omega = 2.5(v_O/r)$.

19–19. 5.89 rad/s $\upharpoonleft$.

19–21. 23.4 rad/s $\downarrow$.

19–22. $y = \frac{2}{3}\ell$.

19–23. $\omega_{BC} = \omega_{AB} = \dfrac{3}{4\sqrt{2}} \dfrac{I}{\ell}$.

19–25. 6.75 rad/s.

19–26. 0.0708 rad/s.

19–27. $\omega_1 = 2.41$ rad/s, $\omega_2 = 1.86$ rad/s.

19–29. 3.56 rad/s.

19–30. 79.8 N · s.

19–31. 0.533 m.

19–33. (a) $\omega_m = (I/I_z)\omega$, (b) $\omega_m = 0$, (c) $\omega_m = (I/I_z)\omega$.

19–34. $h = \frac{7}{5}r$.

19–35. 0.500 ft.

19–37. 6.45 rad/s $\downarrow$.

19–38. (a) $\omega_2 = \frac{1}{4}\omega_1$, (b) $\omega_2 = 1$ rad/s.

19–39. 3.36 ft/s.

19–41. 17.9°.

19–42. 39.8°.

19–43. 0.452 ft/s.

Review 2

R2–1. 20.4 ft/s.

R2–2. $v_C = 25.3$ ft/s, $\measuredangle_{63.4°}$, $a_C = 73.8$ ft/s² $\searrow_{32.5°}$.

R2–3. 4.76 rad/s².

R2–5. 5.19 rad/s².

R2–6. 5.96 ft/s.

R2–7. 3 ft/s.

R2–9. 4.72 m/s.

R2–10. 14.2 ft/s $\uparrow$, 0.25 ft/s² $\downarrow$.

R2–11. 1.80 m.

R2–13. 1.02 m/s $\leftarrow$.

R2–14. 6.14 lb.

R2–15. $a_C = 50.6$ m/s² $\measuredangle_{9.09°}$, $a_B = 8$ m/s² $\uparrow$.

R2–17. 18 m/s $\rightarrow$.

R2–18. 33.3 lb.

R2–19. 68.0 m.

R2–21. 16.1 rad/s $\upharpoonleft$.

R2–22. 195 mm/s.

R2–23. 14.3 rev, 8.5 rad/s, 0.4 rad/s².

R2–25. (a) $B_y = C_y = 7.22$ N; (b) $B_y = C_y = 14.7$ N.

R2–26. 9.15 ft/s.

R2–27. $\alpha_{DC} = 0.708$ rad/s^2 ↙, $a_C = 429$ mm/s^2 ↘$_{7.85°}$.

R2–29. $a_G = 4.90$ m/s^2, $\alpha = 14.7$ rad/s^2.

R2–30. $\omega_{AB} = 0.667$ rad/s ↖, $\alpha_{AB} = 3.08$ rad/s^2 ↙.

R2–31. $v_B = 5.20$ m/s ↘$_{30°}$, $v_C = 6$ m/s →.

R2–33. $A_x = B_x = 0$, $A_y = B_y = 147$ N, $\omega_2 = 66.7$ rad/s ↙.

R2–34. $a_A = a_B = 9.20$ ft/s^2, $\alpha = 18.4$ rad/s^2.

R2–35. 1.64 s, 48.7 N.

R2–37. 9.42 m/s.

R2–38. 48 rad/s ↖.

R2–39. 6.11 ft/s ↘$_{8.13°}$.

R2–41. $v = -r\omega \sin \theta$, $a = -r\omega^2 \cos \theta$.

R2–42. 60 mm/s ↑.

R2–43. 0.714 s.

R2–45. 19.3 ft/s^2, 13.4 lb ∡$_{63.4°}$.

R2–46. $\alpha = 12.1$ rad/s^2, $F = 30.0$ lb.

R2–47. $\alpha = 160$ rad/s^2 ↙, $a_A = 1.0$ m/s^2 →.

R2–49. $\alpha = 12.5$ rad/s^2, $N_A = 88.8$ N.

R2–50. $\alpha = 25.4$ rad/s^2 ↖, $a_B = 5.21$ m/s^2 ↓.

R2–51. 23.7 rad/s ↖.

Chapter 20

20–1. $\boldsymbol{\omega} = \{13.9\mathbf{i} + 8.80\mathbf{k}\}$ rad/s,
$\dot{\boldsymbol{\omega}} = \{-1.73\mathbf{i} + 11.1\mathbf{j} + 11.0\mathbf{k}\}$ rad/s^2.

20–2. $\mathbf{v}_P = \{-1.60\mathbf{i}\}$ m/s, $\mathbf{a}_P = \{-0.32\mathbf{i} - 12\mathbf{j} - 8\mathbf{k}\}$ m/s^2.

20–3. $\mathbf{v}_A = \{0.0828\mathbf{i} - 0.464\mathbf{j} + 0.124\mathbf{k}\}$ m/s,
$\mathbf{a}_A = \{-0.371\mathbf{i} - 0.108\mathbf{j} - 0.278\mathbf{k}\}$ m/s^2.

20–5. $\mathbf{v}_A = \{-1.80\mathbf{i}\}$ft/s, $\mathbf{a}_A = \{-0.720\mathbf{j} - 0.831\mathbf{k}\}$ ft/s^2.

20–6. $\boldsymbol{\omega} = \{-4\mathbf{j}\}$ rad/s, $\boldsymbol{\alpha} = \{16\mathbf{j}\}$ rad/s^2, $v_A = 283$ mm/s,
$a_A = 1600$ mm/s^2.

20–7. $\omega = 11.0$ rad/s, $\dot{\omega} = 44.8$ rad/s^2.

20–9. $\mathbf{v}_A = \{-8.66\mathbf{i} + 8\mathbf{j} - 13.9\mathbf{k}\}$ ft/s,
$\mathbf{a}_A = \{-24.8\mathbf{i} + 8.28\mathbf{j} - 30.9\mathbf{k}\}$ ft/s^2.

20–10. $\boldsymbol{\omega}_P = \{-40\mathbf{j}\}$ rad/s, $\boldsymbol{\alpha} = \{-6400\mathbf{i}\}$ rad/s^2.

20–11. $\mathbf{v}_B = \{12\mathbf{i}\}$ ft/s, $\boldsymbol{\omega} = \{0.980\mathbf{i} - 1.06\mathbf{j} - 1.47\mathbf{k}\}$ rad/s.

20–13. $\mathbf{v}_B = \{10\mathbf{k}\}$ ft/s.

20–14. $\boldsymbol{\omega}_{BD} = \{-2\mathbf{k}\}$ rad/s.

20–15. $\mathbf{v}_B = \{-0.333\mathbf{j}\}$ m/s,
$\boldsymbol{\omega} = \{0.204\mathbf{i} - 0.612\mathbf{j} + 1.36\mathbf{k}\}$ rad/s.

20–17. $\boldsymbol{\omega} = \{1.17\mathbf{i} + 1.27\mathbf{j} - 0.779\mathbf{k}\}$ rad/s,
$\mathbf{v}_B = \{-2.82\mathbf{i} + 3.76\mathbf{k}\}$ ft/s.

20–18. $\boldsymbol{\alpha} = \{-2.78\mathbf{i} - 0.629\mathbf{j} - 2.91\mathbf{k}\}$ rad/s^2,
$\mathbf{a}_B = \{-10.6\mathbf{i} + 14.1\mathbf{k}\}$ ft/s^2.

20–19. $\boldsymbol{\omega} = \{-0.375\mathbf{i} + 0.5\mathbf{j} + 0.625\mathbf{k}\}$ rad/s,
$\mathbf{v}_B = \{-3.75\mathbf{j}\}$ ft/s.

20–21. $\boldsymbol{\omega} = \{-8.66\mathbf{i} - 5\mathbf{j} - 2\mathbf{k}\}$ rad/s,
$\mathbf{v}_C = \{17.3\mathbf{j} - 43.3\mathbf{k}\}$ m/s.

20–22. $\boldsymbol{\omega} = \{0.769\mathbf{i} - 2.31\mathbf{j} + 0.513\mathbf{k}\}$ rad/s,
$\mathbf{v}_B = \{-0.333\mathbf{j}\}$ m/s.

20–23. $\mathbf{v}_A = \{-2.50\mathbf{k}\}$ ft/s, $\mathbf{v}_C = \{4.33\mathbf{i} + 2.17\mathbf{j} - 2.50\mathbf{k}\}$ ft/s.

20–25. $\mathbf{v}_C = \{-1\mathbf{i} + 5\mathbf{j} + 0.8\mathbf{k}\}$ m/s,
$\mathbf{a}_C = \{-28.8\mathbf{i} - 5.45\mathbf{j} + 32.3\mathbf{k}\}$ m/s^2.

20–26. $\mathbf{v}_B = \{-10.2\mathbf{i} - 30\mathbf{j} + 52.0\mathbf{k}\}$ m/s,
$\mathbf{a}_B = \{-31.0\mathbf{i} - 161\mathbf{j} - 90\mathbf{k}\}$ m/s^2.

20–27. $\mathbf{v}_P = \{-3.88\mathbf{i} - 4.54\mathbf{j} + 2\mathbf{k}\}$ m/s,
$\mathbf{a}_P = \{8.41\mathbf{i} - 3.68\mathbf{j} + 65.2\mathbf{k}\}$ m/s^2.

20–29. $\mathbf{v}_P = \{-18\mathbf{i} + 450\mathbf{j} + 4\mathbf{k}\}$ ft/s,
$\mathbf{a}_P = \{-810\mathbf{i} - 16.2\mathbf{j} - 81.0(10^3)\mathbf{k}\}$ ft/s^2.

20–30. $\mathbf{v}_A = \{-7.79\mathbf{i} - 2.25\mathbf{j} + 3.90\mathbf{k}\}$ ft/s,
$\mathbf{a}_A = \{8.30\mathbf{i} - 35.2\mathbf{j} + 7.02\mathbf{k}\}$ ft/s^2.

20–31. $\mathbf{v}_T = \{-9.5\mathbf{i} + 3\mathbf{j} + 1.2\mathbf{k}\}$ m/s,
$\mathbf{a}_T = \{-14.4\mathbf{i} - 4.23\mathbf{j} + 4.20\mathbf{k}\}$ m/s^2.

20–33. $\mathbf{v}_C = \{3\mathbf{i} + 6\mathbf{j} - 3\mathbf{k}\}$ m/s,
$\mathbf{a}_C = \{-13\mathbf{i} + 28.5\mathbf{j} - 10.2\mathbf{k}\}$ m/s^2.

20–34. $\mathbf{v}_C = \{-1.73\mathbf{i} - 5.77\mathbf{j} + 7.06\mathbf{k}\}$ ft/s,
$\mathbf{a}_C = \{9.88\mathbf{i} - 72.8\mathbf{j} + 0.364\mathbf{k}\}$ ft/s^2.

20–35. $\mathbf{v}_P = \{8\mathbf{i} + 32\mathbf{j} - 4\mathbf{k}\}$ ft/s, $\mathbf{a}_P = \{-270\mathbf{i} + 72\mathbf{j}\}$ ft/s^2.

20–37. $\mathbf{v}_P = \{-25.5\mathbf{i} - 13.4\mathbf{j} + 20.5\mathbf{k}\}$ ft/s,
$\mathbf{a}_P = \{161\mathbf{i} - 249\mathbf{j} - 39.6\mathbf{k}\}$ ft/s^2.

20–38. $\mathbf{v}_P = \{-25.5\mathbf{i} - 13.4\mathbf{j} + 20.5\mathbf{k}\}$ ft/s,
$\mathbf{a}_P = \{161\mathbf{i} - 243\mathbf{j} - 33.9\mathbf{k}\}$ ft/s^2.

20–39. $\mathbf{v}_A = \{11.2\mathbf{i} + 15.2\mathbf{j} - 3\mathbf{k}\}$ ft/s,
$\mathbf{a}_A = \{7.43\mathbf{i} + 40.7\mathbf{j} - 11.5\mathbf{k}\}$ ft/s^2.

20–41. $\mathbf{v}_A = \{-8.66\mathbf{i} + 2.26\mathbf{j} + 2.26\mathbf{k}\}$ m/s,
$\mathbf{a}_A = \{-22.6\mathbf{i} - 47.8\mathbf{j} + 4.53\mathbf{k}\}$ m/s^2.

20–42. $\mathbf{v}_A = \{-8.66\mathbf{i} + 2.62\mathbf{j} + 1.91\mathbf{k}\}$ m/s,
$\mathbf{a}_A = \{-38.3\mathbf{i} - 38.7\mathbf{j} + 11.8\mathbf{k}\}$ m/s^2.

Chapter 21

21–2. $I_{yz} = \frac{1}{6}mah$.

21–3. $I_{xy} = \frac{1}{12}ma^2$.

21–5. $I_{xy} = \frac{1}{20}ma^2$.

21–6. $\begin{pmatrix} \frac{13}{24}mr^2 & 0 & \frac{1}{24}mr^2 \\ 0 & \frac{7}{12}mr^2 & 0 \\ \frac{1}{24}mr^2 & 0 & \frac{13}{24}mr^2 \end{pmatrix}$

21–7. 0.148 kg · m².

21–9. $I_x = 4.50$ kg · m², $I_y = 4.38$ kg · m², $I_z = 0.125$ kg · m².

21–10. $I_{Oa} = 1.21$ kg · m².

21–11. $I_{xx} = 0.0292$ slug · ft².

21–13. $I_z = 0.0725$ slug · ft².

21–14. $I_x = I_y = 20.2$ slug · ft², $I_z = 16.3$ slug · ft².

21–15. $I_{z'} = 0.0595$ kg · m².

21–17. $I_x = 1.36$ kg · m², $I_y = 0.380$ kg · m², $I_z = 1.26$ kg · m².

21–21. 0.727 kg · m²/s.

21–22. $\mathbf{H} = \{16.6\mathbf{k}\}$ kg · m²/s, $T = 72.1$ J.

21–23. $\omega_{AB} = 21.4$ rad/s, yes.

21–25. $\boldsymbol{\omega} = \{-0.750\mathbf{j} + 1\mathbf{k}\}$ rad/s.

21–26. 58.4 rad/s.

21–27. 10.2 ft/s ↓.

21–29. 27.2 ft/s.

21–30. 20.2 rad/s.

21–31. $\boldsymbol{\omega} = \{8.71\mathbf{i} - 122\mathbf{j}\}$ rad/s.

21–33. $\Sigma M_x = d/dt(I_x\omega_x - I_{xy}\omega_y - I_{xz}\omega_z)$
$- \Omega_z(I_y\omega_y - I_{yz}\omega_z - I_{yx}\omega_x)$
$+ \Omega_y(I_z\omega_z - I_{zx}\omega_x - I_{zy}\omega_y)$, etc.

21–34. $\Sigma M_x = I_x\dot{\omega}_x - I_{xy}\dot{\omega}_y - I_{xz}\dot{\omega}_z$
$- \Omega_z(I_y\omega_y - I_{yz}\omega_z - I_{yx}\omega_x)$
$+ \Omega_y(I_z\omega_z - I_{zx}\omega_x - I_{zy}\omega_y)$, etc.

21–35. $\Sigma M_x = I_x\dot{\omega}_x - I_y\Omega_z\,\omega_y + I_z\Omega_y\omega_z$.

21–37. $M_x = 0$, $M_y = -0.036 \sin\theta$ N · m, $M_z = 0.003 \sin 2\theta$ N · m.

21–38. $M = 0.0320$ N · m, $A_x = 1.08$ N, $A_y = 13.2$ N, $B_x = 1.08$ N, $B_y = 13.3$ N.

21–39. $B_x = 50$ N, $A_y = B_y = 0$, $A_z = B_z = 24.5$ N.

21–41. $A_x = B_x = 0$, $A_y = 1.09$ kN, $B_y = 1.38$ kN.

21–42. $A_x = 15.6$ lb, $A_y = 51.8$ lb, $A_z = 50$ lb, $B_x = 12.5$ lb, $B_y = 41.4$ lb.

21–43. $B_x = 0$, $A_y = 1.69$ lb, $B_y = 3.90$ lb, $A_z = B_z = 7.5$ lb.

21–45. $T = 23.3$ N, $F_A = 51.6$ N.

21–46. $T = 23.3$ N, $F_A = 51.8$ N.

21–47. $\dot{\omega}_y = 25.9$ rad/s², $B_x = 1.15$ N, $B_z = 12.3$ N, $A_x = 0.102$ N, $A_z = 12.3$ N.

21–49. $T_B = 47.1$ lb, $A_x = 0$, $A_y = 93.2$ lb, $A_z = 57.1$ lb, $M_y = M_z = 0$.

21–50. $A_x = 0$, $A_y = 1.30$ lb, $A_z = 20.2$ lb, $M_y = M_z = 0$, $\theta = 64.1°$.

21–51. 328 N · m.

21–53. 1.31 rad/s (low precession), 255 rad/s (high precession).

21–54. 1.21 rad/s (low precession), 76.3 rad/s (high precession.

21–55. 1.19 rad/s.

21–57. $A_y = 13.7$ kN, $B_y = 13.7$ kN.

21–58. 4.91 rad/s.

21–59. 77.7 lb.

21–61. 107 N.

21–62. 12.8 rad/s, retrograde precession.

21–63. 12.5 Mg · m²/s.

21–66. 2.34 rev/h.

21–67. 0.673°, regular precession.

Chapter 22

22–1. $y = -0.2 \cos (12.7t)$, 0.2 ft, 2.02 Hz.

22–2. $x = -0.126 \sin (3.16t) - 0.09 \cos (3.16t)$, $C = 0.152$ m.

22–3. 7.88 Hz, 0.127 s.

22–5. $\ddot{y} + 56.1y = 0$, 192 mm.

22–6. $x = -0.53 \sin (7.49t) + 0.06 \cos (7.49t)$, $C = 0.533$ m.

22–7. 16.9 kip, 5.42 Hz.

22–9. 16.3°.

22–10. (a) 1.13 Hz, (b) 2.26 Hz.

22–11. $\tau = 2\pi\sqrt{3r/2g}$.

22–13. 0.146 m.

22–14. $\tau = 6.10\sqrt{a/g}$.

22–15. 0.457 m.

22–17. $f = \left(\dfrac{1}{2\pi}\right)\sqrt{\dfrac{(R^2 - (\ell/2)^2)^{1/2}}{(R^2 - \ell^2/6)}}$.

22–18. 170 mm.

22–19. 0.624 Hz.

22–21. 609 kg, 17.4 kN/m.

22–22. 2.88 Mg · m².

22–23. $f = \dfrac{1}{2\pi}\sqrt{\dfrac{d}{\mu g}}$.

22–25. $\tau = 2\pi\sqrt{\dfrac{\ell}{2g}}$.

22–26. $\tau = 2\pi\sqrt{\dfrac{3r}{2g}}$.

22–27. $\tau = 2\pi\sqrt{\dfrac{(k_G^2 + d^2)}{gd}}$.

22–29. 0.624 Hz.

22–30. $\tau = 2\pi\sqrt{\dfrac{2r}{g}}$.

22–31. $\tau = 2\pi\sqrt{\dfrac{m}{4k}}$.

22–33. 2.25 Hz.

22–34. 0.399 s.

22–35. $\ddot{y} + (k_1 + k_2/m)y = 0$.

22–37. $\ddot{\theta} + 117\theta = 0$.

22–38. 1.28 Hz.

22–39. $y = A \sin pt + B \cos pt + \left(\dfrac{F_O/m}{k/m - \omega^2}\right)\cos \omega t$.

22–41. $y = (-16.6 \sin (7.07t) + 70 \cos (7.07t) + 29.4 \sin (4t))$ mm.

22–42. 14.0 rad/s.

22–43. 35.5 mm.

22–45. $x_P = (0.0276 \sin(2t))$ ft.

22–46. $\omega = 6.55$ rad/s.

22–47. 0.0521 ft.

22–49. 0.158 ft.

22–50. 29.5 mm.

22–51. 5.88 mm.

22–53. 4.51 mm.

22–54. 9.89°.

22–55. 0.997.

22–57. $x = (0.119 \cos (6t - 83.9°))$ m.

22–58. 714 N/m.

22–59. $x = (-70.2\, e^{-3.57t} \sin (8.54t))$ mm.

22–61. 18.9 N · s/m.

22–62. 21.1 mm.

22–63. $1.55\ddot{\theta} + 540\dot{\theta} + 200\theta = 0$, $c_c = 0.0653$ lb · s/ft.

22–65. $L\ddot{q} + R\dot{q} + (2/c)q = 0$.

22–66. $L\ddot{q} + R\dot{q} + (1/c)q = 0$.

22–67. $L\ddot{q} + (1/c)q = E_0 \sin \omega t$.
$m\ddot{x} + kx = F_0 \sin \omega t$.

22–69. $\ddot{x} + 16\dot{x} + 4x = 0$, overdamped.

Index

Geometric Properties of Line and Area Elements

Centroid Location	Centroid Location	Area Moment of Inertia

Circular arc segment

Circular sector area

$$I_x = \tfrac{1}{4} r^4 (\theta - \tfrac{1}{2} \sin 2\theta)$$
$$I_y = \tfrac{1}{4} r^4 (\theta + \tfrac{1}{2} \sin 2\theta)$$

Quarter and semicircular arcs

Quarter circular area

$$I_x = \tfrac{1}{16} \pi r^4$$
$$I_y = \tfrac{1}{16} \pi r^4$$

Trapezoidal area

Semicircular area

$$I_x = \tfrac{1}{8} \pi r^4$$
$$I_y = \tfrac{1}{8} \pi r^4$$

Semiparabolic area

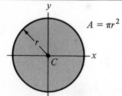

Circular area

$$I_x = \tfrac{1}{4} \pi r^4$$
$$I_y = \tfrac{1}{4} \pi r^4$$

Exparabolic area

Rectangular area

$$I_x = \tfrac{1}{12} bh^3$$
$$I_y = \tfrac{1}{12} hb^3$$

Parabolic area

Triangular area

$$I_x = \tfrac{1}{36} bh^3$$